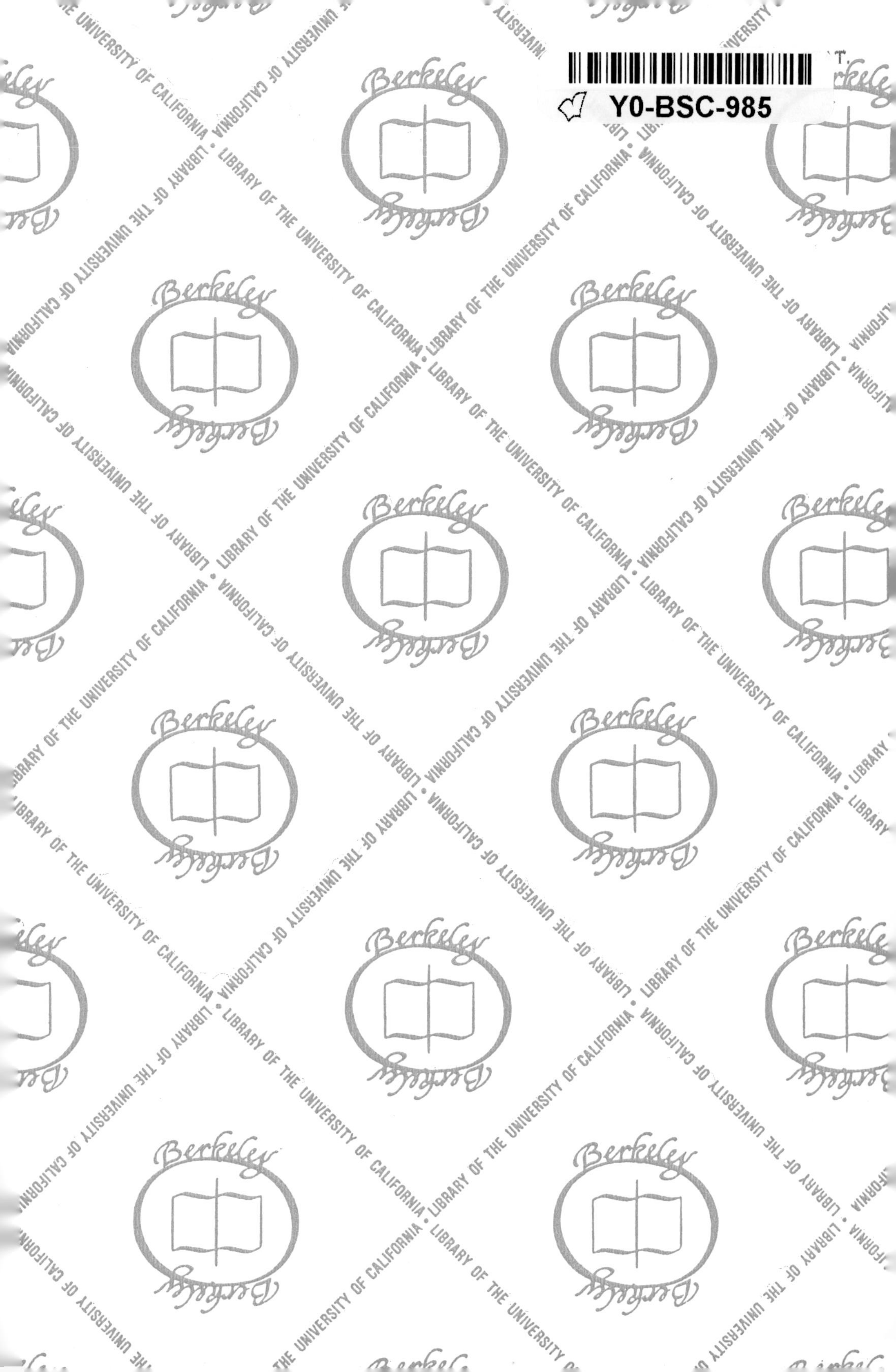
Y0-BSC-985
LIBRARY OF THE UNIVERSITY OF CALIFORNIA
Berkeley

MATH/STAT

COMBINATORIAL DESIGN THEORY

NORTH-HOLLAND MATHEMATICS STUDIES 149
Annals of Discrete Mathematics (34)

General Editor: Peter L. HAMMER
Rutgers University, New Brunswick, NJ, U.S.A.

Advisory Editors
C. BERGE, Université de Paris, France
M. A. HARRISON, University of California, Berkeley, CA, U.S.A.
V. KLEE, University of Washington, Seattle, WA, U.S.A.
J. -H. VAN LINT, California Institute of Technology, Pasadena, CA, U.S.A.
G. -C. ROTA, Massachusetts Institute of Technology, Cambridge, MA, U.S.A.

NORTH-HOLLAND – AMSTERDAM • NEW YORK • OXFORD • TOKYO

COMBINATORIAL DESIGN THEORY

Edited by

Charles J. COLBOURN
Department of Computer Science
University of Waterloo
Waterloo, Ontario
Canada

Rudolf MATHON
Department of Computer Science
University of Toronto
Toronto, Ontario
Canada

1987

NORTH-HOLLAND – AMSTERDAM • NEW YORK • OXFORD • TOKYO

MYLAR MATH

©Elsevier Science Publishers B.V., 1987

All rights reserved. No part of this publication may be reproduced, stored in a retrieval system, or transmitted, in any form or by any means, electronic, mechanical, photocopying, recording or otherwise, without the prior permission of the copyright owner.

ISBN: 0 444 70328 4

Publishers:
ELSEVIER SCIENCE PUBLISHERS B.V.
P.O. BOX 1991
1000 BZ AMSTERDAM
THE NETHERLANDS

Math Sep/ae

Sole distributors for the U.S.A. and Canada:
ELSEVIER SCIENCE PUBLISHING COMPANY, INC.
52 VANDERBILT AVENUE
NEW YORK, N.Y. 10017
U.S.A.

Library of Congress Cataloging-in-Publication Data

Combinatorial design theory / edited by Charles J. Colbourn, Rudolf Mathon.
p. cm. -- (North-Holland mathematics studies ; 149) (Annals of discrete mathematics ; 34)
Festschrift for Alex Rosa.
ISBN 0-444-70328-4 (U.S.)
1. Combinatorial designs and configurations. 2. Rosa, Alexander. I. Colbourn, C. J. (Charles J.), 1953- . II. Mathon, R. A. III. Rosa, Alexander. IV. Series. V. Series: Annals of discrete mathematics ; 34.
QA166.25.C652 1987
511'.6--dc19 87-22311
CIP

SD 2/11/88 Pd

PRINTED IN THE NETHERLANDS

QA166
.25
C6521
1987
MATH

PREFACE

Combinatorial design theory has experienced an explosive growth in recent years, both in theory and in applications. This volume is dedicated to one individual who has played a major role in fostering and developing combinatorial design theory, Alex Rosa.

Although Alex is a young man at the middle of his career, he is truly an individual worthy of such an honour. The strength and diversity of his contributions to the theory of Steiner systems over the past twenty years have directed research into new and exciting avenues, and thereby contributed dramatically to the vibrancy of research on designs found today. His research has encompassed difference methods, colourings, decomposition and partitioning, recursive and direct construction methods, analysis of designs, and relationships with graph theory and with geometry. But Alex's contributions extend beyond this impressive research. Virtually all design theorists have profited from Alex's astonishing ability to organize and disseminate research information. For the past fifteen years, he has maintained a bibliography on Steiner systems, which has proved a great boon to researchers; at the same time, he has unselfishly taken much of his time to help researchers by providing information about a vast range of queries on combinatorial designs. Alex's willingness and ability to help others make him a truly special individual.

In 1987, Alex Rosa is celebrating his fiftieth birthday. We take this opportunity to express our thanks to Alex for his contributions thus-far, and to honour him on the occasion of his fiftieth birthday.

This volume is a collection of forty-one research papers on combinatorial design theory and related topics. They extend the current state of knowledge on Steiner systems, Latin squares, one-factorizations, block designs, graph designs, packings and coverings; they develop recursive and direct constructions, analysis techniques, and computational methods. Hence the collection reflects the current themes in combinatorial design theory, and captures the multifaceted nature of the field. It is not possible here to summarize all of the contributions contained in the papers of this volume, so we simply remark that collectively the papers represent advances on a number of important problems.

We hope that readers of the volume find the individual papers as useful and as enjoyable as we have.

Charles J. Colbourn
Waterloo, Ontario
July, 1987

Rudolf A. Mathon
Toronto, Ontario
July, 1987

ACKNOWLEDGEMENTS

In a project such as the publication of this volume, there are always many people without whom the project would never reach fruition. We want to thank all of the people who provided assistance. Primarily we thank the authors for such an enthusiastic response, and for helping us meet a very tight schedule. We also thank the referees; we list their names here to acknowledge their invaluable assistance: J. Abrham, B.A. Anderson, F.E. Bennett, A.E. Brouwer, J.I. Brown, C.J. Colbourn, J.H. Dinitz, R. Fuji-Hara, M. Gionfriddo, T.S. Griggs, J.J. Harms, A. Hartman, K. Heinrich, P. Hell, M. Jimbo, D. Jungnickel, E.S. Kramer, D.L. Kreher, C.C. Lindner, S.S. Magliveras, R.A. Mathon, W. McCuaig, B.D. McKay, E. Mendelsohn, D.M. Mesner, W.H. Mills, R.C. Mullin, K.T. Phelps, V. Rödl, F.A. Sherk, D.R. Stinson, L. Teirlinck, S.A. Vanstone, and W.D. Wallis.

We also thank all of those who helped with preparing the final manuscript, including many of the authors, but primarily Karen Colbourn and Zoe Kaszas. Finally, we thank Peter Hammer and the mathematics staff at North-Holland for supporting the project and providing excellent technical assistance, as always.

CONTENTS

Annals of Discrete Mathematics 34 (1987) 1—26
© Elsevier Science Publishers B.V. (North-Holland)

The Existence of Symmetric Latin Squares with One Prescribed Symbol in each Row and Column

L.D. Andersen

Department of Mathematics
Institute of Electronic Systems
Aalborg University
Strandvejen 19
DK-9000 Aalborg
Denmark

A.J.W. Hilton

Department of Mathematics
University of Reading
Whiteknights
Reading RG6 2AX
United Kingdom

Visiting:

Division of Mathematics
Auburn University
Auburn
Alabama 36849
U.S.A.

TO ALEX ROSA ON HIS FIFTIETH BIRTHDAY

ABSTRACT

Let P be a partial symmetric $n \times n$ Latin square in which up to $n+1$ entries are specified such that there is at least one specified entry in each row and column. We say exactly when P can be completed to an $n \times n$ symmetric Latin square L. This is the first part of a proof of the symmetric Latin square analogue of the Evans Conjecture.

1. Introduction

A *partial* $n \times n$ *Latin square* P is an $n \times n$ matrix in which some cells may be empty and the non-empty cells will contain exactly one symbol, such that no symbol occurs more than once in any row or in any column. P is called *symmetric* if whenever a cell (i,j) contains some symbol, then cell (j,i) also contains the same symbol.

Our concern in this paper is to characterize those partial symmetric $n\times n$ Latin squares with up to $n+1$ cells occupied, satisfying the restriction that there is at least one occupied cell in each row and in each column, which can be completed to a symmetric $n\times n$ Latin square.

Before we state our main result we recall that in a symmetric Latin square of even side, each symbol occurs an *even number* of times on the diagonal, and in a symmetric Latin square of odd side, each symbol occurs *exactly once* on the diagonal. Both these facts are easy to see. We shall call the diagonal of a partial symmetric Latin square of side n *admissible* if the number of symbols occurring a number of times not congruent to n modulo 2 is less than or equal to the number of empty cells (so that the parity can be made right for every symbol with the 'wrong' parity). Let $t(i)$ denote the number of times that a symbol i occurs on the diagonal. Then, if the symbol set is $\{1,...,n\}$, the condition for admissibility can be written.

$$|\{i : t(i) \neq n \pmod 2\}| \leq n - \sum_{i=1}^{n} t(i).$$

For odd n, this simply means that a diagonal is admissible if and only if no symbol occurs more than once on it. Obviously, a partial symmetric Latin square cannot be completed to a symmetric Latin square if its diagonal is not admissible.

Our main result is the following:

Theorem 1. Let $n \geq 3$ and let P be a partial symmetric Latin square of side n with admissible diagonal; suppose also that there is at least one occupied cell in each row and in each column. Let c be the number of non-empty cells of P.

If $c=n$, then P can be completed to a symmetric Latin square of side n if and only if P is not of the form of any of the Types $E1$, 01 or 02 of Figure 1.

If $c=n+1$, then P can be completed to a symmetric Latin square of side n if and only if P is neither of the form of any of the types $E1$, 01 or 02 with a further diagonal cell filled, nor of the form of the types $E2$ or $L5$ of Figure 1.

Remark. No doubt the meaning of the phrase 'of the form of' is self-evident. But, to be formal, a partial Latin square is of the form of one of these Types if it can be transformed to a partial Latin square of one of these Types by permuting rows, permuting columns the same way, and relabelling the symbols.

It is easy to see that the partial squares in Figure 1 cannot be completed symmetrically. In types $E1$ and 01 the symbol 1 needs an extra diagonal occurrence but cannot get it; in Types $E2$ and 02 it is impossible to make the symbol 1 occur in the first row; finally we leave it to the reader to check that Type $L5$ cannot be completed.

The analogous problem for Latin squares with no symmetry requirement to the problem considered here was settled by Chang [6]; he characterized the possible diagonals of a Latin square. A completely different proof of Chang's result was given by Hilton and Rodger [10].

The result in this paper is a necessary preliminary to a more general result [5] in which we characterize the partial symmetric $n\times n$ Latin squares with up to $n+1$ cells occupied which can be completed. For our proof of this more general result, we need to have the result in this paper proved separately, for the general method of that paper

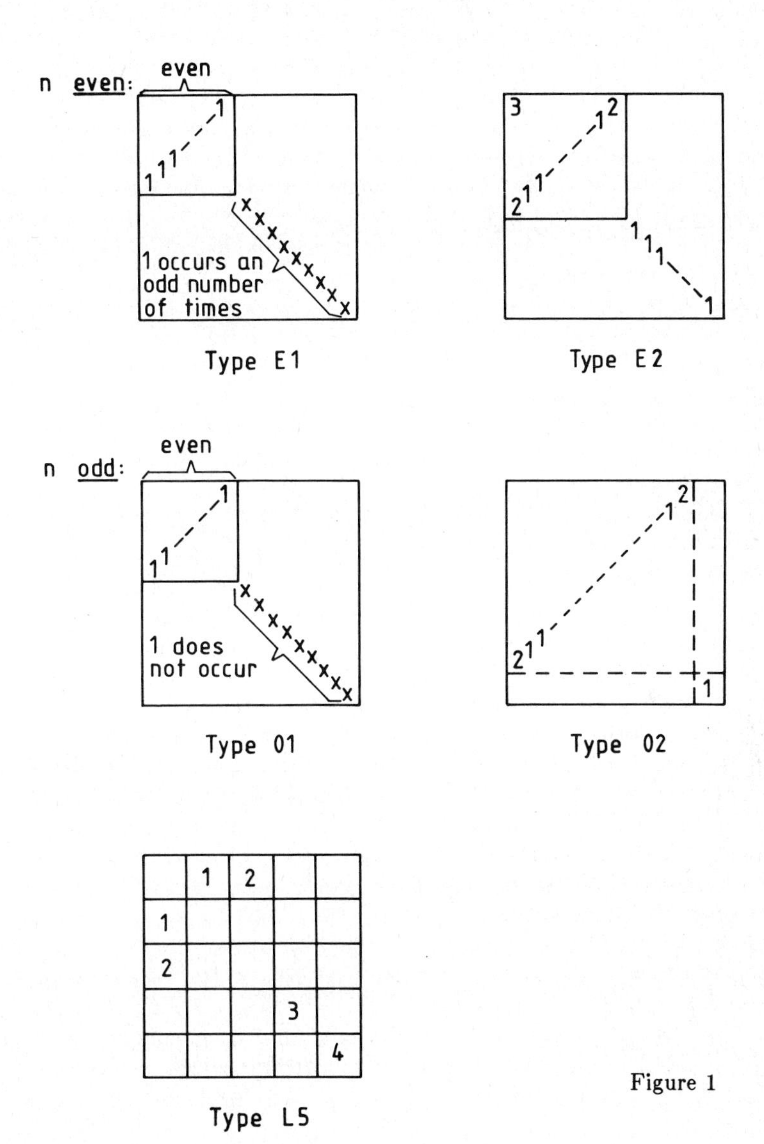
n even:
even
1 occurs an odd number of times
Type E1
Type E2
n odd:
even
1 does not occur
Type 01
Type 02
Type L5

Figure 1

fails when every row and column has an occupied cell. This general question has been known and unsolved for a number of years.

The analogous more general problem for Latin squares with no symmetry requirement was settled by Smetaniuk [12] and, independently, Andersen and Hilton [3]. Smetaniuk only considered the case when $n-1$ cells were occupied; Andersen and Hilton considered the case when n cells were occupied. Damerell [8] extended Smetaniuk's method to cover the case when n cells were occupied. Finally, Andersen [2] dealt with the case when $n+1$ cells were occupied.

2. Edge-Colourings of Graphs

In this section we discuss briefly edge-colourings of graphs, and the relationship between symmetric Latin squares and edge-colourings of complete graphs with loops. The reason for this is that the proof of Theorem 1 is easier to present in terms of edge-colourings.

An *edge-colouring* of a graph G is an assignment of a colour to each edge and loop of G in such a way that all edges and loops incident with the same vertex have distinct colours. The *chromatic index* $CHI'(G)$ of G is the least integer k for which G has an edge-colouring with k colours. Let K_n^ℓ denote the graph obtained from the complete graph K_n by adding a loop at each vertex. It is well-known that $CHI'(K_n^\ell)=n$.

A symmetric Latin square S of side n gives rise to an edge-colouring of K_n^ℓ as follows. Let the vertices of K_n^ℓ be $v_1, \cdots, v_n$. Define the colour of the edge v_iv_j to be the entry of cell (i,j) of S, and the colour on the loop on v_i to be the entry of cell (i,i) of S, $i\neq j$, $1\leq i\leq n$, $1\leq j\leq n$. This clearly defines an edge-colouring of K_n^ℓ with n colours, and it obviously works the other way round as well; it also works for partial symmetric Latin squares and edge-coloured subgraphs of K_n^ℓ. In this connection, we note that if an edge-coloured subgraph of K_n^ℓ corresponds to a partial symmetric Latin square P, then the number of filled cells of P is equal to the degree sum $\sum_{i=1}^{n} d(v_i)=2\,|E(G)|+|L(G)|$, where $E(G)$ is the edge-set of G and $L(G)$ the loop-set of G (a loop is not an edge; a loop on v_i contributes one to the degree $d(v_i)$ of (v_i). We shall say that G has *admissible loop-colouring* if the diagonal of P is admissible.

In this terminology, Theorem 1 becomes:

Theorem 2. Let $n\geq 3$ and let G be an edge-coloured subgraph of K_n^ℓ with an admissible loop-colouring; suppose also that each vertex of G has at least one edge or loop on it, and that G spans K_n^ℓ. Let $t=2\,|E(G)|+|L(G)|$.

If $t=n$, then the edge-colouring of G can be extended to an edge-colouring of K_n^ℓ with n colours if and only if G is not any of the Types $e1$, $o1$ or $o2$ of Figure 2.

If $t=n+1$, then the edge-colouring of G can be extended to an edge-colouring of K_n^ℓ with n colours if and only if G is not any of the Types $e1$, $o1$ or $o2$ with a further loop added, nor of the form of Types $e2$ or $g5$ of Figure 2.

When G is one of the edge-coloured graphs of Figure 2, possibly with a loop added in case $e1$, $o1$ or $o2$ we refer to G as being a *bad case*.

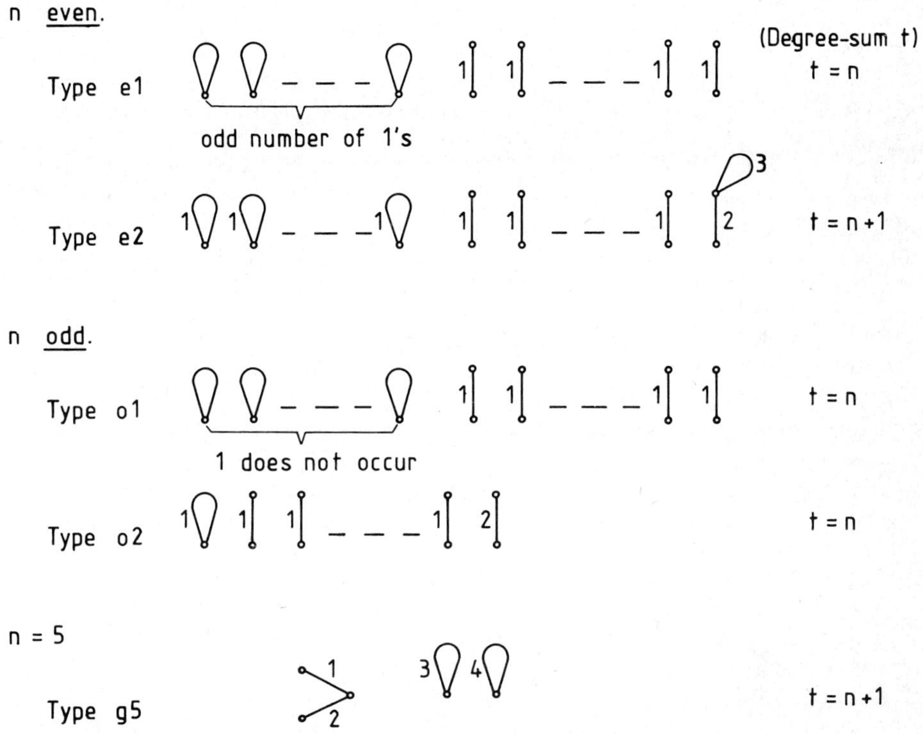

n even. (Degree-sum t)

Type e1 — odd number of 1's — $t = n$

Type e2 — $t = n+1$

n odd.

Type o1 — 1 does not occur — $t = n$

Type o2 — $t = n$

n = 5

Type g5 — $t = n+1$

Figure 2

In G, a loop on a vertex with no edge of G incident with it, is called a *free loop.* Similarly an edge of G which is not adjacent to any other edge or loop of G is called a *free edge.* The letter ℓ denotes the number of free loops of G.

3. Preliminary Result

We make extensive use of the following result due independently to Hoffman [11] and Andersen [1].

Theorem 3. Let G be a subgraph of K_n^ℓ consisting of a K_r^ℓ and possibly some free loops. Let G have an edge-colouring with n colours, $c_1, \ldots, c_n$, and let the loop-colouring be admissible. For $1 \leq k \leq n$ let $\ell(k)$ be the number of free loops of colour k. Then the edge-colouring can be completed to an edge-colouring of K_n^ℓ with $c_1, \ldots, c_n$ if and only if, for $1 \leq k \leq n$, c_k occurs at at least $2r - n + \ell(k)$ vertices of the K_r^ℓ.

A more general result was recently proved by Andersen and Hilton [4]. The special case of Theorem 3 in which G consists of just the K_r^ℓ, and there are no free loops, is often what we need. This special case was due to Cruse [7].

4. Proof of the Main Theorem

In this section we prove Theorem 2 (and so also Theorem 1, as they are equivalent). In places, we leave out some details having certain similarities to other parts of the proof. The full proof is available from the authors.

Proof of Theorem 2. We showed in Section 1 that the exceptional cases of the theorem cannot be completed. Let G be a graph with n vertices, each of which is incident with at least one edge or loop of G, the degree-sum being n or $n+1$; suppose also that G is not a bad case, and that the loop-colouring is admissible. It remains to show that the edge-colouring of G can be extended to an edge-colouring of K_n^ℓ with n colours. We do this by induction on n.

We leave it to the reader to check that, if $n \leq 5$, then such an edge-colouring can be completed.

Now let $n \geq 6$ and assume that the theorem is true for complete graphs with loops which have fewer than n vertices.

It is easy to see that G must be of the form of (i), (ii) or (iii) of Figure 3, as the conditions imply that all vertices of G have degree 1 except possibly for one vertex having degree 2.

If G is of the form of (iii) with n loops and no edges edges it can be completed by Theorem 3, and if it has at least one edge we can add a loop to make it of the form of (ii). As we can easily avoid Type $e2$, the new graph can be assumed not to be a bad case. Henceforth we shall only consider cases (i) and (ii).

Our first move is to add, for each i, $1 \leq i \leq n$, a further coloured loop of colour c_i if c_i occurs on a number of loops incongruent to n modulo 2. Thus if n is odd, each uncoloured loop receives a colour, and if n is even, we obtain an even number of loops of each colour, but some (an even number) of loops may remain uncoloured.

We now show that this can be done. Let s be the number of colours occurring on a number of loops incongruent to n modulo 2, i.e. occurring with the "wrong" parity. If $s=0$, then we have nothing to prove, so suppose that $s>0$ and that the colours in question are $c_1,\ldots,c_s$. For $1 \leq i \leq s$, let W_i be the set of vertices with uncoloured loops, the vertices being incident with no edge of colour c_i. We simply wish to find a system of distinct representatives for the sets W_i. By P. Hall's well-known theorem [9], we must show that

$$\left| \bigcup_{i \in I} W_i \right| \geq |I| \quad (\forall I \subseteq \{1, \ldots, s\}).$$

But if $|I| \geq 2$, $\bigcup_{i \in I} W_i$ contains all vertices with uncoloured loops, except possibly the vertex y in case (i) if $|I|=2$. In case (i) there are at least three uncoloured loops, so in this case, if $|I|=2$, $\bigcup_{i \in I} W_i$ contains at least two elements. Thus we need only consider the case $|I|=1$. But clearly any W_i contains an element, except if all edges have colour c_i. This is impossible in case (i), and in case (ii) it corresponds to G being of

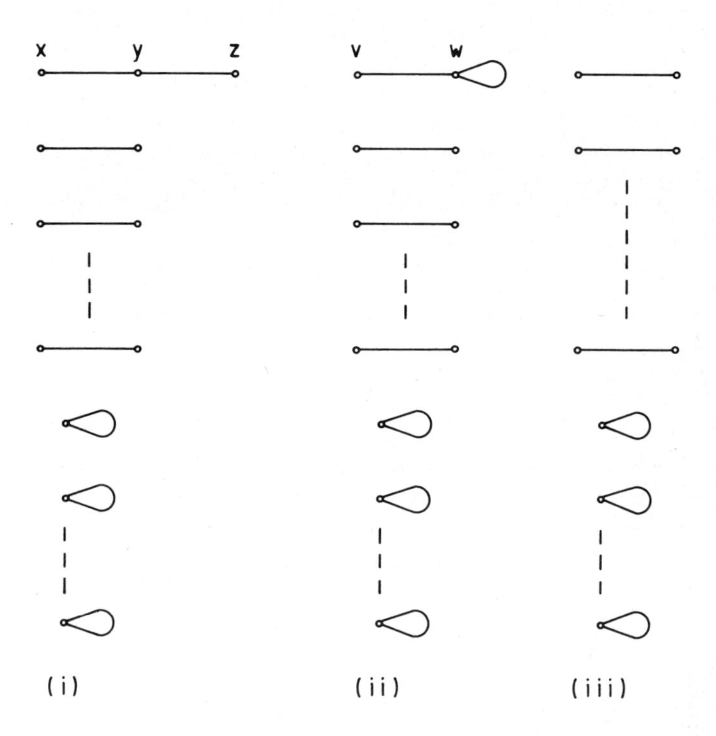

Figure 3

Type $e1$ or $o1$ with a loop added. Thus we can indeed find the required system of distinct representatives, and so we can colour a further loop for each of the colours $c_1, \ldots, c_s$.

We now consider various cases.

Case 1. n even.

Case 1a. $\ell \geq \frac{n}{2}$. Let $K^{\ell}_{n/2}$ be a subgraph of the K^{ℓ}_n containing G, and suppose that all edges of G are inside $K^{\ell}_{n/2}$. Suppose that some further coloured loops have been added as described above, so that the number of loops of each colour is now even (but some loops may remain uncoloured). We now show that, with one exception, we can colour the edges of $K^{\ell}_{n/2}$ so that each colour occurs at least as many times (i.e. with at least as great degree-sum) in $K^{\ell}_{n/2}$ as outside. We do this by colouring the uncoloured edges of $K^{\ell}_{n/2}$ one by one, keeping a proper edge-colouring at each stage, and , at each stage, using a colour which, at the stage before, occurred more times outside $K^{\ell}_{n/2}$ than inside.

Suppose that this process stops because there is some colour, say c_1, which occurs more times outside $K_{n/2}^{\ell}$ than inside, and which cannot be placed on any uncoloured edge of $K_{n/2}^{\ell}$ because at least one end-vertex of each uncoloured edge is incident with an edge or loop coloured c_1.

Let v be the number of vertices of $K_{n/2}^{\ell}$ which are not incident with edges or loops coloured c_1. The number of edges between these v vertices is $\frac{1}{2}v(v-1)$, and the number of precoloured edges joining pairs of these vertices was at most $\frac{1}{2}(v+1)$. Thus the number of edges originally available for colouring was at least $\frac{1}{2}v(v-1)-\frac{1}{2}(v+1)$. Since there are v vertices in $K_{n/2}^{\ell}$ not incident with edges or loops coloured c_1, there are $\frac{1}{2}n-v$ vertices in $K_{n/2}^{\ell}$ which are incident with such edges or loops. Therefore there are at least $(\frac{1}{2}n-v)+1$ loops outside $K_{n/2}^{\ell}$ coloured c_1, and thus the number of loops outside $K_{n/2}^{\ell}$ which were not coloured c_1 is at most $v-1$. Hence at most $v-1$ edges within $K_{n/2}^{\ell}$ have to be coloured with colours other than c_1, and so we have

$$\tfrac{1}{2}v(v-1) - \tfrac{1}{2}(v+1) \leq v-1,$$

from which it follows that $v \leq 3$.

If $v \leq 3$, then the number of loops outside $K_{n/2}^{\ell}$ not coloured c_1 is at most two. But since the number of vertices at which c_1 occurs is even, either $v=3$ and c_1 occurs on $\frac{1}{2}n-1$ loops outside $K_{n/2}^{\ell}$, or $v=2$ and c_1 occurs on all loops outside $K_{n/2}^{\ell}$.

In the first case, let the colour of the last loop outside $K_{n/2}^{\ell}$ and c_2; then the colour c_2 has already been ascribed to some loop of $K_{n/2}^{\ell}$, so all non-prescribed edges coloured during the process have colour c_1, contradicting that the K_3 has no available edges. Therefore we can assume that $v=2$ and that all free loops are coloured c_1. In that case, if we have given the colour c_1 to a non-prescribed edge of $K_{n/2}^{\ell}$, we remove it to have 4 vertices without c_1 on them. Then 2 independent edges in this K_4 are not prescribed and both can be given the colour c_1. A similar argument holds if a loop of colour c_1 was added during the first part of the proof. So we may assume that c_1 is prescribed at all but 2 vertices. Then since G is not the bad case $e2$, both the loops on the two vertices which do not already have a loop or edge incident with them coloured c_1 can themselves be coloured c_1 (this is the exception referred to earlier; it is exceptional as we have here to colour loops c_1 rather than edges).

We now colour the remaining edges of $K_{n/2}^{\ell}$ one by one, and we also colour the loops in pairs with the same colour. Since each vertex with an uncoloured edge or loop has at most $\frac{n}{2}-1$ colours occurring on the loop or the edges incident with it, at most $2(\frac{n}{2}-1)=n-2$ colours are not available for an uncoloured edge or an uncoloured pair of loops, so this can be done greedily; i.e. one edge or loop-pair at a time, in any order, without ever recolouring. When this process is complete, the loop-colouring is admissible.

The conditions of Theorem 3 are now satisfied and so the edge-colouring can be extended to an edge-colouring of K_n^{ℓ}.

Case 1b. $\frac{1}{2}n$ is odd, $\ell=\frac{1}{2}(n-2)$, G is Case (ii) of Figure 3, and the loops of G all have different colours.

In this case, there are $\frac{1}{2}n$ loops, all of different colours, so that when the initial step of colouring uncoloured loops so that each colour appears on an even number of loops is completed, all loops will be coloured. Consider the $K^{\ell}_{\frac{1}{2}(n+2)}$ containing all the edges of G. One colour, say c_1, occurs on two loops of $K^{\ell}_{\frac{1}{2}(n+2)}$. If we show that we can colour all the uncoloured edges of $K^{\ell}_{\frac{1}{2}(n+2)}$ in such a way that each colour apart from c_1 occurs on at least one edge, then, by Theorem 3, we can complete the edge-colouring to K^{ℓ}_n.

The graph $K_{\frac{1}{2}(n+2)}$ is the union of the 1-factor consisting of all the edges of G, and $\frac{n}{2}-1$ further 1-factors, say $F_2, \ldots, F_{n/2}$. Colour the edges of F_2 with some colour different from c_1, not used on the edges or loops of G, say c_n. Let the colours different from c_1 occurring on the loops of G be $c_2, \ldots, c_{n/2}$.

Suppose first that $n \geq 10$. If c_2 does not occur on an edge of G, colour an edge of $K^{\ell}_{\frac{1}{2}(n+2)}$ with c_2. For $3 \leq i \leq n/2$, F_i has at least one uncoloured edge not incident with the loops coloured c_i; if c_i does not occur on an edge of G, colour such an edge with c_i. There are now at least $2(\frac{n}{2}-2)-1=n-5 \geq \frac{n}{2}-1$ further uncoloured edges. For each of the $\frac{n}{2}-1$ remaining colours, if it does not occur on an edge of G, place it one one of the uncoloured edges. Finally, colour all the remaining uncoloured edges of $K^{\ell}_{\frac{1}{2}(n+2)}$ greedily (this can certainly be done since all edges of F_2 are coloured the same (c_n)).

Since $n/2$ is odd, the remaining value of n to be considered is $n=6$. This is left to the reader.

Case 1c. $n/2>\ell$, but Case 1b does not apply.

In this case, before we colour some uncoloured loops so as to make the number of loops of each colour even, we find a free edge of G, and we replace this free edge by two loops of the same colour. We continue to do this until the number of coloured free loops reaches $\frac{1}{2}n$ or $\frac{1}{2}n+1$. When we perform this operation the number of loops of the 'wrong parity' does not change, and, when we stop, the number of loops of the 'wrong parity' is no greater than the number of uncoloured loops; therefore the loop-colouring is still admissible. Also, when we perform this operation, we avoid creating a base case: We cannot create a Type $e2$, as G is not of this Type; and if all edges of G have the same colour, we cannot create a Type $e1$ as G does not contain this Type: finally, if G is case (ii) of Figure 3 and has two edges of distinct colours, we avoid replacing two such edges by loops; in this way we can avoid creating a Type $e1$ unless $n=6$, $\ell=2$ with the free loops having colours c_1 and c_2, say, c_1 also occurring on the edge vw. As Case 1b does not apply, we can assume that the loop on w has colour c_2. An easy application of Theorem 3 completes this special case directly. In all other cases, we then colour some uncoloured loops so as to 'correct the parity' in the usual way.

By Case 1a, this modified edge-colouring can be completed; finally we interchange the colours between each of the extra pairs of coloured loops we introduced and the edge joining them. This gives the required edge-colouring of K_n^ℓ which contains the edge-colouring of G.

Case 2. n odd.

In this case, after we have ensured that each colour appears on a number of loops congruent to n modulo 2, each loop has a colour, and each colour is on exactly one loop.

Case 2i. $\ell \geq \frac{n-1}{2}$.

Let $K^\ell_{\frac{n+1}{2}}$ be a subgraph of K_n^ℓ containing all edges of G. For each colour on a loop outside $K^\ell_{\frac{n+1}{2}}$ and not on an edge of G, colour an edge of $K^\ell_{\frac{n+1}{2}}$. This is easy to do. When this is done, colour the remaining edges greedily. By Theorem 3 this may be extended to an edge-colouring of K_n^ℓ, as required.

Case 2ii. $\ell = \frac{n-3}{2}$.

Since n is odd, either $n \equiv 1$ $(mod\ 4)$, ℓ is odd and Case (ii) of Figure 3 applies, or $n \equiv 3$ $(mod\ 4)$, ℓ is even and Case (i) of Figure 3 applies.

$K^\ell_{\frac{n+3}{2}}$ be the subgraph of K_n^ℓ containing all the edges of G. If we extend the edge-colouring of G so that each edge of $K^\ell_{\frac{n+3}{2}}$ is coloured, in such a way that each colour on a loop of $K^\ell_{\frac{n+3}{2}}$ occurs on at least one edge of $K^\ell_{\frac{n+3}{2}}$, and each colour on a loop outside $K^\ell_{\frac{n+3}{2}}$ occurs on at least two edges of $K^\ell_{\frac{n+3}{2}}$, then, by Theorem 3, this edge-colouring can be completed to an edge-colouring of K_n^ℓ.

Case 2iia. $n \equiv 1$ $(mod\ 4)$, ℓ is odd and G has the form of (ii) of Figure 3.

Case 2iia1. $n \geq 13$. $K^\ell_{\frac{n+3}{2}}$ has a 1-factorization in which one 1-factor consists of the edges of G, and there are $\frac{n-1}{2}$ further 1-factors, say $F_1, \ldots, F_{\frac{n-1}{2}}$. At most $\frac{n+3}{4}$ colours appear on the edges of G; $\frac{1}{2}(n-3)$ colours appear on loops outside $K^\ell_{\frac{n+3}{2}}$. Since $n \geq 13$, it follows that

either (a) there are two colours, say c_1 and c_2, which occur on loops outside $K^\ell_{\frac{n+3}{2}}$ but not on any edge of G, or (b) $n=13$ and the four edges of G are coloured with distinct colours, each such colour occurring on a loop outside $K^{\ell}_{\frac{n+3}{2}}$.

Suppose first that (a) occurs. Colour the edges of F_1 and F_2 with c_1 and c_2 respectively. Each 1-factor contains at least four edges. There are $\frac{n-3}{2}-2$ further colours on loops outside $K^{\ell}_{\frac{n+3}{2}}$, say $c_3, \ldots, c_{\frac{n-3}{2}}$. For $3 \leq i \leq \frac{n-3}{2}$, colour two edges of F_i with colour c_i if c_i appears on no edge of G, and colour one edge of F_i with colour c_i if c_i appears on one edge of G. When this is done, $F_3, \ldots, F_{\frac{n-3}{2}}$ each have at least two uncoloured edges, and $F_{\frac{n-1}{2}}$ has at least four uncoloured edges. Each of the $\frac{n+3}{2}$ colours, say $c_{\frac{n-1}{2}}, \ldots, c_n$, used on loops of $K^{\ell}_{\frac{n+3}{2}}$ which are not used on an edge of G, must be placed on one of these uncoloured edges not adjacent to the loop of that colour. This is easy to do: If no edges of G are coloured with any of $c_{\frac{n-1}{2}}, \ldots, c_n$, then colour two uncoloured edges of F_3 with colours occurring on loops on the other of the two edges, colour four edges of $F_{\frac{n-1}{2}}$ with further colours occurring on loops of some other of the four edges, and finally colour one uncoloured edge of each of $F_4, \ldots, F_{\frac{n-3}{2}}$. If some edges of G are coloured with some of $c_{\frac{n-1}{2}}, \ldots, c_n$, then carry out the process just described ignoring the edges of G, and then remove the colour from any edge which is adjacent to an edge of G of the same colour. Finally colour the remaining uncoloured edges of $K^{\ell}_{\frac{n+3}{2}}$ greedily. This is possible since there are two colours, c_1 and c_2, which occur at every vertex, so at most $2(\frac{n+3}{2}-1)-2=n-1$ colours are forbidden for an uncoloured edge.

Now suppose that (b) occurs. Then $\frac{n+3}{2}=8$. Suppose that the loops outside K^{ℓ}_8 are coloured c_1, c_2, c_3, c_4 and c_5, and that G has an edge of each of the colours c_1, c_2, c_3 and c_4. Let $F_1, \ldots, F_7$ be edge-disjoint 1-factors of K^{ℓ}_8 such that, for $1 \leq i \leq 4$, F_i contains the edge of G coloured c_i. (It is easy to see by direct construction that this is possible, or one could appeal to the recent strong result of Woolbright (private communication) that in any edge-colouring of K_{2n} with $2n-1$ colours, there is a set of n independent edges, each coloured differently). Colour the edges of $F_1, \ldots, F_5$ with colours $c_1, \ldots, c_5$ respectively. $F_6 \cup F_7$ either consists of two disjoint C_4's, or it is a C_8. In either case it is easy to colour the eight edges of $F_6 \cup F_7$ with the colours occurring on the loops of K^{ℓ}_8, say $c_6, \ldots, c_{13}$ so that each colour is on exactly one edge, and so that no loop of any colour is incident with an edge of that colour.

Case 2iia2. $n=9$. Left to the reader.

Case 2iib. $n \equiv 3 (mod\ 4)$, ℓ is even and G has the form of (i) of Figure 3.

$K^{\ell}_{\frac{n+3}{2}}$ has a near 1-factorization with one near 1-factor, say F_0, containing all bar one edges of G, and $\frac{n+1}{2}$ further near 1-factors, say $F_1, \ldots, F_{\frac{n+1}{2}}$, with $F_{\frac{n+1}{2}}$ containing the edge yz of (i) of Figure 3. Note that $\frac{n+1}{2} \geq 4$. If all but one edges of G have the same colour, let F_0 be chosen to contain edges of this colour only.

Let the colours occurring on the loops outside $K^{\ell}_{\frac{n+3}{2}}$ be $c_{\frac{n+5}{2}}, \ldots, c_n$, and let the colours placed at the start on the loops of $K^{\ell}_{\frac{n+3}{2}}$ be $c_1, \ldots, c_{\frac{n+3}{2}}$. Just for this case, we remove the colours $c_1, \ldots, c_{\frac{n+3}{2}}$ from the loops to which they were assigned.

At most $1+½(\frac{n+1}{2})$ colours occur on the edges of G. For $n \geq 7$, $\frac{n+3}{2} \geq 2+(1+½(\frac{n+1}{2}))$, so there are at least two of the colours $c_1, \ldots, c_{\frac{n+3}{2}}$ which do not occur on the edges of G. Suppose c_1 and c_2 are two such colours. For $i=1,2$, colour the edges of F_1 with colour c_i, and also colour the loop on the vertex of $K^{\ell}_{\frac{n+3}{2}}$ not incident with F_i with colour c_i.

We now assign the colours $c_3, \ldots, c_{\frac{n+3}{2}}$ to the uncoloured loops. An argument similar to the earlier one, based on Hall's theorem, can be used to show that this is possible.

For $\frac{n+5}{2} \leq i \leq n$, we now attempt to place c_i on two uncoloured edges of $K^{\ell}_{\frac{n+3}{2}}$ if c_i is not on any edge of G, and on one uncoloured edge of G if c_i is on exactly one edge of G. For $1 \leq i \leq \frac{n-5}{2}$, the near 1-factor F_{2+i} has $\frac{n+1}{4}$ uncoloured edges, and the near 1-factor $F_{\frac{n+1}{2}}$ has $\frac{n-3}{4}$ uncoloured edges. For $n \geq 11$ and $1 \leq i \leq \frac{n-3}{2}$, therefore, if no edge of G is coloured $c_{\frac{n+3}{2}+i}$, then two edges of the near 1-factor F_{2+i} can be coloured $c_{\frac{n+3}{2}+i}$. For $n \geq 11$ and $1 \leq i \leq ½(n-5)$, and for $n \geq 15$ and $i=½(n-3)$, if one edge of G is coloured $c_{½(n+3)+i}$, there are $\frac{n+1}{4}-2 \geq 1$, or $\frac{n-3}{4}-2 \geq 1$, edges available to colour $c_{½(n+3)+i}$. If $n=11$, the colours c_8, c_9, c_{10}, c_{11} can be relabelled to make this work unless each occurs on exactly one edge of G. But then we may assume that the edge yz (of (i) of Figure 3) is coloured c_{11}, and so a second edge of F_6 can be coloured c_{11} without trouble.

Now suppose that $n=7$ and one of the two edges xy and zy (of (i) of Figure 3) is coloured with one of c_6, c_7; we may suppose that yz is coloured c_7. Then, by permuting F_1, F_2, F_3 if necessary, we may colour one or both edges of F_3, as required, with c_6 and the further edge of F_4 with c_7 also. Finally suppose that $n=7$ and that the two edges

xy and yz are both coloured from $\{c_1, \ldots, c_5\}$, say with c_1 and with c_2 respectively. Then there are three possibilities: the further edge of G is coloured either

(a) with one of c_1, c_2, say c_1; or

(b) with one of c_3, c_4, c_5, say c_3; or

(c) with one of c_6, c_7, say c_6.

Then we remove once again the colours from the loops of K_5^ℓ, and we complete the colouring of K_5^ℓ in the required fashion as shown in Figure 4.

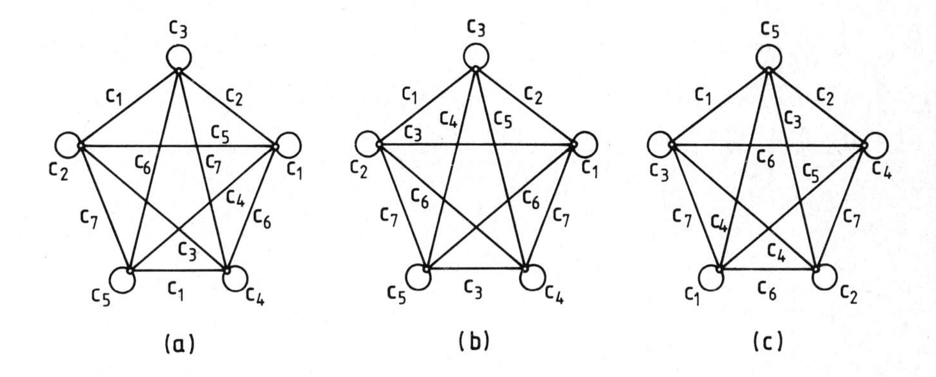

Figure 4

From now on, we shall assume that this case, when $n=7$ and xy and yz are both coloured from $\{c_1, \ldots, c_5\}$, does not occur.

Finally we attempt to show that each of the colours $c_3, \ldots, c_{\frac{n+3}{2}}$, if it does not occur on one or more edges of G, can be placed on an as yet uncoloured edge of $K^\ell_{\frac{n+3}{2}}$. This can almost always be done, and when it cannot be done, we show what to do instead. Recall that c_1 and c_2 have been placed already on $\frac{n+1}{4}$ edges of $K^\ell_{\frac{1}{2}(n+3)}$. The only constraint is that for $3 \leq i \leq \frac{1}{2}(n+3)$, c_i cannot be placed on an uncoloured edge incident with the loop of colour c_i. We perform the colouring, colour by colour; we choose the colour, and the edge to put it on, according to the following procedure.

First suppose there is a vertex v with two or more uncoloured edges incident with it. We choose a colour c which has not so far been used on an edge of $K^{\ell}_{½(n+3)}$ and other than the colour used on the loop on v; we choose an edge incident with v, but not incident with a vertex with a loop of colour c, and we place c on this edge. This is possible unless there is just one colour, say c', left which has to occur on an edge of $K_{½(n+3)}$, and c' is on the loop on v. In this case, either c' can be put on some other edge; or all uncoloured edges are incident with v, and we let vw be any such edge and recolour an edge of colour c_1 or c_2 incident with w with the colour c' - the remaining edges can then be coloured greedily. Now suppose that each vertex with an uncoloured edge incident with it has only one such uncoloured edge. Then a suitable edge and colour can be chosen provided there are at least two uncoloured edges, or, when there is exactly one uncoloured edge, say e, if there is a colour not so far placed on an edge of $K^{\ell}_{\frac{n+3}{2}}$ which does not occur on one of the two loops of e. In any case, if there are at least three colours not placed so far, and just one uncoloured edge, then the uncoloured edge can be coloured. If the final uncoloured edge e cannot be coloured because there are only one or two colours, including, say, c, not placed so far, and these are on loops on the vertices of e, suppose that c is on a loop on the vertex v, and that w is the other end-vertex of e; then recolour an edge coloured c_1 or c_2 (say c_1) on w with the colour c. Similarly, if there is still an unplaced colour, say c', on a loop on w, recolour an edge coloured c_2 incident with v with colour c'. It is easy to see that e can then be coloured greedily. Finally, if there are still some colours in $\{c_3, \ldots, c_{\frac{n+3}{2}}\}$ not placed on any edges, then we choose edges coloured c_1 and c_2, and recolour them, leaving at least one edge of each c_1 and c_2. Clearly this process can be continued until there is only one edge of each colour c_1 and c_2 left.

Let δ be the number of colours $c_3, \ldots, c_{½(n+3)}$ occurring on the edges xy and yz. The number of colours of $c_3, \ldots, c_{½(n+3)}$ which need to be placed on edges of $K^{\ell}_{½(n+3)}$ is not more than $½(n-1)-\delta$. The number of edges which can be coloured, or recoloured, in the process described in the paragraph above is at least

$$\binom{½(n+3)}{2} - \left(\frac{n+1}{4}+1\right) - [2\,\frac{n-3}{2} - (2-\delta)] - 2$$

The inequality

$$\binom{½(n+3)}{2} - \frac{n+1}{4} - 1 - \delta - (n-3) \geq ½(n-1) - \delta$$

reduces to

$$n(n-10) \geq -21$$

and this is true for $n \geq 7$. Thus each of $c_3, \ldots, c_{½(n+3)}$ can be placed on a suitable edge of $K_{½(n+3)}$, as required.

Finally, if there are some edges of $K^{\ell}_{½(n+3)}$ still uncoloured, then they can be coloured greedily, as in Case 2iia.

Case 2iii. $\ell \leq \frac{n-5}{2}$.

In this case we do not, as an initial step, colour the loops of K_n^ℓ not in G.

We remark that, so far, we have not used induction. We do so however in this case.

Case 2iiia. $n \equiv 1 \ (mod\ 4)$. Then $\frac{1}{2}(n+1)$ is odd.

Case 2iiia1. G has the form of (i) of Figure 3.

We may assume that the edges of G are all coloured using $\{c_1, \ldots, c_{\frac{1}{2}(n+1)}\}$. Partition the graph G into disjoint subgraphs L and R of order $\frac{1}{2}(n+1)$ and $\frac{1}{2}(n-1)$ respectively so that R contains all the loops coloured with $c_1, \ldots, c_{\frac{1}{2}(n+1)}$, but no other loops, and so that L contains the two adjacent edges of G. Notice that if the edges of G use $\frac{1}{2}(n+1)$ colours, then G has no loops; the number of loops is even. If at most $\frac{1}{2}(n-1)$ colours are used on the edges and there is at least one loop whose colour is not used on an edge, then, for at least one loop, a choice can be made as to whether to include its colour in $\{c_1, \ldots, c_{\frac{1}{2}(n+1)}\}$ or in $\{c_{\frac{1}{2}(n+1)+1}, \ldots, c_n\}$. Therefore, no parity difficulties arise in performing this partition. Since $\ell \leq \frac{1}{2}(n-5)$, R contains at least $\frac{1}{2}(\frac{1}{2}(n-1)-\ell) \geq \frac{1}{2}(\frac{1}{2}(n-1)-\frac{1}{2}(n-5))=1$ edge. From R form a graph R^* by adjoining an extra vertex v^*, but no extra edges. Then R^* has order $\frac{1}{2}(n+1)$. Let L_e denote the subgraph of L formed by removing any loops there may be from their vertices.

We can complete the edge-colouring of L_e to an edge-coloured $K^\ell_{\frac{1}{2}(n+1)}$, called $H^\ell_{L_e}$, by induction, using the colours $c_1, \ldots, c_{\frac{1}{2}(n+1)}$ (formally, if L_e contains isolated vertices we join them in pairs by edges given arbitrary colours in $\{c_1, \ldots, c_{\frac{1}{2}(n+1)}\}$). Note that L_e cannot be bad with respect to such a completion; also, since L_e has no loops, its loop-colouring is trivially admissible. Let H_{L_e} denote the edge-coloured graph obtained from $H^\ell_{L_e}$ by removing the loops. We similarly complete R^* to an edge-coloured $K^\ell_{\frac{1}{2}(n+1)}$, called $H^\ell_{R^*}$, with $c_1, \ldots, c_{\frac{1}{2}(n+1)}$. Note that R^* contains a vertex (v^*) with no coloured loop or edge incident with it, so R^* cannot be bad (again formally, join v^* to an end-vertex of an edge of R by an edge of a colour, which can be arbitrary except if $\frac{1}{2}(n+1)=5$, where it must be chosen so as not to create a graph of Type g_5). Furthermore, since n is odd and $\frac{1}{2}(n+1)$ is odd, the loop-coiouring of R^* is admissible with respect to a completion to an edge-colouring of $H^\ell_{R^*}$.

Suppose that the vertex sets of $H^\ell_{L_e}$ and $H^\ell_{R^*}$ are $\{v_1, \ldots, v_{\frac{1}{2}(n+1)}\}$ and $\{v_{\frac{1}{2}(n+1)+1}, \ldots, v_{n+1}\}$, respectively, with $v_{n+1}=v^*$. We may suppose that the loops on $v_1, \ldots, v_{\frac{1}{2}(n+1)}$ are coloured $c_1, \ldots, c_{\frac{1}{2}(n+1)}$ respectively, that the edges $v_{\frac{1}{2}(n+1)+1}v^*, \ldots, v_n v^*$ are coloured $c_1, \ldots, c_{\frac{1}{2}(n-1)}$ respectively, and that the loop on v^* is coloured $c_{\frac{1}{2}(n+1)}$; for otherwise we could relabel the vertices or the colours.

Let B be the complete bipartite graph with bipartition $(\{v_1, \ldots, v_{\frac{1}{2}(n+1)}\}, \{v_{\frac{1}{2}(n+1)+1}, \ldots, v_{n+1}\})$, and let F be the 1-factor of B consisting of the edges $v_i v_{\frac{1}{2}(n+1)+i}$ $(i=1, \ldots, \frac{1}{2}(n+1))$. Colour the graph $B \backslash F$ with colours $c_{\frac{1}{2}(n+1)+1}, \ldots, c_n$. Extend this to an edge-colouring of B by colouring the edges $v_i v_{\frac{1}{2}(n+1)+i}$ with the colour c_i $(1 \leq i \leq \frac{1}{2}(n+1))$. From B form a graph B^ℓ by deleting the vertex v^*, placing a loop on each of the vertices $v_1, \ldots, v_{\frac{1}{2}(n+1)}$ and colouring the loop on v_i with the colour of the edge $v_i v^*$ in B.

Let the edge-coloured graph H^{ℓ}_R, be obtained from the edge-coloured $H^{\ell}_{R^*}$ by deleting the vertex v^*. From B^{ℓ}, H_{L_e} and H^{ℓ}_R, now assemble an edge-colouring of K^{ℓ}_n. Possibly after permuting the colours $c_{\frac{1}{2}(n+1)+1}, \ldots, c_n$, this contains the edge-coloured graph G as a subgraph, as required.

2iiia2. G has the form of (ii) of Figure 3.

The number of edges of G is at most $\frac{1}{2}(n-1)$, and so we may assume that the edges and the non-free loop of G are all coloured using $c_1, \ldots, c_{\frac{1}{2}(n+1)}$.

Consider first the case when $\frac{1}{2}(n+1)$ colours are used on the edges and the non-free loop of G, and the one free loop has yet a further colour. Then the edges and the loops all have different colours. Consider the edge-coloured K^{ℓ}_n on vertices $a_{-\frac{1}{2}(n-1)}, \ldots, a_{\frac{1}{2}(n-1)}$ where, for $0 \leq i \leq n-1$, the i-th colour class is the set of edges

$$\{a_{-j+i}\, a_{j+i} : j=1, \ldots, \tfrac{1}{2}(n-1)\},$$

and a loop on the vertex a_i, the indices being read modulo n to be between $-\frac{1}{2}(n-1)$ and $\frac{1}{2}(n-1)$. Then the edges

$$a_{-\frac{1}{4}(n-1)+i}\; a_{\frac{1}{4}(n-1)+1} \qquad (i=0,1, \ldots, \tfrac{1}{2}(n-1)-1)$$

and the loops on $a_{-\frac{1}{4}(n-1)-1}$ and $a_{\frac{1}{2}(n-1)-1}$ all have different colours. The edges and the loop on $a_{-\frac{1}{4}(n-1)-1}$ are mutually non-adjacent, and the loop on $a_{\frac{1}{2}(n-1)}$ is incident with the edge $a_0 a_{\frac{1}{2}(n-1)}$. Thus the edge-coloured K^{ℓ}_n contains an isomorphic copy of our edge-coloured G in this case.

From now suppose that either the edges and the non-free loop use at most $\frac{1}{2}(n-1)$ colours altogether, or that the edges and the non-free loop use $\frac{1}{2}(n+1)$ colours, but the free loop is coloured the same as one of the edges. From G we form a new graph G' by removing the edge from the vertices v and w (referring to (ii) of Figure 3), but retaining v, w and the loop on w. Partition G' into subgraphs L and R of order $\frac{1}{2}(n+1)$ and $\frac{1}{2}(n-1)$ respectively so that L contains the vertex v, R contains the vertex w and all the free loops of G coloured with $c_1, \ldots, c_{\frac{1}{2}(n+1)}$, but no other free loops. It is easy to check that there is no problem with this partition if the edges and non-free loop use $\frac{1}{2}(n+1)$ colours. If the edges and non-free loop use at most $\frac{1}{2}(n-1)$ colours, and if there are free loops not coloured with colours used on the edges, then, for at least one free loop, a choice can be made as to whether to include its colour in $\{c_1, \ldots, c_{\frac{1}{2}(n+1)}\}$ or in $\{c_{\frac{1}{2}(n+1)+1}, \ldots, c_n\}$. Note also that, since $\ell \leq \frac{1}{2}(n-5)$, L contains at least $\lceil \frac{1}{2}(\frac{1}{2}(n+1)-\frac{1}{2}(n-5)) \rceil = 2$ edges, and similarly, R contains at least one edge.

From L form a graph L^* by placing a loop on the vertex v coloured with the colour that vw has in G, say c_1. Then from L^* form a graph L^*_e by removing the loops from all vertices except v. From R form a graph R^* by adjoining a further vertex v^* and join w to v^*; also colour wv^* with the colour c_1. Then both L^*_e and R^* have order $\frac{1}{2}(n+1)$. Assume for the moment that the edge-colourings of L^*_e and R^* are not bad with respect to completing the edge-colourings to edge-colourings of complete graphs with loops of order $\frac{1}{2}(n+1)$, say $H^{\ell}_{L^*_e}$ and $H^{\ell}_{R^*}$, respectively, with colours $c_1, \ldots, c_{\frac{1}{2}(n+1)}$; note that, since n is odd, the loops of G all have different colours, and so, since $\frac{1}{2}(n+1)$ is also odd, the loop-colourings of L^*_e and R^* are both admissible. Then we first complete the edge-colourings of $H^{\ell}_{L^*_e}$ and $H^{\ell}_{R^*}$ by induction.

Suppose that the vertex sets of $H^{\ell}_{L^*_e}$ and $H^{\ell}_{R^*}$ are $\{v_1, \ldots, v_{\frac{1}{2}(n+1)}\}$ and $\{v_{\frac{1}{2}n+1)+1}, \ldots, v_{n+1}\}$ respectively, with $v_1 = v$, $v_{\frac{1}{2}(n+1)+1} = w$ and $v_{n+1} = v^*$. We may suppose that the loops on $v_1, \ldots, v_{\frac{1}{2}(n+1)}$ are coloured $c_1, \ldots, c_{\frac{1}{2}(n+1)}$ respectively, that the edges $v_{\frac{1}{2}(n+1)+1}v^*, \ldots, v_{\frac{1}{2}(n+1)+\frac{1}{2}(n-1)}v^*$ are coloured $c_1, \ldots, c_{\frac{1}{2}(n-1)}$ respectively, and that the loop on v^* is coloured $c_{\frac{1}{2}(n+1)}$. Note that the loop on $v(=v_1)$ and the edge $v_{\frac{1}{2}(n+1)+1}v^*$ $(=wv^*)$ both have the same colour, namely c_1, as required.

Let B be the complete bipartite graph with bipartition $(\{v_1, \ldots, v_{\frac{1}{2}(n+1)}\}, \{v_{\frac{1}{2}(n+1)+1}, \ldots, v_{n+1}\})$, and let F be the 1-factor of B consisting of the edges $v_i v_{\frac{1}{2}(n+1)+i}$ $(i=1,\ldots,\frac{1}{2}(n+1))$. Colour the graph $B \backslash F$ with colours $c_{\frac{1}{2}(n+1)+1}, \ldots, c_n$. Extend this to an edge-colouring of B by colouring the edge $v_i v_{\frac{1}{2}(n+1)+1}$ with the colour c_i $(1 \leq i \leq \frac{1}{2}(n+1))$. For $v' \in V(H^{\ell}_{L^*_e})$, if $v'v^*$ is coloured c_i in B, colour, or recolour, the loop incident with v' with the colour c_i; note that the loop incident with $v_{\frac{1}{2}(n+1)}$ already has the colour $c_{\frac{1}{2}(n+1)}$ on it (i.e. the colour of the edge $v_{\frac{1}{2}(n+1)}v^*$), so that this loop is not actually recoloured. Let the edge-coloured graph R' be obtained from R^* by deleting the vertex v^*. From B, $H_{L^*_e}$ and $H_R{}'$, now assemble an edge-colouring of K^{ℓ}_n. It is easy to check that this edge-colouring is proper; note in particular, that the edge vw is coloured c_1, and that the colour c_1 has been removed from the loop on v_1 and replaced by some other colour, and that the edge on w which was coloured c_1, namely wv^*, has now been removed. Possibly after permuting the colours $c_{\frac{1}{2}(n+1)+1}, \ldots, c_n$, this is the required extension of the edge-colouring of G to an edge-colouring of K^{ℓ}_n.

Suppose that G has r edges of one colour, say y, where $r \geq |E(G)|-2$. If G has 2 edges not coloured y, then their end-vertices and all vertices with free loops of colour distinct from y belong to a K^{ℓ}_p with $p \leq \ell + 4 \leq \frac{n-7}{2} + 4 = \frac{n+1}{2}$, as ℓ is odd in this case. Give this K_p an edge-colouring with colours $\{c_1, \ldots, c_n\}$ with y occurring at every vertex, and with admissible loop-colouring giving no loop the colour of the non-free loop of G if this is on an edge of colour y; further, if $p = \frac{n+1}{2}$, make sure that every colour occurs on a loop or edge of K^{ℓ}_p. This is clearly possible (in the latter case, because $|E(K_{\frac{n+1}{2}})| - \frac{n+1}{4} - 2 \geq n - \frac{n+1}{2}$ as $n \geq 9$). Completing this edge-colouring to an edge-colouring of K^{ℓ}_n, by Theorem 3, gives the required completion of the edge-colouring of G. If G has exactly one edge not coloured y then it has a free loop not coloured y (as it does not contain a Type $o2$ graph), and so the above argument works with $3 \leq p \leq \frac{n-3}{2}$. If all edges of G are coloured y, it is trivial that the edge-colouring of G can be extended to an edge-colouring of K^{ℓ}_n (G does not contain a Type $o1$ graph).

So from now on in this case, we shall make the following supposition.

Supposition 1. If G contains r edges of one colour, then $r \leq |E(G)|-3$

Now suppose that at least one of L^*_e and R^* are bad. We may suppose that if at most $\frac{1}{2}(n+1)$ colours are used on G, then all edges and loops are coloured with $c_1, \ldots, c_{\frac{1}{2}(n+1)}$. It is easy to see that L^*_e can be completed unless all the loops of G are coloured with $c_1, \ldots, c_{\frac{1}{2}(n+1)}$. Suppose for the moment that not all the loops are coloured with $c_1, \ldots, c_{\frac{1}{2}(n+1)}$. Then the number of colours used on the edges of G is at least $(\frac{1}{2}(n+1)+1)-(1+\ell) \geq \frac{1}{2}(n+1)-\frac{1}{2}(n-5)=3$. If R^* has at least three edges, then we can arrange that R^* has edges of at least three colours, and so then R^* cannot be bad. So suppose that R^* has only two edges. Then R^* has $\dfrac{n+1}{2}-4=\dfrac{n-7}{2}$ loops.

Since $\ell \leq \frac{1}{2}(n-7)$, it follows that L has no free loops, so $L^*_e = L^*$. Therefore, from now we may suppose that all the edges and loops of G are coloured with $c_1, \ldots, c_{\frac{1}{2}(n+1)}$.

Suppose that the edge-colouring of L^* is bad. Suppose L^* is of type $o1$ and that all the edges of L^* are coloured c_2 and the loop on v is coloured c_1. In R^*, the edge wv^* is coloured c_1 and, by Supposition 1 there are at least two further edges, say e_1 and e_2, which are not coloured c_2; then, in this case $n \geq 13$. Interchange e_1 and an edge of L^* (coloured c_2). Then L^* is no longer bad. After the change, R^* has at least one edge (wv^*) coloured c_1, at least one edge coloured c_2, and at least one further edge not coloured c_2. R^* could not then be bad unless it consists of one free loop coloured c_1, no further free loops, one edge coloured c_2 and the remaining edges coloured c_1. We may suppose that this also holds if e_2 is removed instead of e_1. But then G must have consisted of a free loop coloured c_1, $\dfrac{1}{4}(n-1)$ free edges coloured c_2, $\dfrac{1}{4}(n-5)$ free edges coloured c_1, the edge vw coloured c_1, as well as a coloured loop on w. We may suppose that the loop on w is coloured with one of $c_2, \ldots, c_{\frac{1}{2}(n-3)}$. We consider the $K_{\frac{1}{2}(n-1)}$ spanned by the $\dfrac{1}{4}(n-1)$ edges coloured c_2. We complete the edge-colouring of this $K_{\frac{1}{2}(n-1)}$ with colours $c_1, c_2, \ldots, c_{\frac{1}{2}(n-3)}$, (so that c_1 occurs at every vertex), and we colour the loops with colours $c_{\frac{1}{2}(n-3)+1}, \ldots, c_{\frac{1}{2}(n-3)+\frac{1}{2}(n-1)}$.

By Theorem 3, this can be completed to an edge-colouring of K_n^ℓ, and, possibly after permuting colours, this edge-colouring of K_n^ℓ contains our edge-coloured G as a subgraph. In any other case, the modified versions of L^* and R^* are not bad, and so the edge-colouring of G can be extended to an edge-coloured K_n^ℓ, as already described.

Next suppose that L^* is Type $o2$ and that the loop on v and all except one edges of L^* are coloured c_1, the one exceptional edge being coloured c_2. Then the edge wv^* of R^* is coloured c_1, and by Supposition 1, R^* has at least two further edges, say e_1 and e_2, which are not coloured c_1. Consequently $n \geq 13$ in this case also. Now interchange an edge coloured c_1 of L^* and the edge e_1 in R^*. Then L^* is no longer bad. The modified version of R^* cannot be bad unless all the edges of R^* are now coloured c_1, except for the edge e_2, and unless R^* has only one free loop, which is coloured c_1. In that case G has three edges not coloured c_1. The K_6^ℓ spanned by these three edges can

be admissibly loop-coloured (so that all the loops receive different colours, but the colour of the free loop (c_1) and the loop on w are not used). Then three independent edges of the K_6^ℓ can be coloured c_1, and finally the remaining edges coloured greedily, all colours used being in $\{c_1, \ldots, c_n\}$. By Theorem 3, this edge-coloured K_6^ℓ can be extended to an edge-coloured K_n^ℓ; such an edge-coloured K_n^ℓ must contain our edge-coloured G as a subgraph. In any other case, the modified L^* and R^* are not bad, and so the edge-colouring of G can be extended to an edge-colouring of K_n^ℓ as described before.

Now suppose that the edge-colouring of R^* is bad. Suppose that R^* is of the Type $o1$. We may assume that all the edges of R^* (including wv^*) are coloured c_1, and that no loops of R^* are coloured c_1. It follows that the loop on v in L^* is coloured c_1 also. By Supposition 1, L^* contains at least three edges not coloured c_1 (and so $n \geq 13$), say e_1, coloured c_2, e_2 and e_3. Interchange e_1 and an edge of R^* (not wv^*) coloured c_1. Then this modified version of L^* cannot be bad. The new R^* has all edges except one coloured c_1, one edge coloured c_2, no free loops coloured c_1, and it has $\frac{1}{2}(n+1) \geq 7$ vertices; it also cannot be bad. We can therefore extend the edge-colouring of G to K_n^ℓ, as described above.

Finally suppose that R^* is of the Type $o2$. We may assume that the edge wv^* of R^* and the loop on v in L^* are both coloured c_1. Then R^* contains just one free loop and either (A) all other edges and the free loop are coloured with some other colour, say c_2, or (B) one edge has some other colour, say c_2, and the remaining edges and the free loop are coloured c_1. In case (A), by Supposition 2, it follows that L^* contains at least two edges not coloured c_2. If L^* contains such an edge e not coloured c_1, then interchange e and an edge of R^* coloured c_2. Then R^* is no longer bad. If all such edges of L^* are coloured c_1, then by Supposition 1, R^* must contain at least two edges coloured c_2. Then interchange an edge coloured c_1 in L^* and an edge coloured c_2 in R^*. Then R^* is no longer bad. Whether L^* is now good or bad, by previous arguments we can complete the edge-colouring of G to an edge-colouring of K_n^ℓ, as required. In case (B), by Supposition 1, it follows that L^* contains at least two edges, say e_1 and e_2, not coloured c_1. If $\frac{1}{2}(n+1) \geq 7$, i.e. $n \geq 13$, then interchange e_1 and an edge other than wv^* coloured c_1. Then R^* is no longer bad, and so, whether or not L^* is bad, we can complete the edge-colouring of K_n^ℓ by previous arguments. If $n=9$ (which is the only remaining possibility here), then, by Supposition 1, neither of the edges of L^* can be colour c_1 or c_2, and they must have different colours, say c_3 and c_4. Then G consists of four edges, vw coloured c_1, and the others coloured c_2, c_3 and c_4, a free loop coloured c_1, and a further loop incident with w coloured, say, c. If $c \in \{c_2, c_3, c_4\}$, then complete the K_6^ℓ spanned by the edges coloured c_2, c_3 and c_4, as shown in Figure 5, and then complete this to a K_9^ℓ by Theorem 3. Then, possibly after a permutation of the colours c_2, c_3 and c_4, the edge-coloured K_9^ℓ contains G as a coloured subgraph. If $c \in \{c_5, \ldots, c_9\}$ the same argument works if the edge xy of Figure 5 is recoloured c_9 and the loop on x is recoloured c_3.

Case 2iiib. $n \equiv 3$ (mod 4). Then $\frac{1}{2}(n-1)$ is odd.

Case 2iiib1. G has the form of (i) of Figure 3.

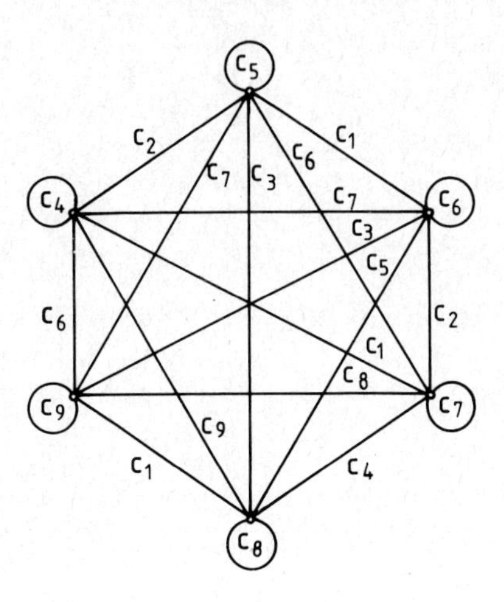

Figure 5

There are at most $\frac{1}{2}(n+1)$ edges of G. Consider first the case when $\frac{1}{2}(n+1)$ colours are used on the edges of G. Then G has no loops and each edge has a different colour.

Consider the edge-coloured K_n^ℓ on vertices $a_{-\frac{1}{2}(n-1)}, \ldots, a_{\frac{1}{2}(n-1)}$ where, for $0 \leq i \leq n-1$, the i-th colour-class is the set of edges

$$\{a_{-j+i}\, a_{j+i} : j=1, \ldots, \tfrac{1}{2}(n-1)\},$$

and a loop on the vertex a_i (where the indices are reduced modulo n to be between $-\frac{1}{2}(n-1)$ and $\frac{1}{2}(n-1)$). Then the edges

$$a_{-\frac{1}{4}(n+1)+i}\; a_{\frac{1}{4}(n+1)+i} \quad (i=0,..,\tfrac{1}{2}(n-1))$$

all have different colours, but induce a graph isomorphic to G. Thus G can be completed to an edge-coloured K_n^ℓ, as required.

From now suppose that the edges of G are coloured with colours in the set $\{c_1, \ldots, c_{\frac{1}{2}(n-1)}\}$. Consider the case where there are $\frac{1}{2}(n-1)$ edges of G, each coloured differently, and two loops, one coloured the same as an edge of G, the other coloured differently from all the edges. Then $n \geq 11$, as $\ell \leq \frac{1}{2}(n-5)$. First we show that the edge-coloured K_n^ℓ just constructed contains an isomorphic copy of the edge-coloured G in the case when a loop has the same colour as one of the adjacent edges.

The edges are:

$$a_{-\frac{1}{4}(n+1)+1}\, a_{\frac{1}{4}(n+1)+i} \quad (i=0,,..,\frac{1}{4}(n+1)-1,\ \frac{1}{4}(n+1)+1,\ \frac{1}{4}(n+1)+2,\ldots,\frac{1}{2}(n-1))$$

and the loops are on the vertices a_0 and $a_{-½(n-1)}$. The edges and the loop on $a_{-½(n-1)}$ all have different colours, and the loop on a_0 has the same colour as the edge $a_{-\frac{1}{4}(n+1)}\, a_{\frac{1}{4}(n+1)}$ which is incident with the edge

$$a_{-\frac{1}{4}(n+1)+½(n-1)}\, a_{\frac{1}{4}(n+1)+½(n-1)},\ \text{since}$$

$$\frac{1}{4}(n+1)+½(n-1)=-\frac{1}{4}(n+1)+n\,.$$

The loops and the remaining edges are mutually non-adjacent. Next we show that the edge-coloured K_n^{ℓ} constructed above also contains an isomorphic copy of the edge-coloured G in the case when a loop has the same colour as a free edge. This time the edges are

$$a_{-\frac{1}{4}(n+1)+i}\, a_{\frac{1}{4}(n+1)+i} \quad (i=0,,\ldots,½(n-1)-2,½(n-1))$$

and the loops are on the vertices $a_{-\frac{1}{4}(n+1)-1}$ and $a_{\frac{1}{4}(n+1)-2}$. The edges and the loop on $a_{-\frac{1}{4}(n+1)-1}$ all have different colours, and the loop on $a_{\frac{1}{4}(n+1)-2}$ has the same colour as the edge $a_{-2}\, a_{½(n+1)-2}$, which is not adjacent to any other edge as $n \geq 11$. The edges as $a_{-\frac{1}{4}(n+1)}\, a_{\frac{1}{4}(n+1)}$ and $a_{-\frac{1}{4}(n+1)+\frac{1}{2}(n-1)}\, a_{\frac{1}{4}(n+1)+\frac{1}{2}(n-1)}$ are adjacent, but the remaining edges and loops are mutually non-adjacent.

From now, suppose that the edges of G are coloured with colours in the set $c_1,\ldots,c_{½(n-1)}$ and that, if G has exactly two loops, then either they are both coloured with colours in $c_1,\ldots,c_{½(n-1)}$ or neither of them is. Notice that if the edges of G use $½(n-1)$ colours, then either G has no loops, or G has two.

The number of free edges that G has is

$$\frac{1}{2}(n-\ell-3)\geq\lceil\frac{1}{2}(n-\frac{1}{2}(n-5)-3)\rceil=\lceil\frac{1}{4}(n-1)\rceil=\frac{1}{4}(n+1).$$

Partition the graph G into subgraphs L and R of order $½(n+1)$ and $½(n-1)$ respectively, so that R contains all the loops coloured with $c_1,\ldots,c_{½(n-1)}$ and no other loops, and so that R contains the two adjacent edges of G. Note that G has at least $\frac{1}{4}(n+1)$ free edges, and furthermore that L contains at least $\frac{1}{4}(n+1)-\frac{1}{2}(\frac{1}{2}(n-1)-3)=2$ free edges. Also note that if the edges of G use fewer than $½(n-1)$ colours, and if there are free loops not coloured with a colour used on the edges, then, for at least one free loop, a choice can be made as to whether to include its colour in $\{c_1,\ldots,c_{½(n-1)}\}$ or in $\{c_{½(n+1)},\ldots,c_n\}$. Therefore no parity difficulties arise in performing this partition.

From L for a graph L^* of order $\frac{1}{2}(n-1)$ from L by replacing an edge e of L between vertices, say $v_{\frac{1}{2}(n-1)}$ and $v_{\frac{1}{2}(n+1)}$ by a loop of the same colour as e on $v_{\frac{1}{2}(n-1)}$, and then deleting $v_{\frac{1}{2}(n+1)}$. Let the remaining vertices of L be $v_1, \ldots, v_{\frac{1}{2}(n-3)}$. Let the vertices of R be $v_{\frac{1}{2}(n+1)+1}, \ldots, v_n$. Let L^*_e denote the graph L^* with all loops removed from their vertices except the loop on $v_{\frac{1}{2}(n-1)}$.

Assume for the moment that the edge-colourings of L^*_e and R are not bad with respect to completing the edge-colourings to edge-colourings of complete graphs with loops of order $\frac{1}{2}(n-1)$, say $H^\ell_{L^*_e}$ and H^ℓ_R respectively; the edge-colouring of L^*_e cannot be bad unless $L^*_e = L^*$ (i.e. there are not free loops in L, so the colours used in G are all in the set $\{c_1, \ldots, c_{\frac{1}{2}(n-1)}\}$), and the the edge-colouring of R cannot be bad unless $\frac{1}{2}(n-1)=5$ (i.e. $n=11$) and R contains no free edges. Note that, since n is odd, the loops of G all have different colours, and so, since $\frac{1}{2}(n-1)$ is also odd, the loop colouring of R is admissible; since L^*_e only has one loop, its loop-colouring is trivially admissible. Then we first complete the edge-colourings of $H^\ell_{L^*_e}$ and H^ℓ_R by induction, using the colours $c_1, \ldots, c_{\frac{1}{2}(n-1)}$. From $H^\ell_{L^*_e}$ we then obtain an edge-colouring of a complete graph without loops, H_{L_e} of order $\frac{1}{2}(n+1)$, by reintroducing the vertex $v_{\frac{1}{2}(n+1)}$, and, for each i, $1 \le i \le 1(n-1)$, replacing the loop on v_i by an edge $v_i v_{\frac{1}{2}(n+1)}$ of the same colour.

Let B be the complete bipartite graph with bipartition $(\{v_1, \ldots, v_{\frac{1}{2}(n+1)}\}, \{v_{\frac{1}{2}(n+1)+1}, \ldots, v_n\})$. Colour the edges of B with colours $c_{\frac{1}{2}(n-1)+1}, \cdots, c_n$. Each of $c_{\frac{1}{2}(n-1)+1}, \ldots, c_n$ will be absent from exactly one of $v_1, \ldots, c_{\frac{1}{2}(n+1)}$, and each such vertex will have exactly one of these colours absent from it. For $1 \le i \le \frac{1}{2}(n+1)$, place a loop on v_i in B coloured with the missing colour, denote the graph obtained by B^ℓ. Finally assemble H_{L_e}, B^ℓ and H^ℓ_R into an edge-coloured K^ℓ_n. Possibly after permuting the colours $c_{\frac{1}{2}(n-1)+1}, \ldots, c_n$, this contains the edge-coloured graph G as a subgraph, as required.

If the edge-colouring of R is bad (implying $n=11$), or if the edge-colouring of L^* is bad, arguments resembling those of the similar situation of Case 2iiia2 are needed. We omit these details here; they can be obtained from the authors.

Case 2iiib2. G has the form of (ii) of Figure 4.

There are at most $\frac{1}{2}(n-1)$ edges in G; we first consider various cases where G has exactly $\frac{1}{2}(n-1)$ edges; then there is one free loop.

Consider first the case when there are $\frac{1}{2}(n-1)$ edges, and the edges and the two loops all have different colours. Consider again the edge-coloured K^ℓ_n on vertices $a_{-\frac{1}{2}(n-1)}, \ldots, a_{\frac{1}{2}(n-1)}$, where, for $0 \le i \le n-1$, the i-th colour class is the set of edges

$$\{a_{-j+i}\ a_{j+i} : j=1, \ldots, \tfrac{1}{2}(n-1)\}$$

and a loop on the vertex a_i (where the indices are reduced modulo n to be between $-\frac{1}{2}(n-1)$ and $\frac{1}{2}(n-1)$). Let S bet the set of edges

$$\{a_{-\frac{1}{4}(n-3)+i}\, a_{\frac{1}{4}(n-3)+i} \;:\; i=0,1,\ldots,\frac{1}{2}(n-3)-2\}.$$

For $n\geq 15$, take S together with the edges $a_{-\frac{1}{4}(n-3)+\frac{1}{2}(n-3)-1}\, a_{-\frac{1}{4}(n-3)-3}$ and $a_{-\frac{1}{4}(n-3)-4}\, a_{-\frac{1}{4}(n-3)-2}$ and the loops on the vertices $a_{-\frac{1}{4}(n-3)-1}$ and $a_{-\frac{1}{4}(n-3)}$. The loop on the vertex $a_{-\frac{1}{4}(n-3)}$ is incident with the edge $a_{-\frac{1}{4}(n-3)}\, a_{\frac{1}{4}(n-3)}$, but the remaining edges and loop are mutually non-adjacent.

The colours of all the loops and edges are distinct. If $n=11$, take the loop on the vertex a_{-1} instead of the loop on the vertex a_{-2}. If $n=7$, take the edges $a_{-2}a_1$, $a_{-3}a_{-1}$, a_2a_3 and loops on a_0 and a_1. Thus in the case when G has ½$(n-1)$ edges and the colours used on G are all distinct, the edge-coloured G can be embedded in an edge-coloured K_n^ℓ, as required.

Next consider the case when the number of colours used on the edges and the non-free loop of G is ½$(n+1)$, but the free loop of G has the same colour as the non-free edge. In this case, together with the set S take the edges $a_{-\frac{1}{4}(n-3)-3}\, a_{-\frac{1}{4}(n-3)-1}$, $a_{-\frac{1}{4}(n-3)-4}\, a_{\frac{1}{4}(n-3)-1}$ and loops on the vertices $a_{-\frac{1}{4}(n-3)-2}$ and $a_{-\frac{1}{4}(n-3)-1}$. The loop on the vertex $a_{-\frac{1}{4}(n-3)-2}$ and the edge $a_{-\frac{1}{4}(n-3)-3}\, a_{-\frac{1}{4}(n-3)-1}$ have the same colour, but the edges and the other loop all have different colours.

We then consider the case when the number of colours used on the edges and the non-free loop of G is ½$(n+1)$, but the free loop has the same colour as a free edge. This is obtained from the above by replacing the loop on $a_{-\frac{1}{4}(n-3)-1}$ by one on $a_{-\frac{1}{4}(n-3)}$.

Now consider the case when the edges all have different colours, the non-free loop has the same colour as one of the edges, and the free loop has a colour not used on any edge. In this case, for $n\geq 11$, take the set S together with the edges $a_{\frac{1}{4}(n-3)-1}\, a_{-\frac{1}{4}(n-3)-3}$ and $a_{-\frac{1}{4}(n-3)-4}\, a_{-\frac{1}{4}(n-3)-2}$ and loops on the vertices $a_{-\frac{1}{4}(n-3)-1}$ and a_0. Then the loop on a_0 and the edge $a_0\, a_{½(n-3)}$ are adjacent, but the remaining edges and loop are independent. The loop on the vertex a_0 and the edge $a_{-\frac{1}{4}(n-3)}\, a_{\frac{1}{4}(n-3)}$ have the same colour, but the remaining edges and loop have distinct colours. For $n=7$, take the edges $a_{-3}a_{-1}$, $a_{-2}a_1, a_2a_3$ and loops on the vertices a_{-1} and a_0.

Next consider the case when the non-free edge has the same colour as a free edge, and the total number of colours used on G is ½$(n+1)$. Take the set S together with the edges $a_{\frac{1}{4}(n-3)-1}\, a_{-\frac{1}{4}(n-3)-4}$ and $a_{-\frac{1}{4}(n-3)-3}\, a_{-\frac{1}{4}(n-3)-2}$ and loops on the vertices $a_{-\frac{1}{4}(n-3)-1}$ and $a_{-\frac{1}{4}(n-3)-3}$. Then the loop on $a_{-\frac{1}{4}(n-3)-3}$ and the edge

$a_{-\frac{1}{4}(n-3)-3}\, a_{-\frac{1}{4}(n-3)-2}$

are adjacent, and the remaining edges and loop are independent. The edges $a_{-\frac{1}{4}(n-3)-3}\, a_{\frac{1}{4}(n-3)-2}$ and $a_{-1} a_{\frac{1}{2}(n-3)-1}$ have the same colour, but the remaining edges and loop have distinct colours.

Finally consider the case when two free edges have the same colour, but the remaining edges and loops all have different colours. In the case above remove the loop from the vertex $a_{-\frac{1}{4}(n-3)-3}$ and place it on the vertex $a_{-\frac{1}{4}(n-3)-4}$.

We then consider two cases where G has exactly three free loops. Then $n \geq 11$. In this case, first suppose that all the edges and loops have different colours. Take the set S together with the edge $a_{-\frac{1}{4}(n-3)-4}\, a_{\frac{1}{4}(n-3)-1}$ and loops on the vertices $a_{-\frac{1}{4}(n-3)-3}$, $a_{-\frac{1}{4}(n-3)-2}$, $a_{-\frac{1}{4}(n-3)-1}$ and $a_{-\frac{1}{4}(n-3)}$.

Finally in this list of special cases, suppose that G has $\frac{1}{2}(n-3)$ edges, that the edges and the non-free loop all have distinct colours, and that exactly 2 of the free loops have colours also used on edges of G. If both these colours occur on free edges, then take the edges

$$a_{-\frac{1}{4}(n+1)+i}\, a_{\frac{1}{4}(n+1)+i} \qquad (i=0,1,\ldots,\tfrac{1}{2}(n+1)-3)$$

and loops on the vertices $a_{-\frac{1}{4}(n+1)-1}$, $a_{-\frac{1}{4}(n+1)}$, $a_{\frac{1}{4}(n+1)-2}$, $a_{\frac{1}{4}(n+1)-1}$.

If the colour of one of the free loops occurs on the non-free edge, move the loop from $a_{-\frac{1}{4}(n+1)}$ to a_{-2}.

We may from now suppose that the edges and non-free loop of G are coloured using the colours $c_1,\ldots,c_{\frac{1}{2}(n-1)}$, and further, that if each of these colours is used then G has an odd number (1 or 3) of free loops coloured using these colours.

We partition the graph G into subgraphs L and R of orders $\frac{1}{2}(n+1)$ and $\frac{1}{2}(n-1)$ respectively, so that R contains the edge vw (referring to (ii) of Figure 3) and all the loops coloured with $c_1,\ldots,c_{\frac{1}{2}(n-1)}$, but no other loops. It is easy to check that if all $\frac{1}{2}(n-1)$ colours are used on the edges and non-free loop of G, then this partition can be made. Furthermore, if fewer than $\frac{1}{2}(n-1)$ colours are used on the non-free loop and the edges, and if yet further colours are used on some free loops, then, for at least one free loop, a choice can be made as to whether to include its colours in $\{c_1,\ldots,c_{\frac{1}{2}(n-1)}\}$ or $\{c_{\frac{1}{2}(n+1)},\ldots,c_n\}$.

Therefore no parity difficulties arise in making this partition. Since $\ell \leq \frac{1}{2}(n-5)$, it follows that G has at least $\lceil \frac{1}{2}(n-\frac{1}{2}(n-5)) \rceil = \frac{1}{4}(n+5)$ edges.

Therefore L has at least $\frac{1}{4}(n+5)-\frac{1}{2}(\frac{1}{2}(n-1)-1)=2$ edges, but R may have only 1 edge.

From L form a graph L^* of order $\frac{1}{2}(n-1)$ by replacing an edge e of L between vertices, say $v_{\frac{1}{2}(n-1)}$ and $v_{\frac{1}{2}(n+1)}$ by a loop of the same colour as e on $v_{\frac{1}{2}(n-1)}$, and then deleting $v_{\frac{1}{2}(n+1)}$. Let the remaining vertices of L be $v_1, \ldots, v_{\frac{1}{2}(n-3)}$. Let the vertices of R be $v_{\frac{1}{2}(n+1)+1}, \ldots, v_n$. Let L^*_e denote the graph L^* with all loops removed from their vertices except the loop on $v_{\frac{1}{2}(n-1)}$.

If L^*_e and R are both good, then we proceed to complete the edge-colourings of L^*_e to $H^\ell_{L^*_e}$ and of R to H^ℓ_R as described in Case 2iiib1. We form the graph B^ℓ as described there, and finally assemble H_{L_e}, H^ℓ_R and B^ℓ into an edge-coloured K^ℓ_n containing G, as described there.

Again, we leave out the details needed for the situation where one L^*_e or R is bad.

□

So the proof of Theorem 2 is complete.

Acknowledgement

Research of the first author was supported by a Niels Bohr Fellowship from the Royal Danish Academy of Sciences.

References

[1] L.D. Andersen, "Embedding Latin squares with prescribed diagonal", *Ann. of Discrete Math.* 15 (1982) 9-26.

[2] L.D. Andersen, "Completing partial Latin squares", *Mat. Fys. Medd. Dan. Vid. Selsk.* 41(1) (1985) 23-69.

[3] L.D. Andersen and A.J.W. Hilton, "Thank Evans!", *Proc. London Math. Soc.* (3), 47 (1983) 507-522.

[4] L.D. Andersen and A.J.W. Hilton, "Extending edge-colourings of complete graphs and independent edges", manuscript.

[5] L.D. Andersen and A.J.W. Hilton, "Symmetric Latin square and complete graph analogues of the Evans conjecture", in preparation.

[6] G.J. Chang, "Complete diagonals of Latin squares", *Canad. Math. Bull.* 22 (1979) 477-481.

[7] A.B. Cruse, "On embedding incomplete symmetric Latin squares", *J. Comb. Theory (A)* 16 (1974) 18-22.

[8] M. Damerell, "On Smetanuik's construction for Latin squares and the Andersen-Hilton Theorem", *Proc. London Math. Soc.* (3) 47 (1983) 523-526.

[9] P. Hall, "On representatives of subsets", *J. London Math. Soc.* 10 (1935) 26-30.

[10] A.J.W. Hilton and C.A. Rodger, "Latin squares with prescribed diagonals", *Canad. J. Math.* 34 (1982) 1251-1254.

[11] D.G. Hoffman, "Completing incomplete commutative Latin squares with prescribed diagonals", *Europ. J. Combinatorics* 4 (1983) 33-35.

[12] B. Smetanuik, "A new construction on Latin squares I. Proof of the Evans conjecture", *Ars. Combin.* 11 (1981) 155-172.

Annals of Discrete Mathematics 34 (1987) 27–42
© Elsevier Science Publishers B.V. (North-Holland)

A Fast Method for Sequencing Low Order Non-Abelian Groups

B.A. Anderson

Department of Mathematics
Arizona State University
Tempe, Arizona 85287
U.S.A.

TO ALEX ROSA ON HIS FIFTIETH BIRTHDAY

ABSTRACT

B. Gordon characterized sequenceable Abelian groups as those Abelian groups with a unique element of order 2. Limited progress has been made on the corresponding question for non-Abelian groups. In this paper, an "algorithm" is described that can be used to search for sequencings in finite groups. The algorithm is successful on all non-Abelian groups of order n, $10 \leq n \leq 30$, for which sequencings have not previously been known. Thus, all non-Abelian groups of these orders are sequenceable. This is at variance with published statements that claim only one of the three non-Abelian groups of order 12 is sequenceable. The algorithm also is used to find sequencings of the dihedral groups D_n, $16 \leq n \leq 50$.

1. Introduction

Suppose G is a finite group of order n with identity e. A **sequencing** of G is an ordering $e, a_2, \ldots, a_n$ of all the elements of G such that the partial products $e, ea_2, ea_2a_3, \ldots, ea_2 \cdots a_n$ are distinct and hence also all of G. If G has order $2n$ and has a unique element z of order 2, then a sequencing $e, a_2, \ldots, a_n, \ldots, a_{2n}$ of G will be called a **symmetric sequencing** if $a_{n+1} = z$ and for $1 \leq i \leq n-1$, $a_{n+1+i} = (a_{n+1-i})^{-1}$.

Sequencings arose in connection with the problem of constructing complete Latin squares [7]. Later [13] it was noticed that sequencings can be used to decompose complete directed graphs into directed Hamiltonian paths (see also [16]). Sequenceable Abelian groups have been characterized [7] as those Abelian groups with a unique element of order 2 and some progress has been made in finding sequenceable non-Abelian groups [1, 6, 8, 9, 10, 11, 12, 17]. In general, however, patterns generating sequencings of non-Abelian groups have been difficult to construct. In [7] Gordon deduced that the non-Abelian groups of orders 6 and 8 are not sequenceable and gave a sequencing of D_5, the dihedral group of order 10. A combination of theory and computation led to the claim [4], later repeated in [3, p.472], that among the non-Abelian groups of order ≤ 12, only D_5 and D_6 are sequenceable. In this paper, specific sequencings will be given for the other two non-Abelian groups of order 12. The reason for checking these things was that the author noticed that recent examples of perfect 1-factorizations on K_{14} constructed by Seah and Stinson [14] imply the dicyclic group Q_6 of order 12 must have a symmetric sequencing. This was explored in [2] where the concept of a 2-sequencing is defined.

Suppose H is a finite group of order n with identify e. A 2-**sequencing** of H is an ordering $e, c_2, \ldots, c_n$ of certain elements of H (not necessarily distinct) such that

(i) The associated partial products $e, ec_2, ec_3, \ldots, ec_2 \cdots c_n$ are distinct and hence all of H,

(ii) if $y \in H$ and $y \neq y^{-1}$, then

$$|\{i: 2 \leq i \leq n \text{ and } (c_i = y \text{ or } c_i = y^{-1})\}| = 2,$$

(iii) if $y \in$ and $y = y^{-1}$, then

$$|\{i: 1 \leq i \leq n \text{ and } c_i = y\}| = 1.$$

Clearly, every sequencing is a 2-sequencing. In [2] the following two results are verified (Q_{2n} is the dicyclic group of order $4n$).

THEOREM 1. Q_{2n} has a symmetric sequencing if and only if D_n has a 2-sequencing.

THEOREM 2. If $n \geq 3$ is an odd positive integer, then D_n has a 2-sequencing.

One would like to show that if $n \geq 4$ is an even positive integer, then D_n has a 2-sequencing. This project started as an attempt to modify the ideas of Dinitz and Stinson's hill-climbing algorithm [5] to this situation. It soon became apparent that this approach actually could be used to discover sequencings.

In [12] Keedwell conjectures that "most non-Abelian groups are sequenceable". The results of this paper support that conjecture. There are 42 non-Abelian groups of order n, $10 \leq n \leq 30$. The method to be described has been applied to all non-Abelian groups in this range not known to be sequenceable (including two groups A_4 and Q_6 of order 12 which [4] states are not sequenceable) and finds sequencings for all of them. It also has been applied to find sequencings of all dihedral groups D_n, $16 \leq n \leq 50$. Previous work on sequencing dihedral groups has concentrated mainly on the case D_n, n an odd prime. These results were obtained on an IBM PC-AT using compiled BASIC. The limits of computation were certainly not reached (the method sequences D_{50} in less than 2 minutes) but there is no compelling reason to go farther at this time. Given the results of this paper, it appears that the only examples known of non-Abelian groups that are not sequenceable are the groups of orders 6 and 8.

2. Description of the method

Suppose G is a finite group with identity e. A **state** P of the method is

(I) a $1-1$ function $\alpha: \{1,2,\ldots,|G|\} = \Gamma \rightarrow G$ such that $\alpha(1) = e$, and

(II) an associated $(|G|-1)$-tuple $D_P = (h_2, h_3, \ldots, h_{|G|})$.

such that if

$$G = \{e, g_2, g_3, \ldots, g_{|G|}\} \text{ then for } 2 \leq j \leq |G|,$$
$$h_j = |\{i: 1 \leq i \leq |G| - 1 \text{ and } \alpha(i) \cdot g_j = \alpha(i+1)\}|.$$

One may think of $\alpha[\Gamma]$ as a directed Hamiltonian path through G that starts at e. Thus $\alpha[\Gamma]$ is a candidate for a partial product sequence associated with a sequencing. Meanwhile, D_P tells how close $\alpha[\Gamma]$ is to being associated with a sequencing. If D_P is a $(|G|-1)$–tuple of all 1s, then $\alpha[\Gamma]$ comes from a sequencing. Given a state P, the **deficiency** of P is

$$DEF(P) = \sum_{j=2}^{|G|} |h_j - 1|.$$

Clearly, $DEF(P)$ is even and the goal is to find P such that $DEF(P)=0$.

Before defining the allowable operations on a state, it will be useful to establish some notation. Suppose P is a state with associated α and D_P. The final element in the path, $\alpha(|G|)$, will be denoted by f. If $1 \leq i \leq |G| - 1$ and $w = \alpha(1)$, then $w^+ = \alpha(i+1)$, the immediate successor of w and if $2 \leq i \leq |G|$, $w^- = \alpha(i-1)$ is the immediate predecessor of w. If $DEF(P) \geq 2$, then there exist i and j such that $h_i = 0$ and $h_j > 1$. The aim usually will be to modify the directed path so that one g_j is lost and one g_i is gained in the new D_P. The notation L_1 and G_1 will be used to denote these elements. Most operations will also lose a second element L_2 and gain a second element G_2. The relation between L_2, G_2 and D_P will determine the progress of the method. For example, if $DEF(P) = 2$ and an operation is performed that has the property $L_2 = G_2$, the process terminates with the sequencing.

Suppose now that a state P with associated α and D_P is given, $DEF(P) \geq 2$ and L_1 and G_1 have been chosen. Find i such that $w = \alpha(i)$ and $w \cdot L_1 = w^+$. There are several important possibilities.

If $f^{-1} = G_1$, then apply OPERATION A.

*A*1: Build a new directed Hamiltonian path

$$w^+ \rightarrow \cdots \rightarrow f \rightarrow e \rightarrow \cdots \rightarrow w$$

*A*2: Left-translate (each element of) this path by $(w^+)^{-1}$.

This gives

$$e \rightarrow \cdots \rightarrow (w^+)^{-1} f \rightarrow (w^+)^{-1} \rightarrow \cdots \rightarrow (w^+)^{-1} w,$$

a new state Q with associated α and D_Q. Note that if

$$\alpha(j) \cdot x = \alpha(j+1)$$

then

$$[(w^+)^{-1}\alpha(j)]x = (w^+)^{-1}\alpha(j+1)$$

so that D_Q is the required modification of D_P. There is no L_2 or G_2 for this operation and clearly

$$DEF(Q) = DEF(P) - 2.$$

The statement that $w \cdot G_1$ **follows** w in the directed Hamiltonian path of P means that if $w = \alpha(i)$ and $w \cdot G_1 = \alpha(j)$, then $i < j$. If $f^{-1} \neq G_1$ and $w \cdot G_1$ follows w in P, then apply OPERATION B.

$B1$: Build a new directed Hamiltonian path

$$e \rightarrow \cdots \rightarrow w \rightarrow w \cdot G_1 \rightarrow \cdots f \rightarrow w^+ \rightarrow \cdots \rightarrow (w \cdot G_1)^- .$$

This gives a new state Q with associated α and D_Q. Note that $w \cdot G_1 \neq w^+$ since $L_1 \neq G_1$ and that

$$L_2 = [(w \cdot G_1)^-]^{-1}(w \cdot G_1) \text{ and } G_2 = f^{-1}w^+.$$

In general, this operation will be one of several attempted on a state before one is chosen. Thus, the results must be saved. There is one case, however, where the results of operation B are ignored.

$B2$: If $L_2 \neq G_2$ and $(L_2 = L_1$ and $G_2 = G_1)$ then do not save the results.

This step is a precaution against the possibility of cycling occurring in the application of the process. Note that if operation B is used, then

$$DEF(Q) \in \{DEF(P), DEF(P)-2, DEF(P)-4\}.$$

The statement that $w \cdot G_1$ **precedes** w in the directed Hamiltonian path of P means that if $w = a(i)$ and $w \cdot G_1 = \alpha(j)$, then $j < i$. If $f^{-1} \neq G_1$, $w \cdot G_1$ precedes w in P and there is a k, $j \leq k < i$ such that $\alpha(k) \cdot f^{-1} = \alpha(k+1)$, then apply OPERATION C.

$C1$: Build a new directed Hamiltonian path

$$\alpha(k+1) \rightarrow \cdots \rightarrow w \rightarrow w \cdot G_1 \rightarrow \cdots \rightarrow \alpha(k) \rightarrow w^+ \cdots \rightarrow f \rightarrow e \rightarrow \cdots (wG_1)^- .$$

There are several special cases embedded in $C1$. In each case, the path modification should be clear. In all cases

$$G_2 = [\alpha(k)]^{-1}w^+.$$

If $w \cdot G_1 \neq e$, then $L_2 = [(w \cdot G_1)^-]^{-1}. (wG_1)$ and if $w \cdot G_1 = e$, then $L_2 = f^{-1}$ and the path terminates at f.

$C2$: Left-translate (each element of) this path by $[\alpha(k+1)]^{-1}$.

In general, this operation will be one of several attempted on a state before one is chosen. Thus, the results must be saved. As in operation B, ignore one case.

$C3$: If $L_2 \neq G_2$ and $(L_2 = L_1$ and $G_2 = G_1)$ then do not save the results.

Note that if operation C is used, then

$$DEF(Q) \in \{DEF(P), DEF(P)-2, DEF(P)-4\}.$$

The next operation will "double" the possibilities for operations A, B and C. Suppose P, α and D_P are given. Apply OPERATION R.

$R1$: Consider a new directed Hamiltonian path

$$f \rightarrow \cdots \rightarrow e.$$

If L_1 and G_1 were chosen for P, then L_1^{-1} and G_1^{-1} can be chosen to fill corresponding roles for the new path, or, if $DEF(P) > 2$ other values can be chosen.

Note that if

$$\alpha(i) \cdot L_1 = \alpha(i+1)$$

then

$$\alpha(i+1) \cdot L_1^{-1} = \alpha(i).$$

This says that when the path is reversed, even if L_1^{-1} and G_1^{-1} are chosen as the designated loss and gain, the "base point" w of operations A, B and C changes from $\alpha(i)$ to $\alpha(i+1)$ and one may expect new things to happen.

$R2$: Left-translate (each element of) this path by f^{-1}.

Let rP denote the new state and note that

$$DEF(P) = DEF(rP).$$

There is a small chance that the application of operations A, B, C and R to a state will yield no choices of possible new states. In this case, apply OPERATION D.

$D1$: Backtrack to the preceding state.

If Operation D is used, DEF can increase.

There is a reasonable chance that the procedure to be described will cycle if only the operations just defined are implemented. It is useful to have available a simple operation that is not easily undone by the other operations and does not drastically increase DEF. OPERATION E seems to work quite well, although it does not guarantee that the algorithm will always find a sequencing if one exists.

$E1$ Modify the path

$$e \rightarrow e^+ \rightarrow \cdots \rightarrow f^- \rightarrow f$$

to

$$e \rightarrow f \rightarrow e^+ \rightarrow \cdots \rightarrow f^-.$$

Under this operation $L_1 = e^+$, $L_2 = (f^-)^{-1}f$, $G_1 = f$ and $G_2 = f^{-1}e^+$. Note that DEF certainly can increase under an application of operation E.

Several comments about notation and terminology will increase the readability of the algorithm to follow. Given a state P, the notation rA will be used to mean "apply operation A to rP. The term "step" will be used to denote a counter for the number of changes in state executed by the algorithm. The phrase "first available possibility" needs to be explained. If the algorithm arrives at a state via the backtrack operation, the algorithm is to proceed by using the first possibility not previously used at that state, if any such possibility exists. In practice, backtracking happens rarely so the program doesn't have to remember very many previous states. If the algorithm arrives at a previous state by methods other than backtracking, this restriction does not apply. This is the reason for operation E. Finally, X, Y and Z are positive constants determined by the size of G. For example if $|G| \leq 20$, $X = 2$, $Y = 20$ and $Z = 100$ work well. For $|G| \leq 100$, $X = 2$, $Y = 80$ and $Z = 500$ are reasonable choices.

A Sequencing Algorithm

1. Input an initial state.
2. IF A or rA can be executed THEN pick the first available possibility and go to (5) ELSE go to (3).
3. Test all possibilities B, rB, C and rC. IF there is an available possibility with $L_2 = G_2$ THEN pick the first available one and go to (5) ELSE IF any possibility exists THEN pick the first available one and go to (5) ELSE go to (4).
4. IF D is possible THEN do it and go to (5) ELSE failure.
5. IF new $DEF = 0$ THEN success ELSE IF new $DEF \leq X$ and Step is a multiple of Y THEN (6) ELSE IF step $\geq Z$ THEN failure ELSE go to (2).
6. Do E and go to (2).

A few observations are in order before proceeding to a listing of results. Suppose G is a sequenceable group and the algorithm has reached a state P such that $DEF(P) = 2$. If G_2 and L_2 are "at random" one would expect the algorithm to find a sequencing quickly. This expectation seems to be justified by actual computation. Except for a few isolated cases, the results described below were obtained using the initial state derived from the directed Hamiltonian path $1,2,3,\ldots,|G|$ or the path $1,|G|,2,3,\ldots,|G|-1$ (with integers $1,2,\ldots,|G|$ used to denote the elements of G). If troubles occurred, initial states were changed in a pseudo-random way. Of course, one could speed up the algorithm by beginning execution with an initial state having small DEF. For dihedral groups at least, this seems relatively easy to do. Modifications of this algorithm certainly are possible. For example, if the aim is to find 2-sequencings, more operations can quickly be defined.

Some examples may be useful. First, consider the reversing process. Suppose $G = D_5$, the dihedral group of order 10 (see Table 1 for the multiplication table). Let P be the state with directed Hamiltonian path

6 7 4 2 10 2 3 9 8

$$P\colon 1 \to 6 \to 5 \to 3 \to 4 \to 8 \to 7 \to 10 \to 2 \to 9.$$

The row of numbers above P shows how P can be constructed by a collection of partial products, that is, $1\cdot 6 = 6$, $6\cdot 7 = 5$, $5\cdot 4 = 3$, etc. From this row it is clear that $DEF(P) = 2$, $L_1 = 2$ and $G_1 = 5$. Since $9^{-1} = 9 \neq G_1$ operation A is not available. Since $3\cdot 5 = 2$ follows 3 and $8\cdot 5 = 9$ both possibilities for operation B are available. In the first case $L_2 = 9$ and $G_2 = 6$ while the second case gives $L_2 = 8$ and $G_2 = 3$. The algorithm would therefore save both these possibilities and consider rP.

8 9 4 5 10 5 3 7 6

$$rP\colon 1 \to 8 \to 5 \to 3 \to 2 \to 6 \to 7 \to 10 \to 4 \to 9.$$

Now $L_1 = 5$ and $G_1 = 2$. Again $9^{-1} = 9 \neq G_1$ and operation A fails. Again $3\cdot 3 = 4$ follows 3 with $L_2 = 7$ and $G_2 = 8$ while $6\cdot 2 = 10$ follows 6 with $L_2 = G_2 = 3$. This means that the last possibility yields a sequencing via operation B.

1 8 9 4 5 10 2 7 6 3

1 8 5 3 2 6 10 4 9 7

The first row above is the sequencing and the second row is the collection of associated partial products.

Next, consider an example of a state P on D_9. With the table for D_9 as in [15] or via the instructions in §3, it can be shown that the algorithm must backtrack from state P.

16 18 4 2 3 5 14 6 8 9 17 11 17 15 7 10 13

$$P\colon 1 \to 16 \to 8 \to 2 \to 3 \to 5 \to 9 \to 13 \to 17 \to 10 \to 11 \to 4 \to 14 \to 7 \to 12 \to 15 \to 6 \to 18.$$

The last part of this section will be devoted to a sample computation on D_5 that illustrates several features of the algorithm.

1	2	3	4	5	6	7	8	9	10
2	3	4	5	1	7	8	9	10	6
3	4	5	1	2	8	9	10	6	7
4	5	1	2	3	9	10.	6	7	8
5	1	2	3	4	10	6	7	8	9
6	10	9	8	7	1	5	4	3	2
7	6	10	9	8	2	1	5	4	3
8	7	6	10	9	3	2	1	5	4
9	8	7	6	10	4	3	2	1	5
10	9	8	7	6	5	4	3	2	1

Table 1. A Multiplication Table for D_5.

Suppose that an initial state P_0 has been input and after a few steps, the process arrives at stage P_1.

4 6 3 5 4 6 7 10 2

$$P_1\colon 1 \to 4 \to 9 \to 7 \to 8 \to 10 \to 5 \to 6 \to 2 \to 3 \qquad DEF(P_1)=4, L_1=6, G_1=9.$$

The implementation I have used always picks the largest available numbers for L_1 and G_1. After testing all possibilities, the algorithm chooses the first available option, which in this case is the first possibility for operation B on P_1.

$$P_2\colon\ 1 \to \overset{4}{4} \to \overset{9}{7} \to \overset{5}{8} \to \overset{4}{10} \to \overset{6}{5} \to \overset{7}{6} \to \overset{10}{2} \to \overset{2}{3} \to \overset{7}{9} \qquad DEF(P_2)=4, L_1=7, G_1=8.$$

$$rP_2\colon\ 1 \to \overset{7}{7} \to \overset{5}{8} \to \overset{10}{4} \to \overset{7}{10} \to \overset{6}{5} \to \overset{3}{2} \to \overset{2}{3} \to \overset{9}{6} \to \overset{3}{9} \qquad DEF(rP_2)=4, L_1=7, G_1=8.$$

After testing all possibilities, the algorithm chooses the first available option which is the first possibility for operation B on rP_2.

$$P_3\colon\ 1 \to \overset{8}{8} \to \overset{10}{4} \to \overset{7}{10} \to \overset{6}{5} \to \overset{3}{2} \to \overset{2}{3} \to \overset{9}{6} \to \overset{3}{9} \to \overset{3}{7} \qquad DEF(P_3)=4, L_1=3, G_1=5.$$

The algorithm chooses the first possibility for operation C on P_3. Note that DEF falls even though $L_2 \neq G_2$.

$$P_4\colon\ 1 \to \overset{6}{6} \to \overset{5}{7} \to \overset{4}{9} \to \overset{2}{8} \to \overset{9}{5} \to \overset{3}{2} \to \overset{3}{4} \to \overset{7}{10} \to \overset{8}{3} \qquad DEF(P_4)=2, L_1=3, G_1=10.$$

The algorithm chooses the first possibility for operation C on P_4 (from the second occurrence of L_1).

$$P_5\colon\ 1 \to \overset{2}{2} \to \overset{9}{10} \to \overset{3}{8} \to \overset{10}{4} \to \overset{5}{3} \to \overset{9}{6} \to \overset{7}{5} \to \overset{8}{7} \to \overset{4}{9} \qquad DEF(P_5)=2, L_1=9, G_1=6.$$

The algorithm chooses the first possibility for operation B on P_5.

$$P_6\colon\ 1 \to \overset{2}{2} \to \overset{6}{7} \to \overset{4}{9} \to \overset{5}{10} \to \overset{3}{8} \to \overset{10}{4} \to \overset{5}{3} \to \overset{9}{6} \to \overset{7}{5} \qquad DEF(P_6)=2, L_1=5, G_1=8.$$

The step that gives P_6 is a multiple of Y (see the algorithm) so the algorithm chooses operation E. Note that DEF increases.

$$P_7\colon\ 1 \to \overset{5}{5} \to \overset{3}{2} \to \overset{6}{7} \to \overset{4}{9} \to \overset{5}{10} \to \overset{3}{8} \to \overset{10}{4} \to \overset{5}{3} \to \overset{9}{6} \qquad DEF(P_7)=6, L_1=5, G_1=8.$$

The algorithm chooses an instance of operation B on P_7 such that $L_{2=G_2}$.

$$P_8\colon\ 1 \to \overset{5}{5} \to \overset{3}{2} \to \overset{6}{7} \to \overset{4}{9} \to \overset{5}{10} \to \overset{3}{8} \to \overset{10}{4} \to \overset{8}{6} \to \overset{9}{3} \qquad DEF(P_8)=4, L_1=4, G_1=7.$$

$$rP_8\colon\ 1 \to \overset{9}{9} \to \overset{8}{2} \to \overset{10}{6} \to \overset{4}{8} \to \overset{2}{7} \to \overset{3}{10} \to \overset{6}{5} \to \overset{4}{3} \to \overset{2}{4} \qquad DEF(rP_8)=4, L_1=4, G_1=7.$$

The algorithm chooses an instance of operation C on rP_8 such that $L_{2=G_2}$.

$$\begin{array}{llllllllllll} & & 6 & 7 & 4 & 2 & 10 & 2 & 3 & 9 & 8 & \\ P_9: & 1 \rightarrow & 6 \rightarrow & 5 \rightarrow & 3 \rightarrow & 4 \rightarrow & 8 \rightarrow & 7 \rightarrow & 10 \rightarrow & 2 \rightarrow & 9 & \quad DEF(P_9)=2, L_1=2, G_1=5. \end{array}$$

$$\begin{array}{llllllllllll} & & 8 & 9 & 4 & 5 & 10 & 5 & 3 & 7 & 6 & \\ rP_9: & 1 \rightarrow & 8 \rightarrow & 5 \rightarrow & 3 \rightarrow & 2 \rightarrow & 6 \rightarrow & 7 \rightarrow & 10 \rightarrow & 4 \rightarrow & 9 & \quad DEF(rP_9)=2, L_1=5, G_1=2. \end{array}$$

Finally, the algorithm terminates with a sequencing by choosing an instance of operation B on rP_9 such that $L_{2=G_2}$.

$$\begin{array}{lllllllllllll} & & 9 & 1 & 4 & 5 & 10 & 2 & 7 & 6 & 3 & \\ P_{10}: & 1 \rightarrow & 8 \rightarrow & 5 \rightarrow & 3 \rightarrow & 2 \rightarrow & 6 \rightarrow & 10 \rightarrow & 4 \rightarrow & 9 \rightarrow & 7 & \quad DEF(P_{10})=0. \end{array}$$

3. Results

The sequencings listed in this section can be verified by looking at the appropriate group table in [15] or, in the case of dihedral groups of order > 32, constructing a table according to the following instructions.

Suppose $n \geq 2$ is a positive integer. The **dihedral group** D_n is the group of order $2n$ defined by

$$D_n = \{a^i b^j : 0 \leq i \leq n-1, 0 \leq j \leq 1, a^n = e, b^2 = e, ba = a^{n-1}b\}.$$

It is easy to see that if j is any integer and D_n is any dihedral group, then $a^j b = ba^{-j}$ and $(a^j b)(a^j b) = e$. The interest then, is in products of the form

$$(a^i b^j)(a^k b^l); \; 0 \leq i, k \leq n-1, 0 \leq j, l \leq 1.$$

Use (i,j) to denote $a^i b^j$. With this notation, a multiplication table for D_n can be constructed by the following rules:

$$(i, 0) \cdot (k, l) = ((i+k)(mod\ n), l)$$
$$(i, 1) \cdot (k, 0) = ((i-k)(mod\ n), 1)$$
$$(i, 1) \cdot (k, 1) = ((i-k)(mod\ n), 0).$$

Now, tables, as in [15], for the dihedral groups are constructed by making the identifications

$$(i, 0) \Leftrightarrow i+1 \text{ and } (k, 1) \Leftrightarrow k+n+1.$$

Here are sequencings for the dihedral groups $D_5 - D_{50}$.

N = 5: 1 10 5 9 6 7 3 2 8 4

N = 6: 1 11 2 7 3 9 12 10 6 5 8 4

N = 7: 1 8 2 10 7 6 9 5 11 4 14 13 12 3

N = 8: 1 13 11 16 4 14 3 5 6 15 8 7 9 12 2 10

N = 9: 1 4 3 16 15 12 6 18 11 2 13 8 9 10 7 5 14 17

N =10: 1 15 4 14 2 20 7 8 10 13 9 12 19 16 18 17 3 11 5 6

11: 1 8 18 15 16 5 10 4 6 13 11 3 19 17 7 9 22 2 14 12
20 21

12: 1 17 4 11 2 8 9 13 10 3 23 24 22 15 14 6 20 18 16 7
21 19 12 5

13: 1 13 7 10 15 23 14 18 6 17 12 2 4 25 24 3 5 11 22 21
26 8 16 19 9 20

14: 1 6 27 28 10 4 2 9 15 7 5 24 25 21 23 18 22 16 3 20
17 26 11 8 14 12 13 19

15: 1 21 24 25 3 16 19 7 4 18 15 20 12 29 2 13 11 30 5 23
8 10 6 26 9 14 28 27 17 22

16: 1 29 13 19 28 20 12 26 5 15 2 11 25 14 3 27 18 22 16 17
9 21 23 7 10 30 31 6 32 8 4 24

17: 1 11 6 9 3 30 13 12 15 17 18 22 33 24 19 21 10 20 2 32
26 34 29 4 23 28 7 27 14 25 5 31 16 8

18: 1 10 18 31 20 26 9 4 30 27 35 33 12 17 32 29 14 8 6 21
11 25 19 36 28 16 7 34 3 5 24 23 22 13 15 2

19: 1 32 15 24 23 8 38 14 22 19 37 34 5 33 36 26 12 25 13 6
28 21 7 29 10 4 20 11 31 18 30 35 3 2 16 17 27 9

20: 1 27 19 35 6 3 37 7 17 16 2 22 8 11 10 33 5 15 14 39
20 29 24 31 13 28 23 12 26 36 32 18 9 34 21 38 25 40 30 4

21: 1 33 23 41 13 36 42 28 9 37 40 25 4 30 24 2 7 34 6 32
39 27 17 3 18 35 20 14 31 38 15 16 22 8 11 19 26 12 10 21
5 29

22: 1 37 15 10 31 35 11 39 41 16 29 24 2 30 40 17 21 8 33 14
19 44 25 7 6 4 22 42 5 9 23 12 43 26 32 13 27 34 36 18
28 38 20 3

23: 1 30 33 29 24 16 8 18 42 45 39 17 26 11 46 40 28 27 36 14
5 3 35 21 6 9 7 43 31 22 44 2 15 4 37 38 32 41 23 13
19 34 10 12 25 20

24: 1 4 6 8 3 29 17 11 34 30 15 10 39 5 31 20 38 32 18 46
16 22 44 7 9 25 12 27 36 14 28 13 47 40 42 24 19 43 45 37
41 48 21 2 35 26 23 33

25: 1 49 45 8 36 3 26 35 7 13 43 32 37 47 28 24 5 23 12 34
25 19 31 50 38 40 39 6 44 9 21 20 48 16 15 10 33 4 18 14
17 41 22 2 46 30 42 29 11 27

26: 1 2 37 3 22 25 36 15 29 20 48 46 14 38 33 45 52 12 7 30
43 41 24 13 21 11 39 40 6 42 10 34 5 27 26 44 31 50 23 8
19 49 16 51 4 35 18 17 47 32 9 28

27: 1 10 13 42 50 52 17 9 23 22 5 51 19 46 32 27 38 11 39 14
54 45 7 18 3 40 43 35 44 47 15 26 34 29 21 30 49 8 41 6
4 53 24 2 20 28 36 37 31 33 12 16 25 48

28: 1 48 34 7 39 37 43 16 33 10 55 5 22 3 12 35 8 15 24 4
23 36 53 41 6 49 21 27 26 25 17 45 2 11 40 30 29 51 19 28
31 20 54 56 46 14 9 18 35 52 32 44 42 47 13 50
29: 1 15 7 25 27 43 9 55 52 46 16 5 39 33 47 26 12 31 28 41
40 37 18 23 6 11 21 42 36 17 50 3 56 49 20 29 58 32 45 19
24 10 51 14 4 57 54 44 35 22 38 48 13 34 2 8 30 53
30: 1 5 56 57 20 32 46 47 53 14 48 19 23 36 45 52 4 16 31 18
35 3 40 2 15 30 38 54 9 44 28 26 59 39 10 37 34 13 33 17
27 55 43 8 49 7 21 29 25 60 41 51 6 11 22 24 58 50 12 42
31: 1 29 19 37 44 38 25 51 4 56 24 47 26 32 20 11 16 62 18 28
21 22 17 52 39 14 42 60 7 45 15 13 35 10 53 31 43 5 41 8
54 48 27 9 58 33 23 50 46 61 49 12 55 59 40 3 6 36 30 2
57 34
32: 1 41 23 8 52 18 50 30 37 28 26 25 10 17 60 47 16 43 59 6
35 15 4 24 55 44 62 38 11 64 32 9 56 14 13 36 58 34 54 33
5 40 3 7 46 53 22 63 49 21 42 48 31 51 19 39 61 2 57 27
29 45 20 12
33: 1 63 31 33 22 20 58 59 9 44 55 28 2 57 23 18 54 5 45 7
56 10 15 46 12 11 8 14 30 19 52 42 25 26 60 61 17 37 3 62
40 13 16 66 27 53 50 43 49 51 47 29 21 36 24 32 6 34 65 35
41 48 4 64 38 39
34: 1 44 52 12 19 64 21 6 4 22 57 60 7 20 41 25 8 9 67 36
40 63 5 48 30 2 58 51 68 61 11 59 65 56 62 39 43 54 15 38
27 35 49 18 17 10 28 42 45 50 3 26 32 34 55 24 46 66 23 37
53 13 47 29 14 31 33 16
35: 1 3 45 10 22 33 11 16 32 54 47 61 43 62 31 12 53 20 67 35
8 46 29 21 7 60 25 39 34 57 64 59 6 55 66 4 38 63 65 51
70 2 13 68 28 37 26 50 30 24 23 58 5 40 27 69 15 48 19 42
56 9 18 36 17 41 44 49 14 52
36: 1 16 46 42 48 27 67 29 28 63 70 23 45 69 71 26 53 60 41 56
32 34 19 72 25 31 62 68 35 55 33 18 49 15 64 20 36 52 10 50
7 22 5 54 11 21 38 24 37 13 2 14 4 12 44 30 59 51 9 40
65 66 47 17 3 57 43 6 39 61 58 8
37: 1 32 47 9 15 6 45 28 49 65 74 67 56 35 31 38 72 62 52 48
7 11 20 24 44 29 33 55 73 22 14 71 60 70 18 3 54 64 36 27
66 13 2 63 68 42 51 69 57 43 4 50 10 34 19 12 17 58 41 40
25 21 39 53 16 5 23 59 46 30 37 61 26 8
38: 1 67 25 45 24 11 10 66 58 16 31 26 54 18 32 28 64 50 2 36
34 4 19 72 15 6 13 61 63 46 17 38 8 69 74 20 9 70 12 39
76 30 65 23 60 33 48 40 55 71 22 35 47 57 3 27 7 56 29 5
42 44 49 52 75 53 62 68 21 73 14 43 37 51 41 59
39: 1 74 23 33 48 56 13 17 63 46 8 34 62 59 65 19 71 78 4 76
53 50 54 31 18 24 58 3 37 30 64 38 51 28 49 36 40 43 61 39
27 67 45 12 2 75 5 69 21 35 47 25 14 22 42 41 60 70 44 6
52 9 68 72 57 55 20 26 10 66 16 73 15 7 29 11 72 32
40: 1 21 41 13 38 61 29 8 11 51 44 20 47 55 25 53 26 30 54 52
78 58 2 48 42 18 10 35 6 23 24 59 32 63 65 79 34 3 9 5

43 68 36 56 45 74 37 75 33 80 39 57 72 69 46 12 4 73 50 76
71 40 22 27 64 66 70 14 60 67 16 28 49 7 15 17 31 19 77 62
41: 1 19 79 4 34 35 47 22 77 67 7 73 2 38 18 20 41 17 48 30
15 64 27 12 21 24 31 3 76 36 54 60 71 51 29 61 26 28 72 5
8 13 16 74 65 81 68 58 53 44 69 66 75 80 33 63 39 32 23 70
78 14 52 46 40 57 45 25 9 82 6 55 42 50 11 43 62 56 59 10
49 37
42: 1 30 77 11 36 24 18 73 76 15 74 64 3 40 55 83 28 25 80 34
66 78 38 21 50 57 68 7 47 75 58 33 12 27 59 51 71 19 16 70
5 45 26 43 63 8 17 84 13 62 2 23 22 48 32 39 44 79 41 53
69 37 52 10 67 14 9 42 35 46 6 72 61 56 54 65 81 82 49 29
60 20 31 4
43: 1 72 5 18 34 35 46 49 51 81 70 73 19 7 47 84 48 86 67 14
50 30 74 57 39 63 59 17 58 56 38 69 20 9 68 26 60 4 44 55
8 28 40 62 27 3 61 31 13 11 12 37 22 82 29 36 64 78 25 2
75 41 43 45 71 15 54 66 52 65 85 32 80 83 53 21 77 16 79 23
24 6 42 76 33 10
44: 1 8 18 51 64 7 78 2 12 21 32 17 24 16 10 37 86 62 13 70
29 81 63 67 50 14 5 57 15 28 53 35 6 34 3 76 38 77 59 54
20 36 44 66 47 75 74 56 83 55 23 49 48 43 19 68 22 41 52 72
79 87 88 25 42 27 61 85 58 65 11 69 84 46 9 39 26 30 40 60
33 80 31 45 71 73 4 82
45: 1 47 72 70 29 45 57 4 55 83 67 69 26 7 5 12 59 71 19 2
16 89 88 20 9 8 75 36 60 44 42 87 34 33 90 14 28 51 63 64
13 73 61 21 54 27 66 3 15 17 22 43 46 79 81 37 74 68 52 65
78 32 39 6 24 40 85 76 84 86 77 38 30 23 53 35 10 58 50 18
80 49 25 62 31 11 56 41 48 82
46: 1 60 65 33 63 44 10 55 15 37 66 88 70 7 13 74 85 67 8 34
28 32 22 90 12 3 5 83 27 50 62 24 81 73 36 25 82 4 30 54
53 29 61 72 59 23 52 77 14 21 45 58 79 75 51 47 18 9 11 43
80 42 56 40 39 41 92 17 71 64 2 86 91 31 87 38 20 19 6 35
68 76 49 16 46 78 26 89 69 57 84 48
47: 1 26 47 34 13 11 42 65 37 51 8 90 43 50 45 38 46 14 86 81
58 53 33 74 39 30 24 20 71 84 66 76 54 36 93 59 25 85 48 73
21 52 15 78 4 64 70 87 69 3 2 16 17 79 5 89 12 77 63 61
7 60 18 88 91 22 10 92 83 56 29 82 57 72 32 19 41 75 31 28
68 67 27 35 55 44 94 49 40 80 6 62 9 23
48: 1 77 41 14 43 9 95 80 6 74 76 44 71 4 51 91 94 63 73 19
33 20 79 92 84 50 2 87 56 49 88 53 69 57 70 15 93 48 12 36
22 62 78 64 86 38 8 96 5 82 10 61 28 17 35 60 68 31 29 89
25 46 47 24 59 81 26 39 21 40 42 58 32 11 13 65 23 30 85 83
7 34 72 52 75 45 16 18 54 66 27 67 55 90 37 3
49: 1 25 93 42 95 41 94 2 69 46 55 49 11 68 24 43 47 22 80 75
77 57 87 9 39 30 6 97 73 4 63 76 18 70 58 35 54 79 20 62
23 45 83 91 13 59 10 74 5 60 96 29 38 33 14 78 17 72 44 66
27 61 56 89 98 84 15 32 67 92 36 37 51 34 3 71 19 12 26 90
28 86 85 31 88 64 40 82 53 8 48 7 16 65 52 81 21 50

50:																			
1	40	31	74	7	76	19	6	12	26	27	13	57	92	78	70	8	17	72	89
82	14	54	50	63	51	96	2	84	45	65	39	59	24	36	32	11	88	42	83
28	79	25	98	66	29	20	90	91	37	64	93	81	41	61	68	62	33	35	48
87	94	44	5	100	34	99	22	75	38	56	58	15	85	80	77	52	69	4	67
21	71	43	23	49	16	46	73	55	95	53	18	30	10	47	97	60	86	9	3

The paper [2] gives symmetric sequencings for all non-Abelian groups of order $n, 10 \leq n \leq 32$ with a unique element of order 2 (there are 9 such groups). The final list of sequencings presented here is labelled according to the number assigned each group in [15]. A sequencing for Q_6, group 12/5 of [15] is given (it is not symmetric) but the other groups sequenced by the methods of [2] are not included. For the convenience of the reader, tables for A_4 and Q_6 are listed.

1	2	3	4	5	6	7	8	9	10	11	12
2	1	4	3	6	5	8	7	10	9	12	11
3	4	1	2	7	8	5	6	11	12	9	10
4	3	2	1	8	7	6	5	12	11	10	9
5	8	6	7	9	12	10	11	1	4	2	3
6	7	5	8	10	11	9	12	2	3	1	4
7	6	8	5	11	10	12	9	3	2	4	1
8	5	7	6	12	9	11	10	4	1	3	2
9	11	12	10	1	3	4	2	5	7	8	6
10	12	11	9	2	4	3	1	6	8	7	5
11	9	10	12	3	1	2	4	7	5	6	8
12	10	9	11	4	2	1	3	8	6	5	7

A Multiplication Table for A_4.

1	2	3	4	5	6	7	8	9	10	11	12
2	3	4	5	6	1	8	9	10	11	12	7
3	4	5	6	1	2	9	10	11	12	7	8
4	5	6	1	2	3	10	11	12	7	8	9
5	6	1	2	3	4	11	12	7	8	9	10
6	1	2	3	4	5	12	7	8	9	10	11
7	12	11	10	9	8	4	3	2	1	6	5
8	7	12	11	10	9	5	4	3	2	1	6
9	8	7	12	11	10	6	5	4	3	2	1
10	9	8	7	12	11	1	6	5	4	3	2
11	10	9	8	7	12	2	1	6	5	4	3
12	11	10	9	8	7	3	2	1	6	5	4

A Multiplication Table for Q_6.

$12/4 = A_4$: 1 8 2 11 10 6 5 7 4 3 9 12
$12/5 = Q_6$: 1 5 2 7 6 3 4 9 8 10 12 11
16/6: 1 14 13 16 10 15 11 6 3 2 12 7 8 5 4 9
16/7: 1 7 12 6 5 8 9 3 14 2 11 4 13 16 15 10
16/8: 1 16 8 12 6 11 9 2 7 13 4 5 14 15 3 10
16/9: 1 3 9 12 15 2 16 10 11 6 13 7 5 4 14 8
16/10: 1 11 7 4 10 2 13 16 9 14 5 6 8 15 3 12
16/11: 1 9 5 10 2 14 15 7 12 8 3 4 13 16 11 6
16/13: 1 13 15 6 5 10 16 9 7 2 4 3 12 8 14 11
18/3: 1 5 9 15 10 4 18 8 7 2 6 17 14 16 3 13 11 12
18/5: 1 3 18 12 10 6 7 14 5 8 4 15 16 17 9 2 13 11
20/5: 1 14 2 7 16 3 11 20 4 13 12 15 19 8 17 18 10 6
5 9
24/4: 1 19 9 14 21 11 4 17 8 22 7 5 2 18 10 13 24 20
23 3 15 16 12 6
24/5: 1 16 2 4 9 23 20 24 18 19 6 13 14 15 17 8 12 3
21 5 10 11 22 7
24/6: 1 13 5 23 6 18 17 11 9 20 21 8 3 22 15 14 12 2
7 4 10 16 24 19
24/7: 1 22 20 18 16 3 10 15 24 5 9 19 14 13 11 23 7 4
6 17 8 21 2 12
24/9: 1 13 6 19 2 14 18 8 7 9 23 5 12 22 16 15 20 3
4 17 10 24 11 21
24/12: 1 15 7 9 21 24 14 2 18 4 5 12 13 20 17 6 10 19
8 22 16 23 11 3
24/15: 1 6 16 2 5 19 12 9 4 23 22 13 20 11 18 8 21 14
10 3 7 15 17 24
27/4: 1 15 2 21 16 23 14 6 10 26 17 22 13 24 12 3 5 25
20 4 19 11 9 8 27 7 18
30/2: 1 11 13 2 15 19 4 24 30 18 14 7 26 21 27 10 28
29 25 8 20 5 12 17 23 3 22 6 9 16
30/3: 1 8 9 18 19 30 14 10 22 24 29 12 21 16 23 7 13
11 15 27 3 6 25 2 26 4 20 17 28 5

References

[1] B.A. Anderson, "Sequencings and Starters", *Pacific J. Math.* 64 (1976) 17-24.

[2] B.A. Anderson, "Sequencings of Dicyclic Groups", *Ars Combinatoria* 23 (1987) 131-142.

[3] J. Dénes and A.D. Keedwell, *Latin Squares and Their Applications, Academic Press,* New York, 1974.

[4] J. Dénes and E. Török", "Groups and Graphs, in *Combinatorial Theory and Its Applications*, North Holland, Amsterdam, (1970), pp. 257-289.

[5] J.H. Dinitz and D.R. Stinson, "A Fast Algorithm for Finding Strong Starters", *Siam J. Alg. Disc. Meth.* 2 (1981) 50-56.

[6] R. Friedlander, "Sequences in Non-Abelian Groups with Distinct Partial Products", *Aequationes Math.* 14 (1976) 59-66.

[7] B. Gordon, "Sequences in Groups with Distinct Partial Products", *Pacific J. Math.* 11 (1961) 1309-1313.

[8] G.B. Hoghton and A.D. Keedwell, "On the Sequenceability of Dihedral Groups", *Annals of Discrete Math.* 15 (1982) 253-258.

[9] D.F. Hsu and A.D. Keedwell, "Generalized Complete Mappings, Neofields, Sequenceable Groups and Block Designs I, II", Pacific J. Math. 111 (1984) 317-332; 117 (1985) 291-312.

[10] A.D. Keedwell, "Some Problems Concerning Complete Latin Squares", in Proc. British Comb. Conf., Aberystwyth, *London Math. Soc.* Lecture Notes 13 (1974) 89-96.

[11] A.D. Keedwell, "Sequenceable Groups: A Survey", in *Finite Geometries and Designs, London Math. Soc.* Lect. Notes 49 (1981), pp. 205-215.

[12] A.D. Keedwell, "On the Sequenceability of Non-Abelian Groups of Order pq", *Discrete Math.* 37 (1981) 203-216.

[13] N.S. Mendelsohn, "Hamiltonian Decompositions of the Complete Directed n-Graph", in Proc. Colloq. Tihany, *Academic Press* (1968), pp. 237-241.

[14] E. Seah and D.R. Stinson, "Some Perfect One-Factorizations of K_{14}", *Annals of Discrete Math.* (elsewhere in this volume).

[15] A.D. Thomas and G.V. Wood, *Group Tables, Shiva Publishing,* (1980).

[16] T.W. Tillson, "A Hamiltonian Decomposition of K^*_{2m}, $2m \geq 8$", *J. Combin. Theory,* B29 (1980) 68-74.

[17] L.L. Wang, "A Test for the Sequencing of a Class of Finite Groups with Two Generators", *Am. Math. Soc. Notices* 20 (1973) 73T-A275.

Annals of Discrete Mathematics 34 (1987) 43–64
© Elsevier Science Publishers B.V. (North-Holland)

Pairwise Balanced Designs with Prime Power Block Sizes Exceeding 7

F.E. Bennett

Department of Mathematics
Mount Saint Vincent University
Halifax, Nova Scotia B3M 2J6
CANADA

TO ALEX ROSA ON HIS FIFTIETH BIRTHDAY

ABSTRACT

In this paper, we construct pairwise balanced designs (PBDs) having block sizes which are prime powers exceeding 7. If we denote by P^* the set of all prime powers exceeding 7 and by $B(P^*)$ the set of orders of PBDs of index unity having block sizes from P^*, then it is shown that $v \in B(P^*)$ for all $v > 2206$ and for many orders less than this value. We also give some applications to the construction of other types of combinatorial designs, such as conjugate orthogonal Latin squares and sets of mutually orthogonal Latin squares (MOLS). For example, it is proved that the spectrum of idempotent Latin squares with distinct and pairwise orthogonal conjugates contains all orders v mentioned above, and some of the PBDs constructed in this paper can be used to obtain more MOLS of certain orders than previously known.

1. Introduction

Let K be a set of positive integers. A *pairwise balanced design* (PBD) of index unity $B(K,1;v)$ is a pair $(X,\mathcal{B})$ where X is a v-set (of *points*) and $\mathcal{B}$ is a collection of subsets of X (called *blocks*) with sizes from K such that every pair of distinct points of X is contained in exactly one block of $\mathcal{B}$. The number $|X| = v$ is called the *order* of the PBD.

We shall denote by $B(K)$ the set of all integers v for which there exists a PBD $B(K,1;v)$. A set K is said to be *PBD-closed* if $B(K) = K$. Pairwise balanced designs have been used extensively in the construction of other types of combinatorial designs, and one is generally interested in constructing PBDs which contain specified block sizes. R. M. Wilson's remarkable theory concerning the structure of PBD-closed sets (see [18,19,20]) often provides us with some form of asymptotic results as follows. Given a set K of positive integers, we define the two parameters

$$\alpha(K) = \gcd\{k-1 : k \in K\},$$
$$\beta(K) = \gcd\{k(k-1) : k \in K\}.$$

Then according to Wilson's result, there exists a constant C (depending on K) such that, for all integers $v > C$, $v \in B(K)$ if and only if $v-1 \equiv 0 \pmod{\alpha(K)}$ and $v(v-1) \equiv 0 \pmod{\beta(K)}$.

In this paper, the main focus of our attention will be the set $B(P^*)$, where P^* is defined to be the set of all prime powers exceeding 7. It is fairly evident that Wilson's theory guarantees the existence of a constant C such that, for all $v > C$, $v \in B(P^*)$. In other words, we have $v \in B(P^*)$ for all sufficiently large values of v, where the term "sufficiently large" remains to be specified. Constructions provided by the author in previous papers [1,2] established a concrete upper bound on C, namely, $C \leq 5074$. It is the object of this paper to improve substantially on this. We shall present some new recursive constructions for PBDs to show that $v \in B(P^*)$ for all $v > 2206$ and for many orders less than this value. As an immediate consequence of this result, we readily obtain that the spectrum L^* of idempotent Latin squares with distinct and pairwise orthogonal conjugates contains the same values of v mentioned above, since $B(P^*) \subseteq L^*$ (see, for example, [1,11]). Moreover, we are able to utilize some of the PBDs constructed in this paper to obtain more mutually orthogonal Latin squares (MOLS) of certain orders than previously known. For example, among the "small" orders, the results contained in [6] indicate $N(520) \geq 7$, $N(522) \geq 6$, and $N(524) \geq 6$, where $N(v)$ denotes the maximum number of MOLS of order v. Here we can actually show that $N(520) \geq 8$, $N(522) \geq 10$, and $N(524) \geq 12$. For more general information on PBDs and related designs, the reader may refer to [9,18-21], and results concerning the existence of MOLS can be found in [6,7,8] and their excellent bibliography.

2. Preliminaries

In this section, we shall define some terminology and state some fundamental results which will be useful later on.

Definition 2.1: Let K and M be sets of positive integers. A *group divisible design* (GDD) $GD(K,1,M;v)$ is a triple $(X,\mathcal{G},\mathcal{B})$, where

(i) X is a v-set (of *points*),

(ii) $\mathcal{G}$ is a collection of non-empty subsets of X (called *groups*) with sizes in M and which partition X,

(iii) $\mathcal{B}$ is a collection of subsets of X (called *blocks*), each with size at least two in K,

(iv) no block meets a group in more than one point, and

(v) each pairset $\{x,y\}$ of points not contained in a group is contained in exactly one block.

The *group-type* (or *type*) of a GDD $(X,\mathcal{G},\mathcal{B})$ is the multiset $\{|G| : G \in \mathcal{G}\}$, and we usually use the "exponential" notation for its description: a group-type $1^i 2^j 3^k \cdots$ denotes i occurrences of groups of size 1, j occurrences of groups of size 2, and so on.

Definition 2.2: A *transversal design* (TD) $T(k,1;m)$ is a GDD with km points, k groups of size m and m^2 blocks of size k, where each block meets every group in precisely one point, that is, each block is a transversal of the collection of groups.

Definition 2.3: Let $(X,\mathcal{B})$ be a PBD $B(K,1;v)$. A *parallel class* in $(X,\mathcal{B})$ is a collection of disjoint blocks of $\mathcal{B}$, the union of which equals X. $(X,\mathcal{B})$ is called *resolvable* if the blocks of $\mathcal{B}$ can be partitioned into parallel classes. A GDD $GD(K,1,M;v)$ is resolvable if its associated PBD $B(K \cup M,1;v)$ is resolvable with M as a parallel class of the resolution.

We wish to remark that it is fairly well-known that the existence of a resolvable TD $T(k,1;m)$ (briefly $RT(k,1;m)$) is equivalent to the existence of a TD $T(k+1,1;m)$, or equivalently $k-1$ MOLS of order m. The following two results can be found in [9,12].

Theorem 2.4: For every prime power q, there exists a $T(q+1,1;q)$.

Theorem 2.5: Let $m = p_1^{k_1} p_2^{k_2} \cdots p_r^{k_r}$ be the factorization of m into powers of distinct primes p_i, then a $T(k,1;m)$ exists, where $k \leq 1+\min\{p_i^{k_i}\}$.

From current results we have concerning the existence of sets of 7 MOLS (see, for example, [6,7]), we can state the following theorem.

Theorem 2.6: There exists a $T(9,1;m)$ in each of the following cases:

(a) for all odd integers $m \geq 9$, where $m \notin$ {15, 21, 33, 35, 39, 45, 51, 55, 63, 69, 75, 77, 85, 87, 93, 95, 111, 119, 123, 159, 175, 183, 291, 295, 303, 335},

(b) for all positive integers $m \equiv 0 \pmod 8$, where $m \neq 24,40,48$,

(c) for $m \in$ {82, 100, 154, 262, 342, 350, 404, 412, 414, 502, 514, 538, 540, 586, 590, 598, 602, 606, 650, 652, 658, 662, 666, 668, 670, 674, 676, 682, 684, 686, 692, 698, 700, 706, 710, 714, 716, 718, 722, 724, 726, 730, 732, 738, 740, 742, 746, 748, 754, 756, 758, 766, 770, 772, 778},

(d) for all integers $m > 780$.

We shall adapt some of the notations used in earler papers by the author (see, for example, [1]). For example, we shall write $B(k,1;v)$ for $B(\{k\},1;v)$ and similarly $GD(k,1,m;v)$ for $GD(\{k\},1,\{m\};v)$. We observe that a PBD $B(k,1;v)$ is essentially a *balanced incomplete block design* (BIBD) with parameters v,k and $\lambda = 1$. In particular, we note that a $B(q+1,1;q^2+q+1)$ is a finite projective plane $PG(2,q)$. Moreover, a $B(q,1;q^2)$ is a finite Euclidean plane $EG(2,q)$ in which the points can be arranged in a $q \times q$ array so that the rows form one parallel class and the columns another parallel class. It is well-known that both $PG(2,q)$ and $EG(2,q)$ exist for all prime powers q. If $k \notin K$, then $B(K \cup \{k^*\},1;v)$ denotes a PBD $B(K \cup \{k\}^*,1;v)$ which contains a unique block of size k and if $k \in K$, then $B(K \cup \{k^*\},1;v)$ is a PBD $B(K,1;v)$ containing at least one block of size k. For convenience, we define $B(k_1,k_2, \cdots ,k_r)$ to be the set of all integers v such that there is a PBD $B(\{k_1,k_2, \cdots ,k_r\},1;v)$. As in [14], we shall refer to a GDD $(X,\mathcal{G},\mathcal{B})$ as a K-GDD if $B \in K$ for every block $B \in \mathcal{B}$; and if $\mathcal{G} = \{G_1,G_2, \cdots ,G_n\}$, we may represent the type of the K-GDD by the ordered n-tuple $T = (m_1,m_2,...,m_n)$, where $|G_i| = m_i$. Of course, where there is no danger of confusion, the type of the K-GDD will be represented as a multiset using the "exponential" notation.

There are some useful PBDs derived from TDs and finite planes, and we shall present some fundamental lemmas below before proceeding to the next section. For the most part, the proofs of these lemmas can be found in [1].

We have the following obvious "Direct Product" construction.

Lemma 2.7: Let K be a set of positive integers and suppose $u,v \in B(K)$. If there exists a TD $T(u,1;v)$, then $uv \in B(K)$.

From truncated TDs or finite planes we get most of what follows.

Lemma 2.8: If a $T(k+1,1;m)$ exists and $1 \leq t \leq m$, then $km+t \in B(k,k+1,m,t^*)$. In particular, the conclusion holds for a prime power $m \geq k$.

Proof: See [1, Lemma 3.3].

Lemma 2.9: If a $T(k+1,1;m)$ exists and $0 \leq t \leq m$, then $km+t+1 \in B(k,k+1,m+1,t+1)$. In particular, the conclusion holds for all prime powers $m \geq k$.

Proof: See [1, Lemma 3.4].

Lemma 2.10: If a $T(k+2,1;m)$ exists and $1 \leq k \leq m-1$, then for $0 \leq u \leq m$ and $0 \leq v \leq m$, we have $km+u+v \in B(k,k+1,k+2,m,u,v)$.

Proof: See [1, Lemma 3.5].

Lemma 2.11: If m is a prime power and $1 \leq k \leq m$, then for $0 \leq t \leq m-k$, we have $km+t \in B(k,k+1,(k+t)^*,m)$.

Proof: See [1, Lemma 3.6].

We also state the following two useful generalizations of Lemma 2.9, which utilize the technique of adding a set of fixed ("infinite") points to a GDD (see, for example, [5,13,21]).

Lemma 2.12: Suppose a $T(k+1,1;m)$ exists, $0 \leq s \leq m$, and $\{k,k+1\} \subseteq B(K)$. If there exist PBDs $B(K \cup \{t^*\},1;m+t)$ and $B(K \cup \{t^*\},1;s+t)$, then $km+s+t \in B(K \cup \{(m+t)^*\})$.

Proof: See [1, Lemma 3.7].

Lemma 2.13: Suppose a $T(k+1,1;m)$ exists, $0 \leq s \leq m$, and $\{k,k+1\} \subseteq B(K)$. If there exists a PBD $B(K \cup \{t^*\},1;m+t)$, then $km+s+t \in B(K \cup \{(s+t)^*\})$.

Proof: See [1, Lemma 3.8].

We remark that both Lemmas 2.12 and 2.13 are special cases of a more general construction which makes use of GDDs with non-uniform group sizes, and the proof of the following lemma is very similar (see, for example, [14]).

Lemma 2.14: Let K be a set of positive integers and $s \geq 0$. Suppose there exists a K-GDD of type $T=(m_1,m_2,\ldots,m_n)$.

(a) If a PBD $B(K \cup \{s^*\},1;m_i+s)$ exists for $1 \leq i \leq n$, then, for each i, $v+s \in B(K \cup \{(m_i+s)^*\})$, where $v = \sum_{1 \leq i \leq n} m_i$.

(b) If a PBD $B(K \cup \{s^*\},1;m_i+s)$ exists for $1 \leq i \leq n-1$, then $v+s \in B(K \cup \{(m_n+s)^*\})$, where $v = \sum_{1 \leq i \leq n} m_i$.

Proof: The proof of (a) is similar to that of Lemma 2.12 and we omit it. The proof of (b) is similar to that of Lemma 2.13 and it is also omitted.

The following two variations of Lemmas 2.8 and 2.9 were earlier overlooked by the author in [1] and perhaps elsewhere. These two lemmas will be of utmost importance in providing essential ingredients for most of our constructions.

Lemma 2.15: If a $T(k+1,1;m)$ exists and $0 \leq t \leq k+1$, then $km+m+t-k-1 \in B(k,k+1,m-1,m,t^*)$.

Proof: We may delete $k+1-t$ points from a particular block of the $T(k+1,1;m)$ to get a PBD $B(\{k,k+1,m-1,m,t^*\},1;km+m+t-k-1)$.

Lemma 2.16: If a $T(k+1,1;m)$ exists and $0 \leq t \leq k+1$, then $km+m+t-k \in B(k,k+1,m,m+1,t^*)$.

Proof: By deleting $k+1-t$ points from a particular block of the $T(k+1,1;m)$, we essentially obtain a GDD $GD(\{k,k+1,t^*\},1,\{m-1,m\};km+m+t-k-1)$. We then adjoin one "infinite" point to the groups of this GDD for our result.

The following construction of Seiden [16] will be quite useful.

Lemma 2.17: For any positive integer n, there exists a resolvable *BIBD* $B(2^{n-1},1;2^{2n-1}-2^{n-1})$.

3. Recursive Constructions

One of the main tools in our recursive constructions of PBDs will be Wilson's Fundamental Construction relating to GDDs (see [21]). We define a *weighting* of a GDD $(X,\mathcal{G},\mathcal{B})$ to be any mapping $w:X \to Z^+ \cup \{0\}$. A brief description of Wilson's construction is the following:

Construction 3.1: (Fundamental Construction). Suppose that $(X,\mathcal{G},\mathcal{B})$ is a "master" GDD and let $w:X \to Z^+ \cup \{0\}$ be a weighting of the GDD. For every $x \in X$, let S_x be $w(x)$ "copies" of x. Suppose that for each block $B \in \mathcal{B}$, a GDD $(\bigcup_{x \in B} S_x, \{S_x : x \in B\}, \mathcal{A}_B)$ is given. Let $X^* = \bigcup_{x \in X} S_x$, $\mathcal{G}^* = \{\bigcup_{x \in G} S_x : G \in \mathcal{G}\}$, and $\mathcal{B}^* = \bigcup_{B \in \mathcal{B}} \mathcal{A}_B$. Then $(X^*,\mathcal{G}^*,\mathcal{B}^*)$ is a GDD.

We shall require several types of GDDs as ingredients for Construction 3.1 above. Our main objective will be to find suitable GDDs for the application of Lemma 2.14 so that we eventually construct PBDs with block sizes from P^*. To this end, we shall give some useful examples of GDDs in what follows. As in the previous section, most of our examples are derived from TDs.

Lemma 3.2: If a $T(k+1,1;m)$ exists and $0 \leq t \leq m-1$, then there is a $\{k,k+1,m\}$-GDD of group-type $k^m t^1$. In particular, the conclusion holds for all prime powers $m \geq k$.

Proof: Suppose there is a $T(k+1,1;m)$ and $0 \leq t \leq m-1$. Then there exists a $RT(k,1;m)$, which has m parallel classes of blocks of size k. We can then adjoin a set of t "infinite" points to t of these parallel classes, one point being adjoined to each block of one of the t parallel classes. From the resulting design, we take the blocks of one of the remaining $m-t$ parallel classes and the block consisting of the t "infinite" points ("block at infinity") as groups for our GDD. By Theorem 2.4, the conclusion holds for all prime powers $m \geq k$.

In the following lemma, we make use of the fact that it is possible to arrange the points of $EG(2,q)$ in a $q \times q$ square array so that the rows form one parallel class and the columns another parallel class.

Lemma 3.3: If m is a prime power and $0 < k \leq m$, then, for $0 \leq t \leq m-k$, there exists a $\{k,k+1,m\}$-GDD of group-type $k^{m-1}(k+t)^1$.

Proof: Since m is a prime power, $EG(2,m)$ exists. Denote its rows by r_i, $i = 1,2, \cdots ,m$, the columns by c_j, $j = 1,2, \cdots ,m$ and by a_{ij} the intersection point of r_i and c_j. If in $EG(2,m)$ we delete the columns $c_1,c_2, \cdots ,c_{m-k}$ but leave t of the points $a_{11},a_{12}, \cdots ,a_{1m-k}$, then we may take the rows of the truncated plane as groups of our desired GDD.

Lemma 3.4: If there exists a $T(k+1,1;m)$, then there is a $\{k+1,m+1\}$-GDD of group-type $k^m m^1$. In particular, the conclusion holds for all prime powers $m \geq k$.

Proof: We first adjoin a new point, say ∞, to each group of the $T(k+1,1;m)$ to obtain a PBD $B(\{k+1,m+1\},1;km+m+1)$. We then delete a point $x \neq \infty$ from this PBD, and consider those blocks from which x has been expunged as groups of our desired GDD.

Lemma 3.5: If a $T(k+1,1;m)$ exists and $0 \leq t \leq k+1$, tnen there exists a $\{k,k+1,t^*\}$-GDD of group-type $m^t(m-1)^{k+1-t}$. In particular, the conclusion holds for all prime powers $m \geq k$.

Proof: Delete $k+1-t$ points from a particular block of the $T(k+1,1;m)$ and take the groups of the truncated TD as groups of our desired GDD.

Lemma 3.6: Suppose there exists a resolvable BIBD $B(k,1;v)$ and let $r = \dfrac{v-1}{k-1}$. If $0 \leq t \leq r-1$, then there is a $\{k,k+1\}$-GDD of group-type $k^{v/k}t^1$.

Proof: In the resolvable $B(k,1;v)$, there are $r = \dfrac{v-1}{k-1}$ parallel classes of blocks. We adjoin a set of t "infinite" points to t of these parallel classes. In the resulting design, we then take the blocks of one of the remaining $r-t$ parallel classes together with the "block at infinity" of size t as groups of our desired GDD.

The following slight variation of Lemma 3.6 will be quite useful.

Lemma 3.7: Suppose there exists a resolvable BIBD $B(k,1;v)$ and let $r = \dfrac{v-1}{k-1}$. Then there exists a $\{k,k+1\}$-GDD of group type $k^{\frac{v}{k}-1}(r-1)^1$.

Proof: First of all, we select a particular parallel class of the resolvable $B(k,1;v)$, say R_1, and delete all of the points from one block of R_1. Next we adjoin a set of $r-1$ "infinite" points, one point being adjoined to each block of one of the remaining $r-1$ parallel classes, where the block sizes are from $\{k-1,k\}$. Finally, in the resulting design, we take the remaining blocks of R_1 and the "block at infinity" of size $r-1$ as groups of our desired GDD.

For convenience and future reference, we shall summarize some important consequences of the preceding lemmas in the following lemma. These special GDDs will be most useful as our input designs for Construction 3.1 (Wilson's Fundamental Construction).

Lemma 3.8: Let $n \geq 8$ be a prime power. Then the following hold:

(a) There exists a $\{8,9,n\}$-GDD of group-type $8^n t^1$ where $0 \leq t \leq n-1$.

(b) There exists a $\{8,9,n\}$-GDD of group-type $8^{n-1}(8+t)^1$, where $0 \leq t \leq n-8$.

(c) There exists a $\{8,9\}$-GDD of group-type $8^{15}t^1$, where $0 \leq t \leq 16$.

(d) There exists a $\{8,9\}$-GDD of group-type $8^{14}16^1$.

Proof: For (a), we put $k = 8$ in Lemma 3.2, and for (b), we put $k = 8$ in Lemma 3.3. For the proof of (c), we first observe that Lemma 2.17 guarantees the existence of a resolvable BIBD $B(8,1;120)$ with $n = 4$. We can then apply Lemma 3.6 with $k = 8$, $v = 120$ and $r = 17$ to obtain our desired result. For (d), we apply Lemma 3.7 with $k = 8$, $v = 120$ and $r = 17$. This completes the proof of the lemma.

In most of what follows, we shall give several applications of Lemma 3.8 in conjunction with Construction 3.1.

Lemma 3.9: If $0 \leq s,t \leq 16$, then there exists a $\{8,9,16,17\}$-GDD of group-type $(128)^{15}(271)^1(8s)^1(8t)^1$.

Proof: We shall first apply Lemma 3.4 to obtain a $\{17,18\}$-GDD of group-type $16^{17}17^1$. Explicitly, we take a TD $T(17,1;17)$ and adjoin a new point ∞ to each group of the TD and then delete a point $x \neq \infty$ from the resulting design. We then take those blocks from which x has been deleted as groups to obtain a $\{17,18\}$-GDD of group-type $16^{17}17^1$. We observe that, in particular, each block of size 17 of this new GDD meets the group of size 17 in exactly one point different from ∞, and each block of size 18 meets the group of size 17 in ∞. We shall give the points of this GDD weights as follows. We give ∞ weight 15 and all other points in the group of size 17 get weight 16 for a total weight of 271 in this group. In all but two of the groups of size 16, we give the points weight 8. In the second last group, give s points weight 8 and the remaining points weight 0. In the last group, give t points weight 8 and the remaining points weight 0. For our proof of the lemma, we require $\{8,9,16,17\}$-GDDs of group-types $8^{14}16^1$, $8^{15}15^1$, $8^{15}16^1$, $8^{16}15^1$, $8^{16}16^1$, and $8^{17}15^1$, all of which come from Lemma 3.8, using $n \in \{16,17\}$.

Lemma 3.10: Suppose there is a $T(18,1;m)$, $0 \leq s,t \leq m$, and $0 \leq x,y,z \leq m$ such that $x+y+z \leq m$, then there exists a $\{8,9,16,17\}$-GDD of group-type $(8m)^{15}(8x+15y+16z)^1(8s)^1(8t)^1$. In particular, the conclusion holds for all prime powers $m \geq 17$.

Proof: In all but three groups of a $T(18,1;m)$, give the points weight 8. In the third last group, give x points weight 8, y points weight 15, z points weight 16, and give the remaining points weight 0. In the second last group, give s points weight 8, and give the remaining points weight 0. In the last group, give t points weight 8 and give the remaining points weight 0. We require $\{8,9,16,17\}$-GDDs of group-types 8^{15}, 8^{16}, 8^{17}, 8^{18}, $8^{15}15^1$, $8^{15}16^1$, $8^{16}15^1$, $8^{16}16^1$, $8^{17}15^1$, $8^{17}16^1$, and these are all obtainable from Lemma 3.8, using $n \in \{16,17\}$.

Lemma 3.11: Suppose there is a $T(14,1;m)$, $0 \leq s,t,w \leq m$, and $0 \leq x,y,z \leq m$ such that $x+y+z \leq m$, then there is a $\{8,9,11,13\}$-GDD of group-type $(8m)^{10}(8x+9y+10z)^1(8s)^1(8t)^1(8w)^1$. In particular, the conclusion holds for all prime powers $m \geq 13$.

Proof: In all groups but four of a $T(14,1;m)$, give the points weight 8. In the fourth last group, give x points weight 8, y points weight 9, z points weight 10, and give the remaining points weight 0. In the third last group, give s points weight 8, and give the remaining points weight 0. In the second last group, give t points weight 8, and give the remaining points weight 0. In the last group, give w points weight 8, and give the remaining points weight 0. We require $\{8,9,11,13\}$-GDDs of group-types 8^{10}, 8^{11}, 8^{12}, 8^{13}, 8^{14}, $8^{10}9^1$, $8^{10}10^1$, $8^{11}9^1$, $8^{11}10^1$, $8^{12}9^1$, $8^{12}10^1$, $8^{13}9^1$, and $8^{13}10^1$. These come from Lemma 3.8 with $n \in \{9,11,13\}$.

Lemma 3.12: Suppose there is a $T(10,1;m)$, $0 \leq s \leq m$, and $0 \leq x,y \leq m$ such that $x+y \leq m$, then there exists a $\{8,9\}$-GDD of group-type $(8m)^8(8x+7y)^1(8s)^1$. In particular, the conclusion holds for all prime powers $m \geq 9$.

Proof: In all groups but two of a $T(10,1;m)$, give the points weight 8. In the second last group, give x points weight 8, y points weight 7, and give the remaining points weight 0. In the last group, give s points weight 8, and give the remaining points weight 0. We require $\{8,9\}$-GDDs of group-types 8^8, 8^9, 8^{10}, 8^87^1 and 8^97^1, which come from Lemma 3.8 with $n \in \{8,9\}$.

In order to apply Lemma 2.14, we shall require some input PBDs. For most of these, we utilize the results mentioned in the previous section.

Lemma 3.13: Suppose there is a $T(9,1;m)$ and $0 \leq t \leq m$, then $8m+t \in B(8,9,m,t^*)$. In particular, the conclusion holds for all positive integers m satisfying Theorem 2.6.

Proof: The proof follows directly from Lemma 2.8 with $k = 8$.

For the exceptional cases $m = 15$ and 18 in Lemma 3.13, we have

Lemma 3.14:

(a) If $0 \leq t \leq 17$, then $120+t \in B(8,9,t^*)$.

(b) If $0 \leq t \leq 9$, then $144+t \in B(8,9,16,17,t^*)$.

Proof: For (a), we adjoin t "infinite" points to a resolvable BIBD $B(8,1;120)$. For (b), we apply Lemma 2.15 with $k = 8$ and $m = 17$.

Lemma 3.15: Let $0 \leq t \leq 9$. Then the following hold:

(a) $271+t \in B(8,9,31,32,t^*) \subseteq B(P^* \cup \{t^*\})$.

(b) $279+t \in B(8,9,31,32,t^*) \subseteq B(P^* \cup \{t^*\})$.

Proof: For (a), we apply Lemma 2.16 with $k = 8$ and $m = 31$. For (b), we apply Lemma 2.15 with $k = 8$ and $m = 32$.

Lemma 3.16: If $v \in \{342, 404, 522, 524, 538, 540\}$, then there exists a PBD $B(\{11, 13, 16, 17, 31, 32, 32^*\},1;v)$, that is, $(v-32)+32 \in B(P^* \cup \{32^*\}) \subseteq B(P^*)$.

Proof: For $v = 342$ and 404, we may apply Lemma 2.9 with $m = t = 31$. We thus obtain $342 = 11\cdot 31+1 \in B(11,32)$ and $404 = 13\cdot 31+1 \in B(13,32)$. For $v \in \{522,524\}$, we apply Lemma 2.16 with $k = 16$ and $m = 31$. Note that $t = 11$ gives $522 \in B(16,17,31,32,11^*)$, and $t = 13$ gives $524 \in B(16,17,31,32,13^*)$. For $v \in \{538,540\}$, we apply Lemma 2.15 with $k = 16$ and $m = 32$. Note that $t = 11$ gives $538 \in B(16,17,31,32,11^*)$, and $t = 13$ gives $540 \in B(16,17,31,32,13^*)$. In each case, it is evident that the PBD contains at least one block of size 32.

For reference, we record the following special case of Lemma 2.8.

Lemma 3.17: If a $T(9,1;m)$ exists, then we have $8m+32 \in B(8,9,m,32,32^*) \subseteq B(P^* \cup \{32^*\}) \subseteq B(P^*)$ provided $m \geq 31$ and $m \in B(P^*)$.

Proof: For $m = 31$, we observe that $8m+32 = 280$, and $280 = 8{\cdot}32+24 \in B(8,9,32,32^*)$ by Lemma 2.11. If $m \geq 32$, then we apply Lemma 2.8, and the result follows directly.

We are now in a position to apply Lemma 2.14 and present some basic lemmas, which will be useful in the construction of PBDs with block sizes from P^*.

Lemma 3.18: Suppose $m = 16$ or a $T(18,1;m)$ exists and $0 \leq s,t \leq m$. Then the following hold for $m \in B(P^*)$:

(a) There exists a $\{8,9,16,17\}$-GDD of group-type $(8m)^{15}(271)^1(8s)^1(8t)^1$.

(b) Suppose that for $1 \leq a \leq 9$, $\{8s+a, 8t+a\} \subseteq B(P^* \cup \{a^*\})$ and we have one of $8m+a$, $8s+a$, $8t+a$ belongs to $B(P^*)$. Then $120m+8s+8t+271+a \in B(P^*)$.

(c) Suppose that for $1 \leq a \leq 9$, $8s+a \in B(P^*)$ and $8t+a \in B(P^* \cup \{a^*\})$. Then $120m+8s+8t+271+a \in B(P^*)$.

Proof: For (a), we observe that if $m = 16$ the result follows from Lemma 3.9. If a $T(18,1;m)$ exists, then $m \geq 17$ and we can make suitable choices of x,y,z in Lemma 3.10 to obtain our desired result. For example, take $x = 0$, $y = 1$, and $z = 16$. For our proof of (b) and (c), we apply Lemma 2.14, using the GDD of (a), and the fact that $8m+a \in B(P^* \cup \{a^*\})$ follows from Lemma 3.13, while $271+a \in B(P^* \cup \{a^*\})$ as a result of Lemma 3.15.

Lemma 3.19: Suppose a $T(18,1;m)$ exists and $0 \leq s,t \leq m$. Then the following hold for $m \geq 19$ and $m \in B(P^*)$:

(a) There exists a $\{8,9,16,17\}$-GDD of group-type $(8m)^{15}(279)^1(8s)^1(8t)^1$.

(b) Suppose that for $1 \leq a \leq 9$, $\{8s+a, 8t+a\} \subseteq B(P^* \cup \{a^*\})$, and we have one of $8m+a$, $8s+a$, $8t+a$ belongs to $B(P^*)$. Then $120m+8s+8t+279+a \in B(P^*)$.

(c) Suppose that for $1 \leq a \leq 9$, $8s+a \in B(P^*)$ and $8t+a \in B(P^* \cup \{a^*\})$. Then $120m+8s+8t+279+a \in B(P^*)$.

Proof: For (a), we apply Lemma 3.10. For example, we can choose $x = 1$, $y = 1$ and $z = 16$. For the proof of (b) and (c), we apply Lemma 2.14, using the GDD of (a) and the additional fact that $\{8m+a, 279+a\} \subseteq B(P^* \cup \{a^*\})$ from Lemmas 3.13 and 3.15.

Lemma 3.20: Suppose a $T(14,1;m)$ exists and $0 \leq s,t,w \leq m$. Then the following hold for $m \geq 29$ and $m \in B(P^*)$:

(a) There exists a $\{8,9,11,13\}$-GDD of group-type $(8m)^{10}h^1(8s)^1(8t)^1(8w)^1$ for $h \in \{271,279\}$.

(b) Let $1 \leq a \leq 9$. If $\{8s+a, 8t+a, 8w+a\} \subseteq B(P^* \cup \{a^*\})$ and we have one of $8m+a$, $8s+a$, $8t+a$, $8w+a$ belongs to $B(P^*)$, then $80m+8s+8t+8w+h+a \in B(P^*)$ for $h \in \{271,279\}$.

(c) Let $1 \leq a \leq 9$. If $8s+a \in B(P^*)$ and $\{8t+a, 8w+a\} \subseteq B(P^* \cup \{a^*\})$, then $80m+8s+8t+8w+h+a \in B(P^*)$ for $h \in \{271,279\}$.

Proof: For (a) we apply Lemma 3.11. For $h = 271$ we may choose $x = 4$, $y = 1$, $z = 23$, and for $h = 279$, we choose $x = 5$, $y = 1$, and $z = 23$. For the proof of (b) and (c) we apply Lemma 2.14 using the GDD of (a) and the fact that $\{8m+a,h+a\} \subseteq B(P^* \cup \{a^*\})$ for $h \in \{271,279\}$.

Lemma 3.21: Suppose a $T(14,1;m)$ exists and $0 \leq s,t,w \leq m$. Then the following hold for $m \geq 31$ and $m \in B(P^*)$:

(a) There exists a $\{8,9,11,13\}$-GDD with group-type $(8m)^{10}(310)^1(8s)^1(8t)^1(8w)^1$.

(b) If $\{8s+32,8t+32,8w+32\} \subseteq B(P^* \cup \{32^*\})$, then $80m+8s+8t+8w+342 \in B(P^*)$.

(c) If $8s+32 \in B(P^*)$ and $\{8t+32,8w+32\} \subseteq B(P^* \cup \{32^*\})$, then $80m+8s+8t+8w+342 \in B(P^*)$.

Proof: For the proof of (a), we apply Lemma 3.11. For example, take $x = y = 0$, and $z = 31$. For the proof of (b) and (c), we adjoin 32 "infinite" points to the GDD of (a), using Lemma 2.14. We require $\{8m+32,342\} \subseteq B(P^* \cup \{32^*\}$, and this comes from Lemmas 3.16 and 3.17.

Lemma 3.22: Suppose a $T(10,1;m)$ exists and $0 \leq s \leq m$. Then the following hold for $m \geq 37$ and $m \in B(P^*)$:

(a) There exists a $\{8,9\}$-GDD of group-type $(8m)^8 h^1 (8s)^1$ for $h \in \{271,279\}$.

(b) Let $1 \leq a \leq 9$. If $8s+a \in B(P^* \cup \{a^*\})$ and either $8m+a$ or $8s+a$ belongs to $B(P^*)$, then $64m+8s+h+a \in B(P^*)$ for $h \in \{271,279\}$.

(c) Let $1 \leq a \leq 9$. If $8s+a \in B(P^*)$, then $64m+8s+h+a \in B(P^*)$ for $h \in \{271,279\}$.

Proof: For (a), we apply Lemma 3.12. For $h = 271$, we may choose $x = 33$, $y = 1$ and for $h = 279$ take $x = 34$, $y = 1$. For the proof of (b) and (c), apply Lemma 2.14 using the GDD of (a). We need $\{8m+a,h+a\} \subseteq B(P^* \cup \{a^*\})$ for $h \in \{271,279\}$, which we get from Lemmas 3.13 and 3.15.

Lemma 3.23: Suppose a $T(14,1;m)$ exists and $0 \leq s,t,w \leq m$. Then the following hold for $m \geq 41$ and $m \in B(P^*)$:

(a) There exists a $\{8,9,11,13\}$-GDD of group-type $(8m)^{10}(372)^1(8s)^1(8t)^1(8w)^1$.

(b) If $\{8s+32,8t+32,8w+32\} \subseteq B(P^* \cup \{32^*\})$, then $80m+8s+8t+8w+404 \in B(P^*)$.

(c) If $8s+32 \in B(P^*)$ and $\{8t+32, 8w+32\} \subseteq B(P^* \cup \{32^*\})$, then $80m+8s+8t+8w+404 \in B(P^*)$.

Proof: The proof is similar to that of Lemma 3.21. To prove (a), we apply Lemma 3.11 with $x = 2$, $y = 4$, $z = 32$, for example. For (b) and (c), we apply Lemma 2.14, using the GDD of (a). We require $\{8m+32,404\} \subseteq B(P^* \cup \{32^*\})$, which comes from Lemmas 3.16 and 3.17.

Lemma 3.24: Suppose a $T(10,1;m)$ exists and $0 \leq s \leq m$. Then the following hold for $m \geq 41$ and $m \in B(P^*)$:

(a) There exists a $\{8,9\}$-GDD of group-type $(8m)^8(310)^1(8s)^1$.

(b) If $8s+32 \in B(P^* \cup \{32^*\})$, then $64m+8s+342 \in B(P^*)$.

(c) If $8s+32 \in B(P^*)$, then $64m+8s+342 \in B(P^*)$.

Proof: To prove (a), we apply Lemma 3.12, using $x = 37$, $y = 2$, for example. For the proof of (b) and (c) we adjoin 32 "infinite" points to the GDD of (a), using Lemma 2.14. Note that we have $\{8m+32,342\} \subseteq B(P^* \cup \{32^*\})$ from Lemmas 3.16 and 3.17.

Lemma 3.25: Suppose a $T(10,1;m)$ exists and $0 \leq s \leq m$. Then the following hold for $m \geq 47$ and $m \in B(P^*)$:

(a) There exists a $\{8,9\}$-GDD of group-type $(8m)^8(372)^1(8s)^1$.

(b) If $8s+32 \in B(P^* \cup \{32^*\})$, then $64m+8s+404 \in B(P^*)$.

(c) If $8s+32 \in B(P^*)$, then $64m+8s+404 \in B(P^*)$.

Proof: The proof of (a) follows from applying Lemma 3.12 with $x = 43$, $y = 4$, for example. For (b) and (c), we adjoin 32 "infinite" points to the GDD of (a), using Lemma 2.14. Note also that we have $\{8m+32,404\} \subseteq B(P^* \cup \{32^*\})$ from Lemmas 3.16 and 3.17.

Lemma 3.26: Suppose a $T(14,1;m)$ exists and $0 \leq s,t,w \leq m$. Then the following hold for $m \geq 53$ and $m \in B(P^*)$:

(a) There exists a $\{8,9,11,13\}$-GDD of group-type $(8m)^{10}h^1(8s)^1(8t)^1(8w)^1$ for $h \in \{490,492,506,508\}$.

(b) If $\{8s+32,8t+32,8w+32\} \subseteq B(P^* \cup \{32^*\})$, then $80m+8s+8t+8w+h+32 \in B(P^*)$ for $h \in \{490,492,506,508\}$.

(c) If $8s+32 \in B(P^*)$ and $\{8t+32,8w+32\} \subseteq B(P^* \cup \{32^*\})$, then $80m+8s+8t+8w+h+32 \in B(P^*)$ for $h \in \{490,492,506,508\}$.

Proof: To prove (a), we apply Lemma 3.11. It is readily checked that, for $h \in \{490,492,506,508\}$, we can find integers x,y,z satisfying $h = 8x+9y+10z$, where $0 \leq x,y,z \leq 53$ and $x+y+z \leq 53$. For the proof of (b) and (c), we apply Lemma 2.14 by adjoining 32 "infinite" points to the GDD of (a). We require $\{8m+32,h+32\} \subseteq B(P^* \cup \{32^*\})$ for $h \in \{490,492,506,508\}$. We obtain this from Lemmas 3.16 and 3.17.

Lemma 3.27: Suppose a $T(10,1;m)$ exists and $0 \leq s \leq m$. Then the following hold for $m \geq 64$ and $m \in B(P^*)$:

(a) There exists a $\{8,9\}$-GDD of group-type $(8m)^8h^1(8s)^1$ for $h \in \{490,492,506,508\}$.

(b) If $8s+32 \in B(P^* \cup \{32^*\})$, then $64m+8s+h+32 \in B(P^*)$ for $h \in \{490,492,506,508\}$.

(c) If $8s+32 \in B(P^*)$, then $64m+8s+h+32 \in B(P^*)$ for $h \in \{490,492,506,508\}$.

Proof: For (a), we apply Lemma 3.12 and suitable choices of x and y can be made for the desired h. To prove (b) and (c), we adjoin 32 "infinite" points to the GDD of (a), using Lemma 2.14. Note that we require $\{8m+32,h+32\} \subseteq B(P^* \cup \{32^*\})$ for $h \in \{490,492,506,508\}$. We get this from Lemmas 3.16 and 3.17.

At this stage, we remark that Lemmas 3.18-3.27 are designed to provide us with some measure of efficiency in our constructions of the next section. Indeed, these lemmas will prove to be very effective in establishing our main result. We also require the following application of Lemma 3.5, which will take care of some special cases.

Lemma 3.28: Suppose there is a $T(k+1,1;m)$, $0 \leq t \leq k+1$, and $a \geq 0$. Further suppose that $\{k,\ k+1,\ t,\ m-1+a,\ m+a\} \subseteq B(P^*)$. If $\{m-1+a,m+a\} \subseteq B(P^* \cup \{a^*\})$, then $km+m+t-k-1+a \in B(P^*)$.

Proof: By Lemma 3.5, there exists a $\{k,k+1,t^*\}$-GDD of group-type $m^t(m-1)^{k+1-t}$. We can then adjoin a "infinite" points to the groups of this GDD and, by applying Lemma 2.14, we obtain the desired result.

4. The Spectrum of $B(P^*)$

In this section we shall prove that $v \in B(P^*)$ for all integers $v > 2206$. In fact, we shall show that $v \in B(P^*)$ for all $v \geq 1989$ with at most 27 possible exceptions in this range, of which the largest is 2206. For most values of $v \leq 1988$, we also find that $v \in B(P^*)$ holds. A complete list of values of v for which $v \in B(P^*)$ remains to be determined is provided in Table 4.3 at the end of the section. We shall accomplish our task in several steps.

For convenience and future reference, we define the set $E = \{15, 21, 33, 35, 39, 45, 51, 55, 63, 69, 75, 77, 85, 87, 93, 95, 111, 119, 123, 159, 175, 183, 291, 295, 303, 335\}$. Note that Theorem 2.6 guarantees the existence of a $T(9,1;m)$ for all odd integers $m \geq 9$, $m \notin E$; and in [1, Theorem 4.3] the author essentially proved that $m \in B(P^*)$ for these identical values of m. For completeness, we give some details.

Theorem 4.1: If $v \geq 9$ is odd and $v \notin E$, then $v \in B(P^*)$.

Proof: The proof of this theorem is essentially contained in the proof of [1, Theorem 4.3] and we shall only sketch the details here. First of all, it is shown that $57 \in B(8)$ from $PG(2,7)$ and $135 \in B(8,9,11,16)$ comes from a truncated $PG(2,16)$. For the values $v = 177$ and 301, which were inadvertently omitted in the proof of [1, Theorem 4.3], we observe $177 = 11 \cdot 16+1 \in B(11,17)$ from Lemma 2.9 ([1, Lemma 3.4]) and $301 = 8 \cdot 37+5 \in B(8,9,13^*,37)$ from Lemma 2.11 ([1, Lemma 3.6]). For the remaining values of $v \leq 1797$ it was shown that either v can be expressed as a product of elements from P^* or v can be written in the form $8m+t$, where $m \geq 8$ is a prime power, $0 \leq t \leq m$ and either $t \in B(P^*)$ or $8+t \in B(P^*)$. For the first case of v, $v \in B(P^*)$ is guaranteed by Lemma 2.7, and in the second case $v = 8m+t \in B(P^*)$ is guaranteed by either Lemma 2.8 or Lemma 2.11. For all odd $v \geq 1791$, it was shown that v can be expressed in the form $8m+t$, where m is relatively prime to 210 and $0 \leq t \leq m$, where $t \in B(P^*)$. Hence we have $v \in B(8,9,m,t) \subseteq B(P^*)$ follows from Lemma 2.8.

Theorem 4.2: For all positive integers $v \equiv 0 \pmod 8$, $v \neq 24, 40, 48, 56$, $v \in B(P^*)$ holds.

Proof: The proof of this theorem is analogous to the proof of [1, Theorem 4.5] and again we only sketch the details. First of all, we obviously have $\{8,16,32,64\} \subseteq B(P^*)$. For the other stated values of v, we may consider the following cases below:

Case 1: $v = 8n$, odd $n \geq 9$, $n \notin E$

Then, by Theorem 4.1, $n \in B(P^*)$. We also have $8 \in B(P^*)$ and there exists a $T(8,1;n)$. Hence $v \in B(P^*)$ follows from Lemma 2.7.

Case 2: $v = 8n+8$, odd $n \geq 9$, $n \notin E$

By Theorem 2.6, there exists a $T(9,1;n)$, and from this we obtain $v = 8n+8 \in B(8,9,n) \subseteq B(P^*)$, using Lemma 2.8.

Case 3: $v = 8n$, n odd, $n \in E$

As we have already seen, $120 \in B(P^*)$ from a resolvable $B(8,1;120)$. For the remaining values of v, we can express v in the form $8m+t$ in such a way that either Lemma 2.8 or 2.11 can be applied to produce $v \in B(P^*)$ (see, for example, [1, Table 4.3]).

Case 4: $v = 8n+8$, n odd, $n \in E$

For the most part, we can apply Lemma 2.7 (see, for example, [1, Table 4.4]) or Lemmas 2.8 and 2.11.

This completes the proof.

We are now left with investigating values of $v \in B(P^*)$ for the remaining cases $v \equiv 2$, 4, and 6 (mod 8). It will be most convenient to treat these cases separately. However, before doing this, we shall mention some special constructions for obtaining $v \in B(P^*)$, where $v \equiv 2$, 4, and 6 (mod 8).

We shall make use of some "small" examples later on. These are provided by the following lemma.

Lemma 4.3: If $v \in \{342, 404, 522, 524, 538, 540, 590\}$, then $v \in B(P^*)$.

Proof: We already have shown that $\{342, 404, 522, 524, 538, 540\} \subseteq B(P^*)$ in Lemma 3.16. We further have $590 = 19 \cdot 31+1 \in B(19,32) \subseteq B(P^*)$ from Lemma 2.9.

The following lemma is a simple application of Lemma 2.8 and Lemma 2.11.

Lemma 4.4: Let $m \geq 31$ be a prime power and $0 \leq t \leq m$. If $t \in B(P^*)$ or $31+t \in B(P^*)$, then $31m+t \in B(P^*)$.

Proof: If $t \in B(P^*)$, we apply Lemma 2.8 to get $31m+t \in B(31,32,m,t^*) \subseteq B(P^*)$. If $31+t \in B(P^*)$, then we apply Lemma 2.11 to get $31m+t \in B(31,32,(31+t)^*,m) \subseteq B(P^*)$.

For $m \in \{31, 37, 41, 43, 47, 49, 53, 59, 61, 64, 67, 71, 73\}$, we give a list of values of $v = 31m+t$ satisfying Lemma 4.4 in Table 4.1 below. We are concerned mainly with values of v where $v \equiv 2$, 4, 6 (mod 8), although the values of $v \equiv 0$ (mod 8) may occur in a specified interval for t. If m is odd, we take t to be odd. If m is even, we take t to be even. Note that Table 4.1 is an extension of [1, Table 5.1].

For some additional values of $v \equiv 2$, 4, 6 (mod 8), which are not covered in the range of Table 4.1, we may apply Lemma 3.28. We have the following useful result.

Lemma 4.5: If $v \in \{1202, 1204, 1338, 1746, 2018, 2026, 2028, 2030, 2034, 2036, 2038, 2042, 2044, 2046, 2050, 2052, 2054, 2154, 2156, 2162, 2164, 2170, 2172, 2178, 2180\}$, then $v \in B(P^*)$.

Proof: We apply Lemma 3.28 as follows. For $\{1202,1204\} \subseteq B(P^*)$, take $k = 16$, $m = 71$, $t \in \{11,13\}$, and $a = 1$. For $1338 \in B(P^*)$, take $k = 16$, $m = 79$, $t = 11$, and $a = 1$. For $1746 \in B(P^*)$, take $k = 16$, $m = 103$, $t = 11$, and $a = 1$. For $\{2018, 2026, 2028, 2030, 2034, 2036, 2038, 2042, 2044, 2046, 2050, 2052, 2054\} \subseteq B(P^*)$, take $k = 31$, $m = 64$, $t \in \{1, 9, 11, 13, 19, 23, 25, 27, 29\}$, and $a \in \{1,9\}$. Note that $72 = 8 \cdot 9$, $73 = 8 \cdot 9+1$. So $\{72,73\} \subseteq B(8,9)$. For $\{2154, 2156, 2162, 2164\} \subseteq B(P^*)$, take $k = 16$, $m = 127$, $t \in \{11,13\}$, and $a \in \{1,9\}$. We require $135 \in B(8,9,11,16) \subseteq B(P^*)$ from Theorem 4.1, and $136 = 8 \cdot 16+8 \in B(8,9,16) \subseteq B(P^*)$ from Lemma 2.8. For $\{2170, 2172, 2178, 2180\} \subseteq B(P^*)$, take $k = 16$, $m = 128$, $t \in \{11,13\}$ and $a \in \{1,9\}$. We also require $136 \in B(8,9,16)$ and $137 = 8 \cdot 16+9 \in B(8,9,16) \subseteq B(P^*)$. This completes the proof.

Table 4.1: Applications of Lemma 4.4

$v = 31m+t$	m	t	$v = 31m+t$	m	t
962	31	1	1886-1888	59	57-59
970-974	31	9-13	1892	61	1
978-980	31	17-19	1900-1904	61	9-13
984-992	31	23-31	1908-1910	61	17-19
1148	37	1	1914-1924	61	23-33
1156-1160	37	9-13	1932-1934	61	41-43
1164-1166	37	17-19	1938-1940	61	47-49
1170-1180	37	23-33	1948-1952	61	57-61
1280-1284	41	9-13	1990	64	6
1288-1290	41	17-19	1994-1996	64	10-12
1294-1304	41	23-33	2002	64	18
1308	41	37	2006	64	22
1334	43	1	2010-2014	64	26-30
1342-1346	43	9-13	2078	67	1
1350-1352	43	17-19	2086-2090	67	9-13
1356-1366	43	23-33	2094-2096	67	17-19
1370	43	37	2100-2110	67	23-33
1374-1376	43	41-43	2114	67	37
1458	47	1	2118-2120	67	41-43
1466-1470	47	9-13	2124-2126	67	47-49
1474-1476	47	17-19	2130	67	53
1480-1490	47	23-33	2134-2138	67	57-61
1494	47	37	2142-2144	67	65-67
1498-1500	47	41-43	2202	71	1
1528-1532	49	9-13	2210-2214	71	9-13
1536-1538	49	17-19	2218-2220	71	17-19
1542-1552	49	23-33	2224-2234	71	23-33
1556	49	37	2238	71	37
1560-1562	49	41-43	2242-2244	71	41-43
1566-1568	49	47-49	2248-2250	71	47-49
1644	53	1	2254	71	53
1652-1656	53	9-13	2258-2262	71	57-61
1660-1662	53	17-19	2266-2268	71	65-67
1666-1676	53	23-33	2272-2276	73	9-13
1684-1686	53	41-43	2280-2282	73	17-19
1690-1692	53	47-49	2286-2296	73	23-33
1830	59	1	2300	73	37
1838-1842	59	9-13	2304-2306	73	41-43
1846-1848	59	17-19	2310-2312	73	47-49
1852-1862	59	23-33	2316	73	53
1866	59	37	2320-2324	73	57-61
1870-1872	59	41-43	2328-2330	73	65-67
1876-1878	59	47-49	2334-2336	73	71-73
1882	59	53			

We may also apply Lemma 2.9 or 2.12, to get $v \in B(P^*)$ for values of $v \equiv 2, 4, 6 \pmod 8$, which are not covered by Lemmas 4.3-4.5. These values are listed in Table 4.2 below, where we have selected $m \in \{31, 71, 79, 103, 127, 143, 149, 151, 199, 271\}$. Some comments are made concerning appropriate "input" PBDs for application of the lemmas.

We shall now consider the case $v \equiv 2 \pmod 8$.

Theorem 4.6: If $v \equiv 2 \pmod 8$, then $v \in B(P^*)$ for all $v \geq 1994$, with the possible exception of $v = 2058, 2066, 2074, 2082, 2098, 2122, 2146, 2186$.

Proof: First of all, if $m \geq 781$ is odd, then there exists a $T(9,1;m)$, and so $v = 8m+522 \in B(8,9,522,m)$ by Lemma 2.8. Hence $v \in B(P^*)$ by Lemma 4.3 and Theorem 4.1. If $m \geq 781$ is even, then $v = 8m+522 \in B(8,9,522,m+1)$ by Lemma 2.9, and consequently $v \in B(P^*)$ by Lemma 4.3 and Theorem 4.1. All in all, we have $v \in B(P^*)$ for all $v \equiv 2 \pmod 8$, $v \geq 6770$. Apart from the possible exceptions mentioned, it remains to fill the interval $1994 \leq v \leq 6762$. For the most part, this is accomplished by applications of Lemmas 3.18-3.20, 3.22, 3.26 and 3.27. For values of v, $1994 \leq v < 2194$, our solution can be found in Lemma 4.5 and Tables 4.1 and 4.2. For values of v, $2194 \leq v \leq 2586$, we illustrate an application of Lemma 3.18 with $m \in \{16,17\}$ and $a = 3$. Observe that $274 \in B(8,9,31,32,3^*)$ by Lemma 3.15. Moreover, $8m+3 \in B(8,9,m,3^*)$, and $8m+3 \in B(P^*)$ for $m = 16$ and 17. With the help of Lemmas 3.13 and 3.14, appropriate choices of s and t can be made to satisfy the conditions of Lemma 3.18 so that $v = 120m+8s+8t+271+a \in B(P^*)$, where $m \in \{16,17\}$ and $a = 3$, holds for all values of v in the interval $2194 \leq v \leq 2586$ with the possible exceptions of the values $v = 2226$ and 2242. Note, for example, that $2194 \in B(P^*)$ comes from $m = 16$, $s = t = 0$, $a = 3$ and the fact that $8m+3 \in B(P^*)$, while $2202 \in B(P^*)$ comes from $m = 16$, $s = 1$, $t = 0$, $a = 3$ and the fact that $8s+3 = 11 \in B(P^*)$, that is, we apply (c) of Lemma 3.18 in this case. The exceptions $v = 2226$ and $v = 2242$ are taken care of in Table 4.1. Thus we have $v \in B(P^*)$ for all v in the interval $2194 \leq v \leq 2586$. For all of the remaining values of v, $2586 < v \leq 6762$, we may combine Lemma 3.18 ($m \geq 19$) with Lemmas 3.19, 3.20, 3.22, 3.26, 3.27 for a solution. For applications of Lemmas 3.19, 3.20, and 3.22, take $a = 3$ as in the case of Lemma 3.18. For applications of Lemmas 3.26 and 3.27, we choose $h \in \{490,506\}$. This completes the proof of the theorem.

We next consider the case $v \equiv 4 \pmod 8$.

Theorem 4.7: If $v \equiv 4 \pmod 8$, then $v \in B(P^*)$ for all $v \geq 1996$, with the possible exception of $v = 2004, 2020, 2076, 2084, 2092, 2116, 2132, 2140, 2148, 2188$.

Proof: First of all, a $T(9,1;m)$ exists for all odd $m \geq 405$. If $m \geq 405$ is odd, then $8m+404 \in B(8,9,m,404)$ by Lemma 2.8. Hence $v \in B(P^*)$ follows from Lemma 4.3 and Theorem 4.1. If $m \geq 525$ is odd, then $v = 8m+524 \in B(8,9,524,m) \subseteq B(P^*)$, using Lemma 4.3 and Theorem 4.1. Consequently, we have $v \in B(P^*)$ for all $v \equiv 4 \pmod 8$, $v \geq 4724$. We must now consider the values of v, $1996 \leq v \leq 4716$, apart from the possible exceptions mentioned. Our approach is similar to that used in the proof of Theorem 4.6. For values of v in the range $1996 \leq v < 2196$, our proof of $v \in B(P^*)$ can be found in Lemma 4.5 and Tables 4.1 and 4.2. For values of v, $2196 \leq v \leq 4716$, we commence with an application of Lemma 3.18. If Lemma 3.18 is applied with $m \in \{16,17\}$ and $a = 5$, we obtain $v \in B(P^*)$ for all v in the range $2196 \leq v \leq 2588$,

Table 4.2: Applications of Lemmas 2.9 and 2.12

Equation for v	m	Comments for input PBDs
$714 = 23 \cdot 31+1$	31	
$782 = 11 \cdot 71+1$	71	$72 = 8 \cdot 9 \in B(8,9)$
$790 = 11 \cdot 71+9$	71	$80 = 8 \cdot 9+8 \in B(8,9)$ by Lemma 2.8
$838 = 27 \cdot 31+1$	31	
$870 = 11 \cdot 79+1$	79	$80 \in B(8,9)$
$900 = 29 \cdot 31+1$	31	
$924 = 13 \cdot 71+1$	71	$72 \in B(8,9)$
$932 = 13 \cdot 71+9$	71	$80 \in B(8,9)$
$1028 = 13 \cdot 79+1$	79	$80 \in B(8,9)$
$1134 = 11 \cdot 103+1$	103	$104 = 8 \cdot 13 \in B(8,13)$
$1142 = 11 \cdot 103+9$	103	$112 = 8 \cdot 13+8 \in B(8,9,13)$ by Lemma 2.8
$1340 = 13 \cdot 103+1$	103	$104 \in B(8,13)$
$1348 = 13 \cdot 103+9$	103	$112 \in B(8,9,13)$
$1398 = 11 \cdot 127+1$	127	$128 = 8 \cdot 16 \in B(8,16)$
$1406 = 11 \cdot 127+9$	127	$136 = 8 \cdot 16+8 \in B(8,9,16)$ by Lemma 2.8
$1414 = 11 \cdot 127+17$	127	$144 = 8 \cdot 17+8 \in B(8,9,17)$ by Lemma 2.8
$1502 = 19 \cdot 79+1$	79	$80 \in B(8,9)$
$1574 = 11 \cdot 143+1$	143	$144 \in B(8,9,17)$
$1582 = 11 \cdot 143+9$	143	$152 = 8 \cdot 17+16 \in B(8,9,16,17)$ by Lemma 2.8
$1634 = 23 \cdot 71+1$	71	$72 \in B(8,9)$
$1642 = 23 \cdot 71+9$	71	$80 \in B(8,9)$
$1818 = 23 \cdot 79+1$	79	$80 \in B(8,9)$
$1926 = 27 \cdot 71+9$	71	$80 \in B(8,9)$
$1956 = 13 \cdot 149+19$	149	$168 = 8 \cdot 19+16 \in B(8,9,16,19)$ by Lemma 2.8
$1958 = 19 \cdot 103+1$	103	$104 \in B(8,13)$
$1964 = 13 \cdot 151+1$	151	$152 \in B(8,9,16,17)$
$1966 = 19 \cdot 103+9$	103	$112 \in B(8,9,13)$
$1972 = 13 \cdot 151+9$	151	$160 = 8 \cdot 19+8 \in B(8,9,19)$ by Lemma 2.8
$2060 = 29 \cdot 71+1$	71	$72 \in B(8,9)$
$2068 = 29 \cdot 71+9$	71	$80 \in B(8,9)$
$2174 = 8 \cdot 271+6$	271	$277 \in B(8,9,31,32,6^*)$ by Lemma 3.15
$2190 = 11 \cdot 199+1$	199	$200 = 8 \cdot 25 \in B(8,25)$
$2236 = 8 \cdot 271+64+4$	271	$275 \in B(8,9,31,32,4^*)$ by Lemma 3.15, and $68 = 8 \cdot 8+4 \in B(8,9,4^*)$ by Lemma 2.8
$2246 = 8 \cdot 271+72+6$	271	$277 \in B(8,9,31,32,6^*)$ by Lemma 3.15, and $78 = 8 \cdot 9+6 \in B(8,9,6^*)$ by Lemma 2.8

with the possible exception of $v = 2212$ and 2236. Note that $8m+5 \in B(P^*)$ for $m \in \{16,17\}$ and we get some help from Lemmas 3.13 and 3.14 regarding suitable choices for s and t. With regards to the exceptional values $v = 2212$ and 2236, we have $2212 \in B(P^*)$ from Table 4.1 and $2236 \in B(P^*)$ from Table 4.2. Hence we have $v \in B(P^*)$ for all values of v in the interval $2196 \leq v \leq 2588$. For the remaining values of v, $2596 \leq v \leq 4716$, our proof of $v \in B(P^*)$ comes readily from the combined applications of Lemmas 3.18-3.20, 3.22, 3.25, and 3.27. Here we use $a = 5$ in Lemmas 3.18-3.20, and 3.22, and choose $h \in \{492,508\}$ in Lemma 3.27. In actual fact, we can obtain the desired result with repeated applications of Lemmas 3.18-3.20, with little or no loss of efficiency. This completes the proof of the theorem.

Finally, we consider the case $v \equiv 6 \pmod 8$.

Theorem 4.8: If $v \equiv 6 \pmod 8$, then $v \in B(P^*)$ for all $v \geq 1990$, with the possible exception of $v = 1998$, 2022, 2062, 2070, 2150, 2158, 2166, 2182, 2206.

Proof: If $m \geq 343$ is odd, then there exists a $T(9,1;m)$. Thus, if m is odd and $m \geq 343$, we have $v = 8m+342 \in B(8,9,342,m) \subseteq B(P^*)$ by Lemma 4.3 and Theorem 4.1. Similarly, if m is odd and $m \geq 591$, we have $v = 8m+590 \in B(8,9,590,m) \subseteq B(P^*)$ by Lemma 4.3 and Theorem 4.1. Hence we have $v \in B(P^*)$ for all $v \equiv 6 \pmod 8$, $v \geq 5318$. It now remains to fill in the interval $1990 \leq v \leq 5310$, apart from the possible exceptions listed in the theorem. For the interval $1990 \leq v < 2198$, our proof of $v \in B(P^*)$ can be found in Lemma 4.5 and Tables 4.1 and 4.2. Next, for values of v in the interval $2198 \leq v \leq 5310$, we first apply Lemma 3.18 with $m \in \{16,17\}$ and $a = 7$ to obtain $v \in B(P^*)$ for all v in the interval $2198 \leq v \leq 2590$, with the possible exception of $v = 2206$, 2230, 2246, and 2254. Here we use the fact that $8m+7 \in B(P^*)$ for $m \in \{16,17\}$ combined with information we have in Lemmas 3.13 and 3.14 relating to suitable choices of s and t in Lemma 3.18. For the exceptional values, we find $\{2230,2254\} \subseteq B(P^*)$ from Table 4.1, and $2246 \in B(P^*)$ from Table 4.2. Consequently, we have $v \in B(P^*)$ for all v in the interval $2198 \leq v \leq 2590$, with the possible exception of $v = 2206$. For all the remaining values of v in the interval $2590 < v \leq 5310$, it can readily be verified that $v \in B(P^*)$ by applying Lemmas 3.18-3.20 with $a = 7$, Lemma 3.21, Lemma 3.22 with $a = 7$, and Lemma 3.24. This completes the proof.

Combining all the results of Theorems 4.1, 4.2, 4.6-4.8, we have

Theorem 4.9: For all $v \geq 1989$, $v \in B(P^*)$ except possibly for $v \in \{1998, 2004, 2020, 2022, 2058, 2062, 2066, 2070, 2074, 2076, 2082, 2084, 2092, 2098, 2116, 2122, 2132, 2140, 2146, 2148, 2150, 2158, 2166, 2182, 2186, 2188, 2206\}$.

In Table 4.3, we list the values of v for which the existence of a $B(P^*,1;v)$ is unknown. However, it is perhaps worth mentioning that if $v \notin P^*$, then a simple counting argument shows that $v \geq 57$ is necessary for $v \in B(P^*)$. If, in addition, $v \not\equiv 0$ or $1 \pmod 8$, then further considerations regarding PBD closure will require that $v \geq 82$ is necessary for $v \in B(P^*)$. We shall therefore list only values of $v \geq 82$ which have not been accounted for already.

Table 4.3: Existence of $B(P^*,1;v)$ unknown

82	84	85	86	87	90	92	93	94	95	98	100	102	106
108	110	111	114	116	118	119	122	123	124	126	130	132	134
138	140	142	146	148	150	154	156	158	162	164	166	170	172
174	175	178	180	182	183	186	188	190	194	196	198	202	204
206	210	212	214	218	220	222	226	228	230	234	236	238	242
244	246	250	252	254	258	260	262	268	270	274	276	278	282
284	286	290	291	292	294	295	298	300	302	303	306	308	310
314	316	318	322	324	326	330	332	334	335	338	340	346	348
350	354	356	358	362	364	366	370	372	374	378	380	382	386
388	390	394	396	398	402	406	410	412	414	418	420	422	426
428	430	434	436	438	442	444	446	450	452	454	458	460	462
466	468	470	474	476	478	482	484	486	490	492	494	498	500
502	506	508	510	514	516	518	526	530	532	534	542	546	548
550	554	556	558	562	564	566	570	572	574	578	580	582	586
588	594	596	598	602	604	606	610	612	614	618	620	622	626
628	630	634	636	638	642	644	646	650	652	654	658	660	662
666	668	670	674	676	678	682	684	686	690	692	694	698	700
702	706	708	710	716	718	722	724	726	730	732	734	738	740
742	746	748	750	754	756	758	762	764	766	770	772	774	778
780	786	788	794	796	798	802	804	806	810	812	814	818	820
822	826	828	830	834	836	842	844	846	850	852	854	858	860
862	866	868	874	876	878	882	884	886	890	892	894	898	902
906	908	910	914	916	918	922	926	930	934	938	940	942	946
948	950	954	956	958	964	966	982	994	996	998	1002	1004	1006
1010	1012	1014	1018	1020	1022	1026	1030	1034	1036	1042	1044	1046	1050
1052	1054	1058	1060	1062	1066	1068	1070	1074	1076	1078	1082	1084	1086
1090	1092	1094	1098	1100	1102	1106	1108	1110	1114	1116	1118	1122	1124
1126	1130	1132	1138	1140	1146	1150	1154	1162	1182	1186	1188	1190	1194
1196	1198	1206	1210	1212	1214	1218	1220	1222	1226	1228	1230	1234	1236
1238	1242	1244	1246	1250	1252	1254	1258	1260	1262	1266	1268	1270	1274
1276	1286	1292	1306	1310	1314	1316	1318	1322	1324	1326	1330	1332	1354
1372	1378	1380	1382	1386	1388	1390	1394	1396	1402	1404	1410	1412	1418
1420	1422	1426	1428	1430	1434	1436	1438	1442	1444	1446	1450	1452	1454
1460	1462	1478	1492	1506	1508	1510	1514	1516	1518	1522	1524	1526	1534
1540	1554	1558	1564	1570	1572	1578	1580	1586	1588	1590	1594	1596	1598
1602	1604	1606	1610	1612	1614	1618	1620	1622	1626	1628	1630	1636	1638
1646	1650	1658	1678	1682	1694	1698	1700	1702	1706	1708	1710	1714	1718
1722	1724	1726	1730	1732	1734	1738	1740	1742	1748	1750	1754	1756	1758
1762	1764	1766	1770	1772	1774	1778	1780	1782	1786	1790	1794	1796	1798
1802	1804	1806	1810	1812	1814	1820	1822	1826	1828	1834	1836	1844	1850
1868	1874	1884	1890	1894	1898	1906	1930	1942	1946	1954	1962	1974	1978
1980	1982	1986	1988	1998	2004	2020	2022	2058	2062	2066	2070	2074	2076
2082	2084	2092	2098	2116	2122	2132	2140	2146	2148	2150	2158	2166	2182
2186	2188	2206											

5. Applications

In this section, we shall apply some of our constructions of PBDs of the previous sections to other types of combinatorial designs, such as conjugate orthogonal Latin squares and sets of mutually orthogonal Latin squares (MOLS). For the formal definitions pertaining to Latin squares and MOLS, the reader is referred to Dénes and Keedwell [8].

It is fairly well-known that the number of distinct conjugates of a Latin square (quasigroup) is always 1,2,3 or 6, and orthogonality relations between pairs of conjugates of Latin squares have been studied quite extensively (see, for example, [1-4,8,10,11,15,17]). The most general and outstanding problem is concerned with the spectrum of Latin squares which have all six conjugates distinct and pairwise orthogonal. We shall denote by L^* the set of all integers n which are orders of idempotent Latin squares having distinct and pairwise orthogonal conjugates. It is known that $B(P^*) \subseteq L^*$ (see, for example, [1,11]). Moreover, constructions by the author in previous papers [1,2] have essentially established that $n \in L^*$ for all $n > 5074$. As an immediate consequence of Theorem 4.9, we now have the following improvement.

Theorem 5.1: For all $n \geq 1989$, $n \in L^*$ except possibly for $n \in \{1998, 2004, 2020, 2022, 2058, 2062, 2066, 2070, 2074, 2076, 2082, 2084, 2092, 2098, 2116, 2122, 2132, 2140, 2146, 2148, 2150, 2158, 2166, 2182, 2186, 2188, 2206\}$.

It is perhaps worth pointing out that applications of the "Singular Direct Product" construction for Latin squares (quasigroups) will further enlarge the spectrum of L^*. However, this is the subject of another paper. Also, if we translate the above result in the context of orthogonal Latin square graphs (see [11] for this concept), then we readily obtain

Theorem 5.2: Idempotent Latin squares whose conjugacies realize K_6, the complete graph on six vertices, exist for all orders $n \geq 1989$, with the possible exception of $n \in \{1998, 2004, 2020, 2022, 2058, 2062, 2066, 2070, 2074, 2076, 2082, 2084, 2092, 2098, 2116, 2122, 2132, 2140, 2146, 2148, 2150, 2158, 2166, 2182, 2186, 2188, 2206\}$.

Our other problem of interest deals with sets of MOLS. Let $N(v)$ denote the maximum number of MOLS of order v. For $v > 1$, we have $N(v) \leq v-1$. It is fairly well-known that $N(q) = q-1$ for prime powers q. However, for non-prime values of v, the problem of determining $N(v)$ remains for the most part unsolved. In order to apply our construction, we shall adapt some terminology from [8]. Suppose that a PBD $B(\{k_1,k_2, \cdots ,k_m\},1;v)$ exists such that there are b_1 blocks of size k_1, b_2 blocks of size k_2, ... , b_m blocks of size k_m. The set of b_i blocks comprising all those containing k_i elements is called the *i-th equiblock component* of the design. A union of equiblock components such that no two of the blocks contained in it have an element in common is called a *clear set* of the equiblock components.

We shall make use of the following (see [8, Theorem 11.3.1]).

Theorem 5.3: If there exists a PBD $B(\{k_1,k_2, \cdots ,k_m\},1;v)$ such that the union of the first, second, ..., r-th equiblock components is a clear set, then

$$N(v) \geq \min\{N(k_1),N(k_2), \cdots ,N(k_r),N(k_{r+1})-1, \cdots ,N(k_m)-1\}.$$

If, in particular, we apply Theorem 5.3 to our PBD construction of Lemma 2.16, then we immediately obtain

Theorem 5.4: Suppose that a $T(k+1,1;m)$ exists, $0 \leq t < k+1 \leq m$, and $v = km+m+t-k$. Then

$$N(v) \geq \min\{N(t),N(k)-1,N(k+1)-1,N(m)-1,N(m+1)-1\}.$$

In Table 5.1 below, we give some examples where Theorem 5.4 produces more MOLS of order v than previously known (see [6]).

Table 5.1: Applications of Theorem 5.4

v	k	m	t	$N(v) \geq$	Old lower bound
520	16	31	9	8	7
522	16	31	11	10	6
524	16	31	13	12	6
2152	16	127	9	8	7
2154	16	127	11	10	7
2156	16	127	13	12	7
4052	31	127	19	18	8
4056	31	127	23	22	8
4060	31	127	27	26	8
4602	16	271	11	10	7
4604	16	271	13	12	8
6508	16	383	13	12	10

We wish to remark that the above results may be of some use in applying other methods. For example, $5764 = 11 \cdot 524$ implies $N(5764) \geq \min\{N(11),N(524)\} \geq 10$. For much larger orders, one might exploit the fact that we now have five consecutive values of $v \in \{521, 522, 523, 524, 525\}$ for which $N(v) \geq 10$. We should also point out that Theorem 5.3 can be applied to the PBD construction of Lemma 2.15. However, this method appears to produce results similar to those already known.

Acknowledgement

The author acknowledges the financial support of the Natural Sciences and Engineering Research Council of Canada under Grant A-5320.

References

[1] F.E. Bennett, "Latin squares with pairwise orthogonal conjugates", *Discrete Math.* 36 (1981) 117-137.

[2] F.E. Bennett, "On conjugate orthogonal idempotent Latin squares", *Ars Combinatoria* 19 (1985) 37-50.

[3] F.E. Bennett, "Conjugate orthogonal Latin squares and Mendelsohn designs", *Ars Combinatoria* 19 (1985) 51-62.

[4] R.K. Brayton, D. Coppersmith and A.J. Hoffman, "Self-orthogonal Latin squares of all orders $n \neq 2$, 3 or 6", *Bull. Amer. Math. Soc.* 80 (1974) 116-118.

[5] A.E. Brouwer, "Optimal packings of K_4's into a K_n", *J. Combinatorial Theory (A)* 26 (1979) 278-297.

[6] A.E. Brouwer, "The number of mutually orthogonal Latin squares - a table up to order 10000", Research Report ZW 123/79, Math. Centrum, Amsterdam, 1979.

[7] A.E. Brouwer and G.H.J. van Rees, "More mutually orthogonal Latin squares", *Discrete Math.* 39 (1982) 263-281.

[8] J. Dénes and A.D. Keedwell, *Latin Squares and Their Applications,* Academic Press, New York and London, 1974.

[9] H. Hanani, "Balanced incomplete block designs and related designs", *Discrete Math.* 11 (1975) 255-369.

[10] A.D. Keedwell, "Circuit designs and Latin squares", *Ars Combinatoria* 17 (1984) 79-90.

[11] C.C. Lindner, E. Mendelsohn, N.S. Mendelsohn and B. Wolk, "Orthogonal Latin square graphs", *J. Graph Theory* 3 (1979) 325-338.

[12] H.F. MacNeish, "Euler squares", *Ann. Math.* 23 (1922) 221-227.

[13] R.C. Mullin, P.J. Schellenberg, S.A. Vanstone and W.D. Wallis, "On the existence of frames", *Discrete Math.* 37 (1981) 79-104.

[14] R.C. Mullin and D.R. Stinson, "Pairwise balanced designs with odd block sizes exceeding 5", preprint.

[15] K.T. Phelps, "Conjugate orthogonal quasigroups", *J. Combinatorial Theory (A)* 25 (1978) 117-127.

[16] E. Seiden, "A method of construction of resolvable BIBD", *Sankhya (A)* 25 (1963) 393-394.

[17] S.K. Stein, "On the foundations of quasigroups", *Trans. Amer. Math. Soc.* 85 (1957) 228-256.

[18] R.M. Wilson, "An existence theory for pairwise balanced designs I", *J. Combinatorial Theory (A)* 13 (1972) 220-245.

[19] R.M. Wilson, "An existence theory for pairwise balanced designs II", *J. Combinatorial Theory (A)* 13 (1972) 246-273.

[20] R.M. Wilson, "An existence theory for pairwise balanced designs III", *J. Combinatorial Theory (A)* 18 (1975) 71-79.

[21] R.M. Wilson, "Constructions and uses of pairwise balanced designs", *Mathematical Centre Tracts* 55 (1974) 18-41.

Annals of Discrete Mathematics 34 (1987) 65–80
© Elsevier Science Publishers B.V. (North-Holland)

Conjugate Orthogonal Latin Squares with Equal-sized Holes

F.E. Bennett

Department of Mathematics
Mount Saint Vincent University
Halifax, Nova Scotia B3M 2J6
Canada

Lisheng Wu and L. Zhu

Department of Mathematics
Suzhou University
Suzhou
The People's Republic of China

TO ALEX ROSA ON HIS FIFTIETH BIRTHDAY

ABSTRACT

In this paper, we shall extend the earlier investigations undertaken by J.H. Dinitz, R.C. Mullin, D.R. Stinson, and L. Zhu of MOLS (mutually orthogonal Latin squares) with holes, and SOLS (self-orthogonal Latin squares) with holes. Here we shall consider the existence of (3,2,1)-conjugate orthogonal Latin squares (briefly COLS) with holes, corresponding to missing sub-COLS, which are disjoint and spanning. We denote by $(3,2,1)-HCOLS(h^n)$ a $(3,2,1)-COLS$ with n holes of equal size h. It is shown that a $(3,2,1)-HCOLS(h^n)$ exists for $h \geq 2$ if and only if $n \geq 4$, except possibly when $n \in \{8,9,12\}$ and $h=2$, and when $n=6$ and $h \in \{5,7,13,61\}$, which is quite similar to an earlier result for $HSOLS(h^n)$ due to Stinson and Zhu. As a consequence, we also have the identical result for $(1,3,2)-HCOLS(h^n)$. It is hoped that these results will be of some importance in the construction of other combinatorial designs, such as Mendelsohn designs.

1. Introduction

A Latin square which is orthogonal to its (3,2,1)-conjugate will be called a (3,2,1)-conjugate orthogonal Latin square (briefly COLS). For more formal definitions and results on $(3,2,1)-COLS$, including COILS (conjugate orthogonal idempotent Latin squares) and $ICOILS(v, n)$ (incomplete conjugate orthogonal idempotent Latin squares of order v missing a subsquare of order n) the reader is referred to [1, 4, 5, 10, 14, 16]. For the formal definitions of MOLS with holes, we refer the reader to [11]. We shall adapt the notation used in [19]. Accordingly, we denote by $HMOLS(h_1^{n_1} h_2^{n_2} \cdots h_k^{n_k})$ a pair of MOLS of order v from which n_i sub-MOLS of order h_i are "missing" ($1 \leq i \leq k$), and in which the latter subsquares are disjoint and spanning, that is, $\Sigma_{1 \leq i \leq k} n_i h_i = v$. The type T of the HMOLS is $h_1^{n_1} h_2^{n_2} \cdots h_k^{n_k}$. An $HSOLS(h_1^{n_1} h_2^{n_2} \cdots h_k^{n_k})$ is defined to be an $HMOLS(h_1^{n_1} h_2^{n_2} \cdots h_k^{n_k})$ in which the two squares are mutual transposes. A $(3,2,1)$–$HCOLS(h_1^{n_1} h_2^{n_2} \cdots h_k^{n_k})$ is defined to be an $HMOLS(h_1^{n_1} h_2^{n_2} \cdots h_k^{n_k})$ in which the two squares are (3,2,1)-conjugates of each other. It is important to note that the *ICOILS (v , n)* described in [4] are essentially $(3,2,1)$–$HCOLS(1^{v-n} n^1)$. For results on HMOLS and HSOLS, the reader is referred to [11, 15, 18, 19, 22]. In particular, it is now known [19] that for any $h \geq 2$, an $HMOLS(h^n)$ exists if and only if $n \geq 4$, and for any $h \geq 2$, an $HSOLS(h^n)$ exists if and only if $n \geq 4$, except possibly when $n=6$ and $h \in \{7,13\}$.

In this paper, we investigate the spectrum of $(3,2,1)$–$HCOLS(h^n)$. It is proved that for any $h \geq 2$, a $(3,2,1)$–$HCOLS(h^n)$ exists if and only if $n \geq 4$, except possibly when $n \in \{8,9,12\}$ and $h=2$, and when $n=6$ and $h \in \{5,7,13,61\}$. Noting that the existence of $(3,2,1)$–*COLS* is equivalent to the existence of $(1,3,2)$–*COLS* (see, for example, [16]), we also have the identical result for $(1,3,2)$–$HCOLS(h^n)$.

2. Preliminaries

For our main result, we shall use both recursive and direct methods of construction. In particular, we shall make extensive use of pairwise balanced designs (PBDs) and related designs, the well-known direct product construction, and a direct method of construction of the starter-adder type. For the convenience of the reader, we describe some of the auxiliary designs to be used in our constructions. However, the interested reader may wish to refer to [12, 21] for more general information on *PBDs* and other related designs.

Definition 2.1. Let K be a set of positive integers. A pairwise balanced design *(PBD)* of index unity *B(K, 1; v)* is a pair $(X, \mathcal{B})$ where X is a v-set (of points) and $\mathcal{B}$ is a collection of subsets of X (called blocks) with sizes in K such that every pair of distinct points of X is contained in exactly one block of $\mathcal{B}$. The number $|X|=v$ is called the order of the *PBD*. A set N of positive integers is said to be *PBD*-closed if $v \in N$ whenever there exists a *PBD B(N, 1; v)*.

Definition 2.2. Let K and M be sets of positive integers. A group divisible design *GDD GD(K, 1, M ; v)* is a triple $(X, \mathcal{G}, \mathcal{B})$ where

(i) X is a v-set (of points),

(ii) $\mathcal{G}$ is a collection of nonempty subsets of X (called groups) with sizes in M which partition X,

(iii) $\mathcal{B}$ is a collection of subsets of X (called blocks) each with size at least two in K,

(iv) no block meets a group in more than one point, and

(v) each pairset $\{x, y\}$ of points not contained in any group is contained in exactly one block.

The group-type of $(X, \mathcal{G}, \mathcal{B})$ will be the multiset $\{|G|: G \in \mathcal{G}\}$.

A weighting of the GDD is a mapping $w: X \rightarrow Z^+ \cup \{0\}$. For the *GDD* $(X, \mathcal{G}, \mathcal{B})$, w a weighting, and $Y \subseteq X$, we let $w(Y)$ denote the multiset $\{w(x): x \in Y\}$.

Definition 2.3. A transversal design *(TD)* $T(k, 1; m)$ is a *GDD* with km points, k groups of size m and m^2 blocks of size k, where each block is a transversal of the collection of groups.

Definition 2.4. Let $(X, \mathcal{B})$ be a *PBD* $B(K, 1; v)$. A parallel class in $(X, \mathcal{B})$ is a collection of disjoint blocks of $\mathcal{B}$, the union of which equals X. $(X, \mathcal{B})$ is called resolvable if the blocks of $\mathcal{B}$ can be partitioned into parallel classes. A *GDD* $GD(K,1,M;v)$ is resolvable if its associated *PBD* $B(K \cup M,1;v)$ is resolvable with M as a parallel class of the resolution.

We remark that it is fairly well-known that the existence of a resolvable *TD* $RT(k,1;m)$ is equivalent to the existence of a *TD* $T(k+1,1;m)$or equivalently $k-1$ *MOLS* of order m. We shall make use of the following three results which can be found in [6, 7, 12, 17].

Lemma 2.5. For every prime power q, there exists a $T(q+1,1;q)$.

Lemma 2.6. Let $m = p_1^{k_1} p_2^{k_2} \cdots p_r^{k_r}$ be the factorization of m into powers of distinct primes p_i, then a $T(k,1;m)$ exists, where $k = 1 + \min p_i^{k_i}$.

Lemma 2.7. A $T(6,1;m)$ exists for all integers $m \geq 5$, except possibly for $m \in \{6,10,14,18,20,22,26,28,30,34,38,42,44,52\}$.

There are some useful PBDs derived from TDs or truncated TDs. We present some lemmas below which will be quite useful in some of our later constructions. We shall denote by $B(K)$ the set of all integers v for which there is a *PBD* $B(K,1;v)$. If $k \notin K$, then $B(K \cup \{k^*\},1;v)$ denotes a *PBD* $B(K \cup \{k\},1;v)$ which contains a unique block of size k, and if $k \in K$, then $B(K \cup \{k^*\},1;v)$ denotes a *PBD* $B(K \cup \{k\},1;v)$ which contains at least one block of size k. For convenience, we shall define $B(k_1, k_2, \ldots, k_r)$ to be the set of all integers v such that there is a *PBD* $B(\{k_1, k_2, \ldots k_r\},1;v)$.

Lemma 2.8. If m is a prime power and $1 \leq k \leq m-1$, then for $0 \leq u \leq m$ and $0 \leq v \leq m$, we have $km+u+v \in B(k, k+1, k+2, u, v)$.

Proof: See [1, Lemma 3.5].

It is now known [8, 20] that a $T(5,1;m)$ exists for all positive integers $m \neq 2,3,6$ with the possible exception of $m=10$. From this result we readily obtain our next lemma (see, for example, [3]).

Lemma 2.9. Suppose $m \neq 2,3,6,10$ and $0 \leq n \leq m$. Then $4m+n \in B(4,5,m,n)$.

Lemma 2.10. Suppose a $T(k+1,1;m)$ exists, $0 \leq u \leq m$, and $\{k,k+1\} \subseteq B(K)$. If there exists a *PBD* $B(K \cup \{v^*\},1;m+v)$, then we have $km+u+v \in B(K \cup \{(u+v)^*\})$.

Proof: See [1, Lemma 3.8]

Our next lemma is a variation of Lemma 2.10, which utilizes a block of size $k+1$ in the *TD*.

Lemma 2.11. Suppose a $T(k+1,1;m)$ exists, $0 \leq u \leq m$, and $\{k,k+1\} \subseteq B(K)$. If there exist a *PBD* $B(K \cup \{(v+1)^*\},1,1;m+v)$ and a *PBD* $B(K \cup \{(v+1)^*\},1;u+v)$, then $km+u+v \in B(K \cup \{(v+k+1^*\})$.

Proof: Let $(X, \mathcal{G}, \mathcal{B})$ be the $GD(\{k,k+1\},1,\{m,u\};km+u)$ obtained by deleting $m-u$ points from a particular group of the $T(k+1,1;m)$. Let $G_1,G_2,\ldots,G_k$ be the k groups of size m and H the group of size u in this *GDD*. Let S be a set of v fixed points disjoint from X and let $B=\{b_1,b_2,\ldots,b_{k+1}\}$ be a block of size $k+1$ in the GDD, where $B \cap G_i = \{b_i,\}$, $i=1,2,\ldots,k$ and $B \cap H = \{b_{k+1}\}$. We now construct a *PBD* $B(K \cup \{(v+k+1)^*\},1;km+u+v)$ on the set $X \cup S$ as follows:

(1) take the blocks $\mathcal{B} \setminus \{B\}$,

(2) for $i=1,2,\ldots,k$, replace each group G_i by a *PBD* $B(K \cup \{(v+1)^*\},1;m+v)$ based on the set $G_i \cup S$ with $S \cup \{b_i\}$ as its unique block of size $v+1$,

(3) replace the group H by a *PBD* $B(K \cup \{(v+1)^*\},1;u+v)$ based on the set $H \cup S$ with $S \cup \{b_{k+1}\}$ as its unique block of size $v+1$, and finally

(4) replace the $k+1$ blocks $S \cup \{b_i\}, i=1,2,\ldots,k+1$, by the single block $S \cup B$ of size $v+k+1$. It is readily checked that the result is a *PBD* $B(K \cup \{(v+k+1)^*\}$, $1;km+u+v)$ and the lemma is proved. □

Before proceeding, we shall make note of an important known result on (3,2,1)–*COILS* (see [2,5]).

Theorem 2.12. There exists a (3,2,1)–*COILS* *(v)* for every positive integer $v \neq 2,3,6$, except possibly for $v=12$.

As a consequence of the above theorem, we immediately get

Corollary 2.13. A (3,2,1)–*HCOLS*(1^n) exists for all $n \geq 4$, $n \neq 6$, except possibly for $n=12$.

We shall also need the following result for our recursive constructions (see [2, 5, 16]).

Theorem 2.14. There exists a (3,2,1)–*COLS* *(v)* for every positive integer $v \neq 2,6$.

With regards to the necessary conditions for the existence of (3,2,1)–*HCOLS*(h^n), we observe the following lemma (see [11, Theorem 1.1]) and its consequence.

Lemma 2.15. If there exists an *HMOLS*(h^n), then $n \geq 4$.

Corollary 2.16. If there exists a (3,2,1)–*HCOLS*(h^n), then $n \geq 4$.

3. Basic Lemmas

For our recursive type of constructions, we shall employ the well-known direct product construction and use PBD closure. Let us denote by (3,2,1)–*COLS*(m) a (3,2,1)–*COLS* of order m.

The following construction is essentially the same as for the usual direct product construction for orthogonal Latin squares.

Lemma 3.1. If a (3,2,1)–$HCOLS(h^n)$ and a (3,2,1)–$COLS(m)$ exist, then there exists a (3,2,1)–$HCOLS((hm)^n)$.

Define $I_2=\{n\geq 4$: there exists a (3,2,1)–$HCOLS(2^n)\}$. The following lemma is the analog for $HSOLS(2^n)$ in [19], and another variation of [11, Lemma 3.1].

Lemma 3.2. I_2 is PBD-closed.

We also have what is essentially [11, Lemma 2.2] and the analog relating to the construction of HSOLS (see [19]).

Lemma 3.3. Suppose that $(X, \mathcal{G}, \mathcal{B})$ is a GDD and let $w:X\rightarrow Z^+\cup\{0\}$ be a weighting. Suppose there exists a (3,2,1)–*HCOLS* of type $w(B)$ for every block $B\in\mathcal{B}$. Then there exists a (3,2,1)–*HCOLS* of type $\{\Sigma_{x\in G}\, w(x):G\in\mathcal{G}\}$.

Using finite fields, we have the following lemma which comes from a direct-method of construction.

Lemma 3.4. Suppose $q\equiv 3(mod\ 4)$ is a prime power exceeding 3. Then $q\in I_2$.

Proof: We shall construct a (3,2,1)–$HCOLS(2^q)$ based on the elements of the set $X=GF(q)\times Z_2$. For convenience, the ordered pair (x, i) of X will be denoted simply by x_i in what follows. Let R denote the set of nonzero squares in $GF(q)$ and choose $c\in R$ such that $c^2-1\notin R$ and $c^4-1\in R$ (the hypothesis of the lemma allows for this). We can then construct the following incomplete orthogonal array based on the set X, where $y\in R$ and the rows are developed by $GF(q)$:

$y_0, 0_0, cy_1, (-y/c)_1$

$y_1, 0_0, -cy_1, (y/c)_0$

$-y_0, 0_0, -cy_0, (-y/c)_0$

$-y_1, 0_0, cy_0, (y/c)_1$

$y_1, 0_1, cy_0, (-y/c)_0$

$y_0, 0_1, -cy_0, (y/c)_1$

$-y_1, 0_1, -cy_1, (-y/c)_1$

$-y_0, 0_1, cy_1, (y/c)_0$

If we now use the first column as row index, the second column as column index, and the third and fourth columns as entries for partitioned incomplete Latin squares L and M, respectively, then it is readily checked that L and M are orthogonal and (3,2,1)–conjugates of each other, that is, we have constructed a (3,2,1)–$HCOLS(2^q)$. □

Example 3.5 For $q=7$, we illustrate the construction in lemma 3.4 with a choice of $c=2$ in fig. 1.

$(3,2,1)$–$HCOLS(2^7)$

	0_0	1_0	2_0	3_0	4_0	5_0	6_0	0_1	1_1	2_1	3_1	4_1	5_1	6_1
0_0		6_0	5_0	4_1	3_0	2_1	1_1		3_1	6_1	2_0	5_1	1_0	4_0
1_0	2_1		0_0	6_0	5_1	4_0	3_1	5_0		4_1	0_1	3_0	6_1	2_0
2_0	4_1	3_1		1_0	0_0	6_1	5_0	3_0	6_0		5_1	1_1	4_0	0_1
3_0	6_0	5_1	4_1		2_0	1_0	0_1	1_1	4_0	0_0		6_1	2_1	5_0
4_0	1_1	0_0	6_1	5_1		3_0	2_0	6_0	2_1	5_0	1_0		0_1	3_1
5_0	3_0	2_1	1_0	0_1	6_1		4_0	4_1	0_0	3_1	6_0	2_0		1_1
6_0	5_0	4_0	3_1	2_0	1_1	0_1		2_1	5_1	1_0	4_1	0_0	3_0	
0_1		3_0	6_0	2_1	5_0	1_1	4_1		6_1	5_1	4_0	3_1	2_0	1_0
1_1	5_1		4_0	0_0	3_1	6_0	2_1	2_0		0_1	6_1	5_0	4_1	3_0
2_1	3_1	6_1		5_0	1_0	4_1	0_0	4_0	3_0		1_1	0_1	6_0	5_1
3_1	1_0	4_1	0_1		6_0	2_0	5_1	6_1	5_0	4_0		2_1	1_1	0_0
4_1	6_1	2_0	5_1	1_1		0_0	3_0	1_0	0_1	6_0	5_0		3_1	2_1
5_1	4_0	0_1	3_0	6_1	2_1		1_0	3_1	2_0	1_1	0_0	6_0		4_1
6_1	2_0	5_0	1_1	4_0	0_1	3_1		5_1	4_1	3_0	2_1	1_0	0_0	

Figure 1

4. Construction of $(3,2,1)$–$HCOLS(2^n)$

The main result of this section will rely heavily on PBD-closure and the use of Lemma 3.2. For this purpose, we shall construct PBDs with specified block sizes. These constructions will prove not only useful here, but no doubt will also be of special interest in other instances. In some of what follows we may tacitly use the fact that if $v \in B(K')$ and $K' \subseteq B(K)$, then $v \in B(K)$.

Our first lemma is for the most part contained in [3, 7].

Lemma 4.1. If $v \geq 4$ and $v \neq 6,8,9,10,12,14,15,18,19,23,26,27,30,38,39,42,50,54,62,66,74,78,90,98,102,110,114,126,158,162,174$, then $v \in B(4,5,7,11)$.

Proof:

If $v \not\equiv 7,10$, or $11 \pmod{12}$ and v satisfies the condition of the lemma, then the proof that $v \in B(4,5,11)$ is contained in [3, Theorems 4.3, 4.4, 4.8, 4.9], noting that if $v \neq 9$, then 9 is not essential for the case $v \equiv 0$ or $1 \pmod 4$. If $v \equiv 7$ or $10 \pmod{12}$ and $v \neq 10,19$, then $v \in B(4,7^*)$ by a result of Brouwer [7]. For the case $v \equiv 11 \pmod{12}, v \geq 35$, we may adjoin seven new points to a resolvable $B(4,1;v-7)$ to get $v \in B(4,5,7^*)$, and our lemma is proved. □

Lemma 4.2. If $v \in \{98,102,110,114,126,158,162,174\}$, then $v \in B(4,5,7,11)$.

Proof:

First of all $\{98,110\} \subseteq B(4,5,7^*)$ follows from the fact that we may adjoint 22 new points to a resolvable $B(4,1;v)$ with $v=76$ and 88 to get $\{98,110\} \subseteq B(4,5,22^*) \subseteq B(4,5,7^*)$. For $v=102$, we use the fact that $22=4 \cdot 5+2 \in B(4,5,2^*)$ from Lemma 2.9 and apply Lemma 2.10 with $k=4$ and $m=u=20$ to get $102 \in B(4,5,22^*) \subseteq B(4,5,7^*)$. If $v=114$, we first observe that $29=4 \cdot 7+1 \in B(4,5,7)$ follows from Lemma 2.9 and use $22 \in B(4,7^*)$ to obtain $114=4 \cdot 23+16+6 \in B(4,5,7,11^*)$ from Lemma 2.11. In a similar manner, we also use the fact that $32=4 \cdot 7+4 \in B(4,5,7)$ with $22 \in B(4,7^*)$ to get $126=4 \cdot 26+16+6 \in B(4,5,7,11^*)$ from Lemma 2.11. Finally $\{158,162,174\} \subseteq B(4,5,7)$ follows from Lemma 2.9 with $158=4 \cdot 34+22$, $162=4 \cdot 35+22$, $174=4 \cdot 35+34$.

Lemma 4.3. If $v \in \{26,27,30,38,39,42,50,54,62,66,74,78,90\}$, then $v \in B(4,5,6,7,11)$.

Proof:

Except for the case $v=27$, we may apply Lemma 2.8 to get our desired result with $k=4$ or 5 and $m \in \{5,7,11,13,17\}$. We may obtain $27 \in B(4,5,6)$ by deleting two points from a group of an $RT(5,1;5)$ and then adjoining four new points to the resulting truncated TD, with one of these new points added to the groups. □

Combining Lemmas 4.1 and 4.2, we have the following theorem:

Theorem 4.4. If $v \geq 4$ and $v \neq 6,8,9,10,12,14,15,18,19,23,26,27,30,38,39,42,50,54,62,66,74,78,90$, then $v \in B(4,5,6,7,11)$.

And combining Theorem 4.4 with Lemma 4.3, we obtain the basis for most of our constructions in this section in the following theorem.

Theorem 4.5. If $v \geq 4$ and $v \neq 8, 9, 10, 12, 14, 15, 18, 19, 23$, then $v \in B(4,5,6,7,11)$.

We shall use PBD-closure (Lemma 3.2) and Theorem 4.5 to obtain the main result for $(3,2,1)$–$HCOLS(2^n)$. However, we first require some input designs in order to apply Theorem 4.5. For the most part we give a direct method of construction of the starter-adder type, where the arrays are generated cyclically in the usual manner (see, for example, [4, 13]). In Figure 2 we present an example of a $(3,2,1)$–$HCOLS(2^4)$, and in Figure 3 we give an example of a $(3,2,1)$–$HCOLS(2^5)$, where the array is essentially of type $2^4 2^1$. Note that this square can be viewed as being completely cyclically generated (*mod* 8) from its first row, given by the vectors $\mathbf{e}=(\phi,2,1,6,\phi,B,3,A)$ and $\mathbf{f}=(7,5)$, and the last two elements of its first column, given by the vector $\mathbf{g}=(6,2)$, where the symbol ϕ is used for empty cells and *A, B* act as "infinity" elements. In Table 1, we give additional cyclically generated $(3,2,1)$–$HCOLS(2^n)$ for $n=6,14,15$, and 18, by specifying the type along with the vectors **e**, **f** and **g** where ϕ is used for empty cells and *A, B, C*, etc., act as "infinity" elements.

Note, for example, that type $2^{10}8^1$ produces a $(3,2,1)$–$HCOLS(2^{14})$ by filling in the hole of size 8 with a $(3,2,1)$–$HCOLS(2^4)$. Moreover, the type $2^{11}10^1$ square will produce a $(3,2,1)$–$HCOLS(2^{16})$, but this is listed here only for future reference as an input design for the constructions of the next section.

Before proceeding to our main result, we have the following from GDD's:

(3,2,1)–$HCOLS(2^4)$

	0	1	2	3	4	5	6	7
0		2	1	6		7	3	5
1	6		3	2	7		0	4
2	5	7		4	3	0		1
3	2	6	0		5	4	1	
4		3	7	1		6	5	2
5	3		4	0	2		7	6
6	7	4		5	1	3		0
7	1	0	5		6	2	4	

Figure 2

(3,2,1)–$HCOLS(2^5)$

	0	1	2	3	4	5	6	7	*A*	*B*
0		2	1	6		*B*	3	*A*	7	5
1	*A*		3	2	7		*B*	4	0	6
2	5	*A*		4	3	0		*B*	1	7
3	*B*	6	*A*		5	4	1		2	0
4		*B*	7	*A*		6	5	2	3	1
5	3		*B*	0	*A*		7	6	4	2
6	7	4		*B*	1	*A*		0	5	3
7	1	0	5		*B*	2	*A*		6	4
A	6	7	0	1	2	3	4	5		
B	2	3	4	5	6	7	0	1		

Figure 3

Lemma 4.6. If $n \equiv 1 \ (mod\ 3)$ and $n \geq 7$, then $n \in I_2$.

Proof:

If $n \equiv 1 (mod\ 3)$ and $n \geq 7$, then there exists a GDD having group-type 2^n and blocks of size 4 (see [9]). We then apply Corollary 2.13 and Lemma 3.3, giving every point of the GDD weight 1, to obtain the desired result. □

We are now in a position to prove the main theorem of this section.

Theorem 4.7. There exists a (3,2,1)–$HCOLS(2^n)$ for all integers $n \geq 4$, except possibly for $n = 8, 9$, and 12.

Proof:

For the proof of the theorem, we combine the examples given in this section with Lemmas 3.4 and 4.6, and Theorem 4.5, using the fact that I_2 is PBD-closed. □

5. Construction of (3,2,1)–$HCOLS(h^n)$

In this section, we shall focus our attention on the existence of (3,2,1)–$HCOLS(h^n)$ for $h \geq 3$. We commence with the following lemma.

Table 1

Type	e	f	g
$2^5 2^1$	$(\phi,7,A,2,B,\phi,$ $8,4,6,3)$	(1,9)	(1,3) (*mod* 10)
$2^{10} 8^1$	$(\phi,E,3,8,A,F,$ $9,G,4,C,\phi,2,7,$ $6,11,D,15,B,5,H)$	(1,12,13,14, 16,17,18,19)	(9,6,18,14,4,12,8, 2) (*mod* 20)
$2^{11} 8^1$	$(\phi,14,5,20,13,A,$ $10,17,15,H,16,\phi,$ $B,C,19,D,E,18,G,$ $F,6,1)$	(2,3,4,7,8,9, 12,21)	(18,20,14,16,19, 15,21,12) (*mod* 22)
$2^{13} 10^1$	$(\phi,5,G,F,25,I,J,$ $18,2,24,7,D,19,$ $\phi,17,C,21,23,16,$ $A,E,22,B,H,8,1)$	(3,4,6,9,10,11, 12,20,14,15)	(8,9,18,25,17,12, 14,22,16,19), (*mod* 26)
$2^{11} 10^1$	$(\phi,21,H,J,G,20,$ $D,A,5,18,4,\phi,19,$ $C,F,E,I,3,2,B,1,$ $17)$	(6,7,8,9,10,12, 13,14,15,16)	(2,4,12,21,14,1, 13,17,5,10) (*mod* 22)

Lemma 5.1. If $n \equiv 0$ or $3 (mod\ 12)$ and $n \geq 12$, then there exists a $(3,2,1)-HCOLS(h^{n/3})$ for $h=3$ and 6.

Proof:

By deleting one point from a $B(4,1;n+1)$ we obtain a GDD with blocks of size 4 and group-type $3^{n/3}$. We then apply Lemma 3.3, giving each weight 1 for the case $h=3$, and weight 2 for the case $h=6$, to obtain our desired result. □

Lemma 5.2. For any integer $n \geq 4$, $n \neq 6,12$, and for $h \geq 3$, there exists a $(3,2,1)-HCOLS(h^n)$.

Proof:

For any $n \geq 4$, $n \neq 6,12$, and $h \geq 3$, $h \neq 6$, we can combine a $(3,2,1)-HCOLS(1^n)$, which exists by Corollary 2.13, and a $(3,2,1)-COLS(h)$, which exists by Theorem 2.14, to obtain a $(3,2,1)-HCOLS(h^n)$ by Lemma 3.1. Next for a $(3,2,1)-HCOLS(6^n), n \neq 6,8,9,12$, we combine a $(3,2,1)-HCOLS(2^n)$ and a $(3,2,1)-COLS(3)$ to obtain a $(3,2,1)-HCOLS(6^n)$ by Lemma 3.1. Finally, for $n=8,9$, we obtain a $(3,2,1)-HCOLS(6^n)$ from Lemma 5.1. This completes the proof of the lemma. □

We are now left with the problem to determine the existence of $(\dot{3},2,1)$–$HCOLS(h^n)$ for $n=6$ and 12, $h \geq 3$. In what follows, we consider the cases separately. First of all, we investigate $(3,2,1)$–$HCOLS(h^6)$. We need some input squares for TDs as the "base" GDDs in what follows.

Lemma 5.3. There exist $(3,2,1)$–$HCOLS(3^6)$ and $(3,2,1)$–$HCOLS(2^5 3^1)$.

Proof:

We give a direct construction of the starter-adder type in both cases, where the squares are generated cyclically. For the $(3,2,1)$–$HCOLS(3^6)$, we obtain a type $3^5 3^1$ square with its first row given by vectors $\mathbf{e}=(\phi,B,C,1,3,\phi,13,11,9,12,\phi,A,14,4,8)$ and $\mathbf{f}=(2,6,7)$, and the last 3 elements of its first column given by vector $\mathbf{g}=(12,8,11)$ and the square is generated $(mod\ 15)$. For the type $2^5 3^1$ square we use $\mathbf{e}=(\phi,8,C,9,A,\phi,7,B,6,2)$, $\mathbf{f}=(1,3,4)$, and $\mathbf{g}=(9,4,2)$ $(mod\ 10)$. It is readily checked that we get the desired result. □

Lemma 5.4. There exists a $(3,2,1)$–$HCOLS(3^k 4^{6-k})$ for $k=0,1,4,5$.

Proof:

We first consider a $B(5,1;25)$ and, by deleting one point from a block B of this design, we obtain a GDD with blocks of size 5 and group-type 4^6. For $k=1,4,5$, we may further delete additional k points contained in another block of this GDD, which is disjoint from B, to obtain a GDD with blocks of size 4 and 5 and group-type $3^k 4^{6-k}$. We now apply Lemma 3.3, giving every point weight 1, to obtain our desired result. □

Lemma 5.5. If there exists a $T(6,1;m)$ then there exist a $(3,2,1)$–$HCOLS$ $((2m+1)^6)$ and $HCOLS((3m+1)^6)$. Moreover, if $m \neq 5$, there exists a $(3,2,1)$–$HCOLS$ $((3m+2)^6)$.

Proof:

The proof considers each of the above three cases separately. First we consider the $T(6,1;m)$ and let B be a block. We give each point of B weight 3 and all other points of the TD are given weight 2. We then apply Lemma 3.3, using as input designs $(3,2,1)$–$HCOLS(2^6)$, $HCOLS(3^6)$ and $HCOLS(2^5 3^1)$ to obtain a $(3,2,1)$–$HCOLS$ $((2m+1)^6)$.

Secondly, for a $(3,2,1)$–$HCOLS$ $((3m+1)^6)$, we proceed as before except we give every point of B weight 4 and all other points of the TD weight 3. Using $(3,2,1)$–$HCOLS(3^6)$, $HCOLS(4^6)$, and $HCOLS(3^5 4^1)$ as input designs, we obtain a $(3,2,1)$–$HCOLS((3m+1)^6)$.

Finally, for a $(3,2,1)$–$HCOLS$ $((3m+2)^6)$, $m \neq 5$, we take a slightly different approach. Here we select two disjoint blocks B and B' of the TD (this is possible provided $m=5$) and give every point of $B \cup B'$ weight 4, and all other points of the TD weight 3. Then using as input designs $(3,2,1)$–$HCOLS(3^6)$, $HCOLS(4^6)$, $HCOLS(3^5 4^1)$, and $HCOLS(3^4 4^2)$, we get a $(3,2,1)$–$HCOLS((3m+2)^6)$. This proves the lemma. □

Lemma 5.6. There exists a $(3,2,1)$–$HCOLS(h^6)$ for all $h \geq 3$, except possibly for $h=5,7,13,61$.

Proof:

We consider the cases h even and h odd separately in our proof which follows:

Case 1: h even. We write $h=2m$ and need only consider $m \geq 3$. If $m \neq 6$, we combine a $(3,2,1)-HCOLS(2^6)$ with a $(3,2,1)-HCOLS(m)$ to obtain a $(3,2,1)-HCOLS((2m)^6)$ by Lemma 3.1. If $m=6$ we combine a $(3,2,1)-HCOLS(4^6)$ with a $(3,2,1)-COLS(3)$ to get a $(3,2,1)-HCOLS(12^6)$. Thus we have a $(3,2,1)-HCOLS(h^6)$ for all even h.

Case 2: h odd. Let $A=\{2,3,4,6,10,14,18,20,22,26,28,30,34,38,42,44,52\}$. We let $h=2m+1$ and need only consider $m \geq 2$. First of all, if $m \notin A$, then there exists a $T(6,1;m)$ by Lemma 2.7, and by applying Lemma 5.5 we obtain a $(3,2,1)-HCOLS((2m+1)^6)$. If $m \in A \setminus \{2,3,6,30\}$, the existence of a $(3,2,1)-HCOLS((2m+1)^6)$ is established by the equations given for $2m+1$ and the appropriate application of Lemmas 3.1 and 5.5 in Table 2. Consequently, we have a $(3,2,1)-HCOLS(h^6)$ for all odd $h \geq 3$, with the possible exception of $h=5,7,13$ and 61, and our lemma is proved. □

Table 2

Equation for $2m+1$	Authority
$9=3.3$	Lemma 3.1
$21=3.7$	Lemma 3.1
$29=3.9+2$	Lemma 5.5
$37=3.12+1$	Lemma 5.5
$41=3.13+2$	Lemma 5.5
$45=3.15$	Lemma 3.1
$53=3.17+2$	Lemma 5.5
$57=3.19$	Lemma 3.1
$69=3.23$	Lemma 3.1
$77=3.25+2$	Lemma 5.5
$85=17.5$	Lemma 3.1
$89=3.29+2$	Lemma 5.5
$105=3.35$	Lemma 3.1

We now turn our attention to the existence of $(3,2,1)-HCOLS(h^{12}), h \geq 3$. We first consider some small values of h.

Lemma 5.7. There exists a $(3,2,1)-HCOLS(h^{12})$ for $h=3$ and 6.

Proof:

We start with a $B(4,1;37)$ and delete one point from a block of this design to obtain a GDD with blocks of size 4 and group-type 3^{12}. If we give every point weight 1, we obtain a $(3,2,1)-HCOLS(3^{12})$, and if we give every point weight 2, we obtain a $(3,2,1)-HCOLS(6^{12})$, by applying Lemma 3.3. □

Lemma 5.8. There exists a (3,2,1)–$HCOLS(h^{12})$ for $h=4,7,8,9$, and 10.

Proof:

For $4 \le h \le 10, h \ne 5,6$, we start with an $RT(h,1;11)$ and then adjoin h new points to this TD, one point to each of h parallel classes of blocks. We then view the resulting design as a GDD with block sizes in $\{11,h,h+1\}$ and group-type h^{12}. Using as input designs (3,2,1)–$HCOLS(1^{11})$, $HCOLS(1^h)$ and $HCOLS(1^{h+1})$, we apply Lemma 3.3 giving every point of this GDD weight 1 and the result follows. □

Lemma 5.9. There exists a (3,2,1)–$HCOLS(h^{12})$ for $h = 5$ and 14.

Proof:

To begin with, there exists a GDD with blocks of size 5 and group-type $(5)^{13}$ (see, for example, [12]). By deleting one group from this GDD, we obtain another GDD with blocks of size 4 and 5 and group-type $(5)^{12}$. We then give every point of this resulting GDD weight 1 and apply Lemma 3.3 to get a (3,2,1)–$HCOLS(5^{12})$. Secondly, there is a GDD with blocks of size 7 and group-type $(7)^{13}$ (see [12]). From this GDD we delete one group to obtain another GDD with blocks of size 6 and 7 and group-type $(7)^{12}$. We then give every point of this resulting GDD weight 2 and apply Lemma 3.3, using (3,2,1)–$HCOLS(2^6)$ and $HCOLS(2^7)$ as input designs, to get a (3,2,1)–$HCOLS(14^{12})$. □

We now need several input designs for our constructions which remain.

Lemma 5.10. There exist (3,2,1)–$HCOLS(1^{10}3^1)$, $HCOLS(1^{10}4^1)$, $HCOLS(1^{11}3^1)$, $HCOLS(1^{11}5^1)$, $HCOLS(2^{11}8^1)$ and $HCOLS(2^{11}10^1)$.

Proof:

For a (3,2,1)–$HCOLS(1^{10}3^1)$, we present a direct construction of the starter-adder type. We essentially construct a square which is cyclically generated from its first row given by the vectors $\mathbf{e}=(0,C,3,9,6,2,5,A,1,B)$ and $\mathbf{f}=(4,7,8)$, and the last three elements of its first column given by the vector $\mathbf{g}=(5,4,8)(mod\ 10)$. For squares of type $1^{10}4^1, 1^{11}3^1, 1^{11}4^1$, and $1^{11}5^1$, the constructions can be found in [4,5]. For types $2^{11}8^1$ and $2^{11}10^1$, the constructions are given in Table 1. □

Lemma 5.11. Let $m>11$ and suppose there exists a $T(12,1;m)$, then there exists a (3,2,1)–$HCOLS(m^{12})$ and $HCOLS((2m)^{12})$.

Proof:

For $m>11$, we may write $m=4x+5y$, where x and y are non-negative integers satisfying $0<x+y<m$. Let $e \in \{1,2\}$ and select a group G of the $T(12,1;m)$. For this particular group G, we shall assign a weight of $4e$ to x points, a weight of $5e$ to y points, and a weight of zero to the remaining points of G. We then assign a weight of e to all the other points of the TD. Using as input designs the squares of type $1^{11}, 2^{11}$ together with those given in Lemma 5.10, we obtain a (3,2,1)–$HCOLS((em)^{12})$ from Lemma 3.3. This proves the lemma. □

Lemma 5.12. There exist (3,2,1)–$HCOLS(11^{12})$ and $HCOLS(22^{12})$.

Proof:

For a $(3,2,1)$–$HCOLS(11^{12})$ we start with a $T(12,1;11)$ and select a group G of this TD. On G we assign a weight of 3 to one point, a weight of 4 to two points and a weight of zero to the remaining 8 points. For all the other points of the TD, we assign a weight of 1. Using as input designs the squares of type $1^{11}, 1^{11}3^1$ and $1^{11}4^1$, we get a $(3,2,1)$–$HCOLS(11^{12})$, by applying Lemma 3.3. For the construction of a $(3,2,1)$–$HCOLS(22^{12})$, we take a different approach. Starting with a $T(12,1;23)$, we first delete one block to obtain a GDD with blocks of size 11 and 12 and group-type $(22)^{12}$. We then select one group G of this GDD and on G we assign a weight of 3 to two points, a weight of 4 to four points, and give weight zero to the remaining 16 points. For all the other points of the GDD, we assign a weight of 1. We then use as input designs the squares of types $1^{10}, 1^{11}$, with those given in Lemma 5.10 to obtain a $(3,2,1)$–$HCOLS(22^{12})$, which proves the lemma. □

We are now in a position to prove

Lemma 5.13. For all $h \geq 3$, there exists a $(3,2,1)$–$HCOLS(h^{12})$.

Proof:

For our proof we consider the cases below with $h \geq 3$.

Case 1: h odd. If $(h,210)=1$, then, by Lemma 2.6, there exists a $T(12,1;,h)$ and we can apply Lemma 5.11 with $h>11$ and Lemma 5.12 to obtain a $(3,2,1)$–$HCOLS(h^{12})$. If $(h,210)\neq 1$, then we can express $h=m \cdot k$ with $m \in \{3,5,7\}$ and k odd. Since a $(3,2,1)$–$HCOLS(m^{12})$ exists by Lemmas 5.7-5.9, and a $(3,2,1)$–$COLS(k)$ exists by Theorem 2.14, we obtain a $(3,2,1)$–$HCOLS(mk)$ by applying Lemma 3.1. Consequently, a $(3,2,1)$–$HCOLS(h^{12})$ exists for all odd $h \geq 3$.

Case 2: h even. If $h=2k$, k odd and $(k,210)=1$, we can apply Lemmas 5.11 and 5.12 to get $(3,2,1)$–$HCOLS((2k)^{12})$. If $h=2k$, k odd and $(k,210)\neq 1$, we can express $h=mk$ with $m \in \{6,10,14\}$ and k odd. Since $(3,2,1)$–$HCOLS((m)^{12})$ exists, by Lemmas 5.7-5.9 and $(3,2,1)$–$COLS(k)$ exists, by Theorem 2.14, we may apply Lemma 3.1 to get a $(3,2,1)$–$HCOLS((mk)^{12})$. Finally, if $4 \mid h$, we can express $h=2^m k$ with $m \geq 2$ and k odd. If $k=1$, we use the fact that we have a $(3,2,1)$–$HCOLS(4^{12})$ and $HCOLS(8^{12})$ and apply Lemma 3.1 as necessary. If $k \geq 3$, we have a $(3,2,1)$–$HCOLS(k^{12})$, from Case 1 above, which can be combined with a $(3,2,1)$–$COLS(2^m)$ to get a $(3,2,1)$–$HCOLS((2^m k)^{12})$ by Lemma 3.1. Thus a $(3,2,1)$–$HCOLS(h^{12})$ exists for all even $h \geq 3$, and the proof of the lemma is complete. □

Combining the results of Lemmas 5.2, 5.6, and 5.13, we obtain the following theorem.

Theorem 5.14. For any $h \geq 3$, a $(3,2,1)$–$HCOLS(h^n)$ exists if $n \geq 4$, except possibly when $n=6$ and $h \in \{5,7,13,61\}$.

In summary, we combine Theorems 4.7 and 5.14 with the necessary condition for existence of $(3,2,1)$–$HCOLS(h^n)$ to obtain our main result:

Theorem 5.15. For any $h \geq 2$, a $(3,2,1)$–$HCOLS(h^n)$ exists if and only if $n \geq 4$, except possibly when $n \in \{8,9,12\}$ and $h=2$, and when $n=6$ and $h \in \{5,7,13,61\}$.

Since the existence of $(3,2,1)$–$COLS(v)$ is equivalent to the existence of $(1,3,2)$–$COLS(v)$, we have the following corollary.

Corollary 5.16. For any $h \geq 2$, a $(1,3,2)$–$HCOLS(h^n)$ exists if and only if $n \geq 4$, except possibly when $n \in \{8,9,12\}$ and $h=2$, and when $n=6$ and $h \in \{5,7,13,61\}$.

6. Concluding Remarks

It is hoped that the results of this paper will be of importance in the construction of other combinatorial designs, such as Mendelsohn designs (see, for example, [2, 14]). We conjecture that $n \geq 4$ is also a sufficient condition for the existence of $(3,2,1)$ (or $(1,3,2)$)–$HCOLS(h^n)$. At present the existence of $(3,1,2)$ (or $(2,3,1)$)–$HCOLS(h^n)$ is under investigation by the authors, and it is hoped that results similar to those mentioned above for the other conjugates can be obtained.

Acknowledgements

The first author acknowledges the financial support of the National Sciences and Engineering Research Council of Canada under Grant A-5320. A portion of this work was carried out while the second author was a visiting professor at Mount Saint Vincent University and he gratefully acknowledges the kind hospitality accorded him.

References

[1] F.E. Bennett, "Latin squares with pairwise orthogonal conjugates", *Discrete Math.* 36 (1981) 117-137.

[2] F.E. Bennett, "Conjugate orthogonal Latin squares and Mendelsohn designs", *Ars Combinatoria* 19 (1985) 51-62.

[3] F.E. Bennett and N.S. Mendelsohn, "On the spectrum of Stein quasigroups", *Bull. Austral. Math. Soc.* 21 (1980) 47-63.

[4] F.E. Bennett and L. Zhu, "Incomplete conjugate orthogonal idempotent Latin squares", *Discrete Math.*, to appear.

[5] F.E. Bennett, Lisheng Wu, and L. Zhu, "Some new conjugate orthogonal Latin squares", *J. Combinatorial Theory (A)*, to appear.

[6] Th. Beth, D. Jungnickel, and H. Lenz, *Design Theory*, Bibliographisches Institut, Zurich, 1985.

[7] A.E. Brouwer, "Optimal packings of K_4's into a K_n," *J. Combinatorial Theory (A)* 26 (1979) 278-297.

[8] A.E. Brouwer, "The number of mutually orthogonal Latin squares - a table up to order 10000", *Research Report ZW 123/79, Mathematisch Centrum,* Amsterdam, June 1979.

[9] A.E. Brouwer, A. Schrijver, and H. Hanani, "Group divisible designs with block-size four", *Discrete Math.* 20 (1977) 1-10.

[10] J. Dénes and A.D. Keedwell, *Latin Squares and Their Applications,* Academic Press, New York and London, 1974.

[11] J.H. Dinitz and D.R. Stinson, "MOLS with holes", *Discrete Math.* 44 (1983) 145-154.

[12] H. Hanani, "Balanced incomplete block designs and related designs", *Discrete Math.* 11 (1975) 255-369.

[13] K. Heinrich and L. Zhu, "Incomplete self-orthogonal Latin squares", *J. Austral. Math. Soc, Ser. A,* to appear.

[14] A.D. Keedwell, "Circuit designs and Latin squares", *Ars Combinatoria 17* (1984) 79-90.

[15] R.C. Mullin and D.R. Stinson, "Holey SOLSSOMs", *Utilitas Math. 25* (1984) 159-169.

[16] K.T. Phelps, "Conjugate orthogonal quasigroups", *J. Combinatorial Theory (A)* 25 (1978) 117-127.

[17] Robert Roth and Matthew Peters, "Four pairwise orthogonal Latin squares of order 24", *J. Combinatorial Theory (A)* 44 (1987) 152-155.

[18] D.R. Stinson and L. Zhu, "On sets of three MOLS with holes", *Discrete Math. 54* (1985) 321-328.

[19] D.R. Stinson and L. Zhu, "On the existence of MOLS with equal-sized holes", *Aequationes Math.,* to appear.

[20] D.T. Todorov, "Three mutually orthogonal Latin squares of order 14", *Ars Combinatoria* 20 (1985) 45-48.

[21] R.M. Wilson, "Construction and uses of pairwise balanced designs", *Mathematical Centre Tracts* 55 (1974) 18-41.

[22] L. Zhu, "Existence of holey SOLSSOMs of type 2^n", *Congressus Numer.* 45 (1984) 295-304.

Annals of Discrete Mathematics 34 (1987) 81–100
© Elsevier Science Publishers B.V. (North-Holland)

On regular packings and coverings

J-C. Bermond, J. Bond, D. Sotteau

U.A. 410 CNRS, L.R.I.
bât 490, Université Paris-Sud, 91405 Orsay Cedex, FRANCE

TO ALEX ROSA ON HIS FIFTIETH BIRTHDAY

ABSTRACT

Let us call a regular packing of K_n with K_k's a set of edge disjoint subgraphs of K_n isomorphic to K_k such that the partial graph G, called the leave of the packing, generated by the edges not covered is regular. Similarly, a regular covering of K_n with K_k's is a set of subgraphs of K_n isomorphic to K_k which cover all the edges of K_n at least once and such that the multigraph H, called the excess of the covering, obtained by deleting K_m from the union of the K_k's, is regular. Here we study maximum regular packings and minimum regular coverings. More exactly we determine the minimum degrees of the leaves and excesses in case of regular packings and coverings of K_n with K_3's and K_4's . We also exhibit minimum regular coverings containing maximum regular packings.

1. Introduction

A classical problem in design theory concerns the existence of (n,k,λ) BIB designs. It is well known that the existence of a $(n,k,1)$ BIB design is equivalent to the existence of a decomposition of the complete graph K_n into K_k's. It is also well known that such a decomposition can exist only if $n(n-1) \equiv 0$ $(mod\ k(k-1))$ and $n-1 \equiv 0$ $(mod\ k-1)$. These conditions have been shown to be sufficient for any k and n large enough (Wilson [18]) and also for $k = 3$ (for a survey on Steiner triple systems, see Doyen and Rosa [7]), $k = 4$ and $k = 5$ (see Hanani [8]). When n does not meet the necessary conditions one can either try to find the maximum cardinality of a packing of K_k's into K_n (that is the maximum number of K_k's included in K_n), or the minimum cardinality of a covering of K_n with K_k's (that is the minimum number of edge disjoint K_k's necessary to cover all the edges of K_n at least once). Such values have been completely determined for $k = 3$ (see Hanani [8]) and for $k = 4$ (see Brouwer [5] and Mills [12,13]).

Here we are interested in a slightly different problem that we call regular packing or covering of K_n with K_k's. This problem was initially motivated by a problem of interconnection networks, called bus networks of diameter 1, modeled by hypergraphs. In such a network there are n processors connected to buses. Each bus contains the same number k of processors. For every pair of processors there must exist at least one bus containing them (diameter 1). These conditions correspond to the classical covering of K_n with K_k's. A further requirement (of regularity) is that each processor belongs to the same number of buses, which gives rise to the notion of regular covering (for details see Bermond, Bond and Sacle [3] and[2]).

A *regular packing* of K_n with K_k's is a set of edge disjoint subgraphs of K_n isomorphic to K_k such that the partial graph G, generated by the edges not covered, is regular of degree d_G. In the literature G is also called the *leave* of the packing (see [6]). In other words we want to find a regular graph G such that $K_n - G$ can be decomposed into K_k's. Note that a regular packing of K_n with K_k's is just a packing with K_k's such that each vertex of K_n belongs to the same number of K_k's. We shall say that a regular packing is *maximum* if d_G is as small as possible.

Similarly, a *regular covering* of K_n with K_k's is a set of subgraphs of K_n isomorphic to K_k which cover all the edges of K_n at least once such that the multigraph H, obtained by deleting K_m from the union of the K_k's, is regular of degree d_H. The multigraph H is also called the *excess* of the covering. In other words we want to find a regular multigraph H of degree d_H such that $K_n + H$ can be decomposed into K_k's. We shall say that a regular covering is *minimum* if d_H is as small as possible.

Furthermore we are interested in finding, when it is possible, a minimum regular covering of K_n with K_k's *containing* a maximum regular packing. That is equivalent to find whether there exist graphs G, d_G-regular, and H, d_H-regular, with d_G and d_H minimum, associated respectively to the packing and covering such that the (d_G+d_H)-regular graph $G + H$ can itself be decomposed into K_k's. In other words, for that purpose, it is sufficient to find a maximum regular packing such that the d_G-regular graph G can be covered by K_k's in such a way that every vertex belongs to $(d_G+d_H)/(k-1)$ graphs K_k's (obviously a covering of a subgraph G of K_n with K_k's is a set of K_k's containing all the edges of G).

The following proposition is immediate

Proposition 1.1

The degree d_G of the regular leave G of a regular packing of K_n with K_k's satisfies the following equalities:

$$n-1-d_G \equiv 0 \pmod{k-1}$$
$$n(n-1-d_G) \equiv 0 \pmod{k(k-1)}.$$

The degree d_H of the regular excess H of a regular covering of K_n with K_k's satisfies the following equalities:

$$n-1+d_H \equiv 0 \pmod{k-1}$$
$$n(n-1+d_H) \equiv 0 \pmod{k(k-1)}$$

From that proposition, one can easily deduce lower bounds on d_G and d_H .

Here we show that these bounds are always reached in the cases $k = 3$ or 4 by constructing maximum regular packings and minimum regular coverings of K_n with K_3's and K_4's for any n (except for n=8 in case of a packing with K_4's). In most of the cases the covering constructed contains the packing.

Let us remark that in some cases the optimum packings or coverings given in the literature ([5,13,14]) are necessarily regular (by counting arguments). However as we want, if possible, to give a covering containing the packing, we give the construction again.

2. Notation and basic lemmas

• K_n will denote the complete graph on n vertices. Usually $V(K_n)$, the vertex set of K_n, will be Z_n, the group of integers modulo n.
• $[x_1,x_2,\ldots,x_r]$ will denote the complete graph on the vertices $x_1,x_2,\ldots,x_r$. In case r=2 it denotes an edge.
• $[x_1,x_2,\ldots,x_r]$ $(mod\ n)$ will denote $\{[x_1+i,x_2+i,\ldots,x_r+i] : i=0,1,\ldots,n-1\}$.
• $K_n - G$ denotes the partial graph obtained from K_n by deleting the edges of G from K_n.
• If G and H are graphs on the same set of vertices V, $G+H$ denotes the multigraph with vertex set V containing the edges of G plus the edges of H (with eventually multiple edges).
• K_{r_1,r_2} denotes the complete bipartite graph with vertex sets of cardinality r_1 and r_2.
• $K_{r_1,r_2,\ldots,r_h}$ denotes the complete multipartite graph. If $r_1 = r_2 = \ldots = r_h = r$ then we denote $K_{r_1,r_2,\ldots,r_h}$ by $K_{h\times r}$.
• Let G be a graph on k vertices. A *parallel class* of G's on $n = pk$ vertices is the vertex disjoint union of p copies of G.

It is well known that the existence of a decomposition of $K_{h\times r}$ into K_h's is equivalent to the existence of $h-2$ orthogonal Latin squares of order r (or also to what is called a $GD(h,1,r;hr)$ or $T(1,r)$ design). Classical results are the following.

Lemma 2.1 (see Hanani [8])

$K_{h\times r}$ can be decomposed into K_3's if and only if $\begin{cases} (h-1)r \equiv 0 \ (mod\ 2) \\ hr^2(h-1) \equiv 0 \ (mod\ 6) \\ h \geq 3 \end{cases}$

Lemma 2.2 (see Brouwer, Hanani, Schrijver [4])
$K_{h\times r}$ can be decomposed into K_4's if and only if

$$\begin{cases} (h-1)r \equiv 0 \ (mod\ 3) \\ hr^2(h-1) \equiv 0 \ (mod\ 12) \\ h \geq 4, \ (h,r) \neq (4,2) \ or \ (4,6) \end{cases}$$

Lemma 2.3 (see Bermond, Huang, Rosa, Sotteau [1])

If $K_{r,r,r}$ and $K_{r,r,r,r}$ can be decomposed into graphs G, then $K_{pr,pr,pr,qr}$ can be decomposed into G for $p \neq 2, 6$, $0 \leq q \leq p$ (and if $q = 0$ for any p).

If $K_{r,r,r,r}$ and $K_{r,r,r,r,r}$ can be decomposed into graphs G , then $K_{pr,pr,pr,pr,qr}$ can be decomposed into G for $p \neq 2, 3, 6, 10$ and $0 \leq q \leq p$.

In [10] Huang, Mendelsohn and Rosa introduce a problem which can be formulated as that of the existence of a decomposition of K_n into graphs all isomorphic to K_k except one isomorphic to K_r. They give the necessary conditions.
The following results have been proved on this problem, which will be used in the proof of the main theorem.

Lemma 2.4 (Brouwer [5])

K_n can be decomposed into one K_7 and K_4's if and only if $n \equiv 7$ or $10 \pmod{12}$, $n \neq 10, 19$.

Lemma 2.5 (Bermond, Bond [3])

K_n can be decomposed into one K_{10} and K_4's if and only if $n \equiv 7$ or $10 \pmod{12}$, $n \neq 7, 19, 22$.

In the section 5 we study a more particular case of this problem, since we want to have the additional property that there exists, in the decomposition, a parallel class of K_k's on $n - r$ vertices .

3. Main results

Theorem 3.1

If $n \geq 3$ there exist a maximum regular packing and a minimum regular covering of K_n with K_3's where the graph G and multigraph H are as follows.

- **o)** if $n \equiv 0 \pmod 6$, *G is 1-regular and H is 1-regular*
- **i)** if $n \equiv 1 \pmod 6$, *G and H are empty*
- **ii)** if $n \equiv 2 \pmod 6$, *G is 1-regular and H is 5-regular*
- **iii)** if $n \equiv 3 \pmod 6$, *G and H are empty*
- **iv)** if $n \equiv 4 \pmod 6$, *G is 3-regular and H is 3-regular*
- **v)** if $n \equiv 5 \pmod 6$, *G is 4-regular and H is 2-regular*

Moreover there exists a minimum covering containing a maximum packing in all cases except case **o**)

Proof

First, from Proposition 1.1, the given values of the degrees of G and H are lower bounds. Now we will prove that these bound are reached.

Case o) ($n=6t$, $t \geq 1$) :
Since K_{6t+1} can be decomposed into K_3's, we get a packing of K_n by deleting a vertex and all the K_3's containing it. The leave of the packing is a perfect matching.

It is known (see Mills [14] for a survey on the subject) that K_n can be covered with $\left\lceil n/3 \left\lceil (n-1)/2 \right\rceil \right\rceil$ K_3's. In this case the covering is regular and the excess H is a perfect matching.

Obviously in this case it is not possible that the minimum covering contains the maximum packing since G and H are necessarily perfect matchings and the sum of two perfect matchings cannot be decomposed into K_3's.

Case i) ($n=6t+1$, $t \geq 1$) and **case iii)** ($n=6t+3$, $t \geq 0$) :
It is known that, in these cases, the graph K_n can be decomposed into K_3's.

Case ii) ($n=6t+2$, $t \geq 1$) :
The packing is obtained exactly as in **o)**, and the leave G is a perfect matching. A regular covering of G with each vertex in three K_3's is obtained by Corollary 4.2 applied to the 3-regular graph formed by the sum of G and any 2-regular graph on the same set of vertices.

Case iv) ($n=6t+4$, $t \geq 0$) :
We distinguish two subcases:

i) $t=2r$, $r \geq 0$
If $r=0$ then $n=4$, $G = K_4$ is itself 3-regular. We can take $H = K_4$ since the multigraph formed by two identical K_4's can be decomposed into K_3's , $[0,1,2]$ $(mod 3)$.
If $r \geq 1$, K_{12r+4} is the edge disjoint union of $3r+1$ vertex disjoint graphs K_4's and a $K_{(3r+1)\times 4}$. By Lemma 2.1 $K_{(3r+1)\times 4}$ can be decomposed into K_3's. Therefore the leave G consists of the $3r+1$ vertex disjoint K_4's. We shall take for the excess H the same graph. As the multigraph formed by two identical K_4's can be decomposed into K_3's, $G + H$ can also be decomposed into K_3's.

ii) $t=2r+1$, $r \geq 0$
If $r=0$ then $n=10$. A maximum regular packing consists of the following K_3's:

$$[0,3,9],\ [1,4,5],\ [2,0,6],\ [3,1,7],\ [4,2,8],$$
$$[0,7,8],\ [1,8,9],\ [2,9,5],\ [3,5,6],\ [4,6,7].$$

The leave G is the Petersen graph which is 3-regular. From Corollary 4.2 it can be covered with each vertex in three K_3's. So we have a covering of K_{10} containing the maximum packing where the excess H is a 3-regular multigraph.
If $r=1$ then $n=22$. The packing is given by the K_3's:

$$[0,1,3],\ [0,4,10] \text{ and } [0,5,13]\ (mod\ 22).$$

The leave G is 3-regular and consists of the edges $[0,7]$ and $[0,11]$ $(mod\ 22)$. It is the union of a hamiltonian cycle and a perfect matching so, by Corollary 4.2, it can be covered with each vertex in three K_3's. Thus we have a covering of K_{22} containing the packing with an excess H which is 3-regular.
If $r \geq 2$, K_{12r+10} can be decomposed into r graphs K_{12}'s on vertex sets X_i, $1 \leq i \leq r$, one K_{10}, on vertex set Y and a $K_{12,\ldots,12,10}$ on vertex set $\bigcup X_i \bigcup Y$.

From above the K_{10} can be decomposed into K_3's plus the 3-regular Petersen graph.

From case **o**) each K_{12} on vertex set X_i can be decomposed into K_3's plus a perfect matching M_i for $1 \leq i \leq r$.

Now $K_{12,\dots,12,10}$ can be obtained from the multipartite graph $K_{(r+1)\times 12}$ by deleting two vertices in the same part. As from Lemma 2.1, $K_{(r+1)\times 12}$ can be decomposed into K_3's, the graph $K_{12,\dots,12,10}$ can be decomposed into K_3's plus two perfect matchings M and M' on $\bigcup X_i$. Now we have a regular packing of K_{12r+10} where the leave G is the 3-regular graph on $12r+10$ vertices which consists of two parts: one is the Petersen graph on 10 vertices, the other one is on $12r$ vertices: $\{\bigcup M_i + M + M'\}$. According to Corollary 4.2, both parts can be covered by K_3's in such a way that each vertex belongs to three K_3's.

Thus we have a covering of K_{12r+10} containing the packing with an excess H which is 3-regular.

Case v) $(n=6t+5,\ t\geq 0)$:

If $t=0$ then $n=5$. We have $G=K_5$ and $H=C_5$, by covering G with the K_3's $[0,1,2]$ $(mod\ 5)$.

If $t=1$ then $n=11$. The packing is given by the K_3's $[0,1,3]$ $(mod\ 11)$. The leave G is the 4-regular formed by the edges $[0,4]$ and $[0,5]$ $(mod\ 11)$. We obtain the covering by adding the K_3's $[0,1,5]$ $(mod\ 11)$. Note that the excess H is a hamiltonian cycle.

If $t\geq 2$, K_{6t+5} is the edge disjoint union of t vertex disjoint K_6's, on vertex sets X_i, $1\leq i\leq t$, one K_5 on vertex set Y and a multipartite graph $K_{6,\dots,6,5}$. According to Lemma 2.1, there exists a decomposition of $K_{(t+1)\times 6}$ into K_3's. By deleting one vertex we obtain a decomposition of $K_{6,\dots,6,5}$ into K_3's and a perfect matching M on the vertex set $\bigcup X_i$.

Each K_6 on vertex set X_i is the edge disjoint union of two vertex disjoint K_3's and a 3-regular graph which consists of a perfect matching M_i and a cycle of length six C_6^i.

K_5 is itself a 4-regular graph on Y. So we have a maximum regular packing of K_{6t+5} with K_3's, where the leave G is the 4-regular graph which consists of two parts: one is $\{\bigcup M_i + M + \bigcup C_6^i\}$, on vertex set $\bigcup X_i$, the other one is a K_5, on vertex set Y. Both parts can be covered with each vertex in three K_3's, the K_5 as in case $n=5$ and the other one as shown in Corollary 4.3.

Thus we have a covering of K_{6t+5} containing the packing with an excess H which is 2-regular. □

Remark 3.2

The referee suggested to mention the following conjecture which seems very difficult.

Conjecture

For every given degree d, all d-regular graphs meeting the necessary conditions are leaves of a packing of K_n with K_3's, with finitely many exceptions.

Note that the case of hamiltonian cycles was solved by E. Mendelsohn [11] and the general case of any 2-regular graph is a particular case of a result of Colbourn and Rosa [6] also obtained by Hilton and Rodgers [9].

Note that the above conjecture is in fact a special case of the following one of Nash-Williams [15].

Conjecture (Nash-Williams)

If G is a graph of order n, $n \geq 15$, such that its number of edges is a multiple of 3, and every vertex has an even degree at least $\frac{3n}{4}$, then G can be decomposed into K_3's.

Remark 3.3

K. Heinrich drew our attention to the following strong recent result of Rees [17] which can be restated as follows:

Theorem (Rees)

K_{6t} can be decomposed into α perfect matchings and β parallel classes of K_3's if and only if $\alpha+2\beta=6t-1$ except for $\alpha=1$ and $t=1$ or $t=2$ (corresponding to the non-existence of Nearly Kirkman triple systems of order 6 or 12).

Using this theorem we can give a shorter proof of the existence of maximum regular packings for the non immediate cases $n=6t+u$ $(u=4,5)$ in the following way. Let $V(K_n) = X \bigcup Y$ with $|X|=6t$, $|Y|=u$. By the theorem above the K_{6t} on X can be decomposed into $2u-1$ perfect matchings on X and $3t-u$ parallel classes of K_3's. Then K_n can be decomposed into the K_u on Y, $u-1$ perfect matchings on X and K_3's formed by the parallel classes of K_3's on X and the $3tu$ triples obtained by joining any vertex of Y to a perfect matching of X. Thus we have a regular packing of K_n with K_3's for which the leave G consists of the K_u on Y plus the $u-1$ perfect matchings on X and is clearly a regular graph of degree $u-1$.

In fact we used only a corollary of the result of Rees. We needed only to decompose K_{6t} into perfect matchings and triangles (Note that such a decomposition solves the conjecture mentioned in Remark 3.2, when the graph is the union of d perfect matchings).

Theorem 3.4

If $n \geq 4$, there exist a maximum regular packing and a minimum regular covering of K_n with K_4's where the leave G and excess H are as follows.

o) if $n \equiv 0$ *(mod 12), G is 2-regular and H is 1-regular*
i) if $n \equiv 1$ *(mod 12), G and H are empty*
ii) if $n \equiv 2$ *(mod 12), G is 1-regular and H is 5-regular*
iii) if $n \equiv 3$ *(mod 12), G is 2-regular and H is 10-regular*
iv) if $n \equiv 4$ *(mod 12), G and H are empty*
v) if $n \equiv 5$ *(mod 12), G is 4-regular and H is 8-regular*
vi) if $n \equiv 6$ *(mod 12), G is 5-regular and H is 1-regular*
vii) if $n \equiv 7$ *(mod 12), G is 6-regular and H is 6-regular*
viii) if $n \equiv 8$ *(mod 12), G is 1-regular and H is 2-regular*
ix) if $n \equiv 9$ *(mod 12), G is 8-regular and H is 4-regular*
x) if $n \equiv 10$ *(mod 12), G is 3-regular and H is 3-regular*
xi) if $n \equiv 11$ *(mod 12), G is 10-regular and H is 2-regular*

except for n=8 where G is 4-regular.

Moreover there exists a minimum regular covering which contains a maximum regular packing in all cases except **o**) *and* **xi**) *and* n=18 (where we don't know whether it exists).

Proof:

First, from Proposition 1.1, the given values of the degrees of G and H are lower bounds (except for n=8). Now we will prove that these bounds are reached.

case o) (n=12t, $t \geq 1$) :
We obtain a decomposition of K_{12t} into K_4's and a parallel class of K_3's by deleting one vertex and the edges containing it from a decomposition of K_{12t+1} into K_4's. Thus we have a maximum regular packing with a leave G which is 2-regular.
A union of K_3's cannot be covered with K_4's with each vertex in only one K_4. In order to be able to find a minimum covering containing the minimum packing we would need to have a decomposition of K_{12t} into K_4's and a parallel class of cycles of length four. Such a decomposition can not exist for n=12. Maybe it does for bigger n,and we can state the following problem.

Problem : *Does there exists a decomposition of K_{12t} into K_4's and a parallel class of C_4's for $t \geq 2$?*

However it is known that there exists a maximum regular packing of K_{12t} with K_4's where the excess H is a perfect matching. Indeed it has been proved in [12] that there exists a covering of K_{12t} with $12t^2$ graphs K_4's. Since each vertex belongs to $4t$ K_4's, the excess H is a perfect matching and the covering is regular.

case i) (n=12t+1, $t \geq 1$) and **case iv)** (n=12t+4, $t \geq 0$) :
It is known that, in these cases, the graph K_n can be decomposed into K_4's.

case ii) (n=12t+2, $t \geq 1$) :
K_{12t+2} is the edge disjoint union of a perfect matching M and a multipartite graph $K_{(6t+1)\times 2}$. From Lemma 2.2, $K_{(6t+1)\times 2}$ can be decomposed into K_4's. Let us take

$G=M=\{[2i,2i+1], 0\leq i\leq 6t\}$. We have a covering of G with each vertex in two K_4's by taking $\{ [2i,2i+1,2i+2,2i+3], 0\leq i\leq 6t \}$.
So we have a covering of K_{12t+2} containing the packing where the excess H is a 5-regular multigraph.

case iii) $(n=12t+3, t\geq 1)$:
K_{12t+3} is the edge disjoint union of a parallel class of K_3's and a multipartite graph $K_{(4t+1)\times 3}$. By Lemma 2.2, $K_{(4t+1)\times 3}$ can be decomposed into K_4's. Let us take for G the parallel class of K_3's which is a 2-regular graph. By Corollary 4.4 in section 4, we know that any 6-regular graph that can be decomposed into K_3's can be covered by K_4's with each vertex in four K_4's . We apply this result with the 6-regular graph formed by G and any two other parallel classes of K_3's.
So we have a covering of K_{12t+3} containing the packing where the excess H is a 10-regular multigraph.

case v) $(n=12t+5, t\geq 0)$:
First we prove the result for $n=5$ and 17.
If $n=5$ then $G=K_5$ and H is the 8-regular multigraph formed by two copies of K_5. Indeed, the graph $G+H$ can be decomposed into K_4's by taking $[0,1,2,4]$ $(mod\ 5)$.
If $n=17$, the packing is given by the following K_4's: $[0,1,4,6]$ $(mod\ 17)$. The leave G is formed by the edges $[0,7]$ and $[0,8]$ $(mod\ 17)$. It is the union of two hamiltonian cycles and can be covered with each vertex in four K_4's by $[0,2,5,9]$ $(mod\ 17)$.
Now let $n=12t+5$, with $t\geq 2$. By Lemma 2.4, K_{12t+7} can be decomposed into one K_7 and K_4's, if $t\neq 1$. We take such a decomposition and delete two vertices of the K_7. We obtain a decomposition of K_{12t+5} into a K_5, two parallel classes of K_3's (on the $12t$ other vertices) and K_4's. So the leave G we obtain consists of two parts, one is a K_5 and the other one is a 4-regular graph on $12t$ vertices, which is the union of two parallel classes of K_3's.
The K_5 can be covered as in the case $n=5$. By Corollary 4.4 in section 4, any 6-regular graph that can be decomposed into K_3's can be covered by K_4's, with each vertex in four K_4's . We apply this result to the 6-regular graph formed by the part of G on $12t$ vertices and any other parallel class of K_3's on the same set of vertices.
So we have a covering of K_{12t+5} containing the packing where the excess H is a 8-regular multigraph.

case vi) $(n=12t+6, t\geq 0)$:
First we prove the result for $n=6$ and 18.
If $n=6$ then $G=K_6$ is itself 5-regular and H is a perfect matching. The covering is given by the K_4's: $[0,1,2,3]$, $[0,1,4,5]$, $[2,3,4,5]$.
If $n=18$, then a regular packing can be obtained by taking $[0,1,3,8]$ $(mod\ 18)$. An optimal covering (which is regular in that case) is given in [12] by Mills. We note that this covering does not contain any maximum regular packing and we don't know if such a covering does exist.
Now let $n=12t+6$, with $t\geq 2$. The graph K_{12t+6} is the edge disjoint union of a parallel class of K_6 and a multipartite graph $K_{(2t+1)\times 6}$. By Lemma 2.2, $K_{(2t+1)\times 6}$ can be decomposed into K_4's. Therefore we obtain a leave G which is a parallel class of K_6's. We cover each K_6 of G as in the case $n=6$ and the excess H is a perfect matching.

case vii) ($n=12t+7$, $t\geq 0$) :
First we prove the result for the cases $n=7$, 19.
If $n=7$ then $G=K_7$ is itself a 6-regular graph. We can take $H=K_7$ since the multigraph formed by two copies of the same K_7 can be decomposed into K_4's as follows: [0,1,2,4] (*mod* 7).
If $n=19$, a maximum packing is given by [0,1,3,8] (*mod* 19) and the leave is a 6-regular graph G which is an edge disjoint union of K_3's : [0,4,10] (*mod* 19). So by Corollary 4.4, G can be covered by K_4's with each vertex in four K_4's.
Now let $n=12t+7$, with $t\geq 2$. By Lemma 2.5, K_{12t+10} can be decomposed into one K_{10} and K_4's, if $t\neq 1$. We take such a decomposition and delete three vertices of the K_{10}. We obtain a decomposition of K_{12t+7} into a K_7, three parallel classes of K_3's (on the other $12t$ vertices) and K_4's. So the leave G we obtain consists of the vertex disjoint union of a K_7 and a 6-regular graph on $12t$ vertices, which can be decomposed into K_3's.
The K_7 can be covered as in the case $n=7$. By Corollary 4.4 in section 4, the part of G which is 6-regular on $12t$ vertices can be covered by K_4's, with each vertex in four K_4's .
So we have a covering of K_{12t+7} containing the packing where the excess H is a 6-regular multigraph.

case viii) ($n=12t+8$, $t\geq 0$) :
First we prove the result for the case $n=8$.
If $n=8$ then there exists no regular packing of K_8 with K_4's with a leave G 1-regular. Indeed, if such a packing exists $K_8 - G$ should be decomposed into four K_4's with each vertex in two K_4's which is easily seen to be impossible. A maximum regular packing of K_8 is obtained with a 4-regular leave G by taking two disjoint K_4's.
A minimum regular covering of K_8 with a 2-regular excess H is given by the following K_4's (found in [13]):

[1,2,3,4],[1,2,5,6],[1,2,7,8],[3,4,5,6],[3,4,7,8],[5,6,7,8]

This covering contains the maximum regular packing since it contains two vertex disjoint K_4's.
Now let $n=12t+8$, with $t\geq 1$. The graph K_{12t+8} is the disjoint union of a perfect matching and a multipartite graph $K_{(6t+4)\times 2}$. From Lemma 2.2, if $t\geq 1$, $K_{(6t+4)\times 2}$ can be decomposed into K_4's. Let G be the matching $\{[2i,2i+1] : i=0,1,...,6t+3\}$. It can be covered with each vertex in one K_4 with the following K_4's : $\{[4i, 4i+1, 4i+2, 4i+3] : i=0,1,...,3t+1\}$. Note that the excess H is the vertex disjoint union of cycles of length 4.

case ix) ($n=12t+9$, $t\geq 0$) :
It is well known that K_{12t+13} can be decomposed into K_4's. If we choose a K_4 and delete its four vertices, we get a decomposition of K_{12t+9} into four parallel classes of K_3's and K_4's. So the leave G is the 8-regular graph formed by four parallel classes of K_3's.
From Corollary 4.5, we have a covering of G with each vertex in four K_4's.
So we have a covering of K_{12t+9} containing the packing where the excess H is a 4-regular multigraph.

case x) $(n=12t+10,\ t\geq 0)$:
If $t=0$ then $n=10$. The packing consists of the following K_4's:

$$[0,2,8,9],\ [1,3,9,5],\ [2,4,5,6],\ [3,0,6,7],\ [4,1,7,8]$$

The leave G is the Petersen graph which is a 3-regular graph.
In order to cover G with each vertex in two K_4's, we add the following K_4's:

$$[0,1,5,7],\ [1,2,6,8],\ [2,3,7,9],\ [3,4,8,5],\ [4,0,9,6]$$

If $t=1$ then $n=22$. Let $V(K_{22}) = Z_2 \times Z_{11}$. A maximum regular packing is given by the following K_4's:

$$\{[(0,i),(0,i+1),(0,i+3),(1,i)],[(0,i+9),(1,i),(1,i+1),(1,i+5)],$$
$$[(0,i),(0,i+5),(1,i+6),(1,i+9)] : 0\leq i\leq 10\}.$$

The leave G is the union of two vertex disjoint cycles of length 11 and a perfect matching joining them. It can be covered by the following K_4's :

$$\{\ [(0,i),(0,i+4),(1,i+5),(1,i+7)] : 0 \leq i \leq 10\ \}$$

where each vertex is in two K_4's.
If $t \geq 2$, the graph K_{12t+10} is the edge disjoint union of one K_{10}, $3t$ vertex disjoint K_4's and a multipartite graph $K_{4,4,\ldots 4,10}$.
As $t\geq 2$, by Theorem 5.1, which will be proved in section 5, there exists a decomposition of the multipartite graph $K_{4,4,\ldots,4,10}$ into K_4's since the number of 4-sets is equal to $3t$. From the case $n=10$, the graph K_{10} is the edge disjoint union of K_4's and a Petersen graph. So the leave G is the vertex disjoint union of a Petersen graph and a parallel class of K_4's.
In order to obtain a minimum regular covering (with an excess H 3-regular) containing the packing given above we add two parallel classes of K_4's on $12t$ points and the K_4's we used to cover the Petersen graph in K_{10}.

case xi) $(n=12t+11\ t\geq 0)$:
First we prove the result for $t=0,1,2$.
If $t=0$ then $G=K_{11}$ is itself 10-regular. A minimum covering is given by $[0,1,4,6]$ (*mod* 11). Note that the excess H is a hamiltonian cycle.
If $t=1$ then $n=23$. A maximum regular packing with G 10-regular is obtained by taking the following K_4's : $[0,1,7,21]$ (*mod* 23) . In order to cover G with each vertex in four K_4's we add the following ones : $[0,1,11,19]$ (*mod* 23). Note that the excess H is again a hamiltonian cycle.
If $t=2$ then $n=35$. A maximum regular packing is given by : $[0,1,4,14]$ and $[0,7,9,15]$ (*mod* 35). A covering containing it is obtained by adding the following K_4's : $[0,1,12,17]$ (*mod* 35). Note that the excess H is a hamiltonian cycle.
Now let $n=12t+11$, with $t\geq 3$, the graph K_{12t+11} is the edge disjoint union of t vertex disjoint K_{12}'s, one K_{11} and the multipartite graph $K_{12,12,\ldots,12,11}$. From Lemma 2.2, the graph $K_{(t+1)\times 12}$ can be decomposed into K_4's for $t\geq 3$. Therefore $K_{12,12,\ldots,12,11}$ can be decomposed into one parallel class of K_3's on $12t$ vertices and K_4's. Each K_{12} is the union of three vertex disjoint K_4's and an 8-regular graph. Let us take for leave G the 10-regular graph which consists of two parts, one is formed on $12t$ vertices by these 8-regular graphs plus the parallel class of K_3's, the other one is the K_{11}. Thus we have a maximum regular packing.
The existence of a minimum regular covering has been proved in [13] where it is proved that there exists a covering with $(t+1)(12t+11)$ K_4's with each vertex in $4t+4$

graphs K_4's. Therefore the excess H is 2-regular and the covering is regular minimum. Notice that in this last case we don't know whether there exists a minimum regular covering which contains a minimum regular packing. □

Remark 3.5

The proof of the theorem suggests a lot of decomposition problems. The most general one would be a problem analogous to the conjecture of remark 3.2, i.e. that all d-regular graphs meeting the necessary conditions are leaves (respectively excesses) of a regular packing (respectively covering) of K_n with K_4's. First interesting cases are when the leaves or excesses are hamiltonian cycles or more generally regular graphs of degree two, which generalize Mendelsohn's [11] or Colbourn and Rosa's [6] result. For example we state the following conjectures.

Conjectures

K_{12t} *minus a hamiltonian cycle*
K_{12t+3} *minus a hamiltonian cycle*
K_{12t+8} *plus a hamiltonian cycle* *can be decomposed into* K_4*'s*
K_{12t+11} *plus a hamiltonian cycle*

One can also ask for a generalization of Rees' theorem.

Conjecture

K_{12t} *can be decomposed into* α *parallel classes of* K_3*'s and* β *parallel classes of* K_4*'s if and only if* $2\alpha+3\beta=12t-1$ *(with a finite number of exceptions).*

If we don't want parallel classes of K_4's we obtain the problem introduced by Huang, Mendelsohn and Rosa [10]. A related problem is considered in section 5.

Note that the problem of decomposing K_n into α perfect matchings and β parallel classes of K_4's for $n\equiv 0$ $(mod\ 4)$ is easy to solve (essentially because a parallel class of K_4's is the union of 3 perfect matchings).

4. Some lemmas

We state here some lemmas used in the proofs of the main theorem. The results are not necessarily best possible, and they are given in the form used in section 3.

Lemma 4.1 :

i) *Let G be an $r(r-1)$-regular graph on n vertices that can be decomposed into K_r's. Then G can be covered with K_{r+1}'s, in such a way that each vertex belongs to $r+1$ graphs K_{r+1}'s.*

ii) *Let G be an $r(r-1)+1$-regular graph on n vertices that can be decomposed into K_r's and a perfect matching. Then G can be covered with K_{r+1}'s, in such a way that each vertex belongs to $r+1$ graphs K_{r+1}'s.*

iii) *Let G be an $r(r-1)+2$-regular graph on n vertices that can be decomposed into K_r's and a 2-regular graph. Then G can be covered with K_{r+1}'s, in such a way that each vertex belongs to $r+1$ graphs K_{r+1}'s.*

Proof

First we remark that in case **iii**) the 2-regular graph is the vertex disjoint union of cycles, and one can give an orientation to the cycles so that every vertex is the origin of only one arc. Any vertex of G is contained in r graphs K_r's of the decomposition , so the total number of K_r's is n. Let us add one vertex of G to each of the K_r's in order to transform it into a K_{r+1}, with each vertex of G added exactly once. Obviously after that transformation we have n graphs K_{r+1}'s, with each vertex of G belonging to $r+1$ of them. In order to do that and to have all the edges of G covered we need a bijection f between $V(G)$ and the set of the n K_r's, such that for any vertex x of G:

- in case **i**): x does not belong to $f(x)$.
- in case **ii**): if $[x,y]$ is an edge of the matching, then $y \in f(x)$.
- in case **iii**): if $[x,y]$ is the arc having x as origin in the 2-regular oriented partial graph of G, then $y \in f(x)$.

In order to show that such a bijection exists we define a bipartite graph, having as stable sets of vertices the K_r's and the vertices of G. We put an edge between a vertex of G and a K_r if the vertex can be added to the K_r. The result is a regular bipartite graph (of degree $n-r$, r and r respectively in cases **i**), **ii**) and **iii**)). A well-known corollary of the König-Hall theorem states that a regular bipartite graph admits a perfect matching. The matching defines the bijection we need. □

We state now some corollaries of this lemma, which are used in section 3, in these terms.

Corollary 4.2 :

Let G be a 3-regular graph containing a perfect matching. Then G can be covered by K_3's, in such a way that each vertex belongs to three K_3's.

Corollary 4.3 :

Let G be a 4-regular graph containing two disjoint perfect matchings. Then G can be covered by K_3's, in such a way that each vertex belongs to three K_3's.

Corollary 4.4 :

Let G be a 6-regular graph that can be decomposed into K_3's. Then G can be covered by K_4's, in such a way that each vertex belongs to four K_4's.

Corollary 4.5 :

Let G be an 8-regular graph that can be decomposed into K_3's and a 2-regular graph. Then G can be covered by K_4's, in such a way that each vertex belongs to four K_4's.

5. A multipartite graph decomposition problem

We propose to study the following problem:

Problem

For what values of h and r can the multipartite graph $K_{4,4,\ldots,4,r}$ on $4h+r$ vertices be decomposed into K_4's ?

If the multipartite graph $K_{4,4,\ldots,4,r}$ on $4h+r$ vertices can be decomposed into K_4's then necessarily

$$\begin{cases} h \equiv 0 \ (mod\ 3) \\ r \equiv 1 \ (mod\ 3) \\ h \geq r/2 + 1 \end{cases}$$

We obtain the first two necessary conditions by divisibility arguments and the third one by counting the edges covered by the K_4's containing an element of the r-set.

We can remark that Lemma 2.2 proves that the above necessary conditions are sufficient for the case $r = 4$ (this case can also be seen as a corollary of the well known theorem of Hanani [8], which says that K_n can be decomposed into parallel classes of K_4's if and only if $n \equiv 4 \ (mod\ 12)$).

Here we will prove that the above necessary conditions are sufficient for the case $r = 10$.

We conjecture that the necessary conditions are sufficient in general, except eventually for a few values (for example it can be shown that $K_{4,4,4,1}$ cannot be decomposed into K_4's).

Theorem 5.1

$K_{4,4,\ldots,4,10}$ can be decomposed into K_4's if and only if h, the number of 4-vertex parts, satisfies

$$\begin{cases} h \equiv 0 \ (mod\ 3) \\ h \geq 6 \end{cases}$$

or equivalently

K_n can be decomposed into one K_{10}, a parallel class of K_4's on the other $n-10$ vertices and K_4's if and only if $n \equiv 10 \ (mod\ 12)$, $n \neq 22$. or also,

K_n can be decomposed into a parallel class of K_4's, 10 parallel classes of K_3's and K_4's if and only if $n \equiv 0 \ (mod\ 12)$, $n \neq 12$.

In what follows we will use the three alternative forms.
One can easily see, on the third formulation of the theorem, that it is a refinement of Lemma 2.5.

To prove the theorem we will need some composition lemmas, for which we will use the following remark:

Remark 5.2

$K_{t,t,t,t}$ can be decomposed into parallel classes of K_4's, for $t \neq 2,3,6,10$.

Proof

It is well known that $K_{t,t,t,t,t}$ can be decomposed into K_5's if and only if $t \neq 2,3,6,10$ (see [8]). Given such a decomposition we can delete the five vertices of a stable set of $K_{t,t,t,t,t}$. Each deleted vertex gives rise to a parallel class of K_4's in $K_{t,t,t,t}$. □

We have the following composition lemmas.

Lemma 5.3

Let $t \geq u$, if K_{12t} and K_{12u} can be decomposed into one parallel class of K_4's, 10 parallel classes of K_3's and K_4's then so do K_{48t} and $K_{48t+12u}$.

Proof

The graph $K_{48t+12u}$ is the edge disjoint union of four K_{12t}'s, one K_{12u} and the multipartite graph $K_{12t,12t,12t,12t,12u}$. The K_{12t}'s and the K_{12u} can be decomposed into 10 parallel classes of K_3, one parallel class of K_4 and K_4's by hypothesis. By Lemma 2.2, $K_{4\times 12}$ and $K_{5\times 12}$ can be decomposed into K_4's. Therefore, according to Lemma 2.3, $K_{12t,12t,12t,12t,12u}$ can be decomposed into K_4's for $u \leq t$. Therefore K_{48t+36} can be decomposed into 10 parallel classes of K_3's , one parallel class of K_4's , and K_4's. □

Lemma 5.4

If K_{12t+9} can be decomposed into 10 parallel classes of K_3's and K_4's then K_{48t+36} can be decomposed into a parallel class of K_4's, 10 parallel classes of K_3's and K_4's.

Proof

K_{48t+36} is the disjoint union of four K_{12t+9} and the multipartite graph $K_{12t+9,12t+9,12t+9,12t+9}$. By hypothesis the K_{12t+9} can be decomposed into 10 parallel classes of K_3's and K_4's. As $12t + 9 \neq 2,3,6,10$, by applying Remark 5.2, there exists a decomposition of $K_{12t+9,12t+9,12t+9,12t+9}$ into parallel classes of K_4's. □

Lemma 5.5

If K_{12t+3} can be decomposed into 7 parallel classes of K_3's and K_4's then K_{48t+12} can be decomposed into a parallel class of K_4's, 10 parallel classes of K_3's and K_4's.

Proof

Let $X = X_1 \bigcup X_2 \bigcup X_3 \bigcup X_4$ with $|X_i| = 12t+4$. By Remark 5.2 the multipartite graph $K_{12t+4,12t+4,12t+4,12t+4}$ on vertex set X can be decomposed into a parallel class of K_4's and K_4's as $12t+4 \neq 2,3,6,10$. Choose a particular K_4, $\{x_1,x_2,x_3,x_4\}$ with $x_i \in X_i$. Then the edges of the K_{48t+12} constructed on $X - \{x_1, x_2, x_3, x_4\}$ can be partitioned into the four K_{12t+3} on vertex set $X_i - x_i$, each decomposable into 7 parallel classes of K_3's and K_4's by hypothesis, the K_4's of $K_{12t+4,12t+4,12t+4,12t+4}$ not containing x_i for any i and the K_3's obtained from the K_4's containing one of the vertices x_i when deleting it. The K_3's of this decomposition can be partitioned into 10 parallel classes of K_3's on $48t+12$ vertices in the following way:

6 parallel classes, each of them being the union of parallel classes on each $X_i - x_i$,
4 parallel classes obtained for $i = 1, 2, 3, 4$ by taking, for each i, the parallel class still unused on $X_i - x_i$ and by adding the parallel class of K_3's on $\bigcup_{j \neq i} (X_j - x_j)$ obtained from the K_4's containing x_i after deletion of x_i. □

Proof of Theorem 5.1.

First the given conditions are necessary as we saw before. To prove that they are sufficient, by using Lemma 5.3, we only have to prove that K_n can be decomposed into a parallel class of K_4's, 10 parallel classes of K_3's and K_4's, for $n=12t$ with $t = 2,3,4,5,6,7,9,11,13,17,21$ in order to finish the proof. From Lemma 5.4 and Lemma 2.5 we get the result with $t = 7,11$. From Lemma 5.5 and Lemma 2.4 we get the result with $t = 5,9,13,17,21$.

In what follows we will give the direct constructions for the remaining cases ($t = 2,3,4,6$). □

The case $t = 2$:

For $n = 24$, the decomposition of K_{24} into a parallel class of K_4's and 10 parallel classes of K_3's is as follows. Let $V(K_{24}) = Z_6 \times Z_4$; the vertices are labeled (i,j) with $i = 0, 1, \ldots, 5$ and $j = 0, 1, 2, 3$. The decomposition is given below:

Classes of K_3's:		
$[(0,j),(1,j),(2,j)]$	$[(3,j),(4,j+1),(5,j)]$	for $j = 0,1,2,3$
$[(0,j),(1,j+1),(3,j)]$	$[(2,j),(4,j),(5,j+1)]$	for $j = 0,1,2,3$
$[(0,j),(1,j+2),(4,j)]$	$[(2,j),(3,j+2),(5,j)]$	for $j = 0,1,2,3$
$[(0,j),(1,j+3),(5,j)]$	$[(2,j),(3,j+3),(4,j+2)]$	for $j = 0,1,2,3$
$[(0,j),(2,j+1),(3,j+1)]$	$[(1,j),(4,j+1),(5,j+3)]$	for $j = 0,1,2,3$
$[(0,j),(2,j+2),(4,j+1)]$	$[(1,j),(3,j+1),(5,j+2)]$	for $j = 0,1,2,3$
$[(0,j),(2,j+3),(5,j+1)]$	$[(1,j),(3,j+2),(4,j)]$	for $j = 0,1,2,3$
$[(0,j),(3,j+2),(4,j+2)]$	$[(1,j),(2,j+1),(5,j)]$	for $j = 0,1,2,3$
$[(0,j),(3,j+3),(5,j+2)]$	$[(1,j),(2,j+2),(4,j+3)]$	for $j = 0,1,2,3$
$[(0,j),(4,j+3),(5,j+3)]$	$[(1,j),(2,j+3),(3,j)]$	for $j = 0,1,2,3$
Class of K_4's:		
$[(i,0),(i,1),(i,2),(i,3)]$		for $i = 0,...,5$

The case $t = 3$:

For $n = 36$ the decomposition of K_{36} into 10 parallel classes of K_3's and 5 parallel classes of K_4's is as follows. Let $V(K_{36}) = Z_4 \times Z_9$; the vertices are labeled (i,j) with $i = 0, 1, 2, 3$ and $j = 0, 1,..., 8$. The decomposition uses the known decomposition of K_9 into 4 parallel classes of K_3's ([16]). We form 8 parallel classes in the following way: for each $t = 0, 1, 2, 3$ we take 2 parallel classes on the 27 vertices (i,j) , $i \neq t$ with two parallel classes of the K_9 constructed on the vertices (t,j). The last two parallel classes are formed with the two unused classes on each K_9.

Classes of K_3's:		
$[(0,j),(1,j+7),(2,j+3)]$	for $j = 0,...,8$	and a class from $(3,j)$
$[(0,j),(1,j+8),(2,j+6)]$	for $j = 0,...,8$	and a class from $(3,j)$
$[(0,j),(1,j+4),(3,j+2)]$	for $j = 0,...,8$	and a class from $(2,j)$
$[(0,j),(1,j+6),(3,j+7)]$	for $j = 0,...,8$	and a class from $(2,j)$
$[(0,j),(2,j+4),(3,j+1)]$	for $j = 0,...,8$	and a class from $(1,j)$
$[(0,j),(2,j+7),(3,j+6)]$	for $j = 0,...,8$	and a class from $(1,j)$
$[(1,j),(2,j+4),(3,j+6)]$	for $j = 0,...,8$	and a class from $(0,j)$
$[(1,j),(2,j+6),(3,j+4)]$	for $j = 0,...,8$	and a class from $(0,j)$
a class from (i,j)	for $i = 0,1,2,3$	
a class from (i,j)	for $i = 0,1,2,3$	
Classes of K_4's:		
$[(0,j),(1,j),(2,j),(3,j)]$	for $j = 0,...,8$	
$[(0,j),(1,j+1),(2,j+2),(3,j+3)]$	for $j = 0,...,8$	
$[(0,j),(1,j+2),(2,j+1),(3,j+5)]$	for $j = 0,...,8$	
$[(0,j),(1,j+3),(2,j+5),(3,j+8)]$	for $j = 0,...,8$	
$[(0,j),(1,j+5),(2,j+8),(3,j+4)]$	for $j = 0,...,8$	

The case $t = 4$:

For $n = 48$ the decomposition of K_{48} into 10 parallel classes of K_3's, a parallel class of K_4's and K_4's is as follows. Let $V(K_{48}) = Z_{48}$.

Classes of K_3's:		Class of K_4's:	
$[i,i+16,i+32]$	for $i = 0,...,15$	$[i,i+12,i+24,i+36]$	for $i = 0,...,11$
$[i,i+1,i+5]$	for $i \equiv 0 \ (mod\ 3)$		
$[i,i+1,i+5]$	for $i \equiv 1 \ (mod\ 3)$		
$[i,i+1,i+5]$	for $i \equiv 2 \ (mod\ 3)$	K_4's:	
$[i,i+2,i+13]$	for $i \equiv 0 \ (mod\ 3)$	$[i,i+3,i+20,i+30]$	for $i = 0,...,47$
$[i,i+2,i+13]$	for $i \equiv 1 \ (mod\ 3)$	$[i,i+6,i+14,i+39]$	for $i = 0,...,47$
$[i,i+2,i+13]$	for $i \equiv 2 \ (mod\ 3)$		
$[i,i+7,i+26]$	for $i \equiv 0 \ (mod\ 3)$		
$[i,i+7,i+26]$	for $i \equiv 1 \ (mod\ 3)$		
$[i,i+7,i+26]$	for $i \equiv 2 \ (mod\ 3)$		

The case $t = 6$:

For $v = 72$ the decomposition of K_{72} into 10 parallel classes of K_3's, a parallel class of K_4's and K_4's is as follows. Let $V(K_{72}) = Z_{72}$.

Classes of K_3's:		Class of K_4's:	
$[i,i+24,i+48]$	for $i = 0,...,23$	$[i,i+18,i+36,i+54]$	for $i = 0,...,17$
$[i,i+1,i+5]$	for $i \equiv 0 \ (mod\ 3)$		
$[i,i+1,i+5]$	for $i \equiv 1 \ (mod\ 3)$		
$[i,i+1,i+5]$	for $i \equiv 2 \ (mod\ 3)$	K_4's:	
$[i,i+2,i+10]$	for $i \equiv 0 \ (mod\ 3)$	$[i,i+3,i+12,i+41]$	for $i = 0,...,71$
$[i,i+2,i+10]$	for $i \equiv 1 \ (mod\ 3)$	$[i,i+11,i+26,i+51]$	for $i = 0,...,71$
$[i,i+2,i+10]$	for $i \equiv 2 \ (mod\ 3)$	$[i,i+14,i+30,i+49]$	for $i = 0,...,71$
$[i,i+7,i+20]$	for $i \equiv 0 \ (mod\ 3)$	$[i,i+6,i+33,i+50]$	for $i = 0,...,71$
$[i,i+7,i+20]$	for $i \equiv 1 \ (mod\ 3)$		
$[i,i+7,i+20]$	for $i \equiv 2 \ (mod\ 3)$		

Acknowledgement

This research was partially supported by P.R.C. Math. Info.

References

1. J-C. Bermond, C. Huang, A. Rosa, and D. Sotteau, Decomposition of complete graphs into isomorphic subgraphs with five vertices, *Ars Combinat.* **10** pp. 211-254 (1980).

2. J-C. Bermond, J. Bond, and J-F. Sacle, Large hypergraphs of diameter one, *in Graph Theory and Combinatorics, Proc. Coll. Cambridge, 1983,* pp. 19-28 (1984).

3. J-C. Bermond and J. Bond, Combinatorial designs and hypergraphs of diameter one, in *Proc. First China-USA Conf. on Graph Theory, Jinan,June 1986,* (1987).

4. A.E. Brouwer, H. Hanani, and A. Schrijver, Group divisible designs with block size four, *Discrete Math.* **20** pp. 1-10 (1977).

5. A.E. Brouwer, Optimal packings of K_4's into a K_n, *Journal of Comb. Th., ser. A* **26** pp. 278-297 (1979).

6. C.J. Colbourn and A. Rosa, Quadratic Leaves of Maximal Partial Triple Systems, *Graphs and Combinatorics 2,* pp. 317-337 (1986).

7. J. Doyen and A. Rosa, An updated bibliography and survey of Steiner systems, pp. 317-349 in *Topics on Steiner systems, C.C. Lindner and A. Rosa ed., Annals of Discrete Math., 7,* (1980).

8. H. Hanani, Balanced incomplete block designs and related designs, *Discrete Math.* **11** pp. 255-369 (1975).

9. A.J.W. Hilton and C.A. Rodger, Triangulating nearly complete graphs of odd order, *in preparation,* (1986).

10. C. Huang, E. Mendelsohn, and A. Rosa, On partially resolvable t-partitions, pp. 169-183 in *Theory and Practice of Combinatorics, A. Rosa, G. Sabidussi and J. Turgeon ed., Annals of Discrete Math., 12,* (1982).

11. E. Mendelsohn, On (Near)-Hamiltonian Triple Systems and related one-factorizations of complete graphs, *Technion Report MT-655,* (1985).

12. W.H. Mills, On the covering of pairs by quadruples. I, *Journal of Comb. Th., ser. A* **13** pp. 55-78 (1972).

13. W.H. Mills, On the covering of pairs by quadruples. II, *Journal of Comb. Th., ser. A* **15** pp. 138-166 (1973).

14. W.H. Mills, Covering designs I: Coverings by a small number of subsets, *Ars Combinat.* **8** pp. 199-315 (1979).

15. C.St.J.A. Nash-Williams, An unsolved problem concerning decomposition of graphs into triangles, *Technical report, University of Waterloo,* (1969).

16. D. K. Ray-Chaudhuri and R. M. Wilson, Solution of Kirkman's schoolgirl problem, pp. 187-204 in *Proc. of Symp. in Pure Math., vol 19 Combinatorics,* Amer. Math. Soc. Providence (1971).

17. R. Rees, Uniformly resolvable pairwise balanced designs with blocksizes two and three, *Preprint, Queen's University, Kingston,* (1986).

18. R.M. Wilson, An existence theory for pairwise balanced designs, III: Proof of the existence conjectures, *Journal of Comb. Th., ser. A* **18** pp. 71-79 (1975).

Annals of Discrete Mathematics 34 (1987) 101–106
© Elsevier Science Publishers B.V. (North-Holland)

An inequality on the parameters of distance regular graphs and the uniqueness of a graph related to M_{23}.

A.E. Brouwer & E.W. Lambeck

Techn. Univ. Eindhoven, The Netherlands

To Alex Rosa on his fiftieth birthday

ABSTRACT

We give an inequality on the parameters of distance regular graphs, and show the geometric consequences of equality. As an application we show that there is a unique distance regular graph with intersection array $\{15,14,12;1,1,9\}$.

0. Introduction.

If Γ is a graph and γ is a vertex of Γ, then let us write $\Gamma_i(\gamma)$ for the set of all vertices of Γ at distance i from γ, and $\Gamma(\gamma) = \Gamma_1(\gamma)$ for the set of all neighbours of γ in Γ. We shall also write $\gamma \sim \delta$ to denote that γ and δ are adjacent.

The graph Γ is called *distance regular* with diameter d and *intersection array* $\{b_0, \cdots, b_{d-1}; c_1, \cdots, c_d\}$ if for any two vertices γ, δ at distance i we have $|\Gamma_{i+1}(\gamma) \cap \Gamma(\delta)| = b_i$ and $|\Gamma_{i-1}(\gamma) \cap \Gamma(\delta)| = c_i$ $(0 \leqslant i \leqslant d)$. Clearly $b_d = c_0 = 0$ (and $c_1 = 1$). Also, a distance regular graph Γ is regular of degree $k = b_0$, and if we put $a_i = k - b_i - c_i$ then $|\Gamma_i(\gamma) \cap \Gamma(\delta)| = a_i$ whenever $d(\gamma,\delta) = i$. We shall also use the notation $k_i = |\Gamma_i(\gamma)|$ (this is independent of the vertex γ). For basic properties of distance regular graphs, see BIGGS [1].

Distance regular graphs of diameter 2 (i.e., strongly regular graphs) are very common, but not so many of larger diameter are known. Often, but not always, such graphs turn out to be uniquely determined (i.e., determined up to isomorphism) by their intersection array.

Our aim in this note is to present an inequality for the parameters of distance regular graphs, and use the geometric information obtained from the proof of this inequality in the case of equality to show that there is a unique distance regular graph with intersection array $\{15,14,12;1,1,9\}$.

Let us first show that there is at least one such graph. Indeed, let Γ be the graph with as vertices the 506 blocks of the Steiner system $S(5,8,24)$ not containing one fixed symbol, where two blocks are adjacent whenever they are disjoint. From the parameters of $S(5,8,24)$ one sees that two blocks have distance 0,1,2,3 in Γ when they meet in 8,0,4,2 points, respectively, and it is easy to check that Γ is distance regular with intersection array $\{15,14,12;1,1,9\}$ and distance distribution diagram

(1) 15 — 1 (15) 14 — 1 (210) 12 — 9 (280) $v = 506$.
 2 6

In fact Γ is distance transitive under the action of M_{23}. (For undefined terminology, see, e.g., BIGGS [1] and CAMERON & VAN LINT [4].) Any two blocks at distance two determine a tetrad (4-

set), and hence a sextet (partition of the set of 24 symbols into six tetrads such that the union of any two of these is a block of $S(5,8,24)$). One of the tetrads of the sextet contains the forbidden symbol, and the pairwise unions of the remaining five tetrads give us ten blocks, forming a Petersen subgraph of Γ. Thus: any two vertices of Γ at distance two are contained in a unique (geodetically closed) subgraph of Γ isomorphic to the Petersen graph.

In this note we show that Γ is the unique distance regular graph with intersection array $\{15,14,12;1,1,9\}$. Our strategy will be, given some graph Γ with these parameters, to first construct Petersen subgraphs, and next embed the graph as a subgraph in the (unique) near hexagon on 759 vertices, the graph on all the blocks of $S(5,8,24)$ (where disjoint blocks are adjacent).

[Added in proof: A.A. Ivanov informs us that he and S.V. Shpectorov had done some work on the characterization of this graph, and in particular also had proved the existence of Petersen subgraphs.]

1. An inequality.

Theorem. *Suppose Γ is a distance regular graph and $a_i \neq 0$ for some i, $1 \leqslant i \leqslant d$. Put $a_{d+1} = 0$. Then*

(i)

$$b_i + c_i \leqslant a_i + \frac{a_{i+1}b_i}{a_i} + \frac{a_{i-1}c_i}{a_i}.$$

Equality holds if and only if the following is true:

For any four vertices α, β, γ, δ of Γ such that $\alpha \sim \beta$, $\gamma \sim \delta$, and $d(\alpha,\gamma) = i$, the three distances $d(\alpha,\delta)$, $d(\beta,\gamma)$, $d(\beta,\delta)$ are not all equal.

In particular, equality implies $a_{i+1} \leqslant a_i$ and $a_{i-1} \leqslant a_i$.

(ii) If $i < d$, then

$$b_i \leqslant a_i + \frac{a_{i+1}b_i}{a_i},$$

and equality implies $a_{i-1} = a_i$.

(iii) If $i > 1$, then

$$c_i \leqslant a_i + \frac{a_{i-1}c_i}{a_i},$$

and, for $i < d$, equality implies $a_{i+1} = a_i$.

Proof. Count 4-tuples $(\alpha,\beta,\gamma,\delta)$ with $\alpha \sim \beta$, $\gamma \sim \delta$, $d(\gamma,\alpha) = d(\gamma,\beta) = i$, $d(\delta,\alpha) \neq i$. Clearly, there are $vk_ia_i(b_i + c_i)$ such 4-tuples. We find the inequality

$$vk_ia_i(b_i + c_i) \leqslant vk_ia_i^2 + vk_{i+1}a_{i+1}c_{i+1} + vk_{i-1}a_{i-1}b_{i-1}$$

by covering this collection of 4-tuples with the three sets

$$\{(\alpha,\beta,\gamma,\delta) \mid \alpha \sim \beta, \gamma \sim \delta, d(\gamma,\alpha) = d(\gamma,\beta) = i, d(\delta,\beta) = i\},$$

$$\{(\alpha,\beta,\gamma,\delta) \mid \alpha \sim \beta, \gamma \sim \delta, d(\delta,\alpha) = d(\delta,\beta) = i+1, d(\gamma,\alpha) = i\},$$

$$\{(\alpha,\beta,\gamma,\delta) \mid \alpha \sim \beta, \gamma \sim \delta, d(\delta,\alpha) = d(\delta,\beta) = i-1, d(\gamma,\alpha) = i\}.$$

This proves the inequality in (i); those in (ii), (iii) follow similarly by restricting $d(\alpha,\delta)$ to be $i+1$ and $i-1$, respectively. Equality in (i) is equivalent with: if $\alpha \sim \beta$, $\gamma \sim \delta$, $d(\alpha,\gamma) = i$, then the three distances $d(\alpha,\delta)$, $d(\beta,\gamma)$, $d(\beta,\delta)$ are not all equal. With $d(\gamma,\beta) = d(\delta,\beta) = i \pm 1$ this implies $a_{i\pm1} \leqslant a_i$ (varying δ). Similarly, we find $a_{i-1} \leqslant a_i$ in case of equality in (ii), but comparing with (i) yields the opposite inequality, so $a_{i-1} = a_i$. Likewise $a_{i+1} = a_i$ in case of equality in (iii). □

Remarks. Clearly, some more geometric information is available in case of equality in (ii) or (iii). A special case of this inequality has already been given by IVANOV, IVANOV & FARADJEV [6]; we have been told that also NOMURA [7] contains a weak form of this result.

2. Construction of Petersen subgraphs.

Lemma. *Let Γ be a graph with $\lambda = 0$, $\mu = 1$, $a_2 = 2$, and suppose that each induced hexagon with at least one pair of vertices at distance 3 contains three such pairs. Then for each vertex γ of Γ, the subgraph $\Gamma_2(\gamma)$ is the disjoint union of $\frac{1}{6}k_2$ hexagons, and hence any two vertices of Γ at distance 2 are contained in a (unique) Petersen subgraph.*

Proof. Fix $\infty \in \Gamma$ and let for $\alpha \in \Gamma_2(\infty)$, $\bar{\alpha}$ be the unique common neighbour of ∞ and α. From $a_2 = 2$ it follows that $\Gamma_2(\infty)$ is a union of polygons; let $\alpha_0 \sim \cdots \sim \alpha_{g-1} \sim \alpha_0$ be such a polygon; clearly $g \geqslant 5$. Since $d(\alpha_i,\alpha_{i+1}) = 2$, the vertex $\overline{\alpha_{i+1}}$ has two neighbours in $\Gamma_2(\alpha_i)$, say ∞ and β_i. Let γ_i be the common neighbour of α_i and β_i. Since $d(\overline{\alpha_i},\alpha_{i+1}) = 2$, it follows from the hypothesis applied to the hexagon $\infty \sim \bar{\alpha}_i \sim \alpha_i \sim \gamma_i \sim \beta_i \sim \overline{\alpha_{i+1}} \sim \infty$ that also $d(\infty,\gamma_i) = 2$. But then $\gamma_i \in \{\alpha_{i-1},\alpha_{i+1}\}$ and $\beta_i \in \{\alpha_{i-2},\alpha_{i+2}\}$. If $\beta_i = \alpha_{i+2}$ then we have the triangle $\overline{\alpha_{i+1}}\alpha_{i+1}\alpha_{i+2}$, contradicting $\lambda = 0$. Thus $\beta_i = \alpha_{i-2}$, so that $\overline{\alpha_{i+1}} = \overline{\alpha_{i-2}}$ $(i \in \mathbf{Z}_g)$. Now $\alpha_0 \sim \overline{\alpha_{3j}} \sim \alpha_{3j+1} \sim \bar{\alpha}_1$ for all $j \in \mathbf{Z}_g$, so that there are only two points $\overline{\alpha_{3j}}$, and $g = 6$. □

Now assume that Γ has intersection array $\{15,14,12;1,1,9\}$. Then $c_3 = 9$, $a_2 = 2$, $a_3 = 6$ and we have equality in (iii) (and hence in (i)) of the theorem in the previous section. The information about the case of equality stated there immediately implies that the hypothesis of the Lemma about hexagons is satisfied, and hence the conclusion of the Lemma holds: any two vertices of Γ at distance two are contained in a unique Petersen subgraph of Γ.

3. The equivalence relation on the edges.

Let Γ be a distance regular graph with intersection array $\{15,14,12;1,1,9\}$, and let P be a Petersen subgraph of Γ. Write $\Gamma_j(P)$ for the set of vertices of Γ at distance j from P, and $\Gamma(P) = \Gamma_1(P)$. Let us study the structure of Γ around P.

1. *For every $\gamma \in \Gamma(P)$ we have $|\Gamma(\gamma) \cap P| = 1$ and $|\Gamma(\gamma) \cap \Gamma(P)| = 0$.*
(For: there are no triangles or 4-gons in Γ; every 5-gon is in a unique Petersen subgraph and P is geodetically closed.)

2. *If $\gamma \in \Gamma_2(P)$, then $\Gamma_2(\gamma) \cap P$ does not contain 2-claws.*
(For suppose $\gamma_1, \gamma_2, \gamma_3 \in \Gamma_2(\gamma) \cap P$ such that $\gamma_1 \sim \gamma_2 \sim \gamma_3$. Then γ and γ_2 determine a unique Petersen subgraph P', which contains $\Gamma(\gamma_2) \cap \Gamma_2(\gamma)$. But now $\gamma_1, \gamma_2, \gamma_3 \in P \cap P'$, contradiction.)

3. *If $\gamma \in \Gamma_2(P)$ then $\Gamma_2(\gamma) \cap P$ is either a 4-coclique or it induces $3K_2$ as a subgraph of P.*
Proof. By 2. there are only 10 possibilities left for the isomorphism type of the subgraph of P induced by $\Gamma_2(\gamma) \cap P$, namely (i) $3K_2$, (ii) $2K_2 + K_1$, (iii) $2K_2$, (iv) $K_2 + 2K_1$, (v) $K_2 + K_1$, (vi) K_2, (vii) $4K_1$, (viii) $3K_1$, (ix) $2K_1$, (x) K_1. (Note that in case (viii) there are two essentially distinct embeddings in P.)
Write $Q = P \setminus (P \cap \Gamma_2(\gamma)) = P \cap \Gamma_3(\gamma)$. For each vertex δ of Q we find $c_3 = 9$ points in $\Gamma(\gamma) \cap \Gamma_2(\delta)$. If we put $e_\delta = |\Gamma(\delta) \cap (P \setminus Q)|$, then e_δ of these points lie in $\Gamma(P)$, and the remaining $9 - e_\delta$ are in $\Gamma_2(P)$. Put $A(\delta) = \Gamma(\gamma) \cap \Gamma_2(\delta) \cap \Gamma_2(P)$.
Now suppose Q contains a path $\delta_1 \sim \delta_2 \sim \delta_3$; then by 2. the set $A(\delta_1) \cap A(\delta_2) \cap A(\delta_3)$ must be empty. In particular, if we put $a = |P \setminus Q|$ so that $|\Gamma(\gamma) \cap \Gamma_2(P)| = 15 - a$, then the inequality $27 - e_{\delta_1} - e_{\delta_2} - e_{\delta_3} \leqslant 30 - 2a$, or, equivalently, $3 + e_{\delta_1} + e_{\delta_2} + e_{\delta_3} \geqslant 2a$ holds. This observation will kill most of our ten possibilities.
Indeed, suppose we have situation (ii). Then we can find a path $\delta_1 \sim \delta_2 \sim \delta_3$ in Q such that $\underline{e} = (e_{\delta_1}, e_{\delta_2}, e_{\delta_3}) = (2,1,2)$, but $a = 5$, contradicting the above inequality.

Similarly, in cases (iii),(iv),(v) we find $a = 4$, $e = (2,1,0)$ and $a = 4$, $e = (1,1,1)$ and $a = 3$, $e = (1,0,0)$, respectively, a contradiction in each case. In case (vi) this argument is not strong enough, but we can find a $K_{1,3}$ in Q, say with vertices δ_0, δ_1, δ_2, δ_3, where δ_0 is the vertex of degree 3, and there must be a pair i, j in 1, 2, 3 such that $A(\delta_0) \cap A(\delta_i) \cap A(\delta_j) \neq \emptyset$, again a contradiction. The cases (viii)-(x) can be ruled out by similar arguments, but since we shall not need the absence of these, we do not give the details. □

We define a relation $/\!/$ on the edges of Γ by: $e_1 /\!/ e_2$ if and only if there is a Petersen subgraph P containing them, and where $e_1 \cup e_2$ induces a subgraph isomorphic to $2K_2$ (i.e., where each vertex of e_1 has distance 2 to each vertex of e_2). Observe that $/\!/$ restricted to a Petersen graph is an equivalence relation with equivalence classes of size 3.

4. *The relation* $/\!/$ *is an equivalence relation.*
Proof. Suppose $e_1 /\!/ e_2$ and $e_2 /\!/ e_3$. If e_1, e_2 and e_3 are in one Petersen subgraph, then by the above observation $e_1 /\!/ e_3$. So assume $e_1, e_2 \subset P_1$ and $e_2, e_3 \subset P_2$, and put $e_1 = \{\gamma_1,\gamma_2\}$. Now it follows from 1. that $d(\gamma_i,P_2) = 2$ $(i = 1,2)$ and $e_2 \subset \Gamma_2(\gamma_i) \cap P_2$ $(i = 1,2)$. By 3. we have $e_3 \subset \Gamma_2(\gamma_i) \cap P_2$ $(i = 1,2)$ and therefore e_1 and e_3 determine a unique Petersen subgraph P_3 containing them, and $e_1 \cup e_3$ induces $2K_2$ as subgraph of P_3. This means that $e_1 /\!/ e_3$. □

Note that there are precisely 253 equivalence classes in Γ, with 15 edges in each equivalence class. (Indeed, each edge is in 7 Petersen graphs, and hence equivalent to 14 other edges.) These equivalence classes will be the missing points of our near hexagon.

4. Embedding in a near hexagon.

A *near polygon* is a partial linear space such that given a point and a line, there is a unique point on the given line closest to the given point. Here 'closest' refers to distance in the collinearity graph. A *near hexagon* is a near polygon of diameter 3. For the theory of near polygons, see BROUWER & WILBRINK [3].

Continuing with the assumptions of the previous section, we construct a partial linear space $(X, \mathfrak{L})$ as follows. The point set X is the union of the set X_0 of vertices of Γ and the set X_1 of equivalence classes of edges. The set of lines $\mathfrak{L}$ consists of the 3-sets $\{\gamma,\delta,E\}$ for each edge $e = \{\gamma,\delta\}$ of Γ, where E is the equivalence class of e. Let us check that this linear space is a regular near polygon of diameter 3, with parameters $(s+1, t_2+1, t+1) = (3,3,15)$.

First of all, we really have a partial linear space: an equivalence class E of edges and a collinear vertex γ determine uniquely an edge $e = \{\gamma,\delta\} \in E$. Next, the partial linear space is connected, since Γ is connected. Note that Γ is an induced subgraph of the collinearity graph of $(X,\mathfrak{L})$, and that X_1 is a coclique.

Let us look at the parameters. We have $|X_0| = 506$, $|X_1| = 253$ and hence $|X| = 759$. Each point is on $t+1 = 15$ lines, and each line contains $s+1 = 3$ points. In order to show that $t_2+1 = 3$, we must show that any two points ξ, η at distance 2 have 3 common neighbours. If $\xi, \eta \in X_0$, then they determine a Petersen subgraph P, and the three common neighbours are their common neighbour in Γ and the two equivalence classes containing edges on them. If $\xi \in X_0$, $\eta \in X_1$, then η contains an edge $\{\alpha,\beta\}$, where $\xi \sim \alpha$. Now α, β and ξ are contained in a Petersen subgraph P, and any common neighbour of ξ and η must be in X_0 (since X_1 is a coclique), and therefore must be one of the three neighbours of ξ on the three edges that η has in P. Finally, if $\xi, \eta \in X_1$, then ξ contains an edge $\{\alpha,\beta\}$ and η contains an edge $\{\alpha,\gamma\}$; again α, β, γ are contained in a Petersen subgraph P, and any common neighbour of ξ and η must be a vertex δ at distance two from each of α, β and γ, and therefore (by 2.) in P; but in P we see three common neighbours of ξ and η. This proves that any two points at distance 2 have precisely three common neighbours.

Now, let us check the near polygon axiom. Let L be a line, say $L = \{\gamma,\delta,E\}$ with $E \in X_1$, and let ξ be a point with $d(\xi,L) = 1$. If $\xi \sim E$, then ξ, γ and δ are contained in a Petersen subgraph of Γ and ξ has distance two to the edge $\{\gamma,\delta\}$. Otherwise, if $\xi \in X_0$, then ξ is adjacent to either γ or δ, but not both, since Γ does not contain triangles, while if $\xi \in X_1$ and $\xi \sim \gamma, \delta$ then $\xi = E$. Thus, the near polygon axiom is satisfied in this case.

Next suppose that $d(\xi,L) = 2$. If $\xi \in X_0$ and $d(\xi,\gamma) = d(\xi,\delta) = 2$, then ξ, γ and δ are contained in a Petersen subgraph, and ξ is on an edge parallel to $\{\gamma,\delta\}$, so that ξ is collinear with E, contrary to our assumption. If $\xi \in X_0$ and $d(\xi,\gamma) = d(\xi,E) = 2$, then E contains an edge $\{\alpha,\beta\}$, where $\xi \sim \alpha$, but now α, β, γ, δ and ξ are all contained in one Petersen subgraph of Γ, and we find $\xi \sim \delta$, contradiction. If $\xi \in X_1$ and $d(\xi,\gamma) = d(\xi,\delta) = 2$, then ξ contains two edges $\{\alpha,\beta\}$ and $\{\zeta,\eta\}$ where $\gamma \sim \alpha$ and $\delta \sim \zeta$. But these two edges are contained in a Petersen subgraph P, and this situation contradicts 1. Finally, if $\xi \in X_1$ and $d(\xi,\gamma) = d(\xi,E) = 2$, then ξ contains two edges $\{\alpha,\beta\}$ and $\{\zeta,\eta\}$ where $\gamma \sim \alpha$ and, for some $\theta \in X_0$, $\{\eta,\theta,E\}$ is a line. But then the edge $\{\alpha,\gamma\}$ is contained in $\Gamma_2(\eta)$, so there is a Petersen subgraph P of Γ containing η, α and γ, and therefore also $\beta, \zeta, \delta, \theta$. It follows that $d(\xi,\delta) \leqslant 2$, contradiction. Thus, the near polygon axiom is also satisfied in case $d(\xi,L) = 2$.

Given a line L, there are 3 points on L, $3.2.(15-1) = 84$ points at distance one to L, and $84.2.(15-3)/3 = 656$ points at distance two from L. But $3 + 84 + 672 = 759$, so $(X,\mathfrak{L})$ has diameter 3 and each point is at distance at most two from any line. This concludes our verification: we do have a regular near polygon of diameter 3 and with the parameters stated. But BROUWER [2] shows that there is a unique such near polygon, namely that with as points the blocks of the Steiner system $S(5,8,24)$, and as lines the triples of pairwise disjoint blocks. Our last task is to identify Γ as a subgraph of the collinearity graph of this near polygon.

Lemma. *The near polygon constructed above contains precisely 24 cocliques of size 253, namely the 24 sets of blocks of $S(5,8,24)$ containing any given symbol.*
Proof. Let C be such a coclique. Since there are 3795 lines, and each point is on 15 lines, it follows that C meets each line in precisely one point. In particular, if Q is any quad, then $C \cap Q$ is an ovoid in Q (for terminology, cf. BROUWER [2]), that is, the set of five blocks containing a fixed tetrad T. Each block in C meets T, for blocks disjoint from T are disjoint from a block containing T. If Q' is another quad, we find another tetrad T'. If $T \cap T' = \emptyset$, then there is a block on T disjoint from T', contradiction. Thus, the set of $1771 = \binom{23}{3}$ tetrads found by varying Q is a linked system, and by a theorem of ERDŐS, KO & RADO [5] this collection of tetrads consists of all tetrads on a fixed symbol. It follows that all blocks in C contain this symbol. □

Thus, starting from an arbitrary graph Γ with intersection array $\{15,14,12;1,1,9\}$ we find Γ as the complement of a 253-coclique in the collinearity graph of the near polygon on 759 points, and we have shown that Γ is uniquely determined.

References

1. Biggs, N.L., *Algebraic Graph Theory*, Cambridge Tracts in Math. 67, Cambridge University Press, Cambridge (1974).

2. Brouwer, A.E., *The uniqueness of the near hexagon on 759 points,* pp. 47-60 in: *Finite Geometries*, Lecture Notes in Pure and Applied Math. 82 (ed. N.L. Johnson, M.J. Kallaher & C.T. Long), Marcel Dekker, New York, 1982. MR 84d:51021 (MR 82j:05038)

3. Brouwer, A.E. and H.A. Wilbrink, *The structure of near polygons with quads*, Geometriae Dedicata **14** (1983) 145-176.

4. Cameron, P.J. and J.H. van Lint, *Graphs, Codes and Designs*, London Math. Soc. Lecture Notes 43, Cambridge Univ. Press, Cambridge (1980).

5. Erdős, P., Chao Ko, and R. Rado, *Intersection theorems for systems of finite sets*, Quart. J. Math. Oxford (2) **12** (1961) 313-320.

6. Ivanov, A.A., A.V. Ivanov, and I.A. Faradjev, **Дистанционно-транзитивные графы цтепени** 5, 6 **и** 7 *(preprint) = Distance-transitive graphs of valency 5, 6 and 7*, Eur. J. Combinatorics (to appear). USSR Comput. Maths. Math. Phys. 24 (1984) pp. 67-76, Žurnal Vičisl. Mat. i Mat. Fiz. 24 (1984) pp. 1704-1718.

7. Nomura, K., *An inequality between intersection numbers of a distance regular graph*, J. Combinatorial Th. (B) (submitted).

Annals of Discrete Mathematics 34 (1987) 107–118
© Elsevier Science Publishers B.V. (North-Holland)

Partitions into indecomposable triple systems

Charles J. Colbourn and Janelle J. Harms

Department of Computer Science
University of Waterloo
Waterloo, Ontario, N2L 3G1
CANADA

TO ALEX ROSA ON HIS FIFTIETH BIRTHDAY

ABSTRACT

The indecomposable partition problem is to partition the set of all triples on v elements into s indecomposable triple systems, where the ith triple system has index λ_i and $\lambda_1 + \cdots + \lambda_s = v-2$. A complete solution for $v \leq 10$ is given here. Extending a construction of Rosa for large sets, we then give a $v \rightarrow 2v+1$ construction for indecomposable partitions. This recursive construction employs solutions to a related partition problem, called indecomposable near-partition. A partial solution to the indecomposable near-partition problem for $v=10$ then establishes that for every order $v = 5 \cdot 2^i - 1$, all indecomposable partitions having $\lambda_i = 1,2$ for each i can be realized.

1. The problem

A *triple system* of *order* v and *index* λ, denoted $TS(v,\lambda)$ is a pair (V,B). V is a set of v elements, and B is a collection of 3-subsets of V called *blocks* or *triples;* every 2-subset of elements appears in precisely λ of these triples. We assume throughout this paper that there are no repeated blocks; that is, blocks are distinct 3-subsets. A triple system is *indecomposable* if there is no proper nonempty subset $B' \subset B$ for which (V,B') is a triple system. The *full* triple system of order v consists of all 3-subsets of a v-set, and hence is the unique $TS(v,v-2)$ without repeated blocks. A *partial* triple system $PTS(v,\lambda)$ relaxes the requirement that each pair appear in exactly λ blocks to a requirement that each pair appear in at most λ blocks.

Mesner [11] asked the following question. Let $\Lambda = \{\lambda_1,...,\lambda_s\}$ be a partition of the integer $v-2$, i.e. $\lambda_1 + \cdots + \lambda_s = v-2$. Can one partition the blocks of the full design $TS(v,v-2)$ into s classes $B_1,...,B_s$ so that for each $1 \leq i \leq s$, (V,B_i) is an indecomposable $TS(v,\lambda_i)$? We call Mesner's problem the *indecomposable partition problem* and write $IPP(v,\Lambda)$ to denote the instance of the problem for the set Λ of indices required. The indecomposable partition problem has been studied by Rosa [14], and some results for small v are reported in Gronau and Rentner [4].

Two restricted cases of the indecomposable partition problem are of special interest and have been widely studied. The problem $IPP(v,\{1,1,...,1\})$ is the *large set problem* for Steiner triple systems, the problem of finding $v-2$ block-disjoint Steiner triple systems of order v. The large set problem is now solved with six possible exceptions: for every $v \equiv 1,3 \pmod 6$, $v \neq 7$, there is a large set of disjoint STS(v) except possibly for $v=$ 141, 283, 501, 789, 1501, and 2365 [9,13]. The problem $IPP(v,\{2,2,2,...2\})$ has been completely settled in the affirmative for all $v \equiv 0,4 \pmod 6$ by Teirlinck [15], as has the problem $IPP(v,\{6,6,6,...,6\})$ for all $v \equiv 2 \pmod 6$. Kramer [7] studied partitions into $TS(v,3)$ designs; subsequently, Teirlinck [17] completely settled the existence of all $IPP(v,\{3,3,...,3\})$ for $v \equiv 5 \pmod 6$ in the affirmative.

The indecomposable partition problem also encompasses the problem of existence of indecomposable triple systems. Existence of indecomposable triple systems is far from settled. For $\lambda=2$, Kramer [6] showed that $v \equiv 0,1 \pmod 3$, $v \neq 7$, and $v>3$ form necessary and sufficient conditions for the existence of an indecomposable $TS(v,2)$. For $\lambda=3$, Kramer [6] showed that $v \equiv 1 \pmod 2$, $v>3$ form necessary and sufficient conditions for the existence of an indecomposable $TS(v,3)$. For $\lambda=4$, Colbourn and Rosa [3] showed that $v \equiv 0,1 \pmod 3$ and $v>9$ form necessary and sufficient conditions for the existence of an indecomposable $TS(v,4)$. Beyond this, little is known. Indecomposable triple systems with repeated blocks exist for all odd indices, as a consequence of the characterization of neighbourhood graphs by Colbourn [1]. However, no indecomposable triple systems with $\lambda>6$ and λ even are presently known, even if repeated blocks are allowed.

The large set problem and the existence of indecomposable designs provide two strong reasons for studying the indecomposable partition problem; the current lack of knowledge about indecomposable triple systems for higher λ leads us to focus on partitions involving only small λ_i's.

In this paper, we present a number of partial results on the indecomposable partition problem. The main result is an extension of Rosa's $v \rightarrow 2v+1$ construction for large sets [12], to produce other indecomposable partitions of the full design $TS(2v+1,2v-1)$. This generalization of Rosa's result requires, in addition to an indecomposable partition of the $TS(v,v-2)$, a different type of partition of the $TS(v+1,v-1)$. Rosa's construction employs the observation that whenever v is the order of a Steiner triple system, the full design $TS(v+1,v-1)$ can be partitioned into $v+1$ Steiner triple systems of order v. This can be done by taking the derived Steiner triple systems of each point in a Steiner quadruple system of order $v+1$, for example (see [8]). Notice that each triple system produced misses precisely one of the $v+1$ points, and each point is missed by one system. To generalize this to higher λ, we define a *pointed triple system* $POTS(v,\lambda)$ to be a pair (V,B); V is a collection of elements, of which λ are distinguished. B is a set of 3-subsets of V, with the balance property that (1) pairs between undistinguished elements appear in λ blocks, (2) pairs between a distinguished and an undistinguished element appear in $\lambda-1$ blocks, and (3) pairs of distinguished elements appear in $\lambda-2$ blocks. When $\lambda=1$, a $POTS(v,1)$ is a Steiner triple system adjoined with an extra element appearing in no blocks. A pointed triple system $POTS(v,\lambda)$ is indecomposable exactly when it has no partition into partial triple systems $PTS(v,\lambda')$ and $PTS(v,\lambda-\lambda')$, with $0<\lambda'<\lambda$. Now letting $\Gamma=\{\lambda_1,...,\lambda_s\}$ be a partition of the integer v into nonzero integer parts, we define the *indecomposable near-partition problem* $INPP(v,\Gamma)$ to be a partition of the full design $TS(v,v-2)$ into classes $B_1,...,B_s$, so that (V,B_i) is an indecomposable $POTS(v,\lambda_i)$. At first glance, this problem seems somewhat artificial, but we see later that it is exactly what we need for the extension to Rosa's theorem.

In section 2, we give a complete solution to the indecomposable partition problem for $v \leq 10$. In section 3, we develop a number of indecomposable near-partitions for order ten, to provide a basis for the recursive construction. Section 4 then develops an easy recursive construction for near-partitions, and section 5 uses this to provide a $v \rightarrow 2v+1$ construction for indecomposable partitions. Finally, in section 6, we combine all of these pieces to show that for $v = 5 \cdot 2^i - 1$, $i \geq 1$ all indecomposable partitions with $\lambda_i = 1$ or 2 are realizable.

2. The partition problem: small orders

For $v = 3,4,5$, and 8, the full design $TS(v,v-2)$ is itself indecomposable, and hence only the indecomposable partition $IPP(v,\{v-2\})$ exists for these orders. For $v=6$, the full design contains a $TS(6,2)$ and hence must partition as $\{2,2\}$. For $v=7$, there is no indecomposable $TS(7,2)$. Moreover, a result of Teirlinck [16] ensures that there is no indecomposable $TS(7,4)$. Since in addition the large set problem $IPP(7,\{1,1,1,1,1\})$ has no solution [13], the only indecomposable partition here is the $IPP(7,\{1,1,3\})$.

The first nontrivial case is $v=9$. Triple systems $TS(9,\lambda)$ have been exhaustively enumerated for $\lambda = 1$, 2, and 3. There is a unique $TS(9,1)$, and there are 13 nonisomorphic $TS(9,2)$'s without repeated blocks [10]. Recently, Harms and Colbourn [5] enumerated the 332 nonisomorphic $TS(9,3)$ designs without repeated blocks. By complementation, there are 332 $TS(9,4)$ designs, 13 $TS(9,5)$ designs, and one $TS(9,6)$ design. Teirlinck's theorem [16] again ensures that the $TS(9,6)$ is decomposable. We have undertaken an exhaustive search by computer, and found decompositions for all of the $TS(9,4)$ and $TS(9,5)$ designs. Hence, the only possible partitions for $v=9$ have maximum index equal to 3. This leaves eight candidate partitions; in appendix one, we give indecomposable partitions for each of these eight. Hence we have

Lemma 2.1: An $IPP(9,\Lambda)$ exists if and only if Λ is one of $\{1,1,1,1,1,1,1\}$, $\{1,1,1,1,1,2\}$, $\{1,1,1,1,3\}$, $\{1,1,1,2,2\}$, $\{1,1,2,3\}$, $\{1,2,2,2\}$, $\{1,3,3\}$, or $\{2,2,3\}$. □

Remarkably, exactly one of the 332 $TS(9,3)$ designs partitions both into $\{1,1,1\}$ and into $\{1,2\}$. Of the 331 remaining, 172 are indecomposable, 8 partition only into $\{1,1,1\}$, and 151 partition only into $\{1,2\}$. The existence of the unique system which decomposes in both ways simplifies the proof of lemma 2.1 substantially, especially since the complementary $TS(9,4)$ of this $TS(9,3)$ contains an isomorphic copy of the original $TS(9,3)$.

Finally, for order ten, the only possible partitions are $\{2,2,2,2\}$, $\{2,2,4\}$, $\{2,6\}$, and $\{4,4\}$. Colbourn, Colbourn, Harms, and Rosa [2] enumerated the 394 nonisomorphic $TS(10,2)$ designs without repeated blocks. We have verified that the 394 complementary $TS(10,6)$ designs are all decomposable, and hence the $\{2,6\}$ partition is ruled out. Colbourn and Rosa [3] found an indecomposable $TS(10,4)$; in fact, both partitions $\{2,2,4\}$ and $\{4,4\}$ are given in appendix two. Together with the $\{2,2,2,2\}$ partition from Teirlinck's construction [15], this gives

Lemma 2.2: An $IPP(10,\Lambda)$ exists if and only if Λ is one of $\{2,2,2,2\}$, $\{2,2,4\}$, or $\{4,4\}$. □

Higher orders become much more difficult, in large part because the existence of indecomposable designs with high λ is open. For v=11, the existence of an indecomposable $TS(11,6)$ is open, for example. For v=13, the existence of an indecomposable $TS(13,\lambda)$ is open for $5 \leq \lambda \leq 9$. Hence we focus on partitions into small indices from here on.

3. The near-partition problem: small orders

It is an easy exercise to see that any indecomposable $POTS(v,2)$ produces an indecomposable $TS(v-1,2)$ by simply identifying the two distinguished elements in the *POTS*. Moreover, one can easily verify that v must be even, since considering triples containing a fixed distinguished element we have $(\lambda-1)v \equiv 0 \pmod 2$, and considering triples with a fixed nondistinguished element we have $\lambda v \equiv 0 \pmod 2$.

Hence the smallest order of any real interest is v=10. Using some clever backtracking techniques, we found numerous near-partitions of the full design $TS(10,8)$. Near-partitions are given in appendix three which establish the following lemma.

Lemma 3.1: There exist $INPP(10,\Lambda)$ for Λ one of $\{1,1,1,1,1,1,1,1,1,1\}$, $\{1,1,1,1,1,1,2,2\}$, $\{1,1,1,1,2,2,2\}$, $\{1,1,2,2,2,2\}$, and $\{2,2,2,2,2\}$. □

It appears reasonable to expect that near-partitions can be obtained in a direct way from partitions; at present, however, we know of no way to do this.

4. Doubling near-partitions

In this section, we describe a simple method for producing indecomposable near-partitions of order $2v$ from indecomposable near-partitions of order v. This doubling construction is based on the standard $v \rightarrow 2v+1$ construction for $TS(v,1)$.

Theorem 4.1: If there exists an $INPP(v,\Lambda)$ and an $INPP(v,\Gamma)$, there exists an $INPP(2v,\Lambda \cup \Gamma)$.

Proof:

Let $T_1,...,T_m$ be an $INPP(v,\Lambda)$ on elements $\{y_1,...,y_v\}$. Let $B_1,...,B_n$ be an $INPP(v,\Gamma)$ on elements $\{x_1,...,x_v\}$. Let $E(B_i)$ $(E(T_i))$ denote the indices of the distinguished elements in the POTS B_i (resp., T_i). Now let $F = (F_1,...,F_{v-1})$ be any 1-factorization of the complete v-vertex graph on $\{y_1,...,y_v\}$ and let $G = (G_1,...,G_{v-1})$ be a 1-factorization on $\{x_1,...,x_v\}$. Finally, let L be a $v \times v$ Latin square on symbols $\{0,...,v-1\}$, with constant zero diagonal.

Now define sets $C_1,...,C_v$ as follows. Whenever the (i,j) position of L is $k \neq 0$, form $C_{ij} = \{\{x_j,a,b\} \mid \{a,b\} \in F_k\}$. Then $C_i = \bigcup_{j \neq i} C_{ij}$. Similarly, whenever the (i,j) position of L is $k \neq 0$, form $D_{ij} = \{\{y_j,a,b\} \mid \{a,b\} \in G_k\}$. Then $D_i = \bigcup_{j \neq i} D_{ij}$.

For each T_i, define $M_i = T_i \cup \bigcup_{j \in E(T_i)} D_j$. For each B_i, define $N_i = B_i \cup \bigcup_{j \in E(B_i)} C_j$. We claim that $M_1,...,M_m,N_1,...,N_n$ is an $INPP(2v,\Lambda \cup \Gamma)$. It is an easy exercise to see that no blocks appear in more than one of the systems; moreover, the construction ensures that each M_i and N_i is a POTS. Since each contains an indecomposable sub-POTS with the required index, each is indecomposable as required. □

Together with the results from section 3, this gives many indecomposable near-partitions. Naturally, our interest is in using these near-partitions to form indecomposable partitions; we pursue this next.

5. Doubling partitions

Rosa [12] gave a $v \rightarrow 2v+1$ construction for large sets. We adapt Rosa's construction here to combine an indecomposable partition of order v with an indecomposable near-partition of order $v+1$ to form an indecomposable partition of order $2v+1$. We are able to use much of Rosa's proof unchanged; hence we review the main ideas of his proof.

Let $(D_1,...,D_{v-2})$ be a large set on the v elements $X = \{x_1,...,x_v\}$; append $v+1$ additional elements $Y = \{y_1,...,y_{v+1}\}$. Triples come in four types:

I. All three elements are from X.

II. Two elements are from X, and one is from Y.

III. One element is from X, and two are from Y.

IV. All three elements are from Y.

All type I triples are exhausted by the $\{D_i\}$. Each D_i is extended using a collection E_i of type III triples, to form a $TS(2v+1,1)$. To get the remaining $v+1$ required designs for the large set, Rosa partitions the type IV triples into $v+1$ triple systems of order v, $B_1,...,B_{v+1}$; he then uses collections C_i of type II triples and the remaining type III triples, so that $B_i \cup C_i$ is a $TS(2v+1,1)$. The resulting $2v-1$ systems are proved to be disjoint in [12].

We adapt Rosa's proof by leaving the sets C_i and E_i of type II and III triples unchanged, and altering the partitioning of the type I and type IV triples.

Theorem 5.1: Given an $IPP(v,\Lambda)$ and an $INPP(v+1,\Gamma)$, there is an $IPP(2v+1,\Lambda \cup \Gamma)$.

Proof:

Let $D_1,...,D_m$ be an $IPP(v,\Lambda)$, with λ_i being the index of D_i. Arbitrarily partition the collection $\{E_j\}$ into m groups $G_1,...,G_m$ so that G_i contains precisely λ_i of the $\{E_j\}$. Let H_i denote the union of all the sets contained in G_i. Now $D_i \cup H_i$ is a $TS(2v+1,\lambda_i)$, which is indecomposable since D_i is. Notice that this exhausts *exactly* the same collection of triples which Rosa's $\{D_i \cup E_i\}$ exhaust.

Now let $(B_1,...,B_n)$ be an $INPP(v,\Gamma)$, with B_i an indecomposable $POTS(v+1,\gamma_i)$ having distinguished points P_i. For each B_i, form $J_i = \bigcup_{j \in P_i} C_j$. Then the system $B_i \cup J_i$ is a $TS(2v+1,\gamma_i)$ which is indecomposable. By construction, the $\{B_i \cup J_i\}$ exhaust the same triples exhausted by Rosa's $\{B_i \cup C_i\}$.

The required partition is obtained by combining these two pieces. Since type I triples only appear in the $\{D_i\}$ and type IV triples only appear in the $\{B_i\}$, we need only ensure that the type II and III triples are partitioned correctly. This follows directly from Rosa's proof. □

A somewhat simpler way to view this process is that we simply follow Rosa's construction; we then "unplug" all type I triples and replace using the $IPP(v,\Lambda)$, and "unplug" all type IV triples and replace by the $INPP(v+1,\Gamma)$. The only complication in the proof is ensuring that the POTS in the indecomposable near-partition become associated with the appropriate sets $\{C_i\}$.

6. An infinite family of indecomposable partitions

Combining the results of sections 2-5, we can observe that there is an interesting infinite family of indecomposable partitions:

Theorem 6.1: Let $v = 5 \cdot 2^i - 1$, $i \geq 1$, and let Λ be a partition of $v-2$ into parts all equal to 1 or 2. Then there is a *IPP*(v,Λ).

Proof:

For $i=1$, this follows from lemma 2.1. By lemma 3.1, we also have indecomposable near-partitions for order 10 for each partition into ones and twos, except the partition with a single two. Using theorem 4.1, we therefore have such indecomposable near-partitions for all orders of the form $5 \cdot 2^i$, $i \geq 1$, again with the exception of the partition having a single two. Finally, using theorem 5.1, the result follows by easy induction. □

We should remark that even if indecomposable near-partitions of order ten were found with all parts of index at most three, theorem 6.1 would not extend to establish the corresponding statement for parts of sizes 1, 2, and 3. The reason for this is simple: for $v=19$, the partition of 17 into $\{2,3,3,3,3,3\}$ cannot be itself partitioned into a partition of 7 and a partition of 10, which would be required. Hence the extension of theorem 6.1 to larger indices requires a new idea.

In conclusion, we observe that the indecomposable partition problem is an interesting question incorporating challenging questions on partitioning and on decomposability. The problem remains, at present, far from a general solution. However, we conjecture that theorem 6.1 holds for all $v \equiv 1,3 \pmod 6$, $v \neq 7$.

Acknowledgements

Thanks to Alex Rosa for suggesting the problem to us, and for suggesting the approach in section 2.

References

[1] C.J. Colbourn, "Simple neighbourhoods in triple systems", submitted for publication, 1986.

[2] C.J. Colbourn, M.J. Colbourn, J.J. Harms, and A. Rosa, "A census of (10,3,2) block designs and of Mendelsohn triple systems of order ten. III. (10,3,2) block designs without repeated blocks", *Proc. Twelfth Manitoba Conf. Num. Math. Computing,* Winnipeg, 1982, pp. 211-234.

[3] C.J. Colbourn and A. Rosa, "Indecomposable triple systems with $\lambda=4$", *Studia Sci. Math. Hung.* 20 (1985) 139-144.

[4] H.D.O.F. Gronau and I. Rentner, "On the decomposition of the set of all k-element subsets of a v-element set into indecomposable t-(v,k,λ) designs", *Rostock. Math. Kolloq.* 28 (1985) 49-54.

[5] J.J. Harms, C.J. Colbourn, and A.V. Ivanov, "A census of (9,3,3) designs without repeated blocks", *Proc. Sixteenth Manitoba Conf. Num. Math. Comput.* Winnipeg, 1986, pp. 147-170.

[6] E.S. Kramer, "Indecomposable triple systems", *Discrete Math.* 8 (1974) 173-180.

[7] E.S. Kramer, "Some triple-system partitions for prime powers", *Utilitas Math.* 12 (1977) 113-116.

[8] C.C. Lindner and A. Rosa, "Steiner quadruple systems - a survey", *Discrete Math.* 21 (1978) 147-181.

[9] J.-X. Lu, "On large sets of disjoint Steiner triple systems VI", *J. Comb. Thy.* A37 (1984) 189-192.

[10] R.A. Mathon and A. Rosa, "A census of Mendelsohn triple systems of order 9", *Ars Combinatoria* 4 (1977) 309-315.

[11] D.M. Mesner, private communications to A. Rosa.

[12] A. Rosa, "A theorem on the maximum number of disjoint Steiner triple systems", *J. Comb. Thy.* A18 (1975) 305-312.

[13] A. Rosa, "Intersection properties of Steiner systems", *Ann. Discrete Math.* 7 (1980) 115-128.

[14] A. Rosa, private communications, 1985-1986.

[15] L. Teirlinck, "On the maximal number of disjoint triple systems", *J. Geometry* 6 (1975) 93-96.

[16] L. Teirlinck, "On making two Steiner triple systems disjoint", *J. Comb. Thy.* A23 (1977) 349-350.

[17] L. Teirlinck, "On large sets of disjoint quadruple systems", *Ars Combinatoria* 17 (1984) 173-176.

Appendix One: $IPP(9,\Lambda)$

We first partition a fixed $TS(9,3)$, #305 from [HC], both as 1+2, and as 1+1+1, as follows:

1+1+1 partition of #305	
λ=1	013 024 057 068 127 148 156 236 258 345 378 467
λ=1	014 023 058 067 125 136 178 247 268 348 357 456
λ=1	012 035 046 078 134 158 167 237 248 256 368 457

1+2 partition of #305	
λ=1	014 023 057 068 127 136 158 248 256 345 378 467
λ=2	012 013 024 035 046 058 067 078 125 134 148 156 167 178 236 237 247 258 268 348 357 368 456 457

Let #305c denote the $TS(9,4)$ complement of #305. Then removing the $TS(9,1)$ on blocks 018, 027, 034, 056, 126, 135, 147, 238, 245, 367, 468, and 578, leaves a $TS(9,3)$ #305i which is isomorphic to #305 and has the following partitions:

1+1+1 partition of #305i	
λ=1	015 026 038 047 128 137 146 234 257 356 458 678
λ=1	016 028 037 045 124 138 157 235 267 346 478 568
λ=1	017 025 036 048 123 145 168 246 278 347 358 567

1+2 partition of #305i	
λ=1	015 026 038 047 124 137 168 235 278 346 458 567
λ=2	016 017 025 028 036 037 045 048 123 128 138 145 146 157 234 246 257 267 347 356 358 478 568 678

The partitions of #305 and #305i combine to give the indecomposable partitions {1,1,1,1,1,1,1}, {1,1,1,1,1,2}, and {1,1,1,2,2}. Next we partition #305c.

1+3 partition of #305c	
λ=1	017 026 038 045 128 135 146 234 257 367 478 568
λ=3	015 016 018 025 027 028 034 036 037 047 048 056 123 124 126 137 138 145 147 157 168 235 238 245 246 267 278 346 347 356 358 458 468 567 578 678

2+2 partition of #305c	
λ=2	015 016 027 028 036 037 045 048 126 128 137 138 145 147 234 235 246 257 346 358 478 567 568 678
λ=2	017 018 025 026 034 038 047 056 123 124 135 146 157 168 238 245 267 278 347 356 367 458 468 578

Partitions of #305 and #305c combine to give {1,1,1,1,3}, {1,1,2,3}, and {1,2,2,2}.

The remaining partitions are obtained by examining an indecomposable *TS*(9,3), #6 in [HC].

Indecomposable #6	
λ=3	012 013 014 023 025 036 045 047 058 067 068 078 124 125 136 138 146 157 158 167 178 237 238 246 247 256 268 278 365 354 347 348 357 458 468 567

The complement *TS*(9,4), #6c, of #6 has two partitions as follows:

1+3 partition of #6c	
λ=1	016 024 038 057 128 137 145 235 267 346 478 568
λ=3	015 017 018 026 027 028 034 035 037 046 048 056 123 126 127 134 135 147 148 156 168 234 236 245 248 257 258 358 367 368 378 456 457 467 578 678

2+2 partition of #6c	
λ=2	015 016 024 028 037 038 046 057 127 128 134 137 148 156 235 236 245 267 346 358 457 478 568 678
λ=2	017 018 026 027 034 035 048 056 123 126 135 145 147 168 234 248 257 258 367 368 378 456 467 578

Partitions of #6 and #6c combine to give the {2,2,3} and {1,3,3} partitions.

Appendix Two: $IPP(10,\Lambda)$

We need only present the {2,2,4} and {4,4} partitions here. Examples follow:

2+2+4 partition	
λ=4	016 017 018 019 026 027 028 029 035 037 038 039 045 046 048 049 056 057 126 127 128 129 134 135 138 139 145 147 148 156 159 167 235 236 237 238 245 246 247 249 258 259 345 346 347 368 369 379 469 478 489 567 568 578 579 589 678 679 689 789
λ=2	012 013 025 034 047 058 067 069 089 123 146 149 157 158 168 179 234 248 256 269 278 279 356 359 367 378 389 457 459 468
λ=2	014 015 023 024 036 059 068 078 079 124 125 136 137 169 178 189 239 257 267 268 289 348 349 357 358 456 458 467 479 569

4+4 partition	
λ=4	016 017 018 019 026 027 028 029 035 037 038 039 045 046 048 049 056 057 126 127 128 129 135 136 138 139 145 147 148 149 156 157 234 235 236 237 245 246 247 258 259 289 345 348 349 367 369 378 467 468 479 568 569 578 579 589 678 679 689 789
λ=4	012 013 014 015 023 024 025 034 036 047 058 059 067 068 069 078 079 089 123 124 125 134 137 146 158 159 167 168 169 178 179 189 238 239 248 249 256 257 267 268 269 278 279 346 347 356 357 358 359 368 379 389 456 457 458 459 469 478 489 567

Appendix Three: $INPP(10,\Lambda)$

1+1+1+1+1+1+1+1+1+1 near partition for v=10
245 780 289 340 237 567 590 260 358 468 479 369
145 356 189 390 137 348 678 160 469 579 580 470
256 467 290 140 127 248 459 789 158 570 680 169
125 367 578 130 238 359 560 890 269 168 179 270
124 236 478 689 349 460 167 190 138 370 279 280
235 347 589 790 450 157 278 120 249 148 380 139
346 458 690 180 123 156 268 389 350 259 149 240
457 569 170 129 234 267 379 490 135 146 360 250
568 670 128 230 345 378 480 150 246 257 147 136
679 178 239 134 456 489 159 126 357 368 258 247

2+2+1+1+1+1+1+1 near partition for v=10
245 289 034 237 026 358 468 369 356 039 137 348 678 579 058 047 467 459 789 057 069 018 156 149
078 567 059 479 145 189 016 469 256 029 014 127 248 158 068 169 346 458 123 389 268 035 259 024
679 178 239 134 456 489 159 126 357 368 258 247
125 367 578 130 238 359 560 890 269 168 179 270
124 236 478 689 349 460 167 190 138 370 279 280
235 347 589 790 450 157 278 120 249 148 380 139
457 569 170 129 234 267 379 490 135 146 360 250
568 670 128 230 345 378 480 150 246 257 147 136

2+2+2+1+1+1+1 near partition for v=10
123 120 135 149 140 168 178 179 237 245 248 269 270 289 346 347 389 380 390 478 490 579 580 670
124 128 134 130 159 150 167 169 236 239 240 256 250 279 349 356 358 370 457 459 468 460 690 890
125 126 136 137 145 148 170 189 234 238 247 259 260 278 345 350 368 379 467 469 480 567 568 578
230 249 258 267 348 357 369 456 470 590 680 789
138 146 157 190 340 359 367 458 479 560 689 780
129 147 158 160 246 257 280 450 489 569 678 790
127 139 156 180 235 268 290 360 378 570 589 679

2+2+2+2+1+1 near partition for v=10
130 149 158 167 230 249 258 267 347 348 357 359 368 369 457 450 468 460 569 560 789 780 790 890
125 126 139 140 156 178 179 180 238 248 256 279 270 290 350 367 459 467 578 570 589 689 680 690
123 127 134 147 150 168 189 190 239 245 240 269 278 280 346 358 378 370 390 479 489 480 579 670
124 128 137 138 146 157 159 160 236 237 246 257 250 289 345 349 356 380 458 478 470 568 678 679
120 135 148 169 234 259 268 360 389 456 490 580
129 136 145 170 235 247 260 340 379 469 567 590

2+2+2+2+2 near partition for v=10
130 149 158 167 230 249 258 267 347 348 357 359 368 369 457 450 468 460 569 560 789 780 790 890
125 126 139 140 156 178 179 180 238 248 256 279 270 290 350 367 459 467 578 570 589 689 680 690
123 127 134 147 150 168 189 190 239 245 240 269 278 280 346 358 378 370 390 479 489 480 579 670
124 120 136 138 148 157 159 169 235 237 246 259 268 289 345 349 360 389 456 478 490 568 580 679
128 129 135 137 145 146 160 170 234 236 247 257 250 260 340 356 379 380 458 469 470 567 590 678

Annals of Discrete Mathematics 34 (1987) 119–136
© Elsevier Science Publishers B.V. (North-Holland)

Cubic Neighbourhoods in Triple Systems

Charles J. Colbourn

Department of Computer Science
University of Waterloo
Waterloo, Ontario, N2L 3G1
CANADA

Brendan D. McKay

Department of Computer Science
Australian National University
Canberra, A.C.T.
AUSTRALIA

TO ALEX ROSA ON HIS FIFTIETH BIRTHDAY

ABSTRACT

In a triple system of index 3, the multiset of pairs appearing in triples with a fixed element form a cubic multigraph called the neighbourhood of the element. We prove that, with precisely three exceptions, every cubic multigraph appears as an element neighbourhood in a triple system. The proof technique involves establishing the existence of certain path factorizations of cubic graphs.

1. Neighbourhoods in Triple Systems

A *triple system* B[3,λ;v] is a v-set of *elements* together with a collection of 3-element subsets called *blocks* or *triples,* with the property that every 2-element subset appears in precisely λ triples. Considering all triples containing a fixed element, say x, and then eliminating x from each triple gives a collection of 2-subsets of elements, called the *neighbourhood* of x, and denoted $N(x)$. In a B[3,λ;v], $N(x)$ can be interpreted as the edges of a multigraph on $v-1$ vertices; in fact, $N(x)$ is a λ-regular multigraph.

The *neighbourhood problem* is to determine which λ-regular multigraphs appear as element neighbourhoods in B[3,λ;v] designs. Easy counting arguments show that an n-vertex λ-regular multigraph can only be a neighbourhood if $\lambda n \equiv 0 \pmod 2$ and $\lambda n(n-1) \equiv \lambda n \pmod 6$; these are the *basic necessary conditions*. For λ=1, the basic necessary conditions are sufficient, as every element in every B[3,1;v] has a 1-factor as its neighbourhood, and these are the only candidate multigraphs. Hence the problem here is settled by observing that B[3,1;v] designs exist for every $v\equiv 1,3 \pmod 6$. For $\lambda=2$, the necessary conditions require a 2-regular multigraph with $n\equiv 0,2 \pmod 3$ vertices; Colbourn and Rosa [4] and Hilton and Rodger [7] showed that

Theorem 1.1: For each 2-regular multigraph G with $n\equiv 0,2 \pmod 3$ vertices, there is a B[3,2;n+1] having G as an element neighbourhood, with two exceptions: $C_2 \cup C_3$, and $C_3 \cup C_3$.

The techniques developed in [4] can be extended to settle the neighbourhood problem for all even λ. Odd λ poses a more serious problem, however. We address the first case, $\lambda=3$, here. In this case, the basic necessary conditions require $3n\equiv 0 \pmod 2$ and $3n(n-1)\equiv 3n \pmod 6$; in other words, they simply require that n be even. Hence every cubic multigraph is a candidate neighbourhood. We will show that with three exceptions, every cubic multigraph is indeed a neighbourhood.

Before delving into technical details, it is important to remark on reasons for interest in the neighbourhood problem. A primary computational reason is that neighbourhoods are used as isomorphism invariants, used as heuristics to certify nonisomorphism of designs (see, for example, [2,8]). In addition, the neighbourhood problem arises as a natural restriction of a very difficult problem in design theory, the leave problem (see [5]). A particular reason for interest in the case $\lambda=3$, is that unlike in the even λ cases, the proof employs a substantial graph-theoretic argument.

2. Path Factorizations of Cubic Graphs

In this section, we develop a number of results on cubic (simple) graphs, which at first seem remote from the neighbourhood problem; the connection will be made in sections 5-7.

A *path factor* in a cubic graph $G=(V,E)$ is a spanning subgraph of G which is a union of paths; a *nontrivial path factor (NTPF)* is a path factor in which each path contains at least one edge.

Theorem 2.1: Every simple cubic graph has an NTPF.

Proof:

Let $H=(V,F)$ be a path factor of $G=(V,E)$ having the minimum number of paths of length zero. If there are none, H is an NTPF. Otherwise, suppose $\{x\}$ is a path of length zero in H. Consider the vertices $\{a,b,c\}$ at distance one from x in G. If any is the endpoint of a path in H, this path can be extended to include x, thereby reducing the number of length zero paths. On the other hand, if one of them, say a, appears in a path of length at least 3, say $...,r,a,s,t,...$, this path can be replaced by $...,r,a,x$ and $s,t,...$, again reducing the number of length zero paths. If neither of these happen, each of a,b,c is a centre of a path of length two, which requires that a,b,c are mutually nonadjacent; we then proceed to consider vertices at distance two from x in G.

Considering vertices at distance i from x, we identify two cases according to the parity of i. If i is even, we proceed as follows. Consider a vertex at distance i. Since we were forced to proceed to distance i, a vertex at this level has degree two in $(V,E-F)$. If, in $(V,E-F)$, two vertices at this level were adjacent, say y,z, find the unique path from y to x in the levels explored thus far, and add $\{z,y\}$ at the beginning of this path. Interchanging the edges in F and those in $E-F$ along this path results in a path factor with fewer paths of length zero. Similarly, if any vertex y at distance i is involved in a (possibly closed) path of three or more edges $(\ldots,r,y,s,t,\ldots)$, we interchange the edges in F and $E-F$ along the path from s through y to x, again reducing the number of length zero paths. The only case which remains is for each vertex at distance i to be the centre of a path of length two in $E-F$, and we proceed to distance $i+1$. The case of odd parity for i is similar with the roles of F and $E-F$ interchanged.

Proceeding in this manner must ultimately provide a means for reducing the number of length zero paths, since the graph is finite. Hence our assumption that H had the minimum number of length zero paths is incorrect. □

Corollary 2.2: For any cubic simple graph $G=(V,E)$, there is a 1-factor (V,F) for which the multigraph $(V,E \cup F)$ is Hamiltonian.

Proof:

Use theorem 2.1 to produce an NTPF for G. In the 1-factor, choose edges which join the paths of the NTPF into a Hamilton cycle, and select the remaining edges arbirarily on the remaining vertices. □

It is interesting to note that if $H=(V,F)$ is an NTPF, then $H'=(V,E-F)$ is a subgraph with maximum degree two. Hence it is possible that H' is also an NTPF; in this case, the graph has a *path factorization (PFZ)* (H,H').

Theorem 2.3: Every simple cubic graph has a PFZ.

Proof:

For a simple cubic graph $G=(V,E)$, let $H=(V,F)$ be an NTPF with the minimum number of paths; let $H'=(V,E-F)$. If H' is acyclic, we are done. Otherwise, H' contains a cycle $(x_1,\ldots,x_k)$. Each x_i is the endpoint of a path in H. But every path has at most two endpoints on the cycle; since $k \geq 3$, this ensures that two neighbours on the cycle $\{x_i,x_{i+1}\}$ are endpoints of different paths. Adding the edge $\{x_i,x_{i+1}\}$ to H therefore reduces the number of paths in H and doesn't create a cycle, which contradicts the assumption that H has the minimum number of paths. □

Corollary 2.4: One can add a 1-factor (V,F) to a simple cubic graph $G=(V,E)$ in such a way that the multigraph $(V,E \cup F)$ has a partition into two Hamilton cycles.

Proof:

Use the theorem to obtain a PFZ (H,H') for G. Select the edges of F to connect the paths of H into a Hamilton cycle, and independently connect the paths of H' into a Hamilton cycle. This can be done since endpoints of paths in H are midpoints of paths in H', and vice versa. □

In the design theory application, we require certain special kinds of path factorizations. An *i-path factorization (i-PFZ)* is a path factorization (H,H') in which one of the path factors has at least i paths of length one. Theorem 2.3 states that every simple cubic graph has a 0-PFZ.

Theorem 2.5: Every simple cubic graph having some component other than K_4 has a 1-PFZ.

Proof:

In view of theorem 2.3, it suffices to show that some component has a 1-PFZ. Consider then some connected component G which is not a K_4. Use theorem 2.3 to produce a PFZ for G, (H,H'). Assume in addition that H has the fewest possible paths. Then all paths in H' have length at most three, for otherwise consider a path in H' with interior vertices $(x_1,...,x_k)$, $k \geq 3$. Each is the endpoint of a path in H, but then some neighbouring pair form endpoints of different paths in H, which would contradict minimality of H.

Then suppose that H' has a path of length three, (a,b,c,d). Since H has the fewest paths, it contains a path with endpoints b and c; consider the neighbour of b on this path, z. Now z cannot be a or c since the graph is simple, but it could be d. If z is not d, remove $\{b,z\}$ from H and add $\{b,c\}$ (then H' contains a path of length one). Otherwise, consider the neighbour y of c. If y is not a, remove $\{y,c\}$ from H and add $\{b,c\}$. Finally we have the case $y=a$ and $z=d$. In this event, consider the third neighbour x of a. It cannot be d, since G is not K_4. Then remove $\{a,x\}$ from H and add $\{b,c\}$.

The remaining possibility is that H' contains only paths of length two. Then H' accounts only for $\frac{2n}{3}$ edges, and hence H accounts for $\frac{5n}{6}$. Thus H contains paths of length at least four. Let $(r,a,s,t,...)$ be such a path in H, and (a,b,c) be a path of H'. Examine the path of H whose endpoint is b. This path may visit both a and c, one of them, or neither. In all cases, we can remove $\{a,s\}$ from H, and add $\{a,b\}$, leaving H' with the length one path $\{b,c\}$. This completes the proof. □

Corollary 2.6: A 1-factor (V,F) can be added to a simple cubic graph on n vertices $G=(V,E)$ so that $(V,E \cup F)$ can be partitioned into a C_n and a $C_{n-2} \cup C_2$ whenever G contains a component other than K_4.

Proof: easy. □

Finally, we will require a characterization of graphs having 2-PFZ's; this is more complicated and requires some preliminary results.

Theorem 2.7: A simple cubic n-vertex graph has an NTPF with at least $\frac{n}{6}$ paths of length one.

Proof:

Choose an NTPF $H=(V,F)$ of $G=(V,E)$ having a *maximum* number of paths. H contains only paths of length one and two, as longer paths can always be broken up. Let m be the number of paths of length one, p the number of length two paths with nonadjacent ends, and q the number of length two paths with adjacent ends. Observe that $2m+3p+3q=n$. Let x be the number of edges joining vertices in a path of length one to vertices in a path of length two; obviously, $x \leq 4m$. Any edge joining two paths of length two must have one end being the midpoint of a path of length two, or one could obtain three paths of length one in the place of the two of length two. But then $x \geq 3(p+q)$. Combining the inequalities, $3p+3q \leq 4m$, whence $n \leq 6m$. □

Corollary 2.8: Every simple cubic graph has a NTPF with at least two paths of length one.

Proof:

This follows from theorem 2.7 for cubic graphs with $n \geq 8$; for $n=4,6$, observe that an NTPF with exactly one path of length one and the remainder of length two cannot exist. □

This theorem and corollary do not immediately give us 2-PFZ's; to obtain these, we first introduce a restricted NTPF. Define a *house* to be the induced graph depicted in figure 2.1.

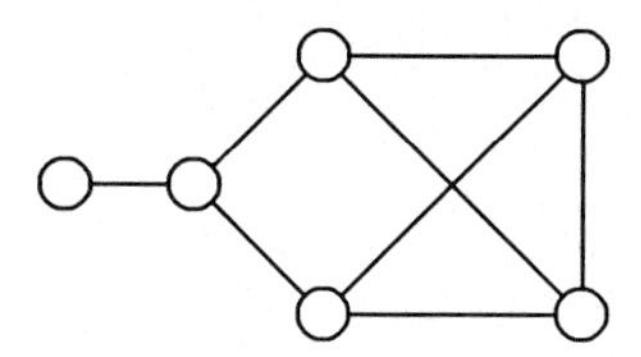

Figure 2.1: A house

We define a *SHNTPF* to be an NTPF with at least two paths of length one, counting at most one in each house. It is an easy matter to verify that

Corollary 2.9: Every cubic graph has a SHNTPF. □

Finally, we are in a position to state the main result:

Theorem 2.10: Every connected simple cubic graph with $n \geq 8$ vertices has a 2-PFZ.

Proof:

Take a SHNTPF $H=(V,F)$ of $G=(V,E)$ in which the number of cycles of $H'=(V,E-F)$ is minimum. If there are no such cycles, we are finished. Otherwise we derive a contradiction. In each case, we either establish a method for reducing the number of cycles, or we produce a smaller graph (still with at least eight vertices) for which the existence of a 2-PFZ ensures the existence of a 2-PFZ for G. So let C be a cycle in H'. Every vertex on C is the endpoint of a path in H. If any two neighbours on C are endpoints of different paths each of length two or more, the edge between these two neighbours can be added to H without reducing the number of paths of length one, while removing one cycle from H' (a contradiction). Similarly, if any two neighbours, say x,y, on C are endpoints of the same path of length at least two, let z be the neighbour of x on this path. Adding $\{x,y\}$ to H, and removing $\{x,z\}$, reduces the number of cycles while retaining all length one paths. If C has nine or more vertices, then C is incident with at least three paths of length one in H. Then there is an edge in C that is incident with one such path that can be removed from H' and placed in H, leaving at least two paths in H and leaving one cycle fewer in H'. This argument ensures that C has at most eight vertices, and eliminates many cases for smaller cycle lengths. We consider the remaining possibilities here.

If C has three vertices, the only case which remains requires that the three vertices of C have three distinct neighbours not on C. Form a graph G' by identifying the three vertices of C. If G' has six vertices, the theorem is easily verified. Otherwise, G' has at least eight vertices but is smaller than G; hence G' has a 2-PFZ, which is easily extended to a 2-PFZ for G.

If C has four vertices, two cases remain. We depict these cases here, drawing edges in H solid and edges in $G-H$ dashed:

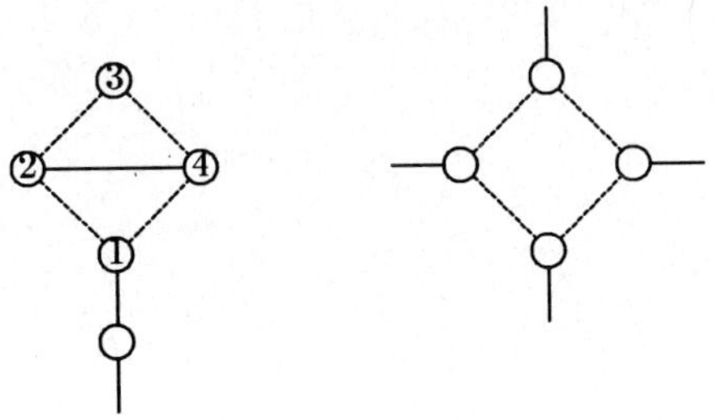

In these two situations, we replace by the configurations drawn below:

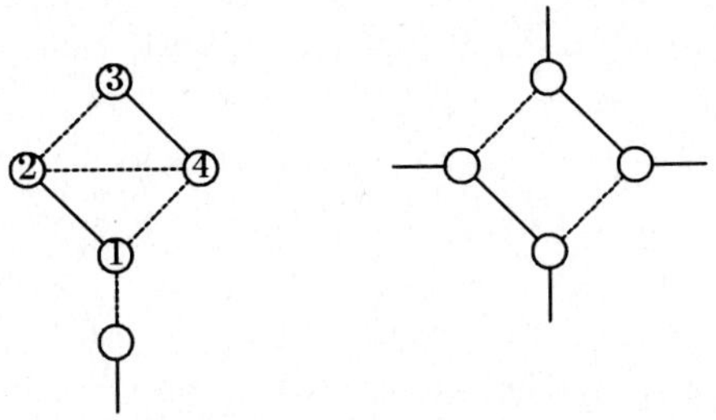

Two remarks are in order. The first transformation applies even when $\{1,2\}$ and $\{3,4\}$ are paths of length greater than one, and hence this handles some of the cases for longer cycle lengths; we refer to this generalized replacement as "handling 1-chorded cycles" (actually, the cycles handled may also have chords between intermediate vertices on the path $\{1,2\}$ and intermediate vertices on the path $\{3,4\}$ in addition to the required chord $\{2,4\}$). When the second transformation is applied, the two required paths of length one appear in $G-H$ rather than H. This necessitates that any cycles in $G-H$ which may remain be destroyed by joining paths in H into longer paths, but since we no longer require any paths of length one in H, this is straightforward.

Next we proceed to consider C having five vertices. Handle the cases with two consecutive long paths, and singly chorded cycles. The only case which remains is that C is the 5-cycle in a house. In this case, H contains two paths of length one in the house, but since H is a SHNTPF, we can reallocate the edges of the house as shown here:

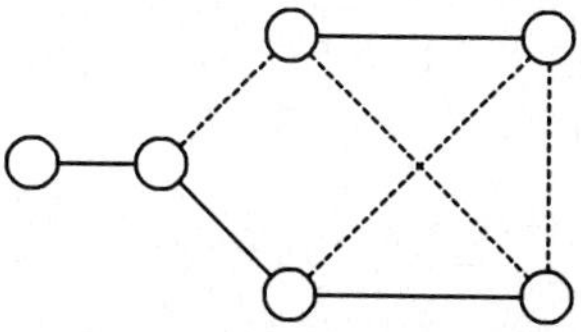

The resulting H still has two paths of length one.

Proceeding to cycles of length six, again handling consecutive long paths and 1-chorded cycles leaves only one case, depicted here:

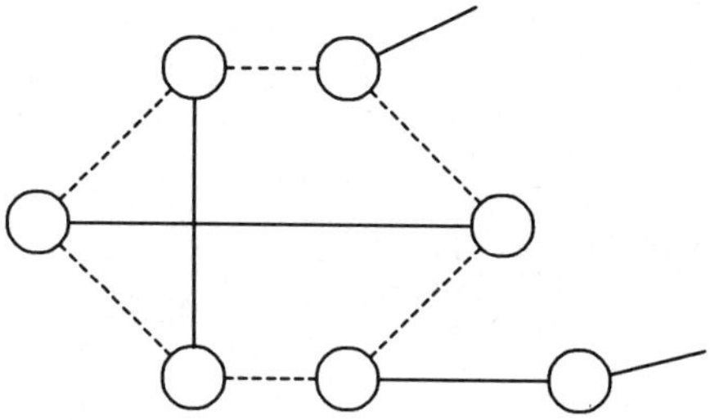

This is replaced by:

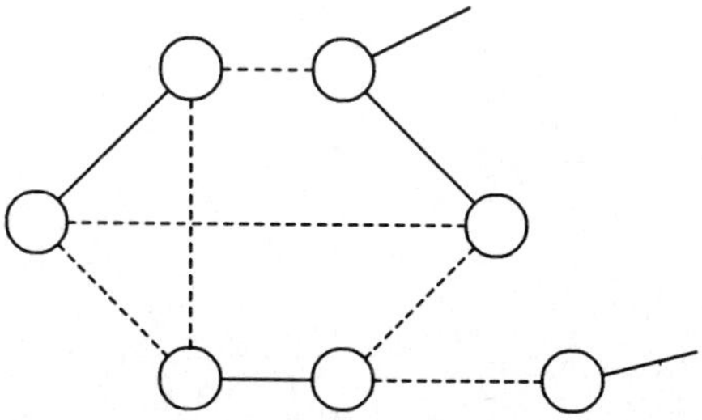

Finally, when C has seven or eight vertices, all cases not handled by consecutive long paths are handled by 1-chorded cycles. □

Corollary 2.11: Every simple cubic graph has a 2-PFZ unless it consists entirely of K_4's, or has one component of size 6 and the remainder consists of K_4's. □

3. Path factorizations of cubic multigraphs

In this section, we extend some of the results from section 2 to handle cubic multigraphs. We extend the results of the previous section, by observing that cubic multigraphs are made from simple cubic graphs by the following three operations:

1. Adding a 3-bond
   Add a new component, a *3-bond,* consisting of two new vertices a and b and three parallel edges from a to b.
2. Adding a 2-bond
   Replace an edge $\{x,y\}$ by

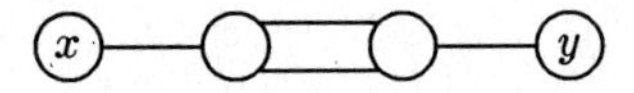

3. Adding a bowtie
   Replace an edge $\{x,y\}$ by

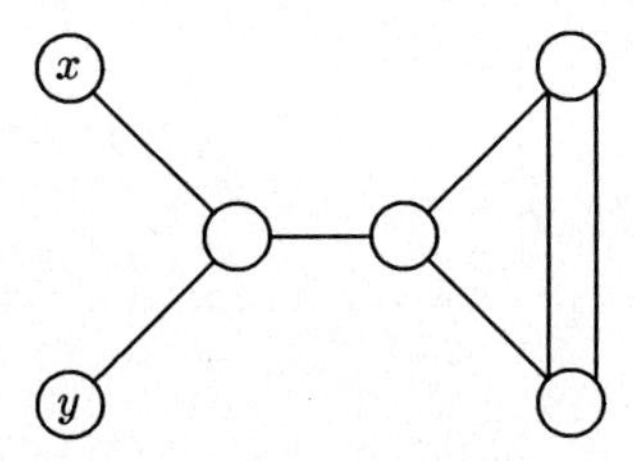

It is easy to verify that every cubic multigraph arises using these operations, by considering all configurations in which a multiple edge can appear. Using these operations, we extend the theorems of section 2.

Theorem 3.1: Every cubic multigraph has an NTPF.

Proof:

We proceed by induction on the number of vertices in G. If G is empty or simple, theorem 2.1 gives the NTPF. Otherwise G is obtained from G' by one of the operations given; let H be an *NTPF* for G'. If G is obtained by adding a 3-bond, add one copy of the edge to H to get an NTPF for G. If G is obtained by adding a 2-bond (x,a,b,y) for x,y, add $\{a,b\}$ to H if $\{x,y\}$ is not in H, and $\{\{x,a\},\{a,b\},\{b,y\}\}$ to H if $\{x,y\}$ is in H. If G is obtained by adding a bowtie, it is again an easy matter to extend the NTPF for G' to one for G. □

Theorem 3.2: Every cubic multigraph having no 3-bond has a PFZ.

Proof:

If G is simple or empty, theorem 2.3 gives the required PFZ. Otherwise, G is formed from a smaller multigraph G' by adding either a 2-bond or a bowtie. The extension of a PFZ for G' is trivial, so we need only consider the cases when G' is itself a 3-bond. Adding a 2-bond or a bowtie to a 3-bond gives graphs having PFZ's (an easy exercise). □

This also extends corollary 2.4, and has a second interesting corollary:

Corollary 3.3: For every cubic multigraph $G=(V,E)$ having no 3-bond, there is a 1-factor (V,F) for which $(V,E \cup F)$ has a partition into two Hamilton cycles. □

Corollary 3.4 (Shannon's theorem, [9]): Every cubic multigraph has a 4-edge-colouring.

Next we extend the result on 1-PFZ's; first, we depict some small exceptional multigraphs:

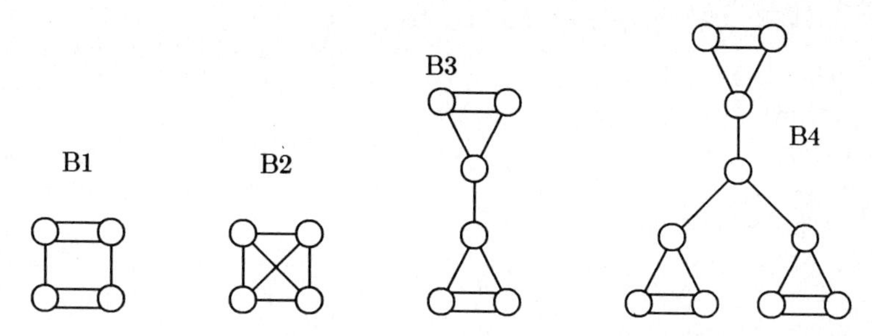

Theorem 3.5: Every cubic multigraph with no 3-bond having some component other than B1,B2,B3, or B4 has a 1-PFZ.

Proof:

Let C be a component of G which is none of B1-B4. Since G has no 3-bond, theorem 3.2 extends a 1-PFZ of C to one of G. If C is simple, it is not K_4, and hence has a 1-PFZ by theorem 2.5. Suppose C is constructed from C' by adding a 2-bond; then it is easy to verify that C has a 1-PFZ if C' has a PFZ at all. By theorem 3.2, B1 is the only graph that can be constructed by adding a 2-bond to a smaller graph and yet does not have a 1-PFZ. Hence we may assume that all 2-bonds of C appear in bowties. Adding a bowtie may reduce the number of paths of length one in the PFZ; however, using the techniques of theorem 2.1, one can ensure that any PFZ in which the replaced edge $\{x,y\}$ is the unique path of length one can be modified to a PFZ in which $\{x,y\}$ appears at the end of a longer path. Extending this modified PFZ to a 1-PFZ for G is straightforward.

Hence the only cases to consider are those graphs obtained by adding bowties to graphs having no 1-PFZ. The proof is completed by observing that adding a bowtie to a 3-bond, or to any of B1-B4, either produces a graph with a 1-PFZ or produces one of B3 or B4. □

Corollary 3.6: For any cubic n-vertex multigraph $G=(V,E)$ having no 3-bond and having some component other than B1-B4, there is a 1-factor (V,F) for which $(V,E \cup F)$ has a partition into C_n and $C_{n-2} \cup C_2$. □

It would be useful to also characterize cubic multigraphs having 2-PFZ's, but this appears to be substantially more complicated. Fortunately, we shall be able to avoid the necessity for such a characterization.

4. Leaves

In this section, we develop some preliminary results in design theory concerning a related (more general) problem, the leave problem. A *partial triple system* PB[3,λ;v] is a collection of 3-subsets (triples) of a v-set of elements, in which each 2-subset appears in at most λ triples. Its *leave* is the multigraph of unordered pairs of λK_v which are not covered by the partial triple system. If there is a B[3,λ;v] having $N(x)=G$, the removal of all triples containing x gives a PB[3,λ;$v-1$] with leave G. Hence the neighbourhood problem is a special case of a much more general *leave problem:* what graphs are leaves of a PB[3,λ;n]? Few sufficient conditions are known, and most involve very sparse graphs; we review the known sufficient conditions to be used later. A *quadratic* graph has all vertices of degree zero or two.

Theorem 4.1 [5]: Let G be an n-vertex e-edge triangle-free quadratic graph with either $n\equiv1,3$ (mod 6) and $e\equiv0$ (mod 3) or $n\equiv5$ (mod 6) and $e\equiv1$ (mod 3). Then G is the leave of a PB[3,1;n] unless $G=C_4\cup C_5$. □

Theorem 4.2 [11]: For each v, there is a PB[3,1;v] whose leave is empty if $v\equiv1,3$ (mod 6), $v/2$ disjoint edges if $v\equiv0,2$ (mod 6), $K_{1,3}$ and $(v-4)/2$ disjoint edges if $v\equiv4$ (mod 6), or C_4 and $v-4$ isolated vertices if $v\equiv5$ (mod 6). □

Theorem 4.3 [4]: Let G be an n-vertex e-edge quadratic multigraph with either $n=e\equiv0$ (mod 3) or $n=e\equiv2$ (mod 3). Then G is the leave of a PB[3,2;n] unless $G=C_2\cup C_3$ or $G=C_3\cup C_3$. □

A related result was first proved in [3], and is also an easy consequence of theorem 4.1:

Theorem 4.4: Let $X=\{x_1,\ldots,x_{2n}\}\cup\{a,b,c\}$ with $n\equiv0,2$ (mod 3). Let F be a 1-factor on $\{x_i\}$ together with $\{b,c\}$, and let F' be a disjoint 1-factor on $\{x_i\}$ together with $\{a,c\}$. Then there exists a B[3,1;$2n+3$] with $N(a)=F$ and $N(b)=F'$.

We require all four of these known theorems on leaves, and prove in addition some closely related results. We assume familiarity with the techniques of [4,5] to avoid repetition here.

Theorem 4.5: Let G be a quadratic multigraph with $n\equiv4$ (mod 6) vertices and $e\equiv0$ (mod 3) edges. Then G is the leave of a B[3,2;n].

Proof:

We consider two cases, in each considering only the parts which differ from the proof of theorem 4.3. First assume $n\equiv4$ (mod 12), and write $n=12t+4$. It suffices to partition the following pure and mixed differences into triples, leaving five 1-factors:

pure(0): 1,2,2,3,3,...,3t-1,3t-1,3t

mixed: 1,2,2,3,3,...,6t-1,6t-1

pure(1): 1,2,2,3,3,...,3t-1,3t-1,3t

To solve this, we employ perfect and hooked Langford sequences [10]. On the pure(1) differences, we place triples from two Langford sequences, chosen according to the class of t (mod 4), as follows:

t	sequences	omitted difference
0 (mod 4)	$(t,1)$ and $(t-1,2)$	$3t-1$
1 (mod 4)	$(t,1)$ and $(t-1,2)$	$3t-1$
2 (mod 4)	$(t,2)$ and $(t-1,1)$	$3t-2$
3 (mod 4)	$(t,1)$ and $(t-1,3)$	2

On the pure(0) and mixed differences, we take the sequences $(3t-1,1)$ and $(3t-2,2)$. Having done this, one pure(0) difference, one pure(1) difference, and three mixed differences remain. This leaves five 1-factors as required in each case.

The second case is $n\equiv 10 \pmod{12}$; write $n=12t+10$. Here we must partition the following pure and mixed differences into triples, leaving five 1-factors:

pure(0): 1,2,2,3,3,...,3t+1,3t+1

mixed: 1,2,2,3,3,...,6t+2,6t+2

pure(1): 1,2,2,3,3,...,3t+1,3t+1

On the pure(1) differences, take two $(t,2)$ Langford sequences; if they are hooked, this creates no problem as $3t+2\equiv-(3t+1) \pmod{6t+3}$. On the pure(0) and mixed differences, take two $(3t,2)$ Langford sequences. The remaining five differences give 5 1-factors by taking any two of the mixed differences which remain as 1-factors; the final three 1-factors are obtained from a 1-factorization of the "prism" induced by the remaining one mixed, one pure(0), and one pure(1) difference. This completes the second case. □

In addition to this extension of theorem 4.3, we require an extension to theorem 4.4 to handle the case $v\equiv 5 \pmod 6$. In this case, there is no B[3,1;v] design, and so we come "as close as possible" by producing instead a covering in which precisely one pair is covered three times:

Theorem 4.6: Let $X=\{a,b,c,d,e\}\cup\{x_1,...,x_{6t}\}$. Let F,F' be disjoint 1-factors on $\{x_i\}$. Then there exists a covering of the pairs of X by triples, in which each pair appears once, except $\{d,e\}$ which appears three times, and moreover, $N(a)=F\cup\{\{b,c\},\{d,e\}\}$ and $N(b)=F'\cup\{\{a,c\},\{d,e\}\}$.

Proof:

Using theorem 4.1, produce a PB[3,1;$6t+3$] on $X-\{a,b\}$ whose leave is the triangle $\{c,d,e\}$ together with the quadratic graph $F\cup F'$. Now add triples $\{a,b,c\},\{a,d,e\},\{b,d,e\},\{c,d,e\}$, and triples $\{\{a,r,s\}|\{r,s\}\in F\}$ and $\{\{b,r,s\}|\{r,s\}\in F'\}$. The resulting covering has the required properties. □

Finally, we need one result on leaves which are not quite quadratic. An *almost quadratic* multigraph is the result of adding two parallel edges between two nonadjacent vertices of a 2-regular bipartite multigraph.

Theorem 4.7: Let G be an almost quadratic multigraph on $n\equiv 4 \pmod 6$ vertices. Then G is the leave of a PB[3,2;n] unless G is

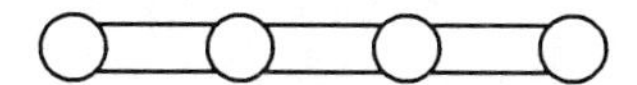

Proof:

The idea is to superimpose two PB[3,1;n]'s to obtain a PB[3,2;n] with leave G. Note that theorem 4.2 shows that there is a PB[3,1;n] whose leave is $K_{1,3}$ together with $(n-4)/2$ disjoint edges; we call this a 1^* *factor*. Notice that G consists of a collection of even cycles together with one "double chord".

Case 1: The double chord $\{a,b\}$ connects vertices on the same cycle at even distance.

In this case, G has a 1^*-factor F so that $G-F$ is also a 1^*-factor. Then superimposing a PB[3,1;n] with leave F and a PB[3,1;n] with leave $G-F$ gives a PB[3,2;n] with leave G.

Case 2: The double chord $\{a,b\}$ connects vertices on the same cycle at odd distance.

In this case, G has no 1^*-factor. Now suppose that a and b are at distance $2s-1$ $(s\geq 2)$ on the cycle; without loss of generality, the cycle contains vertices $(x_0,...,x_{2t-1})$ with double chord $\{x_0,x_{2s-1}\}$. Form G' from G by replacing the double chord with a different double chord $\{x_0,x_{2s-2}\}$. Now use the method in case 1 to form a PB[3,2;n] with leave G'; in addition, ensure that there is an element e for which $\{e,x_{2s-3},x_{2s-1}\}$ and $\{e,x_{2s-2},x_{2s}\}$ are blocks (this is easy to ensure). Remove these two blocks and replace them with $\{e,x_{2s-3},x_{2s-2}\}$ and $\{e,x_{2s-1},x_{2s}\}$. The resulting PB[3,2;n] has leave G.

Case 3: The double chord connects vertices in different cycles.

Suppose that the double chord connects cycles of length $2r$ and $2s$. Form G' by replacing these two cycles by a cycle of length $l=2r+2s$, with the double chord connected vertices at distance $2s$. Without loss of generality, the new cycle is $(x_0,...,x_{l-1})$ with double chord $\{x_0,x_{2s}\}$. Using the method of case 1, form a PB[3,2;n] with leave G' and containing the blocks $\{e,x_0,x_{2s-1}\}$ and $\{e,x_{2s},x_{l-1}\}$. This can always be done if G has more than four vertices. Then remove these two blocks, and add $\{e,x_0,x_{l-1}\}$ and $\{e,x_{2s-1},x_{2s}\}$ to obtain a PB[3,2;n] with leave G. □

Finally, we are in a position to combine the graph-theoretic and design-theoretic preliminaries to address the neighbourhood problem.

5. Cubic multigraphs on $n\equiv 0,2 \pmod 6$ vertices

In this section, we consider the easier two cases. First, we prove a theorem to handle the bulk of the cases:

Theorem 5.1: Let G be a cubic multigraph on $n\equiv 0,2 \pmod 6$ vertices which has at least one component with a 1-PFZ. The G is a neighbourhood in a B[3,3;$n+1$].

Proof:

Let G' be G with all 3-bonds removed, and let (H,H') be a 1-PFZ for G', with $\{x,y\}$ a path of length one in H. Add a 1-factor F to G' so as to obtain a 2-factorization (Q,Q') of the union, in which Q' is Hamiltonian, and Q contains a 2-cycle on $\{x,y\}$ and a single long cycle on the remaining vertices (as in corollary 3.6). Now for each 3-bond $\{a,b\}$ in $G-G'$, add a 2-cycle on $\{a,b\}$ to each of Q and Q'.

Next observe that Q consists of only even cycles, and thus has a partition into 1-factors F_1,F_2. Now using theorem 1.1, let B_1 be a B[3,2;$n+1$] with $N(\infty)=Q'$. Next, using theorem 4.4, let B_2 be a B[3,1;$n+1$] with $N(\infty)=F_1$ and $N(x)=F_2-\{x,y\}\cup\{\infty y\}$.

The union, B, of B_1 and B_2, is a B[3,3;n+1] with $N(\infty)=F_1 \cup Q'$, and $G-N(\infty) \subset N(x)$.

Let $X_{out}=N(\infty)-G$ and $X_{in}=G-N(\infty)$. Observe that X_{out} and X_{in} are 1-factors on the same set of vertices, and hence we can obtain another B[3,3;n+1] by removing triples $\{\{\infty,a,b\} | \{a,b\} \in X_{out}\}$ and $\{\{x,a,b\} | \{a,b\} \in X_{in}\}$, and adding triples $\{\{\infty,a,b\} | \{a,b\} \in X_{in}\}$ and $\{\{x,a,b\} | \{a,b\} \in X_{out}\}$. The result has $N(\infty)=G$ as required. □

This handles all cases, except where each component is a 3-bond or one of B1-B4. The majority of these cases are handled by a simple result:

Theorem 5.2: Let G be a cubic multigraph on $n \equiv 0,2 \pmod 6$ vertices. Then if G has a 1-factor, and G is not B3, G is an element neighbourhood in a B[3,3;n+1].

Proof:

Let F be a 1-factor of G, and let $Q=G-F$. Form a B[3,1;n+1] with $N(\infty)=F$, form a B[3,2;n+1] with $N(\infty)=Q$, and take their union to obtain a B[3,3;n+1] with $N(\infty)=G$. By theorem 4.2, this is always possible unless $Q=C_3 \cup C_3$; the only graph G having only a 1-factor leaving $C_3 \cup C_3$ is B3. □

At this point, it is easy to verify that the only candidate neighbourhoods which remain contain at least one B4, and the remainder is entirely 3-bonds and copies of B1-B4. To handle this last case, we make a small modification in the proof of theorem 5.1:

Lemma 5.3: Let G be a cubic multigraph on $n \equiv 0,2 \pmod 6$ vertices, having one or more B4's, and zero or more copies of 3-bonds and B1-B3. Then G is a neighbourhood in a B[3,3;n+1].

Proof:

Let G' be obtained from G by removing all 3-bonds and one copy of B4 (G' may be empty). Now by theorem 3.2, G' has a PFZ (H,H'); extend this by adding a 1-factor to form two Hamilton cycles (Q,Q') on the vertices of G'. Now for each 3-bond $\{a,b\}$, add the 2-cycle $\{a,b\}$ to each of Q and Q'. Finally, we deal with the B4 which remains. We number the vertices of a B4 for reference:

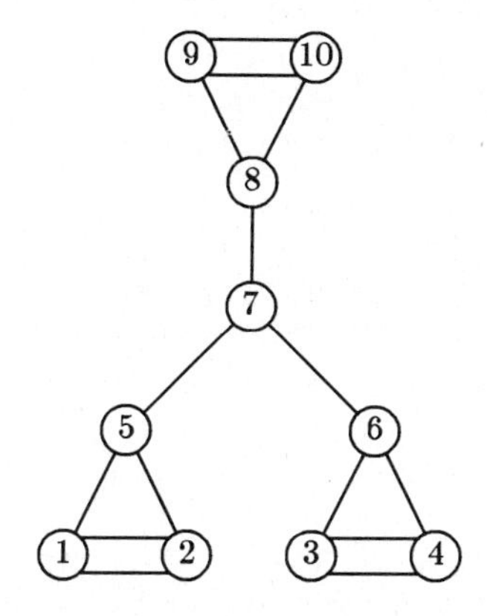

Now add to Q the 2-cycle (5,7) and the 8-cycle (1,2,6,4,3,8,9,10), and add to Q' the 3-cycle (1,2,5) and the 7-cycle (4,3,6,7,8,10,9). Now we proceed just as in theorem 5.1, using (5,7) in the role of $\{x,y\}$. Notice that Q' now doesn't contain only even cycles, but theorem 1.1 doesn't require this. □

In summary, we have proved that

Theorem 5.4: Every cubic multigraph with $n \equiv 0,2 \pmod 6$ vertices is a neighbourhood, except for B3. □

6. Simple cubic graphs, $n \equiv 4 \pmod 6$

The remaining case is more difficult primarily because $n+1 \equiv 5 \pmod 6$ is not an admissible order for a B[3,1;$n+1$] or for a B[3,2;$n+1$]. Nevertheless, the more complicated machinery introduced earlier comes to bear here.

Theorem 6.1: Every simple cubic graph with $n \equiv 4 \pmod 6$ vertices having a 2-PFZ appears as a neighbourhood in a B[3,3;$n+1$].

Proof:

Let (H,H') be a 2-PFZ for G, with two paths of length one in H, $\{x,y\}$ and $\{u,v\}$. Add a 1-factor F to G to transform the 2-PFZ into a pair of 2-factors (Q,Q'), so that Q' is a Hamilton cycle and Q is a 2-cycle on $\{x,y\}$, a 2-cycle on $\{u,v\}$, and an $n-4$-cycle. Since Q has only even cycles, write Q as a union of two 1-factors F_1,F_2.

From Q', form an almost quadratic multigraph Q^* by adding the edge $\{x,y\}$ twice. Use theorem 4.7 to construct a PB[3,2;n] with leave Q^*; appending a point ∞ to this gives a PB[3,2;$n+1$] B_1 having $N(\infty)=Q'$ and having leave C_2 on $\{x,y\}$.

Next use theorem 4.6 to produce a covering with each pair covered once except that $\{x,y\}$ is covered three times, and having $N(\infty)=F_1$ and $N(u)=F_2-\{u,v\}\cup\{\infty v\}$. Adding B_1 to this covering produces a B[3,3;$n+1$] with $N(\infty)=F_1 \cup Q'$. As in theorem 5.1, we now simply exchange edges in $G-N(\infty)$ and $N(\infty)-G$. □

Since in this section we deal only with simple graphs, theorem 2.10 ensures that every cubic graph with a component of at least eight vertices has a 2-PFZ; hence, we need only handle simple cubic graphs with each component of size four or six. Every such cubic graph has a 1^*-factor (a $K_{1,3}$ and $n-4/2$ disjoint edges); we prove a somewhat stronger statement than that needed here:

Lemma 6.2: Let G be a cubic multigraph on $n \equiv 4 \pmod 6$ vertices having a 1^*-factor. Then G is a neighbourhood of a B[3,3;$n+1$].

Proof:

Let F be a 1^*-factor of G and $Q=G-F$. By theorem 4.2, there is a PB[3,1;n] with leave F, and by theorem 4.5 there is a PB[3,2;n] with leave Q; their union is a PB[3,3;n] with leave G. By adding a new element ∞, we get a B[3,3;$n+1$] with $N(\infty)=G$, as required. □

7. Cubic Multigraphs, $n \equiv 4$ (mod 6)

We do not have a characterization of cubic multigraphs having 2-PFZ's, and hence are unable to use theorem 6.1 directly. Nevertheless, we are able to circumvent the need for such a characterization.

Lemma 7.1: Suppose that G is a cubic multigraph on $n \equiv 4$ (mod 6) vertices, and that G is constructed by replacing an edge $\{a,b\}$ not in a 3-bond of G_1 by adding a bowtie. Then G is a neighbourhood in a B[3,3;$n+1$].

Proof:

Form G' by removing all 3-bonds from G_1, and let (H,H') be a PFZ for G' (by theorem 3.2); suppose without loss of generality that $\{a,b\} \in H'$. Now the bowtie added replaces the edge $\{a,b\}$ by edges $\{a,u\},\{b,u\},\{u,v\},\{v,x\},\{v,y\},\{x,y\}$, *and* $\{x,y\}$. Of these edges, place $\{u,v\}$ and both copies of $\{x,y\}$ in H and the remainder in H' (removing $\{a,b\}$ in the process). Now we proceed exactly as in theorem 6.1, with u,v,x,y serving the same roles. □

Next we handle cases resulting from addition of a 2-bond:

Lemma 7.2: Suppose that G is a cubic multigraph on $n \equiv 4$ (mod 6) vertices, constructed from a cubic multigraph G' by adding a 2-bond, and adding any number of 3-bonds. Further suppose that G' has a 1-PFZ. Then G is the neighbourhood of a B[3,3;$n+1$].

Proof:

Let (H,H') be a 1-PFZ for G', and suppose that a 2-bond is added by replacing the edge $\{a,d\}$ by $\{a,b\},\{b,c\},\{b,c\},\{c,d\}$. If $\{a,d\}$ is a path of length one in H, we replace it by $\{a,b\},\{c,d\}$ and add $\{b,c\}$ twice to H'; now we proceed as in theorem 6.1 with a,b,c,d serving the roles of u,v,x,y. If, on the other hand, $\{a,d\}$ appears in the middle of a path of length three of more in H, we replace it by $\{b,c\}$ twice, and add $\{a,b\},\{c,d\}$ to H'. Interchanging the roles of H and H' enables us to proceed as before.

Finally, suppose that $\{a,d\}$ appears at the end of a path of length two or more in H, and that d is an endvertex in this path. Replace $\{a,d\}$ by $\{a,b\},\{c,d\}$ in H, and add $\{b,c\}$ to H' twice. Now since (H,H') is a 1-PFZ, there is a path of length one $\{r,s\}$ either in H or in H'. If it is in H, apply the method in theorem 6.1 with r,s,c,d playing the roles of u,v,x,y. If it is in H', interchange H and H' and apply theorem 6.1 with r,s,b,c playing the roles of u,v,x,y. In all cases, this application is straightforward; one need only ensure that when H' is completed to a quadratic graph, the edge $\{x,y\}$ isn't added, and also that when H is completed to a quadratic graph, the only repeated edges are $\{u,v\}$ and $\{x,y\}$. These conditions are easy to ensure. □

In addition, theorem 6.1 is easily adapted to handle any cubic multigraph having a simple component on eight or more vertices. This, together with lemmas 7.1 and 7.2, leaves only the following as possible components:

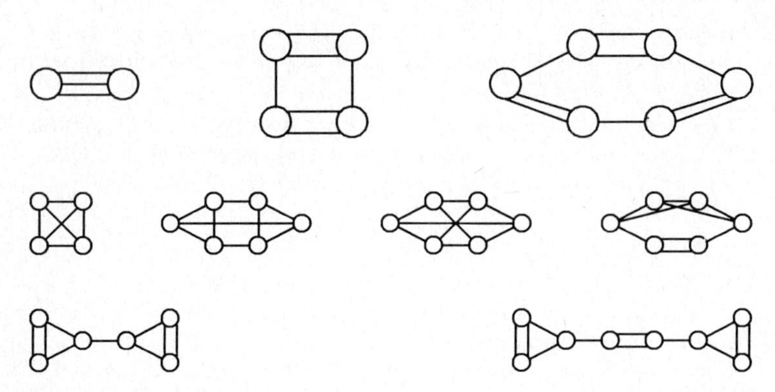

All of these graphs have a 1-factor, and all but the three in the first row have a 1^*-factor. If a cubic multigraph not handled by previous methods has any component other than those in the first row, it is handled by lemma 6.2. What remains is only those cubic multigraphs having components depicted in the first row. All such cubic multigraphs are 3-edge-colourable, and so we establish one more lemma:

Lemma 7.3: Let G be a 3-edge-colourable cubic multigraph with $n \equiv 4 \pmod 6$ vertices having some component other than a 3-bond. Then G is a neighbourhood in a B[3,3;$n+1$] unless G=B1.

Proof:

Let F_1, F_2, F_3 be a 3-edge-colouring of G, with some edge $\{x,y\}$ appearing in F_1, but in neither F_2 nor F_3; this is always possible since G is not entirely 3-bonds. Now let $Q = F_2 \cup F_3 \cup \{\{x,y\},\{x,y\}\}$, and form a B[3,2;$n$] with leave Q using theorem 4.7. This yields a PB[3,2;$n+1$] with $N(\infty) = F_2 \cup F_3$ which leaves $\{x,y\}$ uncovered. Next form a covering with $\lambda=1$ having $N(\infty) = F_1$ and covering $\{x,y\}$ three times. Combining these two yields a B[3,3;$n+1$] with $N(\infty) = G$. □

One final case remains: G contains only 3-bonds. In this case, we use the following lemma of Hanani [6]:

Lemma 7.4: A group divisible design on v elements with blocks of size 3, groups of size 2, and $\lambda=3$, exists whenever $v \equiv 0 \pmod 2$, $v \neq 4$. □

The triples from such a GDD form the blocks of a PB[3,3;v] whose leave is $v/2$ 3-bonds, which gives a B[3,3;$v+1$] with $N(\infty)$ being all 3-bonds, so long as v is not 4.

In the sequence of lemmas developed in sections 5-7, we have dealt with all cubic multigraphs except three: B1, B3, and two 3-bonds. In each of these three cases, the graphs are *not* neighbourhoods, and we prove a simple lemma to show this.

Lemma 7.5: Let G be an n-vertex e-edge graph having a minimal edge cutset of size c separating G into components of size s and $n-s$. If G is a leave of a PB[3,3;n],

$$3\binom{s}{2} + 3\binom{n-s}{2} - (e-c) \geq 2\left[3s(n-s) - c\right]$$

Proof:

The left hand side counts edges which remain "inside" the set of s elements or the set of $n-s$ elements. Now triples use either three of these inside edges, or they use one inside edge and two edges crossing from one set to the other. Hence the number of such edges must exceed twice the number of "cross" edges which remain. □

For B1, we have $n=4, e=6, s=2, c=2$; if B1 is a leave (or a neighbourhood), we require by the lemma $6 + 6 - 4 \geq 2[12 - 2]$, which is not satisfied. For B3, we have $n=6, e=9, c=1, s=3$ and for two 3-bonds we have $n=4, e=6, c=0, s=2$; in both cases, the requirements of the lemma fail.

In summary, we can state the main characterization theorem.

Theorem 7.6: A cubic multigraph G on n vertices is an element neighbourhood of a B[3,3;$n+1$] unless G is B1, B3, or two 3-bonds. □

8. Concluding Remarks

Theorem 7.6 lends strong evidence in support of a conjecture that every λ-regular multigraph meeting numerical conditions and the conditions of lemma 7.5 are realized as neighbourhoods in triple systems; in fact, theorem 7.6 provides the major necessary tool to prove this conjecture for all simple graphs [1]. Perhaps the major reason for interest in the results here is that they provide graph-theoretical tools for addressing problems involving leaves and neighbourhoods.

Acknowledgements

Thanks to Alex Rosa, and to the referee, for helpful comments. The first author also acknowledges the financial support of NSERC Canada.

References

[1] C.J. Colbourn, "Simple neighbourhoods in triple systems", submitted for publication, 1986.

[2] C.J. Colbourn, M.J. Colbourn, J.J. Harms, and A. Rosa, "A complete census of (10,3,2) block designs and of Mendelsohn triple systems of order ten. III. (10,3,2) block designs without repeated blocks", *Proc. Thirteenth Manitoba Conf. Num. Math. Comput.* (1982) 211-234.

[3] C.J. Colbourn, M.J. Colbourn, and A. Rosa, "Completing small partial triple systems", *Discrete Math.* 45 (1983) 165-179.

[4] C.J. Colbourn and A. Rosa, "Element neighbourhoods in twofold triple systems", *J. Geometry,* to appear

[5] C.J. Colbourn and A. Rosa, "Quadratic leaves of maximal partial triple systems", *Graphs and Combinatorics* 2 (1986) 317-337.

[6] H. Hanani, "Balanced incomplete block designs and related designs", *Discrete Math.* 11 (1975) 255-369.

[7] A.J.W. Hilton and C.A. Rodger, "Triangulating nearly complete graphs of odd order", in preparation.

[8] R.A. Mathon and A. Rosa, "A census of Mendelsohn triple systems of order nine", *Ars Combinatoria* 4 (1977) 309-315.

[9] C.E. Shannon, "A theorem on coloring the lines of a network", *J. Math. Phys.* 28 (1949) 148-151.

[10] J.E. Simpson, "Langford sequences: perfect and hooked", *Discrete Math.* 44 (1983) 97-104.

[11] J. Spencer, "Maximal consistent families of triples", *J. Comb. Theory* 5 (1968) 1-8.

Annals of Discrete Mathematics 34 (1987) 137–144
© Elsevier Science Publishers B.V. (North-Holland)

The Geometry of Subspaces of an S(λ;2,3,v)

Michel Dehon

Département de Mathématique
Université Libre de Bruxelles
1050 Brussels
BELGIUM

Luc Teirlinck

Division of Mathematics (Algebra, Combinatorics, and Analysis)
Auburn University
Auburn, Alabama 36849
U.S.A.

TO ALEX ROSA ON HIS FIFTIETH BIRTHDAY

ABSTRACT

We define some linear spaces on the set of all proper subspaces of a triple system S(λ;2,3,v). The connected components of these linear spaces are projective spaces of order 2 and punctured projective spaces of order 3, i.e. projective spaces of order 3 from which a point has been deleted. We show how these connected components can be used to find affine and projective factors in S(λ;2,3,v).

1. Introduction

A *2-covering* is a pair (S,β) where S is a set whose elements are called *points,* and β is a family of subsets of S called *blocks,* such that any two distinct points of S are contained in at least one block and every block contains at least two points. We allow repeated blocks. Where no confusion is possible, we denote (S,β) simply by S. A *subspace* of a 2-covering S is a subset S_1 of S such that any block having at least two points in S_1 is contained in S_1. Clearly, any intersection of a family of subspaces is a subspace. If (S,β) is a 2-covering and if S_1 is a subspace of S, we denote by $\beta|S_1$ the family of all blocks of S contained in S_1. A *connected component* of a 2-covering S is a connected component of the graph obtained on S by joining x and y if and only if x and y are contained in a block B with $|B|\geq 3$. Every connected component is a subspace. A *linear space* is a 2-covering in which any two distinct points x and y are joined by exactly one block, denoted by xy. The blocks of a linear space are usually called *lines.*

Let λ,k and v be integers, $\lambda>0$, $k\geq3$, $v\geq0$. A *2-design* $S(\lambda;2,k,v)$ is a 2-covering (S,β), $|S|=v$, such that every pair of distinct points is contained in exactly λ blocks and every block contains exactly k points. Clearly, if S_1 is a subspace of (S,β), then $(S_1,\beta|S_1)$ is an $S(\lambda;2,k,|S_1|)$. If $\lambda=1$, we write $S(2,k,v)$ instead of $S(1;2,k,v)$.

Let S be an $S(\lambda;2,3,v)$ and let $X(S)$ denote the set of all proper subspaces of S. Let $L_3(S)$ denote the set of all 3-subsets $\{S_1,S_2,S_3\}$ of $X(S)$ such that $S_1\cap S_2=S_1\cap S_3=S_2\cap S_3$ and $S_1\cup S_2\cup S_3=S$. Let $L_2(S)$ denote the set of all 2-subsets of $X(S)$ that are not contained in an element of $L_3(S)$. Put $L(S) = L_2(S)\cup L_3(S)$. Clearly, $(X(S),L(S))$ is a linear space. In section 2, we prove that the connected components of $(X(S),L(S))$ are projective spaces of order 2 (possibly of dimension 0 or 1).

If A and B are sets, $A\Delta B = (A-B)\cup(B-A)$ denotes the symmetric difference. If $A,B\subset S$ for some set S, put $A+_S B = S-(A\Delta B) = (A\cap B)\cup(S-(A\cup B))$.

2. The structure of $L(S)$

Let S be an $S(\lambda;2,3,v)$.

Lemma 1: If S_1 is a subspace of S and if $p\in S-S_1$, then the number of blocks containing p and disjoint from S_1 equals $\lambda(v-2|S_1|-1)/2$.

Proof:

The point p is contained in $\lambda(v-1)/2$ blocks of S. For every point x of S_1, there are λ blocks containing $\{p,x\}$ and every block through p has at most one point in S_1. Thus there are $\lambda|S_1|$ blocks through p intersecting S_1 and $\lambda(v-1)/2-\lambda|S_1| = \lambda(v-2|S_1|-1)/2$ blocks through p disjoint from S_1. □

Lemma 2: If $S_1\in X(S)$, we have $|S_1|\leq(v-1)/2$, and equality holds if and only if every block of S has nonempty intersection with S_1.

Proof:

This is an immediate consequence of Lemma 1. □

Lemma 3: Let $S_1,S_2\in X(S)$, $S_1\not\subset S_2$, $|S_1|=v_1$, $|S_2|=v_2$. Put $S_3 = S_1+_S S_2$ and $|S_3|=v_3$. Then

(1) $v_3\geq v_2$ with equality holding if and only if every block joining a point of S_1-S_2 and a point of $S_3-(S_1\cap S_2)$ has its third point in S_2-S_1.

(2) $|S_1\cap S_2|\geq(v_1+2v_2-v)/2$ with equality holding if and only if $v_2=v_3$.

(3) S_3 is a subspace if and only if $v_1=v_2=v_3$ if and only if $v_1=v_2$ and $|S_1\cap S_2| = (3v_1-v)/2$.

Proof:

Let $p\in S_1-S_2$. Counting, in two different ways, the number of blocks containing p, one point of $S_3-(S_1\cap S_2)$ and one point of S_2-S_1, yields (1).

Clearly $|S_1\cap S_2| = (v_1+v_2+v_3-v)/2$. Together with (1), this yields (2).

If S_3 is a subspace, then (1) yields $v_1=v_2=v_3$.

Assume $v_1=v_2=v_3$. (Obviously, this implies $S_2\not\subset S_1$). We will prove that S_3 is a subspace. As $S_1\cap S_2$ is a subspace, any block containing two points of $S_1\cap S_2$ is contained in $(S_1\cap S_2)\subset S_3$. Any block joining a point in $S_1\cap S_2$ and a point of

$S_3-(S_1 \cap S_2)$ must have its third point in S_3. Indeed, if this third point would lie in S_1 (S_2), the block would have exactly two points in S_1 (S_2), contradicting our assumption that S_1 (S_2) is a subspace. Finally, any block joining two points of $S_3-(S_1 \cap S_2)$ must have its third point in S_3. Indeed, if this third point would lie in S_1-S_2 (S_2-S_1), then by (1) we would have $v_3>v_2$ $(v_3>v_1)$.

Thus S_3 is a subspace if and only if $v_1=v_2=v_3$. Clearly $v_1=v_2=v_3$ if and only if $v_1=v_2$ and $|S_1 \cap S_2| = (3v_1-v)/2$. □

Theorem 1: Every connected component of $(X(S),L(S))$ is a projective space of order 2.

Proof:

We first prove that the relation on $X(S)$ defined by $S_1 \sim S_2$ if and only if $S_1=S_2$ or the line joining S_1 and S_2 has three points, is an equivalence relation. Obviously, $\sim$ is reflexive and symmetric.

Assume that $S_1 \sim S_2 \sim S_3$. By lemma 3, we have $|S_1|=|S_2|=|S_3|$. If $S_2=S_3$, then $S_1 \sim S_3$. Otherwise, let $\{S_2,S_3,S_4\} \in L(S)$. If $S_1 \in \{S_2,S_3,S_4\}$, then $S_1 \sim S_3$. Otherwise, $\{S_1 \cap S_2,\ S_1 \cap S_3,\ S_1 \cap S_4\} \in L(S_1)$. Thus $|S_1 \cap S_3| = |S_1 \cap S_2| = (3|S_1|-v)/2$ and $S_1 \sim S_3$.

Thus if P is a connected component of $(X(S),L(S))$, then all lines of P have three points. This means that P is a subspace of the projective space of order 2 defined on all proper subsets of S by $+_S$. □

3. Generalized Moore-products with projective multiplier

A *transversal design* T(λ;k,l) is a triple (S,β,G) where S is a kl-set, β is a collection of k-subsets of S called *blocks,* and $G=\{A_1,...,A_k\}$ is a partition of S into l-subsets called *groups,* such that

(i) $|B \cap A_i| = 1$ for all $B \in \beta$ and $i=1,...,k$, and

(ii) if $x \in A_i$, $y \in A_j$, $1 \le i < j \le k$, then $\{x,y\}$ is contained in exactly λ blocks.

It is well known that a T(λ;3,l) exists for all $l>0$.

Construction 1: Let W be an S(λ;2,3,w) and let $Y_1,...,Y_n$ be S(λ;2,3,w') containing W as a subsystem, such that $Y_i \cap Y_j=W$ for all $1 \le i < j \le n$. Let Q be an S(2,3,n) whose set of points is $\{Y_1, \ldots ,Y_n\}$. For each block B of Q, let B' be a T(λ;3,$w'-w$) whose groups are the sets $Y-W$ for $Y \in B$. Then we can construct an S(λ;2,3,$w+n(w'-w)$) on $S = Y_1 \cup \cdots Y_n$ whose blocks are:

(1) The blocks of W;

(2) The blocks of Y_i, $1 \le i \le n$, not contained in W;

(3) The blocks of B', B a block of Q.

We call an S(λ;2,3,$w+n(w'-w)$) constructed in the above way, a *generalized Moore-product* with *distinguished set* W, *components* $Y_1,...,Y_n$ and *multiplier* Q.

If P is a set of subsets of a set S, we denote by $\delta(P)$ the lattice of all intersections of elements of P, ordered by inclusion. We put the intersection of zero elements equal to S, so that $S \in \delta(P)$.

Theorem 2: If P is any connected subspace of $(X(S),L(S))$, $\dim P=n$, S an $S(\lambda;2,3,v)$, then:

(1) $\delta(P)$ is the lattice of all subspaces of a projective space P^* of order 2, $\dim P^*=\dim P$, and $\left|\bigcap_{p\in H\in P} H - \bigcap_{H\in P} H\right| = \dfrac{(v-|\bigcap_{H\in P} H|)}{(2^{n+1}-1)}$ for all $p\in S-\bigcap_{H\in P} H$.

(2) S is a generalized Moore-product with distinguished set $\bigcap_{H\in P} H$, components $\bigcap_{p\in H\in P} H$, $p\in S-\bigcap_{H\in P} H$, and multiplier P^*.

Proof:

Claim (1) is obvious, for instance, from a combination of Lemma 3 (3), Theorem 1, and the corollary to Theorem 3 of [3]. Claim (2) is an obvious consequence of (1). □

4. Some results on the large subdesigns of $S(\lambda;2,3,v)$

In sections 4 through 7, S will be an $S(\lambda;2,3,v)$.

Lemma 3 shows that if S_1 and S_2 are in the same connected component of $(X(S),L(S))$, then $|S_1|=|S_2|$. Thus the set $X(S,v')=\{S_1\in X(S): |S_1|=v'\}$ is a union of connected components of $(X(S),L(S))$. We will call S *r-projective* if $X(S,(v-r)/2)$ is a projective space of order 2. (In particular, this projective space may be empty, a single point, or a single line.)

Lemma 4: Let S_1, $S_2\in X(S,v')$, $S_1\neq S_2$, and $|S_1\cap S_2|=s$. Denote by λ' (λ'', respectively) the number of blocks of S (S_2, respectively) containing a given point $p\in S-S_1$ ($q\in S_2-S_1$, respectively) and disjoint from S_1 ($S_1\cap S_2$, respectively). We have

(a) $2(\lambda'-\lambda'')(v'-s)\leq(v-2v'+s)\lambda'$.

(b) $(2r-a+1)v\leq 10r^2-6ra+a^2-4r+a$, where $r=v-2v'$ and $a=v-4s$.

Proof:

Lemma 1 shows that λ' and λ'' are independent of p and q. Counting in two different ways the number of ordered pairs (x,B), $x\in S-(S_1\cup S_2)$, B a block, $x\in B$, $|B-(S_1\cup S_2)|=2$, $|B\cap(S_2-S_1)|=1$, we get $2(\lambda'-\lambda'')|S_2-S_1|\leq|S-(S_1\cup S_2)|\lambda'$, which yields (a).

We now prove (b). We have $v'=(v-r)/2$ and $s=(v-a)/4$. By Lemma 1, we have $\lambda'=\lambda(r-1)/2$ and $\lambda''=\lambda(a-r-2)/4$. Plugging these values of v', s, λ', and λ'' into (a) and simplifying, yields (b). □

Let S_1, $S_2\in X(S,(v-r)/2)$, $S_1\neq S_2$. Put $s=|S_1\cap S_2|$. Lemma 2 shows that $s\leq[(v-r)/2-1]/2=(v-r-2)/4$. Lemma 4b shows that, if v is large enough with respect to r, we have $s\leq(v-2r-1)/4$. Lemma 3 shows that $s\geq[(v-r)/2+(v-r)-v]/2=(v-3r)/4$. Combining Lemma 3 and Theorem 1 shows that S_1 and S_2 are in the same connected component of $(X(S),L(S))$ if and only if $s=(v-3r)/4$. A well known necessary and sufficient condition for the existence of a $S(\lambda;2,3,v)$, $v\geq3$, is $\gcd(6,v-2)\,|\,\lambda$. Applying this condition to v, $(v-r)/2$ and s is, for certain small r and congruences modulo 6 of v and λ, sufficient to exclude all s with $(v-3r)/4<s\leq(v-2r-1)/4$ and thus to prove the r-projectivity of S. This works, at

least for some congruences modulo 6 of v and λ, up to $r=8$ if r is even and up to $r=15$ if r is odd. It would take too much space to list all of the cases in which the above technique works. In the following theorem, we list only those cases in which, at least for some λ, the method works for all but a finite number of v.

Theorem 3:

(a) Every $S(\lambda;2,3,v)$ is 1-projective. (This is a known result; see, for instance, [1,2,3]).

(b) Every $S(\lambda;2,3,v)$ with $v\neq 4$ is 2-projective.

(c) If S is an $S(\lambda;2,3,v)$ with $v\notin\{5,9\}$, then either S is 3-projective or $v\equiv 3 \pmod 4$ and $\lambda\equiv 0 \pmod 6$.

(d) Every $S(\lambda;2,3,v)$ with $\lambda\equiv 1,5 \pmod 6$ and $v\notin\{7,19\}$ is 5-projective.

5. Affine Factors of S(λ;2,3,v)

If $Y\in X(S)$, let $\tilde{Y}$ denote the connected component of Y in $(X(S),L(S))$. If $\dim \tilde{Y}=1$, put $V(Y)=\bigcap_{Z\in\tilde{Y}} Z$. We define an equivalence relation $\mathcal{R}$ on the set of all $Y\in X(S)$ with $\dim\tilde{Y}=1$, by putting $Y_1\mathcal{R}Y_2$ if and only if $V(Y_1)=V(Y_2)$. Obviously $\tilde{Y}\subset\mathcal{R}(Y)$ for all $Y\in X(S)$, $\dim\tilde{Y}=1$.

Theorem 4: Let $Y\in X(S)$, $\dim\tilde{Y}=1$. For each $x\in S-V(Y)$, put $\Omega_x(Y)=\bigcap_{Z\in\mathcal{R}(Y),x\in Z} Z$. Put $Aff(Y)=\{\Omega_x(Y):x\in S-V(Y)\}$ and $L_{Aff}(Y)=\{\{\Omega_x(Y),\Omega_y(Y),\Omega_z(Y)\}\subset Aff(Y): \{x,y,z\}\subset S-V(Y),\ |\{\Omega_x(Y),\Omega_y(Y),\Omega_z(Y)\}|=3,\ \{x,y,z\}\in\beta\}$. (Note that we consider $L_{Aff}(Y)$ as a set, i.e. as containing no repeated elements.) Then

(1) $(Aff(Y),L_{Aff}(Y))$ is an affine space of order 3.

(2) We have $|\Omega_x(Y)|=|\Omega_y(Y)|$ for all $x,y\in S-V(Y)$. Moreover, S is a generalized Moore-product with distinguished set $V(Y)$, components $\Omega_x(Y)$ for $x\in S-V(Y)$ and multiplier $(Aff(Y),L_{Aff}(Y))$.

Proof:

If $x,y\in S-V(Y)$, $\Omega_x(Y)\neq\Omega_y(Y)$, then clearly $\Omega_x(Y)\cap\Omega_y(Y)=V(Y)$. As $\lambda>0$, there is at least one block $\{x,y,z\}\in\beta$. We have $\{\Omega_x(Y),\Omega_y(Y),\Omega_z(Y)\}\in L_{Aff}(Y)$. Thus $(Aff(Y),L_{Aff}(Y))$ is a 2-covering. Moreover, for each $Y_1=\{Y_1,Y_2,Y_3\}$, $Y_1\in\mathcal{R}(Y)$, the sets $Y'_i=\{\Omega_x(Y):x\in Y_i-V(Y)\}$, $i=1,2,3$, are affine hyperplanes of $(Aff(Y),L_{Aff}(Y))$ in the sense of [3]. Thus, (1) follows from [3, section 3, corollary 3]. Claim (2) is an easy consequence of (1). □

6. Some further results on the equivalence relation $\mathcal{R}$

If $Y\in X(S)$ and $\dim\tilde{Y}\geq 1$, then obviously $|Y|\geq v/3$ and, by Lemma 2, Y is maximal in $X(S)$.

Lemma 5: Let $A_1,B_1\in X(S)$, $\tilde{A}_1\neq\tilde{B}_1$, $\dim\tilde{A}_1\geq 1$, $\dim\tilde{B}_1\geq 1$. Then the following are equivalent.

(1) $\dim\tilde{A}_1=\dim\tilde{B}_1=1$ and $A_1\mathcal{R}B_1$.

(2) $|A_1|=|B_1|$ and $|A_1-B_1| = |S-(A_1\cap B_1)|/4$.

Proof:

As an easy consequence of Theorem 4, (1) $\rightarrow$ (2).

We now prove that (2) $\rightarrow$ (1). Let $\{A_1,A_2,A_3\}\in L_3(S)$ and $\{B_1,B_2,B_3\}\in L_3(S)$. Obviously, $\{B_1\cap A_1, B_1\cap A_2, B_1\cap A_3\}\in L_3(B_1)$. Thus $|B_1\cap A_1| = |B_1\cap A_2| = |B_1\cap A_3|$ and $|A_1-B_1| = |A_2-B_1| = |A_3-B_1|$. We have $3|S-(A_1\cap B_1)|/4 = |S-B_1| = |A_1-B_1| + |A_2-B_1| + |A_3-B_1| - 2|(A_1\cap A_2)-B_1| = 3|S-(A_1\cap B_1)|/4 - 2|(A_1\cap A_2)-B_1|$. It follows that $(A_1\cap A_2)-B_1 = \emptyset$. Thus $A_1\cap A_2\subset B_1$. As B_1 and B_2 play symmetrical roles, we have also that $A_1\cap A_2\subset B_2$. Thus $A_1\cap A_2\subset B_1\cap B_2$. By symmetry, $B_1\cap B_2\subset A_1\cap A_2$ and thus $A_1\cap A_2=B_1\cap B_2$.

Assume that $A_4\in\tilde{A}_1$, $A_4\notin\{A_1,A_2,A_3\}$. We get $A_1\cap A_2=B_1\cap B_2 = A_1\cap A_4$ and $A_4=(A_2\cap A_4)\cup(A_3\cap A_4)$. Thus A_4 is a union of two of its proper subspaces, a contradiction. This means that no such A_4 exists and $\dim \tilde{A}_1=1$. Similarly, $\dim \tilde{B}_1=1$.

As $\dim \tilde{A}_1=\dim \tilde{B}_1=1$ and $A_1\cap A_2=B_1\cap B_2$, we have $A_1 \mathcal{R} B_1$. □

Lemma 6: Let $\{A_1,A_2,A_3\}\in L_3(S)$, $C\in X(S)$, $A_1\cap A_2\subset C$, $|C|=|A_1|$, $C\notin\{A_1,A_2,A_3\}$. Then $|A_1-C|=|S-(A_1\cap C)|/4$.

Proof:

We have $|A_1-C|=|A_2-C|=|A_3-C|$. As $|A_1|=|C|$, we have $|A_1-C|=|C-A_1|$. Thus $|S-(A_1\cap C)|=|A_1-C| + |A_2-C| + |A_3-C| + |C-A_1| = 4|A_1-C|$. □

Lemma 7: Let $A,B\in X(S)$, $\tilde{A}\neq\tilde{B}$, $\dim \tilde{A}\geq 1$, $\dim \tilde{B}\geq 1$, $A_2\in\tilde{A}$, $A_2\neq A$. Then the following are equivalent.

(1) $\dim \tilde{A}=\dim \tilde{B}=1$ and $A \mathcal{R} B$.

(2) $|A| = |B|$ and $A\cap A_2\subset B$.

Proof:

Lemma 7 follows immediately from Lemmas 5 and 6. □

Lemma 8: Let $A,B\in X(S)$, $\dim \tilde{A}\geq 1$, $\dim \tilde{B}\geq 1$, $\tilde{A}\neq\tilde{B}$, $C\in X(S)$, $C\notin\{A,B\}$, $A\cap B\subsetneq C$. Then $A\cap B=B\cap C=C\cap A$, $\dim \tilde{A}=\dim \tilde{B}=\dim \tilde{C}=1$ and $A \mathcal{R} B \mathcal{R} C$.

Proof:

We have $A\cap B\in X(A)$ $(X(B))$ and $\dim \widetilde{A\cap B}\geq 1$ in $X(A)$ $(X(B))$. Thus $A\cap B$ is a maximal element of both $X(A)$ and $X(B)$ and $A\cap B=B\cap C=C\cap A$. Let $\{A,A_2,A_3\}\in L_3(S)$. We have $|A\cap B|=|A_2\cap B| = |A_3\cap B| = |A_2\cap C| = |A_3\cap C|$ and $|B| = |A\cap B\cap A_2| + 3(|A\cap B| - |A\cap B\cap A_2|) = 3|A\cap B| - 2|A\cap B\cap A_2| = 3|A\cap C| - 2|A\cap C\cap A_2| = |C|$. Thus $|B|=|C|$. Similarly, $|A|=|C|$.

As $\tilde{A}\neq\tilde{B}$, we have $S-(A\cup B\cup C)\neq\emptyset$. As $|A_2|=|A_3|$ and $|A_2\cap B|=|A_3\cap B| = |A_2\cap C|=|A_3\cap C|$, we have $|A_2-(A\cup B\cup C)| = |A_3-(A\cup B\cup C)|\neq 0$. Thus there is a point $p\in A_2-(A\cup B\cup C)$. The third point of any block joining point p and a point $q\in(A\cap B)-A_2$ lies in $A_3-(A\cup B\cup C)$. This yields $|A_3-(A\cup B\cup C)| \geq |(A\cap B)-A_2|$. We have $|A_3| = |(A\cap A_2)-B| + |A\cap B\cap A_2| + (|A_3\cap B|-|A\cap B\cap A_2|) + (|A_3\cap C|-|A\cap B\cap A_2|) + |A_3-(A\cup B\cup C)|\geq |(A\cap A_2)-B| + 3|A\cap B| - 2|A\cap B\cap A_2|$

$= |(A\cap A_2)-B| + |B|$. As $|A_3| = |B|$, this implies $A\cap A_2 \subset B$.

By Lemma 7, $\dim \tilde{A} = \dim \tilde{B} = 1$ and $A\mathcal{R}B$. Thus there are $D,E \in X(S)$, $\dim \tilde{D} = \dim \tilde{E} = 1$, with $ARB\mathcal{R}D\mathcal{R}E$ such that $A,B \subset D,E$ and $\{A-(A\cap B), B-(A\cap B), D-(A\cap B), E-(A\cap B)\}$ is a partition of $S-(A\cap B)$. Clearly we have $C \in \{D,E\}$. This finishes the proof of the lemma. □

7. Another linear space on $X(S)$

Let $L_4(S)$ denote the set of all 4-subsets $\{S_1,S_2,S_3,S_4\}$ of $X(S)$ such that $S_i\cap S_j = S_1\cap S_2$ for all $i,j\in\{1,2,3,4\}$, $i\neq j$, and such that $\{S_i-(S_1\cap S_2): i=1,2,3,4\}$ is a partition of $S-(S_1\cap S_2)$. (Note that, as we assume all elements of a partition to be non-empty, this implies that $S_1\cap S_2 \notin \{S_1,S_2,S_3,S_4\}$.) Let $L'_2(S)$ denote the set of all 2-subsets of $X(S)$ that are neither contained in an element of $L_3(S)$ nor in an element of $L_4(S)$. Put $L'(S) = L'_2(S)\cup L_3(S)\cup L_4(S)$.

Theorem 5: $(X(S),L'(S))$ is a linear space.

Proof:

It is obvious that two distinct points of $X(S)$ are contained in at most one element of $L_3(S)$. If $\{A_1,A_2,A_3\}\in L_3(S)$ and $\{A_1,A_2,A_4,A_5\}\in L_4(S)$, then $A_3 = (A_3\cap A_4)\cup(A_3\cap A_5)$. This yields a contradiction, as no $S(\lambda;2,3,v)$ is the union of two proper subspaces.

For a similar reason, no two distinct elements of $X(S)$ can be contained in two distinct elements of $L_4(S)$. □

Let $P(S)$ denote the set of all $Y\in X(S)$ with $\dim \tilde{Y} \geq 1$.

Theorem 6: $P(S)$ is a subspace of $(X(S),L'(S))$. Moreover, if $A,B\in P(S)$, then $\{A,B\}$ is contained in an element of $L_4(S)$ if and only if $\dim \tilde{A} = \dim \tilde{B} = 1$, $\tilde{A} \neq B$ and $A\mathcal{R}B$.

Proof:

The first claim follows from Lemma 8. The second is a consequence of Lemma 8 and Theorem 4. □

Theorem 7:

(1) If $Y\in P(S)$, $\dim \tilde{Y} \geq 2$, then the connected component of Y in $L'(S)|P(S)$ is $\tilde{Y}$. Moreover, $L'(S)|\tilde{Y} = L(S)|\tilde{Y}$.

(2) If $Y\in P(S)$, $\dim \tilde{Y} = 1$, then the connected component of Y in $L'(S)|P(S)$ is $\mathcal{R}(Y)$. Moreover, $(\mathcal{R}(Y), L'(S)|\mathcal{R}(Y))$ is a punctured projective space of order 3, i.e. a projective space of order 3 from which a point has been deleted.

Proof:

Theorem 7 is an easy consequence of the preceding lemmas and theorems. In particular, claim (2) is an obvious corollary of Theorem 4. □

Although Theorem 7 completely describes the structure of $(P(S),L'(S)|P(S))$, we were unable to obtain good descriptions of $(X(S),L'(S))$. We believe that the structure of $(X(S),L'(S))$ may be relatively wild.

8. Applications to other structures

In [2,3], projective and affine spaces are associated to certain classes of subspaces of more general structures than $S(\lambda;2,3,v)$. These structures are 2-coverings in which all blocks have 3 points [3], 3-coverings in which all blocks have 4 points [3], and $S(\lambda;t,t+1,v)$ with $2 \leq t \leq 5$ [2]. In the case of $S(\lambda;2,3,v)$, the classes of subspaces correspond to $X(S,(v-1)/2)$ [2,3] and to the set of all $Y \in X(S)$ with $\dim \tilde{Y}=1$ and $V(Y)=\emptyset$ [3]. Moreover, in [3], some general results are proved, but these are only valid for sets of subspaces satisfying an extra condition. Except for the subspaces already studied in [2,3], most results in this paper seem difficult to extend to objects other than $S(\lambda;2,3,v)$. However, sometimes structures which seem more general than $S(\lambda;2,3,v)$ are not really more general and thus, the results of this paper can be applied to them

An $OD(\lambda;2,3,v)$ is a pair (S,β) where S is a v-set and β is a family of ordered triples (x_1,x_2,x_3) of distinct elements of S such that, for any pair of distinct elements $a,b \in S$ and for any $1 \leq i < j \leq 3$, there are exactly λ triples (x_1,x_2,x_3) of β with $x_i=a$ and $x_j=b$. A *subspace* of an $OD(\lambda;2,3,v)$ is a subset S_1 of S such that if $(x_1,x_2,x_3) \in \beta$ and at least two of the components are contained in S_1, then the third is also. It is easy to check that if (S,β) is an $OD(\lambda;2,3,v)$ and if β' denotes the family of all 3-subsets of S obtained by replacing each $(x_1,x_2,x_3) \in \beta$ by $\{x_1,x_2,x_3\}$, then (S,β') is an $S(6\lambda;2,3,v)$. Moreover, the subspaces of (S,β) are exactly the subspaces of (S,β'). Thus the results of this paper can be applied to $OD(\lambda;2,3,v)$.

If $(Q,\cdot)$ is an idempotent quasigroup and if $\beta = \{(x_1,x_2,x_3): x_1 \cdot x_2 = x_3, x_1 \neq x_2\}$, then (Q,β) is an $OD(1;2,3,|Q|)$. The subquasigroups of $(Q,\cdot)$ are exactly the subspaces of (Q,β). Thus the results of this paper can be applied to the subquasigroups of an idempotent quasigroup.

References

[1] J. Coupland, "On the construction of certain Steiner systems", Ph.D. thesis, University College of Wales, 1975.

[2] M. Dehon, "Designs et Hyperplans", *J. Comb. Theory* A23 (1977) 264-274.

[3] L. Teirlinck, "On projective and affine hyperplanes", *J. Comb. Theory* A28 (1980) 290-306.

Annals of Discrete Mathematics 34 (1987) 145—152
© Elsevier Science Publishers B.V. (North-Holland)

On 3-Blocking Sets in Projective Planes

Marialuisa J. de Resmini

Dipartimento di Matematica
Universitá di Roma "La Sapienza"
00185 Rome
ITALY

TO ALEX ROSA ON HIS FIFTIETH BIRTHDAY

ABSTRACT

A 3-blocking set in a projective plane Π is a set of points in Π met by any line in at least three points and containing no line. Lower and upper bounds are found for 3-blocking sets and examples are given of sets whose sizes attain those bounds. A good lower bound is the size of the union of three disjoint Baer subplanes; it is shown that any Hughes plane contains a set with the same parameters as such a union, which does not split into three Baer subplanes.

1. Introduction

A 3-blocking set K, $|K|=k$, in a projective plane Π of order q (q not necessarily a prime power) is a set of points met by any line of Π in at least three points and containing no line. Thus, 3-blocking sets are a specialization of 2-blocking sets. The latter were introduced by A.A. Bruen in [3] and under a weaker assumption (in the sense that lines are allowed to sit on the set) by R. Laskar and F.A. Sherk in [15]; such sets were further investigated by the author in [6].

A 3-blocking set is called *irreducible* whenever $K-\{x\}$ is not a 3-blocking set for any $x \in K$. Unless otherwise stated, by a 3-blocking set we always mean an irreducible one.

One of the first problems on 3-blocking sets is to find lower and upper bounds on their sizes. An obvious lower bound is provided by $4q$ when q is odd, and by $4q-1$ when q is even. Indeed, take four lines in Π, no three of them concurrent, say $\{l_1,l_2,l_3,l_4\}$, and look at the dual complete quadrangle they give rise to. When q is odd, the three diagonal lines d_i are non-concurrent so that the points l_j, $j=1,2,3,4$, plus two out of the three points $d_i \cap d_h$, $i \neq h$, $i,h \in \{1,2,3\}$, form a set met by any line in three points at least. We call such a set a *quasi-3-blocking* set. When q is even, the diagonal lines may be concurrent (they always are when the plane is Desarguesian), so it is enough to add one point to the points on the l_j's. The sizes of these two quasi-3-

blocking sets yield lower bounds for the sizes of 3-blocking sets which in some small planes turn out to be the best possible (section 5).

On the other hand, if q is a square and Π contains three mutually disjoint Baer subplanes, then their union is a 3-blocking set whose size provides a better lower bound than the previous ones for $q>13$. Whenever Π is a Desarguesian plane of square order $q>4$, we can construct such a 3-blocking set since the plane has a partition into Baer subplanes [13]. This raises the question of whether a set with the parameters (i.e., size and intersection numbers) of the union of three mutually disjoint Baer subplanes does split into three Baer subplanes. The answer is negative, as it will be shown that in any Hughes plane there is such a set which contains one Baer subplane at most (section 4).

The union K of three pairwise disjoint Baer subplanes is of type $(3,3+\sqrt{q})$, i.e. its intersection numbers $|K \cap l|$, l a line in Π, are 3 and $3+\sqrt{q}$. Thus we investigate sets of type $(3,n)$ and show that, with just two exceptions, $n=3+\sqrt{q}$ maximizes (minimizes) k. If $n = 3+\sqrt{q}$, then the size of the set is either $k = 3(q+\sqrt{q}+1)$ or $k = q\sqrt{q}+2q-2\sqrt{q}+3$. The latter value for k yields a good upper bound for the size of a 3-blocking set when $q>8$. Examples are given of sets with those sizes (section 3). Moreover, 3-blocking sets are constructed in all small planes (section 5) whose sizes either attain the considered lower bounds or approach them as much as possible. Finally, it is shown that in the Ostrom-Rosati plane or order nine there exists a 3-blocking set on 36 points.

2. Preliminaries

We always denote by Π a projective plane of order q (q not necessarily a prime power), and by K a set of points in Π, $|K|=k$. Let J be a set of nonnegative integers. K is of class J if its intersection numbers $|K \cap l|$, l a line of Π, all belong to J; in particular, K is of *type* J if its intersection numbers are precisely the elements of J. When $|K \cap l| = j \in J$, l is called a *j-secant* of K. Denoting by t_j the number of j-secants of K, then [13,19]:

$$\sum_{j\in J} t_j=q^2+q+1, \quad \sum_{j\in J} jt_j=k(q+1), \quad \sum_{j\in J} j(j-1)t_j=k(k-1). \tag{2.1}$$

In [6] it was observed that the minimum and maximum sizes for sets of class $[m,m+1,...,n-1,n]$, $0<m<n\leq q$, are attained by sets of type (m,n) and for such sets k is a root of the following equation [13,19]:

$$x^2-x((m+n)(q+1)-q) + mn(q^2+q+1) = 0. \tag{2.2}$$

(See [4,6,7,8,9,12] for examples of such sets.) Notice that the left hand side of (2.2) is less than zero for all sets of class $[m,m+1,...,n-1,n]$ and not of type (m,n).

This suggests that good bounds for the sizes of 3-blocking sets might be provided by the sizes of type $(3,n)$ sets. Before investigating such sets, we recall that recently G. Tallini [18] proved the following (see also [6]). Assume that a set exists of type (m,n), $2\leq m<n\leq q-1$; if $\gcd(m-1,n-1)=1$ and $\gcd(m,n)=1$, then $n-m=\sqrt{q}$.

Therefore, if K is a set of type $(3,n)$ and $n\equiv 2,4 \pmod 6$, then $n=3+\sqrt{q}$ and, by (2.2), either $k=3(q+\sqrt{q}+1)$ or $k=q\sqrt{q}+2q-2\sqrt{q}+3$. On the other hand, nothing can be said about sets of type $(3,n)$ when $n\equiv 0,1,3,5 \pmod 6$. If K is a set of type

$(3,n)$, then an easy counting argument shows that each point of K lies on the same number of 3-secants. This suggests to look for the values of n which maximize k. With two exceptions, such a value for n turns out to be $n=3+\sqrt{q}$ (section 3) which implies that q must be a square.

3. The sets of type (3,n)

To investigate the sets of type $(3,n)$, we look for those of maximum possible sizes.

Proposition 3.1: Assume that K is a set of type $(3,n)$ with a unique 3-secant on each of its points. Then either $q=18$, $n=9$, and $k=3\cdot 49$, or $q=3$ and K is PG(2,3) with one point deleted.

Proof:

Under these assumptions, k is obviously maximum. Counting the points of K on the lines through one of its points, $k=(n-1)q+3$. Since k must be a solution of (2.2) with $m=3$,

$$n^2-4n-3q+9 = 0 \tag{3.2}$$

and the discriminant $\Delta = 3q-5$ of (3.2) must be a square. Hence $3q-5 = x^2$, so that $q = 1+(x^2+2)/3$. Since q must be an integer, either (i) $x=3s+1$, or (ii) $x=3s+2$. If (i) holds, then $n=3s+3$. On the other hand, $(n-3)\,|q$ is a necessary existence condition for a type $(3,n)$ set [19]. Thus $3s\,|2s+2$ and this occurs if and only if $s=2$ which implies $q=18$, $n=9$, and $k=3\cdot 49$. Next assume that (ii) holds, so that $q=3s^2+4s+3$, $n=3s+4$. In this case, $(n-3)\,|q$ implies that $3s+1\,|2$ which occurs if and only if $s=0$. Hence $q=3$, $n=4$, and K is PG(2,3) with one point deleted. □

To get a lower bound for the size of a 3-blocking set in a possible plane of order 18, compute the other solution of (2.2) with $m=3$, $n=9$ and $q=18$; this yields $k=63$. On each point of such a set there are fifteen 3-secants and four 9-secants. Furthermore, the 9-secants form a 28-set of type (1,4) in the dual plane. Similarly, the 3-secants of a set with $k=3\cdot 49$ form a 49-set of type (1,4) in the dual plane. A related problem is the following. Looking for possible embeddings of small Steiner systems $S(2,n,v)$ in projective planes, it is easy to find that an $S(2,4,28)$ can be embedded without external lines only in a plane of order either 9 or 18 [11]. Such a result is also a special case of a more general theorem on sets of type $(1,n)$ [18].

In what follows we assume that $q\neq 3,18$.

By proposition 3.1, the maximum size for a set of type $(3,n)$ could be attained if there are precisely two 3-secants on each of its points.

Proposition 3.3: No set of type $(3,n)$ exists with precisely two 3-secants on each of its points.

Proof:

Under these hypotheses, by the same argument as in the proof to proposition 3.1,

$$k = nq-q-n+6. \tag{3.4}$$

Next compute the number y of 3-secants on a point off K; obviously,

$$y = 2+\frac{q}{n-3}, \tag{3.5}$$

implying

$$q = (n-3)(y-2). \tag{3.6}$$

Taking into account that (3.4) must be one root of equation (2.2) with $m=3$, the other root is

$$\bar{k} = 3q+2n-3. \tag{3.7}$$

Consequently, by (3.4) and (3.6), $k\bar{k} = 3n(q^2+q+1)$ implies that

$$(y-2)(2(n-3)^2+n+3) = (n-3)(3(y-2)^2+2). \tag{3.8}$$

If $(y-2,n-3)=1$, then $y-2\,|2$, i.e. $y=4$ so that n is not an integer. Therefore assume that $(y-2,n-3)=d\neq 1$. Write $n-3 = ad$ and $y-2 = bd$, $(a,b)=1$. Thus (3.8) becomes

$$b(2a^2d^2+ad+6) = a(3b^2d^2+2), \tag{3.9}$$

so that $b\,|2$ and $a\,|6$. Direct calculations show that none of the eight possible cases can occur. □

By proposition 3.3, the maximum size for a set of type $(3,n)$ could be attained if there are precisely three 3-secants on each of its points. Under this assumption, by the same counting as before,

$$k = nq-q-2n+9. \tag{3.10}$$

Again, since this value for k must be a root of equation (2.2) with $m=3$, we get

$$n^2-6n-(q-9) = 0, \tag{3.11}$$

whence

$$n = 3+\sqrt{q}. \tag{3.12}$$

Hence, taking into account propositions 3.1 and 3.3, the following result has been proved.

Proposition 3.13: Assume that K is a set of type $(3,n)$ and maximum (minimum) size in Π. Then one of the following occurs.

(i) $q=3$ and K is PG(2,3) with one point deleted.

(ii) $q=18$, K is of type (3,9) and either $k=63$ or $k=3{\cdot}49$.

(iii) q is a square, $n=3+\sqrt{q}$, and either

$$k = 3(q+\sqrt{q}+1), \text{ or} \tag{3.14}$$

$$k = q\sqrt{q}+2q-2\sqrt{q}+3. \quad \square \tag{3.15}$$

Notice that the assumption that K has the maximum (minimum) possible size cannot be dropped. Indeed, sets might exist in a plane of order 32 of type (3,7) and size 147 or 151 [10] (see also [18]).

Thus, (3.14) and (3.15) yield bounds for the size of a 3-blocking set. When $q<8$,the lower bound is given by (3.15) and the upper bound by (3.14); when $q>8$, (3.14) provides a lower bound and (3.15) an upper bound. In case $q=8$, $k=36$ can be taken as an upper bound, whereas a lower bound is given by $4q-1=31$. Suppose $q<8$ and compare the lower bounds given by (3.15) and $4q$ or $4q-1$. In case $q=4$, we get the same lower bound, namely 15; when $q=5$, (3.15) provides the same lower bound as $4q=20$. Finally, if $q=7$, a better lower bound is $4q=28<31$. Next suppose that $q>8$ and compare (3.14) with $4q$ and $4q-1$. This shows that (3.14) yields a good lower bound for a 3-blocking set provided $q>13$.

Even if from the blocking set point of view, the 3-blocking sets of minimum possible size are the most interesting, it seems worthwhile to mention that some sets of type $(3,3+\sqrt{q})$ and size (3.15) were constructed; namely, a 42-set of type (3,6) in all of the known non-Desarguesian planes of order nine, a 91-set of type (3,7) in the Desarguesian plane of order 16 [7] (see also [2]). To the author's best knowledge, these are the only known sets of type $(3,3+\sqrt{q})$ and size (3.15).

4. 3-blocking sets in the Hughes planes

Throughout this section, Π denotes the Hughes plane of order q^2, q an odd prime power, and Π_0 its real subplane, i.e. the subplane over the field GF(q) [5,6,16,17]. The points in $\Pi-\Pi_0$ and the tangents to Π_0 are the complex points and lines of Π.

In [6] it was shown that Π contains a set with the same parameters as the union of two disjoint Baer subplanes which does not split into two Baer subplanes (nor contains such a subplane). More precisely, recall that Π is constructed from PG($2,q^2$), the plane over the field GF(q^2), introducing an alternative condition of collinearity for the points not in Π_0 = PG($2,q$) [16,17]. Furthermore, PG($2,q^2$) has a partition into q^2-q+1 Baer subplanes, one of them being Π_0 [13]. All these subplanes but Π_0 come in pairs of conjugate subplanes, say Π_j and $\overline{\Pi}_j$, under the automorphism $a \to a^q$ of GF(q^2),$a \in$GF(q^2). Moreover, each subplane is an orbit under σ^{q^2-q+1}, σ the Singer cycle of PG($2,q^2$) [1,13]. In Π the points in a pair of such conjugate subplanes Π_j and $\overline{\Pi}_j$ are distributed in a different way on the complex lines and $\Pi_j \cup \overline{\Pi}_j$ has the same parameters as the union of two disjoint Baer subplanes, but neither Π_j nor $\overline{\Pi}_j$ has the intersection numbers of such a subplane [6,16,17]. In addition, no Baer subplane is contained in $\Pi_j \cup \overline{\Pi}_j$. Notice that Π_j ($\overline{\Pi}_j$) is the orbit of a complex point under the Singer cycle σ_0 of Π_0 [6,17,20].

As a consequence, we can take s pairs of conjugate orbits of complex points under σ_0 and look at their union; this union yields a set of size $2s(q^2+q+1)$ and type $(2s,2s+q)$, $s = 1,2,\ldots,(q^2-q)/2$ [6]. When $s = (q^2-q)/2-1$, the complement of such a set is the union of Π_0 and a pair of conjugate orbits Π_j and $\overline{\Pi}_j$ and has the parameters of the union of three mutually disjoint Baer subplanes, i.e. is a set of type $(3,3+q)$ and size $3(q^2+q+1)$. This set does not split into three Baer subplanes, but contains precisely one Baer subplane, namely Π_0 [6].

Therefore, any Hughes plane contains a 3-blocking set whose size attains the lower bound in section 3.

In the special case $q^2=9$, it is possible to construct two nonisomorphic 39-sets of type (3,6). One of them is provided by the above construction and contains a Baer subplane, whereas the second one contains no Baer subplane and comes from gluing together three orbits of complex points under σ_0, no two of them being conjugate. This set was constructed in [6] and its existence is a consequence of the full collineation group of Π (of order nine) being $G\times Sym_3$ where G is the full collineation group of PG(2,3) and Sym_3 if the symmetric group on three letters [20]. Notice that the 39-set of type (3,6) containing Π_0 is the orbit under σ_0 of any type α subplane of order 2 (Denniston's notation [5]); a typical type α subplane consists of the points $A,D,L,N6,P3,S3,X6$ [5].

By the argument in section 3, 39 is not the best lower bound for the size of a 3-blocking set in a plane of order 9. This raises the question of finding a smaller 3-blocking set (containing no line); in section 5, some results will be shown.

5. 3-blocking sets in small planes

In this section some examples are given of 3-blocking sets in planes of orders $q\leq 9$ whose sizes approach the previously given lower bounds.

When $q=4$, no 3-blocking set exists as it would be a non-existent set of type (3,4). The best possible lower bound is 15 (see also [15]) and a quasi-3-blocking set of this size is provided by the complement of a hyperoval.

Next, assume that $q=5$. The best lower bound in section 3 is 20 and some easy arithmetic shows that there is no 3-blocking set on 20 points in PG(2,5). A quasi-3-blocking set of this size contains lines, two at least and five at most. Furthermore, the 6-secants form an arc in the dual plane. Take PG(2,5) on the points $\{0,1,...,29,30\}$ with first line 0 1 3 8 12 18 [1]; the following points yield a quasi-3-blocking set with $t_6=5$, $t_5=1$, $t_4=10$, and $t_3=15$:

0 1 2 3 4 5 7 8 9 10 12 13 15 16 18 19 22 23 25 30.

The construction in section 1 gives a 20-set with $t_6=4$. Actually, no 20-set exists in PG(2,5) with fewer 6-secants. In fact, $t_6=2$ is ruled out by counting the secants on a point off K and taking into account that $t_6=2$ implies that $t_4=1$. Next $t_6=3$ is ruled out by the configuration of the 4-secants as the 6-secants form an arc (the plane should contain a PG(2,2)).

Suppose that $q=7$. The two lower bounds for a 3-blocking set are 28 and 31 (section 3) and the former comes from a set containing lines. Assume PG(2,7) on the points $\{0,1,...,55,56\}$ is given by the first line 0 5 6 8 18 37 41 48 [1]. The following thirty points form a 3-blocking set

0 3 6 9 12 15 18 21 24 27 30 33 36 39 42 45 48 51 54 2 5 8 20 23 26 29 38 41 44 56.

For this set, $t_7=4$, $t_6=7$, $t_5=6$, $t_4=t_3=20$. It might be possible to find a smaller set.

In case $q=8$, the best lower bound in section 3 is 31. On the other hand, the best result the author was able to get is a 3-blocking set of size 34. Take PG(2,8) on the points $\{0,1,...,71,72\}$ with first line being 0 1 3 7 15 31 36 54 63 [1]; the points

0 1 3 7 9 12 14 19 23 25 27 28 30 31 35 36 39 46 48 49 50 53 54 55 56 57 59 61 63
64 65 68 69 72

form the set and $t_8=3$, $t_7=4$, $t_6=7$, $t_5=6$, $t_4=23$, and $t_3=30$.

Finally, suppose that $q=9$. By the arguments in sections 3 and 4, both in the Desarguesian and in the Hughes plane there is a 3-blocking set on 39 points which is of type (3,6). On the other hand, the $4q$ bound gives 36 and this raises the question of whether it is possible to find a set of this size containing no line. The answer is affirmative. More precisely, the Ostrom-Rosati plane of order nine contains a 3-blocking set of size 36 which is of type (3,6,9) with a unique 9-secant, $t_6=27$, and $t_3=63$. Recall that the Ostrom-Rosati plane of order nine is the derived plane of the Hughes plane of order nine. Take as line at infinity a real line l_∞ and as derivation set $l_\infty \cap \Pi_0$; in what follows $l_\infty = n$: A B D J N3 N4 N5 N6 N7 N8 [5]. Using Denniston's notation [5], the following points yield the set:

C E F G H I K L M N3 N4 N5 Q6 Q7 Q8 S3 S4 S5 S6 S7 S8 T3 T4 T5 U3 U4
U5 V3 V4 V5 X6 X7 X8 Z3 Z4 Z5.

Since the translation plane of order nine is both the dual plane of the Ostrom-Rosati plane and the derived plane of the Desarguesian plane, it is likely that a similar result holds in that plane too. On the other hand, trying to lower the 39 bound in the Hughes plane of order nine, the best result the author was able to get is a quasi-3-blocking set of size 38 containing just one line. The conjecture that both in the Desarguesian and in the Hughes plane of order nine the best lower bound for the size of a 3-blocking set is 39 makes sense.

Finally, we observe the existence of a nice 3-blocking set in PG(2,11) of size 57 and type (3,4,7) having precisely 19 3-secants and consisting of the points $\{7j, 7j+1, 7j+5 : j=0,1,\ldots,18\}$. Such a set is constructed by gluing together the orbits under σ^7, σ the Singer cycle of PG(2,11), of three suitable points. (See [1,14] for the first line and [14] for some properties of PG(2,11).)

References

[1] L.D. Baumert, *Cyclic Difference Sets,* Lecture Notes in Mathematics 182, Springer Verlag, Berlin-Heidelberg-New York, 1971.

[2] A.E. Brouwer, "A series of separable designs with applications to pairwise orthogonal Latin squares", *European Journal of Combinatorics* 1 (1980) 39-41.

[3] A.A. Bruen, "Arcs and multiple blocking sets", *Proc. Conf. "Finite Geometries",* I.N.d.A.M., Rome, Italy, May 1983; *Symposia Mathematica* 28 (1986).

[4] M. de Finis, "On k-sets of type (m,n) in projective planes of square order", in: *Finite Geometries and Designs* London Math. Society Lecture Notes Series 49 (1981) 98-103.

[5] R.H.F. Denniston, "Subplanes of the Hughes plane of order 9", *Proc. Cambridge Phil. Soc.* 64 (1968) 589-598.

[6] M.J. de Resmini, "On 2-blocking sets in projective planes", *Ars Combinatoria* 20B (1985) 59-69.

[7] M.J. de Resmini, "An infinite family of type (m,n) sets in PG(2,q^2)", *J. Geometry* 20 (1983) 36-43.

[8] M.J. de Resmini, "On k-sets of type (m,n) in a Steiner system S(2,l,v)", in: *Finite Geometries and Designs,* London Math. Soc. Lecture Notes Series 49 (1981) 104-113.

[9] M.J. de Resmini, "A 35-set of type (2,5) in PG(2,9)", *J. Comb. Theory A,* to appear.

[10] M.J. de Resmini, "On admissible sets with two intersection numbers in a projective plane", *Proc. Conf. "Combinatorics '86",* Passo della Mendola, Italy, 1986; *Annals of Discrete Math.,* to appear.

[11] M.J. de Resmini, "Embedding small Steiner systems S(2,n,v) in projective planes", unpublished.

[12] M.J. de Resmini and G. Migliori, "A 78-set of type (2,6) in PG(2,16)", *Ars Combinatoria* 22 (1986) 73-75.

[13] J.W.P. Hirschfeld, *Projective Geometry over Finite Fields,* Clarendon Press, Oxford, 1979.

[14] J.W.P. Hirschfeld and A.R. Sadeh, "The projective plane over the field of eleven elements", *Mitt. Math. Sem. Giessen* 164 II (1984) 245-256.

[15] R. Laskar and F.A. Sherk, "Generating sets in finite projective planes", in: *Finite Geometries* (C. Baker and L.M. Batten, editors) Dekker, Lecture Notes in Pure and Applied Mathematics 103 (1985) 183-198.

[16] T.G. Room, "Veblen-Wedderburn hybrid planes", *Proc. Roy. Soc. London* A309 (1969) 157-170.

[17] T.G. Room, "The combinatorial structure of the Hughes plane", *Proc. Cambridge Phil. Soc.* 68 (1970) 291-301.

[18] G. Tallini, "Some new results on sets of type (m,n) in projective planes", *J. Geometry,* to appear.

[19] M. Tallini Scafati, "{k,n}-archi di un piano grafico finito, con particolare riguardo a quelli con due caratteri", *Rend. Acc. Naz. Lincei* 40 (1966) 812-818, 1020-1025.

[20] G. Zappa, "Sui gruppi di collineazioni dei piani di Hughes", *Boll. Unione Mat. Ital.* (3) 12 (1957) 507-516.

Annals of Discrete Mathematics 34 (1987) 153–164
© Elsevier Science Publishers B.V. (North-Holland)

Star sub-Ramsey numbers

P. Fraisse

U.A. 410 CNRS, L.R.I., bat 490,
Université Paris-Sud
91405 Orsay Cedex
FRANCE

G. Hahn

Département d'I.R.O.
Université de Montréal
C.P. 6128, Succ. A,
Montréal, P.Q., H3C 3J7
CANADA

D. Sotteau

U.A. 410 CNRS, L.R.I., bat 490,
Université Paris-Sud
91405 Orsay Cedex
FRANCE

TO ALEX ROSA ON HIS FIFTIETH BIRTHDAY

ABSTRACT

Let $sr(G, k)$ be the least m such that in any colouring of the edges of K_m using no colour more than k times there is a (isomorphic copy of) G with all edges of distinct colours. Continuing previous work of one of the authors we determine good bounds on $sr(K_{1,s}, k)$ and use them to find exact values for large classes of parameters.

I. Introduction and notation

Let G be a graph (simple, undirected; for terminology and notation not explained see [3]) and k a positive integer. A (G, k)-colouring of a complete graph K_m is an assignment of colours to the edges of K_m in such a way that

(1) no colour is used more than k times

(2) each (subgraph of K_m isomorphic to) G has at least two edges of the same colour.

Clearly the subgraph G is not induced unless it is a complete graph.

The **sub-Ramsey number** $sr(G, k)$ is the least m such that K_m has no (G, k)-colouring. In other words, $sr(G, k)$ is the least m such that no matter how the edges of K_m are coloured while (1) is satisfied, there always is a (subgraph isomorphic to) G all of whose edges have distinct colours.

Sub-Ramsey numbers were introduced by G. Hahn in [6] , [7] following F. Galvin's (unpublished) work on set partitions. Indeed, we have a sort of a "dual" to Ramsey-type problems: instead of partitioning into k classes we limit the size of each class to k . It is not difficult to see that $sr(G, k) \leq r(G; k)$, the Ramsey number for G and k classes. A more technical proof gives $sr(G, k) \leq (k-1)n(n-1)(n-2)/4 + 3$ where n is the number of vertices of G . Since $r(G; k) \geq r(G; 2) \geq 2^{\frac{n}{2}}$ (see [5]), one would think that determining the exact values of $sr(G, k)$ is much easier than for $r(G; k)$. Unfortunately, this is not the case.

Next to nothing is known for the natural starting point, namely $sr(K_n, k)$, except the above mentioned upper bound and a lower bound of $k(n-1) + 1$, plus a few exact values: $sr(K_3, k) = k + 2$, $sr(K_4, 2) = 7$ and $sr(K_5, 2) \geq 10$, see [6] . The second relatively easy case is that of paths and cycles. If G_n is a path or a cycle on n vertices and if $n \geq ck^3$ then $sr(G_n, k) = n$. This is shown in [8] ; we conjecture that this holds already for $n \geq ck$. Thus we are led to what seems to be the simplest case, that of "stars". The reason for the relative simplicity of finding the values or bounds of $sr(K_{1,s}, k)$ is that when two edges of the same colour appear in a $K_{1,s}$, we know *where* .

The purpose of the present note is to provide good bounds for $sr(K_{1,s}, k)$ (by proving a conjecture of [7]), to give exact values, and to indicate why even here the exact numbers are difficult to obtain.

Throughout this note we will assume that $s \geq 2$, $k \geq 2$ and $k = \binom{p}{2} + r$, $0 \leq r < p$ (that is, p is the largest integer such that $\binom{p}{2} \leq k$).

If K_m is $(K_{1,s}, k)$-coloured we will denote by e_c and m_c the number of edges and the number of vertices, respectively, of the monochromatic subgraph induced by the edges of colour c . Clearly $e_c \leq \binom{m_c}{2}$.

II. Bounds

In [7] the following conjecture is stated.

Conjecture

For any $s \geq 2$ *and any* $k \geq 2$

$$sr(K_{1,s}, k) \leq \frac{1}{2}(s-1)(-1 + \sqrt{1+8k}\,) + 2$$

It is also shown there that the bound can be achieved (for example for $k = 3$ and $s \equiv 1$ *or* 2 (mod 3) , see [6]). In this section we prove the conjecture, improving the bound somewhat for some of the cases where the inequality is strict and show that equality holds almost always for $p \leq 5$ for half the values of r .

The main result is this.

Theorem

Let $s \geq 2$, $2 \leq k = \binom{p}{2} + r$, $0 \leq r < p$. *Then*

$$sr(K_{1,s}, k) \leq \begin{cases} (s-1)(p-1) + 2 & \text{if } r \leq (p-1)/2 \\ (s-1)\dfrac{p(p-1) + 2r}{p+1} + 2 & \text{if } r > (p-1)/2 \end{cases}$$

with equality

- *for* $r < (p-1)/2$ *if and only if* $K_{(s-1)(p-1)+1}$ *can be decomposed into* K_p*'s,*
- *for* $r = (p-1)/2$ *if and only if* $K_{(s-1)(p-1)+1}$ *can be decomposed into* K_p*'s and graphs of order* $p+1$ *with* k *edges in such a way that every vertex belongs to* $s-1$ *graphs of the decomposition,*
- *for* $r > (p-1)/2$ *if and only if* K_M *can be decomposed into graphs on* $p+1$ *vertices with* k *edges in such a way that every vertex belongs to* $s-1$ *graphs of the decomposition (where* $M = (s-1)\dfrac{p(p-1) + 2r}{p+1} + 1$ *).*

Proof: Let K_m be $(K_{1,s}, k)$-coloured, let C be the set of colours used and let $c \in C$. With the convention that the sums and the maxima are over C unless otherwise indicated we have:

(1) $\sum e_c = m(m-1)/2$ since each edge appears in exactly one monochromatic subgraph and all edges are coloured.

(2) $m(s-1) \geq \sum m_c$ since no vertex lies in more than $s-1$ monochromatic subgraphs.

(3) $$m-1 \leq 2(s-1)\frac{\sum e_c}{\sum m_c}$$ from (1) and (2).

(4) $\dfrac{\sum e_c}{\sum m_c} \leq \max \dfrac{e_c}{m_c}$ with equality if and only if $\dfrac{e_c}{m_c} = \max \dfrac{e_c}{m_c}$ for all $c \in C$.

(5) $\max \dfrac{e_c}{m_c} \leq \max \left\{ \dfrac{p-1}{2}, \dfrac{p(p-1)+2r}{2(p+1)} \right\}$ since for each c, e_c can be written as $\binom{s_c}{2} + t_c$, $0 \leq t_c < s_c$, and since $m_c \geq \begin{cases} s_c & \text{if } t_c=0 \\ s_c+1 & \text{if } t_c>0 \end{cases}$, it suffices to show that for each c,

$$\max\left\{\frac{s_c-1}{2}, \frac{s_c(s_c-1)+2t_c}{2(s_c-1)}\right\} \leq \max\left\{\frac{p-1}{2}, \frac{p(p-1)+2r}{2(p+1)}\right\}.$$

But this is easy: recall that $e_c \leq k = \binom{p}{2} + r$ and $s_c \leq p$. Then consider two cases; for each c:

(a) $s_c = p$ or $\max\left\{\dfrac{s_c-1}{2}, \dfrac{s_c(s_c-1)+2t_c}{2(s_c+1)}\right\} = \dfrac{s_c-1}{2}$

(b) $s_c < p$ and $\max\left\{\dfrac{s_c-1}{2}, \dfrac{s_c(s_c-1)+2t_c}{2(s_c+1)}\right\} = \dfrac{s_c(s_c-1)+2t_c}{2(s_c+1)}$.

(6) $$m \leq \begin{cases} (s-1)(p-1)+1 & \text{if } 0 \leq r \leq (p-1)/2 \\ (s-1)\dfrac{p(p-1)+2r}{p+1} + 1 & \text{if } (p-1)/2 < r < p \end{cases}$$ from (1)-(5) and the fact that $\dfrac{(p-1)}{2} \geq \dfrac{p(p-1)+2r}{2(p+1)}$ if and only if $0 \leq r \leq (p-1)/2$

This proves the first part.

In order to have equality in (6) we must have equality in (2), (4) and (5).

If $0 \leq r < (p-1)/2$, equality holds in (4) and (5) if and only if $\dfrac{e_c}{m_c} = \dfrac{p-1}{2}$ and $m_c = p$ for all $c \in C$. What this says is that each colour appears as often as possible in a subgraph on p vertices, that is, the graph K_m is decomposed into monochromatic complete graphs on p vertices (and hence it is better to use each colour $\binom{p}{2}$ times in a complete graph than to use it $\binom{p}{2} + r$ times otherwise, if the goal is to get a $(K_{1,s}, k)$-coloured graph of maximum size).

If $r = (p-1)/2$ we have equality in (4) if and only if $\dfrac{e_c}{m_c} = \dfrac{p-1}{2} = \dfrac{p(p-1)+2r}{2(p+1)}$ for all c in C , i.e. if and only if $K_{(s-1)(p-1)+1}$ can be decomposed into K_p's and graphs on $p+1$ vertices and $\dfrac{p(p-1)}{2} + r$ edges.

If $r > (p-1)/2$ we have equality in (4) if and only if $\dfrac{e_c}{m_c} = \dfrac{p(p-1)+2r}{2(p+1)}$ for all c in C, i.e. if and only if K_M can be decomposed into graphs of order $p+1$ with $\dfrac{p(p-1)}{2} + r$ edges.

In order to have equality in (6) we must also have equality in (3) and thus in (2) which says that every vertex lies in exactly $s-1$ graphs of the decomposition. □

Corollary 1

For any $s \geq 2$ and $k \geq 2$,

$$sr(K_{1,s}\,,\,k) \leq \frac{1}{2}(s-1)(-1+\sqrt{1+8k}\,)+2$$

Proof: In the proof of the Theorem we have from (1)-(4)

$$m-1 \leq 2\,(s-1)\max\frac{e_c}{m_c}$$

Now $e_c \leq m_c(m_c-1)/2$ for every c, and, therefore, $m_c^2 - m_c - 2\,e_c \geq 0$ This implies that $m_c \geq (1+\sqrt{1+8e_c}\,)/2$ and $\dfrac{e_c}{m_c} \leq \dfrac{-1+\sqrt{1+8k}}{4}$ as e_c is always less than or equal to k. □

Corollary 2

Let $s \geq 2$, $2 \leq k = \binom{p}{2} + r$, $0 \leq r \leq \dfrac{p-1}{2}$. *If (i)* $p > 5$ *and* s *is big enough, or (ii)* $p \leq 5$, *then*

$$sr(K_{1,s}\,,\,k) = (s-1)(p-1)+2 \quad \text{if} \quad (s-1)(s-2) \equiv 0 \pmod p$$

(and **only if** for $r \neq \dfrac{p-1}{2}$ *).*

Proof: The conditions necessary for the existence of a decomposition of K_m into K_p's are well known :

$$m(m-1) \equiv 0 \pmod{p(p-1)}$$
$$m-1 \equiv 0 \pmod{p-1}$$

Indeed such a decomposition yields a Steiner p-system.

Now Wilson's theorem [11] says that the necessary conditions are also sufficient if m is "big enough". In particular it is known that "big enough" is "at least p " for $p \leq 5$.

Letting $m = (s-1)(p-1)+1$, the necessary conditions reduce to

$$(s-1)(s-2) \equiv 0 \pmod p$$

and " m big enough " becomes " s big enough " if k is fixed. □

Corollary 3

Let $s \geq 2$,

$$sr(K_{1,s}, 3) \begin{cases} = 2s & \text{if and } \textit{only} \text{ if } s \equiv 1 \textit{ or } 2 \pmod 3 \\ \leq 2s-1 & \textit{otherwise.} \end{cases}$$

$sr(K_{1,s}, 4) \leq 2s$
with equality if $s \equiv 1$ *or* $2 \pmod 3$.

$sr(K_{1,s}, 5) \leq \frac{5}{2}(s-1) + 2$ *with equality if and only if* $K_{5(s-1)/2+1}$ *can be decomposed into graphs on* 4 *vertices and* 5 *edges with each vertex in* $s-1$ *graphs.*

$$sr(K_{1,s}, 6) = sr(K_{1,s}, 7) \begin{cases} = 3s-1 & \text{if and } \textit{only} \text{ if } s \equiv 1 \textit{ or } 2 \pmod 4 \\ \leq 3s-2 & \textit{otherwise.} \end{cases}$$

$sr(K_{1,s}, 8) \leq \frac{16}{5}(s-1) + 2$ *with equality if and only if* $K_{16(s-1)/5+1}$ *can be decomposed into graphs on* 5 *vertices with* 8 *edges with each vertex in* $s-1$ *graphs.*

$sr(K_{1,s}, 9) \leq \frac{18}{5}(s-1) + 2$ *with equality if and only if* $K_{18(s-1)/5+1}$ *can be decomposed into* K_5 *minus one edge with every vertex in* $s-1$ *graphs.*

$$sr(K_{1,s}, 10) = sr(K_{1,s}, 11) \begin{cases} = 4s-2 & \text{if and } \textit{only} \text{ if } s \equiv 1 \textit{ or } 2 \pmod 5 \\ \leq 4s-3 & \textit{otherwise.} \end{cases}$$

$sr(K_{1,s}, 12) \leq 4s-2$ *with equality if* $s \equiv 1$ *or* $2 \pmod 5$.

This is an immediate consequence of the Theorem and Corollary 2 for $p \leq 5$. □

Proposition

(1) $sr(K_{1,s}, 3) = 2s-1$ *for* $s \equiv 0 \pmod 3$.

(2) $sr(K_{1,s}, 4) = 2s$ *for all* s .

(3) $sr(K_{1,s}, 5) = \frac{5}{2}(s-1) + 2$ *if and only if* s *is odd.*
$\frac{5s}{2} - 2 \leq sr(K_{1,s}, 5) \leq \frac{5s}{2} - 1$ *if* s *is even.*

(4) $sr(K_{1,s}, 6) = sr(K_{1,s}, 7) = 3s-2$ *for* $s \equiv 0$ *or* $3 \pmod 4$.

(5) $sr(K_{1,s}, 8) = \frac{16}{5}(s-1) + 2$ *if and only if* $s \equiv 1 \pmod 5$.

Proof

(1) is proven in [6]

(2) For $k = 4$, if $s \equiv 0 \pmod 3$ then $2s-1 \equiv 5 \pmod 6$. It is known (see [10]) that K_{6t+5} can be decomposed into K_5 's and K_3 's. Therefore K_{2s-1} can be decomposed into C_4 's (cycle of length 4) and K_3 's (since each K_5 can be

decomposed into two K_3 's and one C_4). If we colour each graph K_3 or C_4 of this decomposition of K_{2s-1} with a distinct colour we get a $(K_{1,s}, 4)$-colouring. This, together with Corollary 2, proves the claim.

(3) For $k = 5$, if $s \equiv 1$ *or* $3 \pmod 4$ then $\frac{5}{2}(s-1) + 1 \equiv 1 \pmod 5$. In [1] it is shown that K_{5t+1} can be decomposed into graphs isomorphic to Q (where Q is a K_4 minus one edge) in such a way that each vertex has degree 2 (resp. 3) in exactly $(s-1)/2$ graphs of the decomposition. This and the Theorem give the result.

If s is even, then from Corollary 3 we get the upper bound on $sr(K_{1,s}, 5)$. To get the lower bound it is sufficient to notice that we always have $sr(K_{1,s}, k) > sr(K_{1,s-1}, k)$. Here $s-1$ is odd so we have $sr(K_{1,s-1}, 5) = \frac{5}{2}(s-2) + 2$ and therefore $sr(K_{1,s}, 5) \geq \frac{5s}{2} - 2$.

(4) For $k = 6$, if $s \equiv 0$ *or* $3 \pmod 4$ we know already from Corollary 2 that $sr(K_{1,s}, 6) \leq 3s-2$.

If s = 4t+3, $t \geq 0$, we have $3s - 3 = 12t + 6$ and it is known that K_{12t+6} plus a perfect matching on the same set of vertices can be decomposed into K_4's (see [9]) in such a way that every vertex belongs to $4t+2 = s-1$ graphs K_4 of the decomposition. Colour all these K_4's with different colours and then delete the additional perfect matching. Now no colour has been used more than six times in this colouring of K_{3s-3} and only $s-1$ colours appear at each vertex, so we have a $(K_{1,s}, 6)$-colouring. Thus $sr(K_{1,s}, 6) \geq 3s-2$, and we have equality.

If s = 4t+4, $t \geq 0$, consider the complete graph G' of order $3s-2 = 12t+10$. Brouwer showed [4] that if $t \geq 1$ then G' can be decomposed into graphs all isomorphic to K_4 except one isomorphic to K_7.
In this decomposition every vertex belongs to $s-1$ graphs K_4 except seven vertices which belong each to $s-3$ graphs K_4 and one K_7. Colour each K_4 with a different colour and consider the graph G obtained by deleting one of the seven special vertices together with the inherited colouring. Since K_6 plus a perfect matching can be decomposed into K_4's (see [9]) the colouring of G can be completed by adding a perfect matching M to the complete graph induced by the six special vertices, colouring each K_4 of such a decomposition by a distinct colour and deleting the perfect matching M. It is easy to see that the result is a $(K_{1,s}, 6)$-colouring of K_{3s-3}, which implies that $sr(K_{1,s}, 6) \geq 3s-2$ if $s \equiv 0 \pmod 4$, $s \neq 4$.

In the case $s = 4$, a $(K_{1,4}, 6)$-colouring of K_9 can be obtained as follows. Let the vertices of K_9 be the elements of Z_9 and colour each of the following K_4 with a different colour : (1,3,6,8), (2,4,6,0), (3,5,7,0). Consider, for $i = 1, 4, 7$, the three graphs with vertices $i, i+1, i+2, i+3, i+7$ and 6 edges $(i,i+1)$, $(i+1,i+2)$, $(i+2,i+3)$, $(i+3,i)$, $(i+3,i+7)$, $(i+7,i+1)$ and colour each of them with a different new colour. It is easy to check that only three colours appear at each vertex and that the whole graph is coloured. So $sr(K_{1,4}, 6) \geq 10$.

Since any (G, k)-colouring is also a (G, k')-colouring for $k' > k$, the above shows that $sr(K_{1,s}, 7) \geq 3s-2$ when $s \equiv 0$ *or* $3 \pmod 4$ and we have equality in that case as well.

(5) For $k = 8$, let $s \equiv 1 \pmod 5$ (necessary condition for $\frac{16}{5}(s-1) + 2$ to be an integer); then $\frac{16}{5}(s-1) + 1 \equiv 1 \pmod{16}$. It is known that for any t , K_{16t+1} can be decomposed into K_5 minus two non adjacent edges in such a way that each vertex belongs to $5t$ graphs of the decomposition (see [2]). This with Corollary 3 proves the claim. □

Remark 1: Using the Theorem, Corollary 2 and the fact that we have

$$sr(K_{1,s+1}, k) \geq sr(K_{1,s}, k) + 1$$

we can get the following bounds for $sr(K_{1,s}, k)$.

Let $s \geq 2$, $2 \leq k = \binom{p}{2} + r$, $0 \leq r \leq \frac{p-1}{2}$ *and* $s \equiv q \pmod p$, $q \geq 2$. *Then, if (i)* $p > 5$ *and* s *is big enough, or (ii)* $p \leq 5$, *we have*

$$(s-q+1)(p-1) + q \leq sr(K_{1,s}, k) \leq \begin{cases} (s-1)(p-1) + 1 & \text{if } r < (p-1)/2 \\ (s-1)(p-1) + 2 & \text{if } r = (p-1)/2 \end{cases}$$

Indeed, $sr(K_{1,s}, k) \geq sr(K_{1,s-q+2}, k) + q-2$, and, as $s - q + 2 \equiv 2 \pmod p$, we have, from Corollary 2, $sr(K_{1,s-q+2}, k) = (s-q+1)(p-1) + 2$. This gives the lower bound. The upper bound is already in the Theorem and Corollary 2.

For clarity and completeness we collect the values or bounds of $sr(K_{1,s}, k)$ for $k \leq 12$ in Figure 1.

k	Lower bound	Equality	Upper bound	Values of s
3		$2s$		$s \equiv 1$ *or* $2 \pmod 3$
		$2s - 1$		$s \equiv 0 \pmod 3$
4		$2s$		*any* s
5	$\frac{5}{2}s - 2$		$\frac{5}{2}s - 1$	$s \equiv 0 \pmod 2$
		$\frac{5}{2}(s-1) + 2$		$s \equiv 1 \pmod 2$
6, 7		$3s - 1$		$s \equiv 1$ *or* $2 \pmod 4$
		$3s - 2$		$s \equiv 0$ *or* $3 \pmod 4$
8		$\frac{16}{5}(s-1) + 2$		$s \equiv 1 \pmod 5$
	$3s - 2$		$\frac{16}{5}s - \frac{7}{5}$	$s \not\equiv 1 \pmod 5$
9	$3s - 2$		$\frac{18}{5}(s-1) + 2$	*any* s
10, 11		$4s - 2$		$s \equiv 1$ *or* $2 \pmod 5$
	$4s - 5$		$4s - 3$	$s \equiv 3 \pmod 5$
	$4s - 8$		$4s - 3$	$s \equiv 4 \pmod 5$
	$4s - 11$		$4s - 3$	$s \equiv 0 \pmod 5$
12		$4s - 2$		$s \equiv 1$ *or* $2 \pmod 5$
	$4s - 5$		$4s - 2$	$s \equiv 3 \pmod 5$
	$4s - 8$		$4s - 2$	$s \equiv 4 \pmod 5$
	$4s - 11$		$4s - 2$	$s \equiv 0 \pmod 5$

Figure 1 : Values or bounds for $sr(K_{1,s}, k)$ *for* $k \leq 12$

Remark 2 : We take this opportunity to correct a minor omission in [7] . Proposition 1 of that paper proves that

$$sr(K_{1,3}, k) = \left[\frac{(3 + \sqrt{1+24k})}{2} \right]$$

except that $sr(K_{1,3}, 2) = 4$ and $sr(K_{1,3}, 7) = 7$.

The proof provides colourings of K_m for $m = \left[\frac{(3 + \sqrt{1+24k})}{2} \right] - 1$ in all other cases except the cases $k = 8$ and $k = 9$. These two cases should fall in case b), where $m = 3n+1$; the construction, however, needs $n \geq 3$. The missing colouring is given in Figure 2. Clearly if K_7 can be coloured with $k = 8$ then it is coloured for $k = 9$ as well.

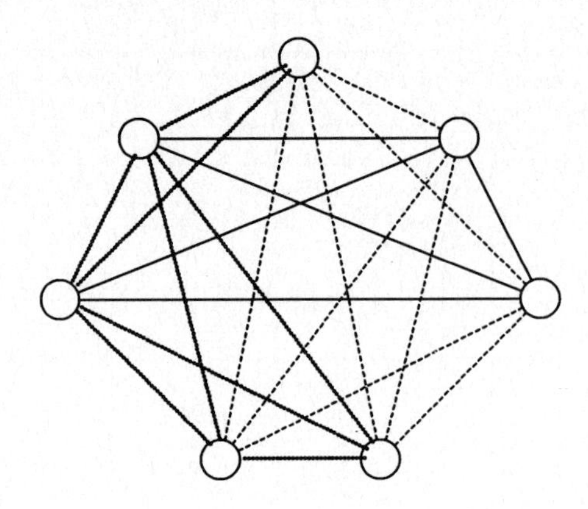

Figure 2

Acknowledgements

Research of the first and third authors is supported by P.R.C. Math. Info.; research of the second author is supported by NSERC Canada under grant number A0199.

References

1. J.C. BERMOND and J. SCHONHEIM, G-decomposition of K_n , where G has four vertices or less, *Discrete Mathematics* **19** pp. 113-120 (1977).
2. J.C. BERMOND, C. HUANG, A. ROSA, and D. SOTTEAU, Decomposition of complete graphs into isomorphic subgraphs with five vertices, *Ars Combinatoria* **10** pp. 211-254 (1980).
3. J.A. BONDY and U.S.R. MURTY, *Graph Theory with Applications.* 1976.
4. A.E. BROUWER, Optimal Packings of K_4 's into a K_n, *Journal of Combinatorial Theory A* **26** pp. 278-297 (1979).
5. P. ERDOS, Some remarks on the theory of graphs, *Bull. Amer. Math. Soc.* **53** pp. 292-294 (1947).
6. G. HAHN, Some Star Anti-Ramsey Numbers, *Proc. 8th S-E Conference on Combinatorics, Graph Theory and Computing, Congressus Numerantium XIX, Utilitas Mathematica*, pp. 303-310 (1977).
7. G. HAHN, More Star Sub-Ramsey Numbers, *Discrete Mathematics* **34** pp. 131-139 (1981).
8. G. HAHN and C. THOMASSEN, On path and cycle sub-Ramsey numbers, *Discrete Mathematics* **62** pp. 29-33 (1986).

9. W.H. MILLS, On the Covering of Pairs by Quadruples I., *Journal of Combinatorial Theory A* **13** pp. 55-78 (1972).

10. R.M. WILSON, Construction and Uses of Pairwise Balanced Designs, *Combinatorics Part I, Math. Centre Tracts 55, Math Centrum Amsterdam,* pp. 18-41 (1974).

11. R.M. WILSON, Decompositions of Complete Graphs into Subgraphs Isomorphic to a Given Graph, *Proc. 5th Brit. Comb. Conference 1975, Congressus Numerantium XV, Utilitas Mathematica,* pp. 647-659 (1976).

Annals of Discrete Mathematics 34 (1987) 165–178
© Elsevier Science Publishers B.V. (North-Holland)

Colored packing of sets

P. Frankl

CNRS
Paris, FRANCE

Z. Füredi

AT&T Bell Laboratories
Murray Hill, NJ, 07974
U.S.A.

TO ALEX ROSA ON HIS FIFTIETH BIRTHDAY

ABSTRACT

Let $\mathcal{H}$ be a family of t-sets on $\{1,2, \ldots, k\}$. A family $\mathcal{F}$ of k-sets on v elements is called a $(v,k,\mathcal{H})$-packing if for all $F \in \mathcal{F}$ there is a copy of $\mathcal{H}$, $\mathcal{H}_F$ such that the t-sets of F corresponding to $\mathcal{H}_F$ are covered only by F. Clearly, $|\mathcal{F}| \leq \binom{v}{t} / |\mathcal{H}|$, and if $\mathcal{H}$ is the complete t-hypergraph then we obtain the usual definition of the (partial) Steiner-system. The main result of this paper is that for every fixed k and $\mathcal{H}$ the size of the largest $\mathcal{H}$-packing is $(1-o(1)) \binom{v}{t} / |\mathcal{H}|$, whenever $v \longrightarrow \infty$

1. Preliminaries. Packings and near perfect matchings

Let X be a v-element set, $X = \{1,2, \ldots, v\}$. For an integer k, $0 \leq k \leq v$ we denote the collection of all k-subsets of X by $\binom{X}{k}$, while 2^X denotes the power set of X. A family of subsets of X is just a subset of 2^X. It is called *k-uniform* if it is a subset of $\binom{X}{k}$. A *Steiner-system* $\mathcal{S} = S(v,k,t)$ is an $\mathcal{S} \subset \binom{X}{k}$ such that for every $T \in \binom{X}{t}$ there is exactly one $B \in \mathcal{S}$ with $T \subset B$. Obviously, $|\mathcal{S}| = \binom{v}{t} / \binom{k}{t}$ holds. A $\mathcal{P} \subset \binom{X}{k}$ is called a (v,k,t)*-packing* if $|P \cap P'| < t$ holds for every pair $P, P' \in \mathcal{P}$. Rödl

[16] proved that

$$\max\left\{|\mathcal{P}|: \mathcal{P} \text{ is a } (v,k,t)\text{—packing}\right\} = (1-o(1))\binom{v}{t}\Big/\binom{k}{t} \qquad (1)$$

holds for every fixed k,t whenever $v \to \infty$

This theorem was generalized by Frankl and Rödl [13]. To state it, we recall some definitions. For a family of finite sets $\mathcal{F}$ and an arbitrary set A the *degree* of A is defined by $deg_{\mathcal{F}}(A) =: |\{F \in \mathcal{F}: A \subset F\}|$. For $A = \{a\}$ we set $deg_{\mathcal{F}}(a) =: deg_{\mathcal{F}}(\{a\})$, the usual definition of the degree of an element. A *matching* $\mathcal{M}$ of $\mathcal{F}$ is a subfamily of pairwise disjoint members, $\mathcal{M} \subset \mathcal{F}$, $M \cap M' = \emptyset$ for all $M, M' \in \mathcal{M}$. The largest cardinality of a matching is denoted by $\nu(\mathcal{F})$. Clearly, for $\mathcal{F} \subset \binom{X}{k}$ we have

$$\nu(\mathcal{F}) \leq v/k. \qquad (2)$$

(Frankl and Rödl [13]) For every $\epsilon > 0$ and k there exists a $\delta > 0$ and a $v_0 = v_0(k,\epsilon)$ such that if $\mathcal{F} \subset \binom{X}{k}$, and every degree of $\mathcal{F}$ is almost d (i.e., $|deg_{\mathcal{F}}(x) - d| \leq \epsilon d$ holds for every $x \in X$) and for every $x,y \in X$ we have $deg_{\mathcal{F}}(\{x,y\}) < d/(\log v)^3$ then

$$\nu(\mathcal{F}) > \frac{v}{k}(1-\delta)$$

holds for $v > v_0$.

For a family of sets $\mathcal{G} \subset 2^X$ the subset $A \subset X$ is an *own* part of $G \in \mathcal{G}$ if $A \subset G$ and $deg_{\mathcal{G}}(A) = 1$, i.e., A is contained only in G. Hence $\mathcal{G} \subset \binom{X}{k}$ is a (v,k,t)-packing if and only if every $G \in \mathcal{G}$ has $\binom{k}{t}$ own t-subsets. The aim of this paper is to construct such families $\mathcal{F}$ in which for every $F \in \mathcal{F}$ the family of own t-subsets of F is isomorphic to a given t-uniform hypergraph $\mathcal{H}$. Such a family is called an $\mathcal{H}$-*packing*. If $\mathcal{H}$ is the complete t-hypergraph on k elements, an $\mathcal{H}$-packing is just the usual (v,k,t)-packing. The existence of large $\mathcal{H}$-packings is proved in Chapter 2 and 5. The proof is probabilistic, the main tool for the construction is (2). In Chapter 3 we give an application solving (at least asymptotically) the question: what is the maximum size of a family $\mathcal{F} \subset \binom{X}{k}$ such that none of the members is contained in the union of r others.

2. $\mathcal{H}$-packings and colored $\mathcal{H}$-packings

Let $\mathcal{H}$ be a family of t-sets over k elements. Suppose that $\mathcal{F} \subset \binom{X}{k}$ where $|X| = v$ and for every $F \in \mathcal{F}$ there exists a copy of $\mathcal{H}$ on F (i.e., $\mathcal{H}_F \subset \binom{F}{t}$, $\mathcal{H}_F \approx \mathcal{H}$). If every t-set $T \in \mathcal{H}_F$ is covered only by F (i.e., $deg_{\mathcal{F}}(T) = 1$) then we call $\mathcal{F}$ a $(v,k,\mathcal{H})$*-packing* (or, briefly, $\mathcal{H}$-packing). Clearly,

$$|\mathcal{F}| \leq \binom{n}{t} / |\mathcal{H}|. \tag{3}$$

E.g., the following family $\mathcal{F}$ is a $(v,4,C_4)$-packing of size $(v^2/8) + O(v)$. $\mathcal{F} = \left\{\{2i-1,2i\} \cup \{2j-1,2j\}: 1 \leq i < j \leq v/2\right\}$.

Definition 2.1. Let $\mathcal{H} \subset \binom{K}{t}$, $|K| = k$, $c = \binom{k}{t} - |\mathcal{H}|$ and fix a partition $\binom{K}{t} = \mathcal{H} \cup \{T_1\} \cup \cdots \cup \{T_c\}$. In other words, this is a coloring χ: $\binom{K}{t} \rightarrow \{0,1,\ldots,c\}$ with $\chi(T) = 0$ for $T \in \mathcal{H}$. The family $\mathcal{F} \subset \binom{X}{k}$, $|X| = v$ is called a *colored* $(v,k,\mathcal{H})$*-packing* if

(i) $|F \cap F'| \leq t$ holds for every two $F, F' \in \mathcal{F}$, and

(ii) there exists a coloring of the t-sets of X with $c+1$ colors $\binom{X}{t} = \mathcal{C}_0 \cup \mathcal{C}_1 \cup \cdots \cup \mathcal{C}_c$ such that for every $F \in \mathcal{F}$ the induced coloring of $\binom{F}{t}$ is isomorphic to χ, especially $\mathcal{C}_0 \cap \binom{F}{t} \approx \mathcal{H}$. (I.e., there exists an injection π_F: $F \rightarrow K$ such that for $T \in \binom{F}{t}$ we have $T \in \mathcal{C}_{\chi(\pi_F(T))}$.) E.g., the following family $\mathcal{F}$ is a colored $(v,4,C_4)$-packing of size $(v^2/8) + O(v)$: $\mathcal{F} = \left\{\{2i-1,2i\} \cup \{2j,2j+1\}: 1 \leq i < j < v/2\right\}$. We prove that the upper bound in (3) for the size of $\mathcal{F}$ is essentially the best possible.

Theorem 2.2. For every given k and $\mathcal{H}$ the size of the largest colored $(v,k,\mathcal{H})$-packing is $(1-o(1)) \binom{v}{t} / |\mathcal{H}|$ when $v \rightarrow \infty$

In the proof we will use the following

Lemma 2.3. Suppose $n > n_0(k)$, $k > t \geq 1$, $D \leq n^{2/3}$. Then there exists a family of k-sets $\mathbf{S}$ on the n-element set N such that

(i) $|S \cap S'| \leq t$ holds for every two $S, S' \in \mathbf{S}$

and

(ii) $|deg_{\mathbf{S}}(T) - D| < 2\sqrt{kD \log n}$ holds for every $T \in \binom{N}{t}$. In other words, there exists a "near" D–(n,k,t)-design. Of course, for $D < 4k \log n$ (ii) is meaningless.

Let $f(v,k,\mathcal{H}) = \max\{|\mathcal{F}|: \mathcal{F}$ is a $(v,k,\mathcal{H})$-packing$\}$ and $f_c(v,k,\mathcal{H}) = \max\{|\mathcal{F}|: \mathcal{F}$ is a colored $(v,k,\mathcal{H})$-packing$\}$. Then $f_c \leq f$. If $|\mathcal{H}| < \binom{k}{t}$ then we cannot expect equality in (3). The following two results are consequences of a theorem of Bollobás [2].

(4) ([2]) If $|\mathcal{H}| = 1$, then $f(v,k,\mathcal{H}) = \binom{v-k+t}{t}$.

(5) ([7]) If $|\mathcal{H}| = \binom{s}{t}$ a complete hypergraph then for $v > v_0(k)$ we have $f(v,k,\mathcal{H}) \leq \binom{v-k+s}{t} / \binom{s}{t}$, where equality holds if and only if there exists a Steiner system $S(v-k+s,\, s, t)$.

Another special case of Theorem 2.2 (when $\mathcal{H}$ is a star) was proved in [7]. Theorem 2.2 says that $\left| f(v,k,\mathcal{H}) - \binom{v}{t} / |\mathcal{H}| \right| = o(v^t)$. In the cases (4) and (5) $\binom{v}{t} / |\mathcal{H}| - f(v,k,\mathcal{H}) > O(v^{t-1})$. In general, the gap may be much larger: Let K be the graph with vertex-set $\{a,b,c,d\}$ and edges $\{a,d\}$, $\{b,c\}$, $\{b,d\}$, $\{c,d\}$.

Proposition 2.4. Suppose that $\mathcal{H}$ is a graph, and that it has an induced subgraph isomorphic to K. Then

$$\binom{v}{2} / |\mathcal{H}| - f(v,k,\mathcal{H}) > \frac{1}{2\sqrt{2}\,|\mathcal{H}|}\, v^{3/2}.$$

3. r-cover-free families

We call the family of sets $\mathcal{F}$ *r-cover-free* if $F_0 \not\subset F_1 \cup \cdots \cup F_r$ holds for all $F_0, F_1, \ldots, F_r \in \mathcal{F}$ ($F_i \neq F_j$ for $i \neq j$). Let us denote by $f_r(n,k)$ the maximum cardinality of an r-cover-free family $\mathcal{F} \subset \binom{N}{k}$, $|N| = n$. Let us set $t = \lceil k/r \rceil$ (upper integer part). An (n,k,t)-packing is r-cover-free, hence by (1)

$$f_r(n,k) \geq (1-o(1)) \binom{n}{t} \Big/ \binom{k}{t}.$$

On the other hand every $F \in \mathcal{F}$ has an own t-subset. Indeed, F can be covered by r t-sets, $F = T_1 \cup \cdots \cup T_r$, $T_i \in \binom{F}{t}$. If for every i $deg_{\mathcal{F}}(T_i) > 1$ then F is covered by r others, which is a contradiction. This yields

$$f_r(n,k) \leq \binom{n}{t}.$$

Proposition 3.1. ([7]) For fixed k and r

$$\lim_{n\to\infty} f_r(n,k) \Big/ \binom{n}{t} = \lim_{n\to\infty} sup\ f_r(n,k) \Big/ \binom{n}{t} =: c_r(k)$$

exists and is positive.

In the next theorem we determine the value of $c_r(k)$, or at least we show that the calculation of it is a finite problem depending only on k.

Definition 3.2. Let k,t,l be positive integers $k \geq t(l+1)$, and $m(k,t,l) = \max\{|\mathcal{N}|: \mathcal{N} \subset \binom{K}{t}, |K| = k, \mathcal{N}$ does not contain $l+1$ pairwise disjoint members$\}$.

For $k < t(l+1)$ define $m(k,t,l) = \binom{k}{t}$ and $m(k,t,0) = 0$. Considering all the t-sets intersecting a given l-set we obtain that $m(k,t,l) \geq \binom{k}{t} - \binom{k-l}{t}$ holds. In several cases this is best possible:

(6) (Erdős, Ko and Rado [9]) $m(k,t,l) = \binom{k-1}{t-1}$ for $k \geq 2t$,

(7) (Erdős [5]) $m(k,t,l) = \binom{k}{t} - \binom{k-l}{t}$ for $k > k_0(t,l)$.

Later $k_0(t,l) < 2t^3 l$ was established by Bollobás, Daykin and Erdös [3]. We can prove $k_0(t,l) < 2t^2 l$ (cf. [12]). The case $t=2$ was solved completely by Erdös and Gallai [8] in 1959:

$$m(k,2,l) = \max\left\{\binom{2l+1}{2}, \binom{k}{2} - \binom{k-l}{2}\right\} \tag{8}$$

for $k \geq 2l+1$ and the only extremal graphs are either K_{2l+1} or $K_k - K_{k-l}$.

Conjecture 3.3. (Erdös [5]) $m(k,t,l) = \max\left\{\binom{tl+t-1}{t}, \binom{k}{t} - \binom{k-l}{t}\right\}$ for all $k \geq (t+1)l$.

A general upper bound was given by Frankl (see, e.g., in [10] or [11])

$$m(k,t,l) \leq l\binom{k-1}{t-1}.$$

For k and r let $k = r(t-1) + l+1$ where $0 \leq l < r$ (i.e., $t = \lceil k/r \rceil$).

Theorem 3.4. $c_r(k) = \left(\binom{k}{t} - m(k,t,l)\right)^{-1}$.

Other results and additional constructions can be found in [6] and [7] where exact results are proved for the case $r=2$ and the cases (6)-(8), and also for $k = 2r$.

4. Proof of Lemma 2.3

We are going to use the following consequence of Chernoff inequality [4] (originally, the Bernstein's improvement of Chebysheff inequality, see, e.g., Rényi [15]).

(9) Let $Y_1, \ldots, Y_m$ be independent random variables with $Prob(Y_i=1) = q$, $Prob(Y_i=0) = 1-q$, then

$$Prob(|\sum Y_i - mq| > \alpha\sqrt{mq}) < 2e^{-\alpha^2/2}.$$

For every $F \in \binom{N}{k}$ let Y_F be a random variable

$$Prob(Y_F{=}1) = D \Big/ \binom{n-t}{k-t}, \tag{10}$$

$$Prob(Y_F{=}0) = 1 - Prob(Y_F{=}1)\ .$$

Let $\mathcal{F}$ be the random family, defined by $\mathcal{F} = \left\{F \in \binom{N}{k} : Y_F = 1\right\}$. For $T \in \binom{N}{t}$ define $Y_T = \sum \left\{Y_F \colon T \subset F\right\}$. Then

$$E(Y_T) = D\ .$$

As Y_T is a sum of $\binom{n-t}{k-t}$ independent random variables (9) gives that for every fixed T

$$Prob(\,|Y_T - D\,| > (2k\,D \log n)^{1/2}) < \frac{2}{n^k}\ .$$

Hence

$$Prob(\exists T \in \binom{N}{t} \text{ with } |Y_T - D\,| > (2k\,D \log n)^{1/2}) < \frac{2\binom{n}{t}}{n^k} = o(1)\ . \tag{11}$$

Proposition 4.1. $Prob(\exists U \in \binom{N}{t+1}$ with $deg_{\mathcal{F}}(U) \geq 3k) < \frac{1}{t+1}$.

Proof: We can choose a set U in $\binom{n}{t+1}$ distinct ways. Then we can choose $3k$ k-sets through U in $\binom{\binom{n-t-1}{k-t-1}}{3k}$ ways. The probability of the appearance of such a configuration is $(D/\binom{n-t}{k-t})^{3k}$. Altogether, the probability in the left hand side is not larger than

$$\binom{n}{t+1}\binom{\binom{n-t-1}{k-t-1}}{3k}\left(\frac{D}{\binom{n-t}{k-t}}\right)^{3k} < \frac{n^{t+1}}{(t+1)!}\left(\frac{\binom{n-t-1}{k-t-1}}{\binom{n-t}{k-t}}\right)^{3k}\frac{D^{3k}}{(3k)!}$$

Using $\binom{n-t-1}{k-t-1}/\binom{n-t}{k-t} = (k-t)/(n-t) < k/n$ and $D \leq n^{2/3}$ we obtain 4.1. □

Proposition 4.2. Let $s = 6\left\lfloor k^{-2.5}(D\log n)^{0.5}\right\rfloor$. Then the probability that there exists a $T \in \binom{N}{t}$ and $2s$ distinct members of $\mathcal{F}$ such that $F_1, \ldots, F_s \in \mathcal{F}$ with $T \subset F_i$ and $F_{s+i} \in \mathcal{F}$ with $|F_i \cap F_{s+i}| > t$ $(1 \leq i \leq s)$ is less than $1/k^2$.

Proof: It is analogous to the 4.1. We can choose such a configuration in at most

$$\binom{n}{t}\binom{\binom{n-t}{k-t}}{s}\left(\binom{n-t-1}{k-t-1}\binom{k}{t+1}\right)^{s}$$

distinct ways. The probability of the appearance of each of these configuration in $\mathcal{F}$ is $(D/\binom{n-t}{k-t})^{2s}$. For n sufficiently large (e.g., $n > \exp(k^{10}4^k)$) an easy calculation gives 4.2. □

The proof of 2.3. Choose a family $\mathcal{F}$ at random as in (10). Then the sum of the probabilities in (11), 4.1 and 4.2 is $o(1) + 1/(t+1) + 1/k^2 < 1$. Hence there exists a family $\mathcal{F}$ without the configurations described in 4.1 and 4.2 and for which

$$|deg_{\mathcal{F}}(T) - D| < (2kD\log n)^{1/2} \tag{12}$$

holds for every $T \in \binom{N}{t}$.

Now call a set $F \in \mathcal{F}$ *bad* if there exists an $F' \in \mathcal{F}$, $F \neq F'$ with $|F \cap F'| > t$. Let $\mathcal{B} = \{F \in \mathcal{F}\colon F \text{ is bad}\}$, and define $\mathcal{S} = \mathcal{F} - \mathcal{B}$. We claim that $\mathcal{S}$ fulfils the constraints of 2.3. Obviously (i) holds. If we prove that for every $T \in \binom{N}{t}$

$$deg_{\mathcal{B}}(T) < (2-\sqrt{2})\sqrt{kD\log n}$$

holds, then we are ready by (12). Suppose on the contrary, that for some T we have $deg_{\mathcal{B}}(T) > 3k^3 s$. Then by 4.1 we can find $B_1, \ldots, B_{sk} \in \mathcal{B}$, $B_i \supset T$ such that $(B_i - T) \cap (B_j - T) = \emptyset$ for $1 \leq i < j \leq sk$. There exists $B_i' \in B$ with $|B_i' \cap B_i| > t$. Then we can choose a subsequence of B_j's such that $B_{i_1}, B_{i_2}, \ldots, B_{i_s}$ and

$B'_{i_1}, \ldots, B'_{i_s}$ are 2s distinct members. This contradicts 4.2. □

5. Proof of Theorem 2.2

To avoid trivialities suppose that $t \geq 2$. To construct a colored $(v,k,\mathcal{H})$-packing we begin with a family $\mathbf{S} \subset \binom{X}{k}$, $|X| = v$, given by Lemma 2.3 with $D = \sqrt{v}$. In the following calculations we suppose that h is a small but fixed positive real depending only on k (e.g., $h = 4^{-k}$). Furthermore, we suppose that $v > v_0(k)$. Let $p = v^{-h}$ and let Z_T be a random variable for every $T \in \binom{X}{t}$ with distribution

$$Prob(Z_T = i) = p \quad \text{for} \quad i = 1,2,\ldots,c \text{ and}$$

$$Prob(Z_T = 0) = 1 - cp\ .$$

In other words we color randomly and independently the t-sets of X. Recall that $c = \binom{k}{t} - |\mathcal{H}|$. Let $\mathbf{C}_i =: \{T\colon Z_T = i\}$, then for $i \geq 1$

$$E(|\mathbf{C}_i|) = p\binom{v}{t},$$

and by (9)

$$Prob\left(\left||\mathbf{C}_i| - p\binom{v}{t}\right| > v^{t-1}\right) < 2e^{-v^t}\ . \tag{13}$$

Call a set $S \in \mathbf{S}$ *well-colored* (with respect to the coloration $\{\mathbf{C}_0, \mathbf{C}_1, \ldots, \mathbf{C}_c\}$) if the restriction of the coloration to $\binom{S}{t}$ is isomorphic to χ. Denote the set of well-colored k-sets by $\mathbf{W}$. If g is the number of non-isomorphic embeddings of χ into $\binom{S}{t}$ then

$$Prob(S \text{ is well-colored}) = g\, p^c (1-p)^{\binom{k}{t}-c}\ .$$

Define an $|\mathcal{H}|$-uniform hypergraph $\mathcal{A}$ with vertex-set $\mathbf{C}_0$, $\mathcal{A} = \left\{\binom{S}{t} \cap \mathbf{C}_0\colon S \in \mathbf{S}\ \text{well-colored}\right\}$. Let $d =: gp^c(1-p)^{\binom{k}{t}-c-1}|\mathcal{H}|D/\binom{k}{t}$. By the choice of p and D we have

$$d > v^{0.4} \tag{14}$$

Proposition 5.1. Let $T \in \mathcal{C}_0$, then

$$Prob(|deg_{\mathcal{A}}(T) - d| > v^{0.3}) < e^{-v^h}.$$

Proof: Consider the sets $S_1, S_2, \ldots, S_l \in \mathcal{S}$ with $T \subset S_i$. As $T \in \mathcal{C}_0$, $Prob(S_i \in \mathcal{W}) = gp^c(1-c)^{\binom{k}{t}-c-1} |\mathcal{H}| / \binom{k}{t}$. Moreover these events are independent, hence by (9)

$$Prob(|deg_{\mathcal{A}}(T) - l| > v^h \sqrt{l}) < 2e^{-v^{2h}} < e^{-v^h}.$$

However $|deg_{\mathcal{A}}(T) - d| > v^{0.3}$ implies $|deg_{\mathcal{A}}(T) - l| > v^h \sqrt{l}$. □

This proposition and (13) yield that there exists a choice of $\{\mathcal{C}_0, \mathcal{C}_1, \ldots, \mathcal{C}_c\}$ for which

$$|\mathcal{C}_i| < p\binom{v}{t} + v^{t-1} < v^{t-h}$$

holds for all $i \geq 1$, and for every $T \in \mathcal{C}_0$

$$|deg_{\mathcal{A}}(T) - d| \leq v^{0.3}.$$

But by (14) $d > v^{0.4}$ and for every $T_1 \neq T_2$, $T_1, T_2 \in \binom{X}{t}$ we have $deg_{\mathcal{X}}(\{T_1, T_2\}) \leq 1 < d/(\log v)^3$. Thus we can apply (2) to $\mathcal{A}$. This gives that whenever $v \longrightarrow \infty$

$$v(\mathcal{A}) \geq (1-o(1)) \, |\mathcal{C}_0| / |\mathcal{H}| = (1-o(1)) \binom{v}{t} / |\mathcal{H}|.$$

Finally, a matching $\mathcal{M}$ in $\mathcal{A}$ gives a colored $\mathcal{H}$-cover, $\mathcal{F} = \{S \in \mathcal{S}: \binom{S}{t} \cap \mathcal{C}_0 \in \mathcal{M}\}$. □

6. Proof of 2.4

Let $\mathcal{F}$ be a $(v,k,\mathcal{H})$-packing and let $\mathcal{G}$ denote the set of "crowded" edges, i.e., $\mathcal{G} = \{\{i,j\}$: *either* $\{i,j\}$ *is uncovered or* $deg_{\mathcal{F}}(\{i,j\}) \geq 2\}$. For a vertex $x \in X$ denote by f_x the number of $F \in \mathcal{F}$ for which the vertex corresponding to a in F is x. Then

$$\sum_{x \in X} f_x = |\mathcal{F}| = \frac{1}{|\mathcal{H}|}\left(\binom{v}{2} - |\mathcal{G}|\right). \tag{15}$$

The main point is that

$$f_x \leq \binom{deg_{\mathcal{G}}(x)}{2}. \tag{16}$$

Indeed, if $x = a_F$ and $\{a_F, b_F\} \notin \mathcal{H}_F$, $\{a_F, c_F\} \notin \mathcal{H}_F$ then, clearly $\{b_F, c_F\} \neq \{b_{F'}, c_{F'}\}$. Moreover,

$$f_x \leq n-1 \tag{17}$$

holds for all $x \in X$ because $\{a_F, d_F\}$ is an own edge. Finally (15)-(17) give $|\mathcal{G}| \geq v^{3/2}/2\sqrt{2}\,|\mathcal{H}|$.

7. Proof of Theorem 3.4

Construction. Let $\mathcal{N}$ be a family of t-sets over the k-element set K such that $\mathcal{N}$ does not contain $l+1$ pairwise disjoint members and suppose that $|\mathcal{N}|$ is maximal, i.e., $|\mathcal{N}| = m(k,t,l)$. Let $\mathcal{H} = \binom{K}{t} - \mathcal{N}$ and let $\mathcal{F}$ be a colored $(n,k,\mathcal{H})$-packing with size

$$|\mathcal{F}| = (1-o(1))\binom{n}{t}/|\mathcal{H}|.$$

We claim that $\mathcal{F}$ is an r-cover-free family. Suppose on the contrary, that $F_0 \subset F_1 \cup \cdots \cup F_r$. As $|F_0 \cap F_i| \leq t$ and $|F_0| = r(t-1) + l+1$, we have at least $l+1$ F_i such that $|(F_i \setminus (F_1 \cup \cdots \cup F_{i-1})) \cap F_0| \geq t$. Then $|F_0 \cap F_i| = t$ must hold and thus we have at least $l+1$ disjoint sets $F_0 \cap F_i$ such that $F_0 \cap F_i \in \mathcal{N}$, a contradiction.

Upper bound. Let $\mathcal{F}_0$ be an r-cover-free family. Let $\mathcal{F}_0$ be the sets with small own parts, i.e., $\mathcal{F}_0 = \{F \in \mathcal{F}: \exists U \subset F,\ |U| \leq t-1$ such that $deg_{\mathcal{F}}(U) = 1\}$. Clearly $|\mathcal{F}_0| \leq \binom{n}{t-1}$. Consider an $F \in \mathcal{F} - \mathcal{F}_0$ and let $\mathcal{N}_F$ be the non-own parts of F with t elements, i.e., $\mathcal{N}_F = \{T \in \binom{F}{t}: \exists F' \in \mathcal{F}$ such that $F \cap F' \supset T\}$.

Proposition 7.1. If $F \in \mathcal{F} - \mathcal{F}_0$ then $\mathcal{N}_F$ does not contain $l+1$ pairwise disjoint members.

Proof: Suppose for contradiction that $T_1, \ldots, T_{l+1} \in \mathcal{N}_F$ with $|\bigcup T_i| = (l+1)t$. Let $\mathcal{P} = \{T_1, \ldots, T_{l+1}, S_1, \ldots, S_{r-l-1}\}$ be a partition of F such that $|S_i| = t-1$. Then for each $P \in \mathcal{P}$ there exists an $F_P \in \mathcal{F}$, $F_P \neq F$ with $P \subset F_P$. Hence $F \subset \bigcup \{F_P: P \in \mathcal{P}\}$, a contradiction. □

Now Proposition 7.1 implies that $|\mathcal{N}_F| \leq m(k,t,l)$, i.e., every $F \in \mathcal{F} - \mathcal{F}_0$ contains at least $\binom{k}{t} - m(k,t,l)$ own t-subsets. Hence

$$|\mathcal{F}| = |\mathcal{F}_0| + |\mathcal{F}-\mathcal{F}_0| \le \binom{n}{t-1} + \binom{n}{t} \Big/ \left(\binom{k}{t} - m(k,t,l)\right). \quad \square$$

A slightly more complicated argument gives that for $n > n_0(k)$

$$|\mathcal{F}| \le \binom{n}{t} \Big/ \left(\binom{k}{t} - m(k,t,l)\right).$$

8. Final Remarks

Actually, using the argument in Chapter 5 we can prove the following stronger statement.

Theorem 8.1. Let $\mathcal{H}$ be a family of t-sets on $\{1,2,\ldots,k\}$. There exists a family $\mathcal{F} \subset \binom{X}{k}$, $|X| = v$ of size

$$|\mathcal{F}| = (1-o(1)) \binom{v}{t} \Big/ |\mathcal{H}|$$

(whenever $v \to \infty$) with the following properties:

(i) $|F \cap F'| \le t$ for all distinct $F, F' \in \mathcal{F}$

(ii) For every $F \in \mathcal{F}$ there is a permutation of its elements $F = (x_1, \ldots, x_k)$ such that whenever $|F \cap F'| = t$ and $F' = (y_1, \ldots, y_k)$ and $F \cap F' = \{x_{i_1}, \ldots, x_{i_t}\}$ then $x_{i_\alpha} = y_{i_\alpha}$ for $1 \le \alpha \le t$, and $\{i_1, \ldots, i_t\} \notin \mathcal{H}$. It remains open whether we can suppose in this theorem that the orderings on each $F \in \mathcal{F}$ can be obtained as a restriction of an ordering of X.

Another open problem arises from the fact that our proof is probabilistic. It would be interesting to give other ("real") constructions. It is not necessarily hopeless, e.g., N. Alon [1] pointed out that an exponentially large r-cover-free family can be obtained using a recent explicit construction of J. Friedman [14] of certain generalized Justensen codes.

References

[1] N. Alon, Explicit construction of exponential sized families of k-independent sets, Discrete Math. 58 (1986), 191-193.

[2] B. Bollobás, On generalized graphs, Acta Math. Acad. Sci. Hungar. 16 (1965), 447-452.

[3] B. Bollobás, D. E. Daykin and P. Erdős, On the number of independent edges in a hypergraph, Quart. J. Math. Oxford (2) 27 (1976), 25-32.

[4] H. Chernoff, A measure of asymptotic efficiency for tests of an hypothesis based on the sum of observations, Ann. Math. Statist. 23 (1952), 493-507.

[5] P. Erdős, A problem of independent r-tuples, Ann. Univ. Budapest 8 (1965), 93-95.

[6] P. Erdős, P. Frankl and Z. Füredi, Families of finite sets in which no set is covered by the union of two others, J. Combin. Th. A 33 (1982), 158-166.

[7] P. Erdős, P. Frankl and Z. Füredi, Families of finite sets in which no set is covered by the union of r others, Israel J. Math. 51 (1985), 79-89.

[8] P. Erdős and T. Gallai, On maximal paths and circuits of graphs, Acta Math. Acad. Sci. Hungar 10 (1959), 337-356.

[9] P. Erdős, C. Ko and R. Rado, An intersection theorem for finite sets, Quart. J. Math. Oxford (2) 12 (1961), 313-320.

[10] P. Erdős and E. Szemerédi, Combinatorial properties of a system of sets, J. Combin. Th. A 24 (1978), 308-311.

[11] P. Frankl, A general intersection theorem for finite sets, Ann. Discrete Math. 8 (1980), 43-49.

[12] P. Frankl and Z. Füredi, Extremal problems concerning Kneser-graphs, J. Combin. Th. B 40 (1986), 270-284.

[13] P. Frankl and V. Rödl, Near perfect coverings in graphs and hypergraphs, Europ. J. Combin. 6 (1985), 317-326.

[14] J. Friedman, Constructing $O(n \log n)$ size monotone formulae for the k-th elementary symmetric polynomial of n Boolean variables, Proc. 25th Symp. on Foundation of Comp. Sci., Florida 1984.

[15] A. Rényi, Probability theory, North-Holland, Amsterdam 1970.

[16] V. Rödl, On a packing and covering problem, Europ. J. Combin. 6 (1985), 69-78.

Annals of Discrete Mathematics 34 (1987) 179–188
© Elsevier Science Publishers B.V. (North-Holland)

Balanced Room Squares from Finite Geometries and their Generalizations

R. Fuji-Hara

Institute of Socioeconomic Planning
University of Tsukuba
Tsukuba, JAPAN

S.A. Vanstone

Department of Combinatorics and Optimization
University of Waterloo
Waterloo, Ontario, N2L 3G1
CANADA

TO ALEX ROSA ON HIS FIFTIETH BIRTHDAY

ABSTRACT

Room squares have been extensively studied and their existence has been completely settled. The balanced Room square problem appears to be much more difficult. The existence problem for these designs is far from complete. In this paper we show how to construct a certain class of balanced Room squares from finite geometries and use recursive constructions for the geometries to produce infinitely many new balanced Room squares. The paper generalizes the concept of a balanced Room square to Kirkman squares with larger block size. Using finite geometries infinitely many balanced Kirkman squares are constructed.

1. Introduction.

A Kirkman square $KS_k(v)$ is an $r \times r$ array ($r = \frac{v-1}{k-1}$) defined on a r-set V such that

(1) each cell of the array is either empty or contains a k-subset of V.

(2) each element of V is contained in precisely one cell of each row and column of the array.

(3) the subsets in the non-empty cells form the blocks of a $(v,k,1)$-BIBD.

The spectrum for $KS_k(v)$ has not yet been completely settled. In fact, we are at present far from the complete solution. The following results are known.

Theorem 1.1. (Mullin and Wallis [7]). *There exists a $KS_2(v)$ if and only if v is a positive integer, $v \equiv 0$ (mod 2) and $v \neq 4$ or 6.*

Theorem 1.2. (Rosa and Vanstone [9]). *There exists an integer v_0 such that for all $v > v_0$ and $v \equiv 3$ (mod 6) there is a $KS_3(v)$.*

It is conjectured that for all $v \geq 27$, $v \equiv 3$ (mod 6) there exists a $KS_3(v)$.

In this paper we are interested in $KS_k(v)$'s with additional properties. These require several more definitions.

A $KS_k(v)$ is said to be ordered if each cell containing a k-subset K of the point set V is replaced by an ordered k-tuple on the elements of K. We denote such an array by $OKS_k(v)$. Let D be an $OKS_k(v)$. For each row (column) of D we form k blocks $B_1, B_2, \ldots, B_k$ where B_i consists of all elements in the i-th coordinate of each k-tuple in this row (column). The set of all blocks taken over all rows (columns) of D is called the row (column) design of D. Let D^* be a $K_k(v)$. If there exists an $OKS_k(v)$ D formed from D^*, as the underlying array, such that the row (column) design of D is a BIBD then D^* is called a row (column) balanced Kirkman square and is denoted $RBKS_k(v)$ ($CBKS_k(v)$). If D^* is both row and column balanced then D^* is denoted $RCBKS_k(v)$. If D^* is a $RBKS_k(v)$ and a $CBKS_k(v)$ and both have the same underlying $OKS_k(v)$ then D^* is called a strongly balanced Kirkman square and is denoted $SBKS_k(v)$.

A $RBKS_k(v)$ or $CBKS_k(v)$ in the case $k = 2$ is commonly referred to as a balanced Room square and denoted $BRS(v-1)$. A number of articles (see for example [8], [10]) have been concerned with the existence of $BRS(v-1)$ but the spectrum is far from determined. The concepts of $RCBKS_2(v)$ and $SBKS_2(v)$ have not been considered before. In this paper we approach the problem through the use of finite geometries. The next section introduces the required geometrical ideas.

We conclude this section with an example of a $SBKS_2(8)$.

(0,8)	(2,6)	(4,5)		(1,3)		
	(1,8)	(3,0)	(5,6)		(2,4)	
		(2,8)	(4,1)	(6,0)		(3,5)
(4,6)			(3,8)	(5,2)	(0,1)	
	(5,0)			(4,8)	(6,3)	(1,2)
(2,3)		(6,1)			(5,8)	(0,4)
(1,5)	(3,4)		(0,2)			(6,8)

ROW DESIGN

{0,2,4,1}
{8,6,5,3}
{1,3,5,2}
{8,0,6,4}
{2,4,6,3}
{8,1,0,5}
{4,3,5,0}
{6,8,2,1}
{5,4,6,1}
{0,8,3,2}
{2,6,5,0}
{3,1,8,4}
{1,3,0,6}
{5,4,2,8}

The row and column designs are (8,4,3)-BIBDs.

2. Skew and Hyperplane Skew Resolutions.

A skew class in $AG(n,q)$, the affine geometry of dimension n and order q, is a set of lines, no two distinct lines are parallel, which partitions the point set of the geometry. A skew resolution of $AG(n,q)$ is a set of skew classes which partitions the line set of $AG(n,q)$. Skew resolutions were introduced in [5] for an approach to constructing packings in $PG(n,q)$. In order to see how skew resolutions can be used to construct Kirkman squares we require another definition.

Let D be a $(v,k,1)$-BIBD and let $\mathbf{R}$ and $\mathbf{R}'$ be two resolutions of the block set. $\mathbf{R}$ and $\mathbf{R}'$ are said to be orthogonal if $|R \cap R'| \leq 1$ for every resolution class $R \in \mathbf{R}$, $R' \in \mathbf{R}'$. It is not difficult to see that a $(v,k,1)$-BIBD with two orthogonal resolutions implies the existence of a $KS_k(v)$ and conversely.

The points and lines of $AG(n,q)$ constitute a $(v,k,1)$-BIBD where $v = q^n$, $k = q$. If we denote the parallel resolution by $\mathbf{P}$ and if $AG(n,q)$ has a skew resolution $\mathbf{S}$ then $\mathbf{P}$ and $\mathbf{S}$ are orthogonal resolutions of the BIBD.

Section 3 will also require the concept of a hyperplane skew resolution. Let H_∞ be the hyperplane at infinity for $AG(n,q)$ so that $AG(n,q)$ plus H_∞ gives $PG(n,q)$. If $\mathbf{S}$ is any set of lines in $AG(n,q)$ we let $\Gamma(S)$ be the set of points in H_∞ which are not incident with any line of S. A skew class T of $AG(n,q)$ is said to be a hyperplane skew class if $\Gamma(T)$ is the point set of a hyperplane of H_∞. A skew resolution $\mathbf{S} = \{S_1, S_2, \ldots, S_r\}$ is said to be a hyperplane skew resolution (HSR) of $AG(n,q)$ if

(i) S_i is a hyperplane skew class for $1 \leq i \leq r$.

(ii) $\Gamma(S_i) \neq \Gamma(S_j)$, for all $i \neq j$.

Hyperplane skew resolutions exist [6]. It is known ([4]) that any skew resolution in $AG(3,q)$ is an HSR.

3. Balanced Kirkman Squares from Finite Geometries.

In this section we apply the concepts of the previous section in order to construct various types of balanced Kirkman squares.

Theorem 3.1. *If there exists a skew resolution in $AG(n,q)$ then there exists a $RBKS_q(q^n)$.*

Proof. Consider the $(q^n,q,1)$-BIBD associated with the points and lines of $AG(n,q)$. Let $\mathbf{P} = \{P_1,P_2,\ldots,P_r\}$ and $\mathbf{S} = \{S_1,S_2,\ldots,S_r\}$, $r = (q^n-1)/(q-1)$, be the parallel and skew resolutions respectively. Form an $r \times r$ array A with the rows indexed by the parallel classes P_i and the columns by the skew classes S_i. In cell (P_i,S_j) place the subset $P_i \cap S_j$ which is either the empty set or a line of the geometry. Clearly, A is a $KS_q(q^n)$. Associated with the parallel class of lines P_i is a parallel class of q hyperplanes $H_i = \{H_i^1,H_i^2,\ldots,H_i^q\}$ such that H_i^j contains precisely one point from each line in P_i and the H_i's are distinct. There are many ways to associate H_i with P_i. The easiest way is the following: Let ℓ be the line of P_i which is incident with the origin, in the sense of linear algebra. Then there is a unique linear subspace F which is the orthogonal complement of ℓ. We can take the parallel class of hyperplanes containing F as H_i. Order a k-subset in row P_i of A so that the j-th coordinate point is on the hyperplane H_i^j. The resulting ordered $KS_q(q^n)$ is a row balanced Kirkman square whose row design is the design formed from the points and hyperplanes of $AG(n-1,q)$.

Theorem 3.2. *If there exists a hyperplane skew resolution in $AG(n,q)$ then there exists a $RCBKS_q(q^n)$.*

Proof. Since a HSR is also a skew resolution we can apply the construction given in the proof of Theorem 3.1 to get a $RBKS_q(q^n)$. We now prove that the array A is also column balanced but perhaps with a different ordering placed on the k-subsets. To do this we need only prove that for each skew class S_i there exist q parallel hyperplanes $H_i^1, H_i^2, \ldots, H_i^q$ such that each H_i^j contains precisely one point from each line of S_i.

Let $\ell_1, \ell_2, \ldots, \ell_t$ $(t = q^{n-1})$ be the lines of a hyperplane skew class S_i and let H_∞ be the hyperplane at infinity for $AG(n,q)$. Let

$$Q = \{x \in H_\infty : \; \ell \cap H_\infty = x \text{ for some } \ell \in S_i\}.$$

The points in $H_\infty \backslash Q$ are the points of a $PG(n-2,q)$ which we denote by T. There are precisely $q+1$ hyperplanes of $PG(n,q)$ which contain T; of course, H_∞ is one of these hyperplanes. Let H' be any one of the remaining q hyperplanes. Each ℓ_i, $1 \leq i \leq t$, meets H' in a point which is not in T. This completes the proof.

In order to apply the above results we need to know about the existence of skew resolutions and HSRs in $AG(n,q)$.

Theorem 3.3. (Fuji-Hara and Vanstone [4]). *There exists a HSR in $AG(3,q)$ for all q a prime or prime power.*

Corollary 3.3. *There exists a $RCBKS_q(q^3)$ for all q a prime or prime power.*

Theorem 3.4. (Fuji-Hara and Vanstone [6]). *There exists a HSR in $AG(2^i-1,q)$ for all q a prime or prime power and for all integers $i \geq 2$.*

Corollary 3.4. *There exists a $RCBKS_q(q^t)$ $t = 2^i - 1$ for all q a prime or prime power and for all integers $i \geq 2$.*

More existence results for skew resolutions will be given in the following section.

4. Balanced Room Squares.

The existence of $BKS_k(v)$ for $k = 2$ is of particular interest since it is the first case considered by other researchers and its spectrum is not yet settled. As indicated in the introduction, these arrays are commonly referred to as balanced Room squares. We again make use of the concepts from section 2.

The following result on skew resolutions in $AG(n,2)$ is originally due to R.D. Baker [2] and independently proved by B.A. Andersen [1] although neither phrased the result in terms of skew resolutions.

Theorem 4.1. *There exists a skew resolution in $AG(2m+1,2)$ for all integers $m \geq 1$.*

Corollary 4.1. *There exists a $RBKS_2(2^{2m+1})$ for all integers $m \geq 1$.*

It was previously unknown whether or not a skew resolution exists in an affine geometry of even dimension. We answer this question in the affirmative.

Theorem 4.2. *There exists a skew resolution in $AG(4,2)$.*

Proof. It is enough to display one skew class in $AG(4,2)$. Instead we list all skew resolutions in $AG(4,2)$ which admit an automorphism of order 15.

We associate the points of $AG(4,2)$ with the points of $GF(2^4)$ generated by the primitive irreducible polynomial $f(x) = x^4 + x^3 + 1$. Let α be a root of $f(x)$. If $\{\alpha^i, \alpha^j\}$ is a line of $AG(4,2)$ we abbreviate the notation to $\{i,j\}$. Also, we define $\alpha^\infty = 0$. A parallel class of lines in $AG(4,2)$ is

$$P_0 = \{\{\alpha^i, 1+\alpha^i\}: i = \infty \text{ or } 0 \leq i \leq 14\}.$$

The other 14 parallel classes are

$$P_j = \{\{\alpha^{i+j}, \alpha^j(1+\alpha^i)\}: i = \infty \text{ or } 0 \leq i \leq 14\},\ 1 \leq j \leq 14.$$

Explicitly,

$$P_0 = \{\{0,\infty\},\{1,12\},\{2,9\},\{3,4\},\{5,10\},\{6,8\},\{7,13\},\{11,14\}\}.$$

A skew class is represented by a list of 7 distinct integers $(a_1, a_2, \ldots, a_7)$ such that $1 \leq a_i \leq 14$ and

$$\begin{aligned}
\{1,12\} + a_1 &= \{1+a_1, 12+a_1\} \\
\{2,9\} + a_2 &= \{2+a_2, 9+a_2\} \\
\{3,4\} + a_3 &= \{3+a_3, 4+a_3\} \\
\{5,10\} + a_4 &= \{5+a_4, 10+a_4\} \\
\{6,8\} + a_5 &= \{6+a_5, 8+a_5\} \\
\{7,13\} + a_6 &= \{7+a_6, 13+a_6\} \\
\{11,14\} + a_7 &= \{11+a_7, 14+a_7\}
\end{aligned}$$

is a partition of the non-zero residues modulo 15. Such a class will generate a skew resolution under the action of the cyclic automorphism of order 15.

a_1	a_2	a_3	a_4	a_5	a_6	a_7
1	9	5	11	4	7	8
1	14	2	9	3	5	11
2	3	5	1	11	9	14
2	3	6	11	5	10	8
2	11	6	7	13	9	12
4	1	3	9	5	10	13
4	5	14	1	2	6	13
4	10	6	12	5	1	7
6	3	7	4	13	10	2
7	1	10	11	14	4	13
8	1	10	7	13	9	12
9	7	14	3	6	13	8

It is a simple matter to check that there is no HSR in $AG(4,2)$ which admits an automorphism of order 15. Recently, the following results in $AG(5,2)$ were obtained.

Theorem 4.3. (Stinson and Vanstone [11]). *In* $AG(5,2)$ *there exist* 179 *inequivalent skew resolutions which admit an automorphism of order* 31 *and* 26 *HSRs admitting an automorphism of order* 31.

Corollary 4.3. *There exists a* $RCBRS(31)$.

Another recent result is stated in the next theorem.

Theorem 4.4. (Stinson and Vanstone [12]). *There exists a skew resolution in* $AG(n,2)$ *for* $n = 6,8,10$ *and* 14.

These give rise to new balanced Room squares.

Corollary 4.4. *There exist* $BRS(2^n-1)$ *for* $n = 6,8,10$ *and* 14.

The following recursive constructions give rise to infinitely many new balanced Room squares.

Theorem 4.5. (Fuji-Hara and Vanstone [5]). *If there exists a skew resolution in* $AG(m,q^n)$ *and a skew resolution in* $AG(n,q)$ *then there exists a skew resolution in* $AG(mn,q)$.

Theorem 4.6. (Fuji-Hara and Vanstone [5]). *If there exists a skew resolution in* $AG(m+1,q^n)$ *and if there exists a skew resolution in* $AG(n,q)$ *then there exists a skew resolution in* $AG(mn+1,q)$.

For hyperplane skew resolutions we have the following result.

Theorem 4.7. (Fuji-Hara and Vanstone [6]). *If there exists a HSR in* $AG(m,q^n)$ *and a HSR in* $AG(n,q)$ *then there exists a HSR in* $AG(mn,q)$.

5. Strongly Balanced Room Squares.

Strongly balanced Kirkman squares were defined in section 1 and an example of a strongly balanced Room square was displayed in the same section. There has so far been very little work done on $SBRS(v-1)$'s and few results on their existence are known.

Theorem 5.1. *If q is a prime power and $q \equiv 3 \pmod 4$ then there exists a $SBRS(q)$.*

Proof. Index the rows and columns of a $q \times q$ array A with the elements of $GF(q)$. Let α be a generator for the nonzero elements of $GF(q)$. Consider the set of ordered pairs $\{(\alpha^{2i},\alpha^{2i+1})\colon 0 \leq i \leq \frac{q-3}{2}\}$. In row g, column $g + \alpha^{2i} + \alpha^{2i+1}$, $0 \leq i \leq \frac{q-3}{2}$, place the ordered pair $(\alpha^{2i}+g,\alpha^{2i+1}+g)$ for all $g \in GF(q)$. In row g, column g place the ordered pair (∞,g) where ∞ $GF(q)$ for all $g \in GF(q)$. A is an $OBRS(q)$. It is easily checked that A is in fact a $SBRS(q)$.

Theorem 5.1 shows the existence of a $SBRS(31)$. An alternate way to construct such a design is through finite geometries. We illustrate the technique.

For notation and a detailed explanation the reader is referred to [11]. A base class for the parallel resolution in $AG(5,2)$ is

$$P_0 = \{2^i\{\{5,2\},\{3,29\},\{26,28\}\}\colon\ 0 \leq i <+ 4\} \cup \{\{0,\infty\}\}.$$

A base class for a hyperplane skew resolution is

$$S_0 = \{2^i\{\{26,27\},\{5,19\},\{3,16\}\}\colon\ 0 \leq i \leq 4\} \cup \{0,\infty\}.$$

If we order the pairs as given in P_0 and S_0 and observe that

$$H_0 = \{2^i\{5,3,26\}\colon 0 \leq i \leq 4\} \cup \{0\}$$

is a base block for a hyperplane then the associated $BRS(31)$ is a $SBRS(31)$.

Clearly, the existence of a $SBRS(v-1)$ implies the existence of a $RCBRS(v-1)$.

It is unlikely that the converse is true.

We conclude this section with the following result concerning strong balanced Room squares. We consider $AG(2k+1,2)$ and associate the points of the geometry with the elements of $GF(2^{2k+1})$. Let α be a generator of $GF(2^{2k+1})$ and let $\alpha^{\infty} = 0$. The automorphism $\delta\colon \alpha^i \longrightarrow \alpha^{i+1}$ generates all lines of the geometry cyclically from q^{n-1} initial lines. Each cycle contains $q^n - 1$ lines which are generated from an initial line. Let H_0 be a hyperplane of $AG(2k+1,2)$ containing α^{∞} but α^0 H_0. The set of lines

$$P_0 = \{(\alpha^a,1+\alpha^a)\colon\ \alpha^a \in H_0\}$$

is a parallel class of lines in the geometries.

The following lemma was proven first by B.A. Anderson [1] but not in the context of skew resolutions.

Lemma 5.1. *Let H_0 be a hyperplane of $AG(2k+1,2)$ such that $\alpha^{\infty} \in H_0$ α^0 H_0 then $S_0 = \{(\alpha^{-a},(1+\alpha^a)^{-1})\colon \alpha^a \in H_0\}$ is a skew class and δ generates a skew resolution from S_0.*

Proof. It is easy to see that the set of lines of S_0 partitions the points of $AG(2k+1,2)$. Since the pair $\{\alpha^{-a},(1+\alpha^a)^{-1}\}$ is in the same cycle generated from $\{\alpha^a,(1+\alpha^1)\}$, we show here that integer j such that $\{\alpha^a\cdot\alpha^j,$ $(1+\alpha^a)\alpha^j\} = \{\alpha^{-a},(1+\alpha^a)^{-1}\}$ is distinct for each $\alpha^a \in H_0$. When $\alpha^a = \alpha^\infty$, $j = 0$. Consider, for $\alpha^a \in H_0\backslash\alpha^\infty$,

$$\alpha^j = (\alpha^{-a}/(1+\alpha^a) = (\alpha^a+\alpha^{2a})^{-1}$$

If the class contains two parallel classes, then the following two cases are considered:

(i) there exists α^a in $H_0\backslash\alpha^\infty$ such that $\alpha^a + \alpha^{2a} = 1$.

(ii) there exists distinct elements α^a and α^b in $H_0\backslash\alpha^\infty$ such that $\alpha^a + \alpha^{2a} = \alpha^b + \alpha^{2b}$

In case (i), since $x^2 + x + 1 = 0$, has no root in H_0 it is impossible.

In case (ii), which implies that $(\alpha^a+\alpha^b) = (\alpha^a+\alpha^b)^2$. It holds only if $\alpha^a + \alpha^b = \alpha^\infty$ or $\alpha^a + \alpha^b = 1$. Since $\alpha^1 \neq \alpha^b$, $\alpha^a + \beta^b \neq \alpha^\infty$. Since H_0 is an additive group, $\alpha^a + \alpha^b$ is an element of H_0 but $1 = \alpha^0$ is not the element of H_0.

Theorem 5.2. *For any integer $k \geq 1$, there exists an $RCBRS(2^{2k+1}-1)$.*

Proof. We define a $t \times t$ array A, $t = 2^{2k+1} - 1$. Rows of A are corresponding to parallel classes $P_0P_1 \cdots P_{t-1}$ and columns are corresponding to skew classes constructed in Lemma 5.1 $S_0S_1 \cdots S_{t-1}$. Since the P_i's are parallel classes, the array is a row balanced room square. The entries of the array are represented by power indices of the primitive element i.e. $G = \{\infty,0,1,...,q^{2k-1}-2\}$. Let H_0^* be the index set of H_0 and let $\alpha^{f(a)} = 1 + \alpha^a$, $1 \in G\backslash H_0$. Consider the difference family $\{E_0,F_0\}$, where

$$E_0 = \{-a;\ a \in H_0^*\}$$
$$F_0 = \{-f(a);\ a \in H_0^*\}.$$

E_0 and F_0 have the same set of differences as H_0^* and $G\backslash H_0^*$. So we have a $(2^{2k+1},2^{2k},2^{2k-1})$-BIBD whose blocks are $E_0,E_1 \cdots E_{t-1}$, $F_0,F_1 \cdots F_{t-1}$.

The initial blocks of the row design and the column design constructed above have the same family of differences, but, in general, they are distinct families and the row and column designs generated from them are different.

The following corollary shows that a stronger result may be possible.

Corollary 5.4. *If H_0 is an additive group of order 2^{2k} in $GF(2^{2k+1})$ such that x^{-1} H_0 for any $x \in H_0\backslash\alpha^\infty$, then there exists a hyperplane skew resolution in $AG(2k+1,2)$ and a $SBRS(2^{2k+1}-1)$ exists.*

Proof. Since there is a cyclic automorphism δ: $\alpha^j \to \alpha^{j+1}$ in $GF(2^{2k+1})$, we can assume that H_0 does not contain α^0 without loss of generality. We order each pair in P_0 as follows:

$$P_0 = \{(\alpha^a,1+\alpha^a):\ \ \alpha^a \in H_0\}$$

then we have the following ordered pairs in the 1-st column as:

$$S_0 = \{(\alpha^\infty,1)\} \cup \{((1+\alpha^a)^{-1},\alpha^{-a}): \ \alpha^a \in H_0 \backslash \alpha^a\}.$$

If the hypothesis is satisfied, we can say that there exist $\alpha^b \in H_0 \backslash \alpha^\infty$ such that $\alpha^{-a} = 1 + \alpha^b$ for any $\alpha^a \in H_0 \backslash \alpha^\infty$. Therefore, the set of elements consisting of all first components of ordered pairs from S_0

$$\{\alpha^\infty\} \cup \{(1+\alpha^a)^{-1}: \ a \in H_0 \backslash \alpha^\infty\}$$

is exactly the same as H_0. So we have an *SBRS* and $\{S_0 S_1 \cdots S_t\}$ is a hyperplane skew resolution. □

We have, displayed in section 1, an example of an *SBRS* which is generated in $GF(2^3)$ by the primitive irreducible polynomial $x^3 + x + 1$.

6. Conclusion.

In this paper we have shown the connection between balanced Room squares, balanced Kirkman squares and finite geometries. It remains an open question whether or not a skew resolution exists in every affine geometry of dimension greater than 2. We conjecture that such a resolution does exist.

References

[1] B.A. Anderson, *Hyperplanes and Balanced Howell Rotations,* Ars Combinatoria Vol. 13 (1981), 163-168.

[2] R.D. Baker, *Partitioning the planes of* $AG_{2^n}(2)$ *into 2-designs,* Discrete Math. 15 (1974), 205-211.

[3] A. Beutelspacher, *On parallelisms in finite projective spaces,* Geometriae Dedicata 3 (1974), 35-40.

[4] R. Fuji-Hara and S.A. Vanstone, *Affine geometries obtained from projective planes and skew resolutions on* $AG(3,q)$, Ann. Discrete Math. 18 (1983), 355-376.

[5] R. Fuji-Hara and S.A. Vanstone, *Recursive constructions for skew resolutions in affine geometries,* Aequationes Math. 23 (1981), 242-251.

[6] R. Fuji-Hara and S.A. Vanstone, *Hyperplane skew resolutions and their Applications,* J. of Comb. Th. Series A (to appear).

[7] R.C. Mullin and W.D. Wallis, *The existence of Room square,* Aequationes Math. 1 (1975) 1-8.

[8] E.T. Parker and A.M. Mood, *Some balanced Howell rotations for duplicate bridge sessions,* Amer. Math. Monthly, 62 (1955) 714-716.

[9] A. Rosa and S.A. Vanstone, *A starter-adder approach to Kirkman squares and cubes of small side,* Ars Combinatoria 14 (1982) 199-212.

[10] P.J. Schellenberg, *On balanced Room squares and complete balanced Howell rotations,* Aequationes Math. 9 (1973) 75-90.

[11] D.R. Stinson and S.A. Vanstone, *Orthogonal packings in* $PG(5,2)$, Aequationes Math. 31 (1986) 159-168.

[12] D.R. Stinson and S.A. Vanstone, *A few more balanced Room squares,* Journal of the Australian Math. Soc. 39 (1985) 344-352.

Annals of Discrete Mathematics 34 (1987) 189–196
© Elsevier Science Publishers B.V. (North-Holland)

On the Number of Pairwise Disjoint Blocks in a Steiner System

Mario Gionfriddo

Dipartimento di Matematica
Universitá di Catania
Viale A. Doria 6
95125 Catania
ITALY

TO ALEX ROSA ON HIS FIFTIETH BIRTHDAY

1. Introduction

An $S(k,k+1,v)$ *Steiner system* is a pair (S,β) where S is a finite set of size v (called the *order* of the Steiner system) and β is a collection of $(k+1)$-element subsets of S (called *blocks)* such that every k-element subset of S belongs to *exactly one* block of β. An $S(2,3,v)$ Steiner system is called a *triple system* (STS(v)) and an $S(3,4,v)$ Steiner system is called a *quadruple system* (SQS(v)). It is well known that a triple system of order v exists if and only if $v\equiv 1,3 \pmod 6$ (see [4]) and in 1960 Hanani [2] proved that a quadruple system of order v exists if and only if $v\equiv 2,4 \pmod 6$. Very little is known about $S(k,k+1,v)$ Steiner systems for $k\geq 4$. In the theory of Steiner systems there are many open problems regarding the *parallelism* among the blocks.

If (S,β) is a Steiner system $S(k,k+1,v)$, by a *partial parallel class* of blocks of the Steiner system (S,β) is meant a collection Π of pairwise disjoint blocks of β. If the blocks of Π partition S, then Π is called a *parallel class* of blocks. Two blocks b',b'' are called *parallel* if $b'\cap b''=\emptyset$ (i.e., they are disjoint).

In this paper we give a result on the number of pairwise parallel blocks in a Steiner system $S(k,k+1,v)$.

2. Short survey on the parallelism in a Steiner system

In [3], C.C. Lindner and K.T. Phelps have proved the following general result:

Theorem 1: Let (S,β) be an $S(k,k+1,v)$ Steiner system with $v\geq k^4+3k^3+k^2+1$. Then β has a partial parallel class containing at least $\dfrac{v-k+1}{k+2}$ blocks. □

For the systems STS(v) and SQS(v) this result implies that:

Theorem 1.1: In every $S(2,3,v)$ with $v\geq 45$, there are at least $(v-1)/4$ pairwise parallel triples.

Theorem 1.2: In every S(3,4,v) with $v \geq 172$ there are at least $(v-2)/5$ pairwise parallel quadruples.

In another theorem of [3], C.C. Lindner and K.T. Phelps have proved that in the statement of Theorem 1.1 (for STS(v)) we can replace the inequality $v \geq 45$ by $v \geq 9$, except possibly the cases $v=19$ and $v=27$.

Theorem 1.3: A Steiner triple system of order $v \geq 9$ has a partial parallel class containing at least $(v-1)/4$ blocks, except possibly for $v=19$ and $v=27$.

In [5] and [6], G. Lo Faro has proved that the statement of Theorem 1.1 is true also for v=19 and v=27.

Theorem 2 (G. Lo Faro): In every STS(19) there are at least five pairwise parallel blocks, and there exist Steiner triple systems of order 19 with a partial parallel class containing exactly five blocks. □

Regarding the Steiner systems STS(v), after C.C. Lindner and K.T. Phelps, D.E. Woolbright [7] and A.E. Brouwer [1] have obtained the following results.

Theorem 3 (D.E. Woolbright): In every STS(v) there are at least $\frac{3v}{10}-7$ parallel blocks. □

This improves the result of Lindner-Phelps for $v \geq 139$.

Theorem 4 (A.E. Brouwer): In every STS(v) there are at least $\frac{(v-5v^{2/3})}{3}$ parallel blocks. □

This result applies for every admissible $v \geq 127$. It improves the result of Woolbright for large values of v (from 9500-10000).

In what follows, for any number a, $\lfloor a \rfloor$ denotes the maximum integer $\leq a$ and $\lceil a \rceil$ denotes the least integer $\geq a$.

3. New results

In this section we prove that in every Steiner system S(k,k+1,v) with $k \geq 3$ there exist at least $\lfloor (v+2)/2k \rfloor$ pairwise disjoint blocks. This result is general, and for $v<k^4+3k^3+k^2+1$ is currently the best result in the literature.

Theorem 5: In every S(k,k+1,v) with $k \geq 3$ there exist at least $\lfloor (v+2)/2k \rfloor$ parallel blocks.

Proof:

Let (S,β) be a Steiner system S(k,k+1,v) with $k \geq 3$. Let Π be a family of parallel blocks of this system, such that if $P = \bigcup_{b \in \Pi} b$ then $|S-P| \geq (k-1)|\Pi|+2(k-1)$. This implies that $v \geq 2k|\Pi|+2(k-1)$.

In what follows, let $|\Pi|=t$.

We prove that in the system there exists a family Π' of parallel blocks, such that $|\Pi'|>|\Pi|$. The result is immediate if there exists a block contained in $S-P$. Therefore we suppose that for each $b \in \beta$, $b \not\subseteq S-P$. We observe that, for a fixed set of $k-1$ distinct points $A_1,\dots,A_{k-1} \in S-P$, if $R=(S-P)-\{A_1,\dots,A_{k-1}\}$, then there exists an injection $\psi:R\to P$ defined as follows: for all $x\in R$, $\psi(x)$ is the element of P such that $\{A_1,\dots,A_{k-1},x,\psi(x)\}\in\beta$.

Now let

$$\{a_{i1},a_{i2},\dots,a_{i,k+1}\}\in\Pi, \quad \text{for } i=1,2,\dots,r,$$

such that $\{a_{i1},\dots,a_{ik}\}\subseteq\psi(R)$;

$$\{b^j_{i1},b^j_{i2},\dots,b^j_{i,k+1}\}\in\Pi, \quad \text{for } j=1,\dots,k-1 \text{ and } i=1,2,\dots,p_j,$$

such that $\{b^j_{i1},\dots,b^j_{ij}\}\subseteq\psi(R)$ and $\{b^j_{i,j+1},\dots,b^j_{i,k+1}\}\cap\psi(R)=\emptyset$;

$$\{c_{i1},\dots,c_{i,k+1}\}\in\Pi \quad \text{for } i=1,\dots,h,$$

such that $\{c_{i1},\dots,c_{i,k+1}\}\cap\psi(R)=\emptyset$.

Necessarily,

$$(k+1)r+\sum_{i=1}^{k-1}p_i\cdot i \geq |\psi(R)| = |S-P|-(k-1)\geq(k-1)t+k-1.$$

Since $t=r+\sum_{i=1}^{k-1}p_i+h$, it follows that

$$(k+1)r+\sum_{i=1}^{k-1}p_i\cdot i\geq(k-1)r+(k-1)\sum_{i=1}^{k-1}p_i+(k-1)h+k-1,$$

and hence

$$r\geq\frac{1}{2}\left[\sum_{i=1}^{k-2}p_i(k-1-i)+h(k-1)+k-1\right].$$

Let $X_{ij}\in R$ such that $\psi(X_{ij})=a_{ij}$ and let $Y^j_{iu}\in R$ such that $\psi(Y^j_{iu})=b^j_{iu}$.

Case 1: At first, suppose that $a_{i,k+1}\notin\psi(R)$ for each $i=1,\dots,r$. It follows that

$$|S-P|-(k-1)=\sum_{i=1}^{k-1}p_i\cdot i+kr.$$

Since $|S-P|\geq(k-1)t+2(k-1)$ and $t=h+r+\sum_{i=1}^{k-1}p_i$, in this case it follows:

$$\sum_{i=1}^{k-1}p_i\cdot i+kr\geq(k-1)t+k-1=(k-1)h+(k-1)r+(k-1)\sum_{i=1}^{k-1}p_i+k-1,$$

and hence

$$r\geq\sum_{i=1}^{k-2}p_i(k-i-1)+h(k-1)+(k-1).$$

Consider the injection $\phi:R'\to P$, where $R'=\{X_{ij}\in R: i\neq 1\}$, such that for all $X_{ij}\in R'$, $\phi(X_{ij})$ is the element of $\psi(R)$ satisfying the condition $\{X_{11},\dots,X_{1,k-1},X_{ij},\phi(X_{ij})\}\in\beta$.

If Γ is the family of the blocks $\{X_{11},...,X_{1,k-1},X_{ij},\phi(X_{ij})\}$ and $L = \{c_{ij}: i=1,...,h, j=1,...,k+1\} \cup \{b_{i1}^1 : i=1,...,p_1\} \cup \{a_{1k}\}$, it follows that $|\Gamma| = k(r-1)$ and $|L| = (k+1)h+p_1+1$, with

$$|\Gamma| = k(r-1) = kr-k \geq$$
$$\sum_{i=1}^{k-2} p_i(k-i-1)+hk(k-1)+k^2-2k >$$
$$(k+1)h+p_1+1 = |L|,$$

where we have considered that for each $k \geq 3$

$$r \geq \sum_{i=1}^{k-2} p_i(k-i-1)+h(k-1)+(k-1)$$
$$hk(k-1) > h(k+1)$$
$$k^2-k > 1.$$

Then it is possible to find an element $x \in P-L$ such that $\{X_{11},...,X_{1,k-1},\phi^{-1}(x),x\} \in \beta$. Further, there exists at least an element $y \in \psi(R)$, $y \neq x$, with x and y belonging to the same block b_{xy} of Π. If

$$\Pi' = \Pi - \{b_{xy}\} \cup \{\{X_{11},...,X_{1,k-1},\phi^{-1}(x),x\},\{A_1,...,A_{k-1},\psi^{-1}(y),y\}\}$$

then Π' is a family of parallel blocks of β with $|\Pi'| > |\Pi|$.

Case 2: Now suppose that there is at least one element $a_{i,k+1}$ such that $\{a_{i1},...,a_{i,k+1}\} \subseteq \psi(R)$.

We can suppose that

$$\{a_{i1},...,a_{i,k+1}\} \subseteq \psi(R)$$

for each $i=1,...,r'$, and

$$\{a_{i1},...,a_{ik}\} \subseteq \psi(R)\} \text{ and } a_{i,k+1} \notin \psi(R)$$

for each $i=r'+1,...,r$.

If $r \geq 2$, consider the injection $\mu: R'' \rightarrow P$, where

$$R'' = \{X_{ij} \in R : (i,j) \neq (1,1),...,(1,\lceil\tfrac{k-1}{2}\rceil),(2,1),...,(2,\lceil\tfrac{k-1}{2}\rceil)\}$$

such that for all $X_{ij} \in R''$, $\mu(X_{ij})$ is the element of $\psi(R)$ satisfying the condition $\{X_{11},...,X_{1,\lceil\frac{k-1}{2}\rceil}, X_{21},...,X_{2,\lfloor\frac{k-1}{2}\rfloor}, X_{ij},\mu(X_{ij})\} \in \beta$.

If Γ' is the family of blocks

$$\{X_{11},...,X_{1,\lceil\frac{k-1}{2}\rceil},X_{21},...,X_{2,\lfloor\frac{k-1}{2}\rfloor},X_{ij},\mu(X_{ij})\}$$

and

$$L' = \{c_{ij}: i=1,...,h,\, j=1,...,k+1\} \cup \{b_{i1}^1 : i=1,...,p_1\}$$

if follows that

$$\begin{aligned}|\Gamma'| &\geq (k-1)t - \sum_{i=1}^{k-1} i \cdot p_i \\ &= (k-1)(r + \sum_{i=1}^{k-1} p_i + h) - \sum_{i=1}^{k-1} i \cdot p_i \\ &= (k-1)r + (k-1)h + \sum_{i=1}^{k-2} (k-i-1)p_i \\ &\geq \frac{k+1}{2} \sum_{i=1}^{k-2} p_i(k-1-i) + \frac{h(k^2-1)}{2} + \frac{(k-1)^2}{2} \\ &> (k+1)h + p_1 + 1 = |L'| + 1\end{aligned}$$

where we have considered that

$$\begin{aligned}t &= r + h + \sum_{i=1}^{k-1} p_i \\ r &\geq \frac{1}{2}\left[\sum_{i=1}^{k-2} p_i(k-i-1) + h(k-1) + (k-1)\right] \\ k &\geq 3.\end{aligned}$$

So it is possible to find at least two distinct elements x', x'' belonging to two distinct blocks of Γ'

$$b' = \{X_{11},...,X_{1,\lceil\frac{k-1}{2}\rceil}, X_{21},...,X_{2,\lfloor\frac{k-1}{2}\rfloor}, \mu^{-1}(x'), x'\}$$

$$b'' = \{X_{11},...,X_{1,\lceil\frac{k-1}{2}\rceil}, X_{21},...,X_{2,\lfloor\frac{k-1}{2}\rfloor}, \mu^{-1}(x''), x''\}$$

such that $x', x'' \in P - L'$.

Since $x' \neq x''$, we can suppose that $x' \neq a_{2,\lceil\frac{k-1}{2}\rceil}$. In this situation, it is possible to find an element $y \in \psi(R)$, $y \neq x'$, with x' and y belonging to the same block b_{xy} of Π.

It follows that there exists a family Π' of parallel blocks with $|\Pi'| = |\Pi| + 1$.

If $r=1$, then $r'=r=1$. Since $r \geq \frac{1}{2}\left[\sum_{i=1}^{k-2} p_i(k-1-i) + h(k-1) + k-1\right]$, then $\sum_{i=1}^{k-2} p_i(k-1-i) + h(k-1) + k-1 \leq 2$.

It follows necessarily: $k=3$, $h=0$, $p_1=0$.

Hence $t = p_2+1$, $|S-P| = 2p_2+6$, and $v = 6p_2+10$. If $p_2=0$, then $v=10$, $t=1$, and it is well-known that in the unique STS(10) there are two parallel blocks.

If $p_2 \geq 1$, consider the blocks

$$b' = \{X_{11}, \psi^{-1}(b_{11}^2), X_{12}, x'\}$$
$$b'' = \{X_{11}, \psi^{-1}(b_{11}^2), X_{13}, x''\}$$

where $x', x'' \in \psi(R)$.

Since $x' \neq x''$, we can suppose $x' \neq b_{12}^2$. This permits to apply the same technique of the previous cases to find a family Π' of parallel blocks with $|\Pi'| = |\Pi| + 1$.

It is therefore proved that in any $S(k,k+1,v)$ (S,β) with $k \geq 3$, if Π is a family of parallel blocks such that for $|\Pi| = t$, $P = \bigcup_{b \in \Pi} b$, and $|S - P| \geq (k-1)t + 2(k-1)$, then there exists always in the system a family Π' of parallel blocks having size $|\Pi'| > |\Pi|$.
□

It follows that, if $t = \lfloor \frac{v-2(k-1)}{2k} \rfloor$, then there exists a family of parallel blocks having size $t' = t+1 = \lfloor \frac{v+2}{2k} \rfloor$.

From Theorem 5 it follows that in every Steiner quadruple system there are at least $\lfloor \frac{v+2}{6} \rfloor$ pairwise disjoint blocks. Therefore we have the following inequalities:

SQS(v)	
v (small values)	number of parallel blocks
4	1
8	2
10	2
14	≥ 2
16	≥ 3
20	≥ 3
22	≥ 4
26	≥ 4
28	≥ 5
32	≥ 5
34	≥ 6
38	≥ 6
40	≥ 7

In particular, for the systems SQS(16) we have the following interesting open problem: is it true that in every Steiner quadruple system of order 16 there are four pairwise disjoint blocks, or that there exists an SQS(16) with exactly three pairwise disjoint blocks?

Acknowledgements

Lavoro eseguito nell'ambito del GNSAGA e con contributo del M.P.I. (40%, 1985)

References

[1] A.E. Brouwer, "On the size of a maximum transversal in a Steiner triple system", *Canadian J. Math.* 33 (1981) 1202-1204.

[2] H. Hanani, "On quadruple systems", *Canadian J. Math.* 12 (1960) 145-157.

[3] C.C. Lindner and K.T. Phelps, "A note on the partial parallel classes in Steiner systems", *Discrete Math.* 24 (1978) 109-112.

[4] C.C. Lindner and A. Rosa, "Steiner quadruple systems: a survey", *Discrete Math.* 22 (1978) 147-181.

[5] G. Lo Faro, "On the size of partial parallel classes in Steiner systems STS(19) and STS(27)", *Discrete Math.* 45 (1983).

[6] G. Lo Faro, "Partial parallel classes in Steiner systems S(2,3,19)", *J. Infor. Optimization Sci.* 6 (1985) 133-136.

[7] D.E. Woolbright, "On the size of partial parallel classes in Steiner systems", *Ann. Disc. Math.* 7 (1980) 203-211.

Annals of Discrete Mathematics 34 (1987) 197–206
© Elsevier Science Publishers B.V. (North-Holland)

On Steiner Systems $S(3, 5, 26)$

M.J. Grannell, T.S. Griggs and J.S. Phelan

School of Mathematics and Statistics
Lancashire Polytechnic
Preston, PR1 2TQ
UNITED KINGDOM

TO ALEX ROSA ON HIS FIFTIETH BIRTHDAY

1. Introduction

Denote a Steiner system by $S(t,k,v)$ where the parameters have their usual meanings. In contrast to the situations where $t=2$ and $k=3,4,5,6$ or $t=3$ and $k=4$ where the spectrum of v for which a Steiner system exists is either completely or substantially determined, very little seems to be known concerning systems $S(3,5,v)$. In this case the necessary admissibility conditions give $v=2,5,17,26,41,50$ $(mod\ 60)$. The system $S(3,5,5)$ is of course trivial and it has long been known [7] that $S(3,5,17)$ is unique. For $v=26$ there are two realisations in the literature. One was constructed by Hanani [5], and the other occurs as a derived subsystem of the $S(5,7,28)$ given by Denniston [2], all such derived subsystems being isomorphic due to the double transitivity of the automorphism group. There appears to have been no published investigation as to whether the Hanani and Denniston systems are isomorphic. The only other known result is a construction also due to Hanani [4] which produces an $S(3,5,4v-3)$ from an $S(3,5,v)$. Hence there exists an $S(3,5,65)$ but the existence of an $S(3,5,v)$ for the intermediate cases $v=41,50,62$ as well as for an infinite number of other values of v is still undetermined.

In this paper we present a method by which with the aid of a computer we hope that further Steiner systems $S(3,5,v)$ may be constructed. We illustrate the method by applying it to produce a third realisation of an $S(3,5,26)$. Finally we determine that the three $S(3,5,26)'s$ are isomorphic thus raising the question of whether this is the unique Steiner system with these parameters.

2. Theory

A Steiner system $S(t,k,v)$ which has an automorphism of order v is said to be cyclic. In this case the base set V is usually represented by the set of residue classes modulo v and the automorphism by the mapping $i \rightarrow i+1$ $(mod\ v)$. We will follow this throughout the remainder of this paper. The Steiner system may be constructed from orbits under the cyclic group C_v generated by the above mapping and each orbit may be characterised by a cyclically ordered difference k-tuple $< d_1, d_2, \ldots, d_k >$ where $d_1+d_2+\ldots+d_k=v$. In the same realisation any orbit which is stabilized by the mapping $i \rightarrow -i \ (mod\ v)$ is said to be symmetric and a cyclic Steiner system, all of whose orbits are symmetric is said to be S-cyclic. In [3], two of the present authors proved that if $S(t,k,v)$ is S-cyclic then $t=3$ and if in addition k is odd then $v=k^2-2k+2$. Hence when $k=5$ then $v=17$ and indeed the unique $S(3,5,17)$ is S-cyclic [3]. For $v=26$, the system $S(3,5,v)$ can not be S-cyclic but the theory of such systems may still be utilized to advantage. All orbits of 5-element sets under C_{26} are of length 26 and precisely 10 of these are required to construct an $S(3,5,26)$. Each cyclic 5-element set orbit contains 10 cyclic 3-element set orbits. If a cyclic 5-element set orbit is symmetric then its characteristic difference quintuple may be written in the form $<a,b,c,b,a>$ where $2a+2b+c=26$. Two of the cyclic 3-element set orbits contained in this have difference triples $<a,a,2b+c>$ and $<a+b,a+b,c>$. Since there are precisely 12 difference triples of the form $<x,x,26-2x>$ i.e. for $x=1,2,....12$ there are at most 6 symmetric cyclic orbits contained in $S(3,5,26)$. We call a cyclic Steiner system in which the maximum number of cyclic orbits are also symmetric, a maximally S-cyclic system or MS-cyclic. The computational advantage of constructing an MS-cyclic Steiner system over an ordinary cyclic system is that the total number of symmetric cyclic orbits is far less than the number of all cyclic orbits. This facilitates the partial construction of the system using symmetric cyclic orbits only which may then be able to be completed with non-symmetric cyclic orbits. We now construct an MS-cyclic $S(3,5,26)$ and prove that this is the unique MS-cyclic $S(3,5,26)$.

3. Construction of partial systems

The Appendix lists all difference quintuples of symmetric cyclic 5-element set orbits under C_{26}. For use as an orbit in the construction of a Steiner system the 10 difference triples of the cyclic 3-element set orbits contained in each of the 5-element set orbits must be distinct. In the case of the system $S(3,5,17)$ if the difference quintuple was of the form $<a,b,c,b,a>$ it was shown in [3] that this implies that all of the following conditions must hold.

(i)	$a \neq b$
(ii)	$b \neq c$
(iii)	$2a+b \neq c$
(iv)	$a \neq c$
(v)	$a \neq b+c$
(vi)	$2a \neq b+c$
(vii)	$b \neq 2a$

In the case of $S(3,5,26)$ the same conditions apply as well as one other namely (viii) $c \neq 2a$ which is automatically satisfied when $v=17$ because then c must be odd. Those orbits listed in the first column of the Appendix which are unsuitable because they do not satisfy all of the above have the reason given alongside and the others, 36 in all, are numbered. For the latter orbits the second column gives the 2 values x, namely a and $a+b$, of the difference triples $<x,x,26-2x>$ of the cyclic 3-element set orbits of this form which they contain. The third column gives the difference triples of 4 of the other cyclic 3-element set orbits contained, the other 4 being obtained as the reverse triples.

The next stage is to find collections of 6 of these orbits which form a partial cyclic $S(3,5,26)$. Although this is feasible by hand the tedious nature of the task makes it more suitable for computer. Programs were written by two of the present authors independently and the results obtained showed that there are only 2 such collections. These are given below in the same format as in the Appendix. Also given are the difference triples of the cyclic 3-element set orbits which are not contained in any of the 6 symmetric cyclic 5-element set orbits and remain to be contained in non-symmetric cyclic 5-element set orbits if either of the partial systems are to be completed.

Partial System 1

Difference quintuple	Difference triples <x,x,26-2x>		Other difference triples (together with their reverse triples)			
<1,3,18,3,1>	1	4	<1,3,22>	<3,5,18>	<1,4,21>	<2,3,21>
<2,5,12,5,2>	2	7	<2,5,19>	<5,9,12>	<2,7,17>	<4,5,17>
<3,9,2,9,3>	3	12	<3,9,14>	<2,9,15>	<3,11,12>	<6,9,11>
<5,1,14,1,5>	5	6	<1,5,20>	<1,11,14>	<5,6,15>	<1,10,15>
<8,3,4,3,8>	8	11	<3,8,15>	<3,4,19>	<7,8,11>	<3,7,16>
<9,1,6,1,9>	9	10	<1,9,16>	<1,6,19>	<7,9,10>	<1,7,18>

Difference triples remaining

<1,2,23>	<2,8,16>	<4,6,16>	<5,7,14>
<1,8,17>	<2,10,14>	<4,7,15>	<5,8,13>
<1,12,13>	<2,11,13>	<4,8,14>	<5,10,11>
<2,4,20>	<3,6,17>	<4,9,13>	<6,7,13>
<2,6,18>	<3,10,13>	<4,10,12>	<6,8,12>

together with reverse triples of all the above.

Partial System 2

Difference quintuple	Difference triples <x,x,26-2x>		Other difference triples (together with their reverse triples)			
<1,5,14,5,1>	1	6	<1,5,20>	<5,7,14>	<1,6,19>	<2,5,19>
<2,7,8,7,2>	2	9	<2,7,17>	<7,8,11>	<2,9,15>	<4,7,15>
<3,5,10,5,3>	3	8	<3,5,18>	<5,10,11>	<3,8,15>	<5,6,15>
<4,1,16,1,4>	4	5	<1,4,21>	<1,9,16>	<4,5,17>	<1,8,17>
<7,3,6,3,7>	7	10	<3,7,16>	<3,6,17>	<7,9,10>	<3,9,14>
<11,1,2,1,11>	11	12	<1,11,14>	<1,2,23>	<3,11,12>	<1,3,22>

Difference triples remaining

<1,7,18>	<2,6,18>	<3,10,13>	<5,8,13>
<1,10,15>	<2,8,16>	<4,6,16>	<5,9,12>
<1,12,13>	<2,10,14>	<4,8,14>	<6,7,13>
<2,3,21>	<2,11,13>	<4,9,13>	<6,8,12>
<2,4,20>	<3,4,19>	<4,10,12>	<6,9,11>

together with the reverse triples of all the above.

4. Completion of partial systems

It remains to ascertain whether either of the partial systems can be completed to form a cyclic $S(3,5,26)$. We deal with each in turn.

Partial System 1

Since difference triple <1,2,23> remains, one of the 4 nonsymmetric cyclic 5-element set orbits must have a difference quintuple of the form <1,2,c,d,e>. By considering other remaining difference triples of the form <1,x,y> it then follows that $c=6,10,11,15$ or 21. But the latter two can be rejected since the difference triples <2,15,9> and <2,21,3> are already contained in symmetric orbits. Similarly $c=11$ is impossible because the difference triple <3,11,12> is contained in the orbit with difference quintuple <3,9,2,9,3>. Hence $c=6$ or 10. Similarly $e=2,8,12,13,17$ or 23. But by considering the remaining difference triples of the form <3,x,y> the possibilities reduce to $e=13$ or 17. Since $c+d+e=23$ and $d\geq 1$, this leaves the orbit having difference quintuple <1,2,6,4,13> as the only possibility.

By a similar argument the second of the 4 non-symmetric cyclic 5-element set orbits has the reverse difference quintuple <2,1,13,4,6>. The remaining 2 orbits can now be deduced in a similar manner using the difference triples still remaining and it is easy to show that they must have difference quintuples $<2,4,7,5,8>$ and $<4,2,8,5,7>$.

Partial System 2

The argument follows the same reasoning as for the first partial system. By considering difference triple $<1,7,18>$ which remains there must be a non-symmetric cyclic 5-element set orbit with difference quintuple $<1,7,c,d,e>$. Further consideration of the remaining difference triples of the form $<1,x,y>$ gives $c=3,5,6,8$ or 11 but the remaining difference triples of the form $<7,x,y>$ eliminate all of these possibilities except $c=6$. Hence $d+e=12$ and again by considering the remaining difference triples of the form $<1,x,y>$ there are just two possibilities namely $e=7$ and $d=5$ or $e=10$ and $d=2$. However the difference quintuples $<1,7,6,5,7>$ and $<1,7,6,2,10>$ can be eliminated as in both cases the orbits contain the cyclic 3-element set orbit with difference triple $<7,8,11>$ which is already contained in the orbit with difference quintuple $<2,7,8,7,2>$. Hence this partial system can not be completed to a cyclic $S(3,5,26)$.

Summarising we have proved that there exists a unique *MS*-cyclic $S(3,5,26)$. It may be constructed from the set of residue classes modulo 26 under the action of the cyclic group C_{26} generated by the mapping $i \rightarrow i+1 \ (mod\ 26)$ and consists of 10 orbits, 6 of which are symmetric and 4 non-symmetric. The orbits are characterized by the following difference quintuples.

A : $<1,3,18,3,1>$	B : $<2,5,12,5,2>$
C : $<3,9,2,9,3>$	D : $<5,1,14,1,5>$
E : $<8,3,4,3,8>$	F : $<9,1,6,1,9>$
G : $<1,2,6,4,13>$	H : $<2,1,13,4,6>$
I : $<2,4,7,5,8>$	J : $<4,2,8,5,7>$

Finally we note that a primitive of 26 is 7 and that the action of the mapping $i \rightarrow 7i \ (mod\ 26)$ on the above orbits is as follows.

$$A \rightarrow B \rightarrow C \rightarrow D \rightarrow F \rightarrow E \rightarrow A$$

$$G \rightarrow J \rightarrow H \rightarrow I \rightarrow G$$

Thus the system is stabilized by the group of mappings $i \rightarrow ai+b \ (mod\ 26)$ where $b=0,1,2,\ldots.25$ and $(a,26)=1$, being formed by two orbits of this group, one consisting of the 6 symmetric cyclic orbits and the other of the 4 non-symmetric cyclic orbits.

5. Isomorphism testing

In [1], Cameron states the following theorem.

Theorem: Suppose a Steiner system $S(3,k,v)$ has the property that given any block B of the system, the blocks which intersect B in precisely two elements form an $S(2,k-2,v-k)$ on the complement of B. Then $k=5, v=26$ or $k=23, v=5084$ or $k=105, v=557026$.

He continues by saying that no systems satisfying the above property are known. For the *MS*-cyclic $S(3,5,26)$ constructed above a program was written to determine whether it had the property. For each block of the system there are 70 blocks which intersect it in two elements. Removing these two elements leaves 70 triples which give rise to 210 pairs of elements. The program counted, for each block of the system, how many pairs occurred once and how many occurred twice among the total 210 pairs. It is easy to see that no pair can occur three times and to satisfy Cameron's property, there must be 210 pairs occurring precisely once for each block considered. The results obtained are that for each of the 156 blocks in a symmetric cyclic orbit, 126 pairs occur once and 42 occur twice whilst for the remaining 104 blocks, 132 pairs occur once and 39 occur twice. Hence the *MS*-cyclic $S(3,5,26)$ does not have the required property nor indeed seems to have any similar structure. However, the figures given above are an invariant of the system which can be used in helping to determine whether it is isomorphic to the $S(3,5,26)$'s constructed by Denniston and Hanani. In [2], Denniston constructs a Steiner system $S(5,7,28)$ as the union of 2 orbits under the group $PSL_2(3^3)$ of 7-element subsets of $GF(3^3) \cup \{\infty\}$ with $x^3 = x + 2$ as the irreducible polynomial. He gives starter blocks for the two orbits as

$$\{\infty, 0, 1, 2, x, x+1, x+2\} \quad \text{and}$$

$$\{\infty, 0, 1, x^2+2x, x^2+2x+2, 2x^2, 2x^2+2x\}$$

By determining all blocks of this system which contain the pair $\{\infty, 0\}$, another task for the computer, a derived $S(3,5,26)$ was obtained and found to have the same invariant numbers with respect to Cameron's property as the *MS*-cyclic $S(3,5,26)$. The element x is a primitive in $GF(3^3)$ and further investigation leads to establishing that the mapping $i \rightarrow x^i$ from the set of residue classes modulo 26 to $GF(3^3) \setminus \{0\}$ with the above irreducible polynomial is an isomorphism from the system constructed in this paper to the derived system of Denniston's $S(5,7,28)$ through $\{\infty, 0\}$.

Hanani [5] constructs an $S(3,5,26)$ on the Cartesian product $Z(2) \times Z(13)$ where $Z(n)$ denotes the set of residue classes modulo n.

He gives the following starter blocks

$$\{(0,0), (1,0), (1,2^x), (1,2^{x+4}), (1,2^{x+8})\},\ x = 0,1,2,3, \quad \text{and}$$

$$\{(0,2^y), (0,2^{y+6}), (1,0), (1,2^{y+2}), (1,2^{y+8})\},\ y = 0,1,2,3,4,5$$

which under the mapping $(i,j) \rightarrow (i+1, j+1)\ (mod\ (2,13))$ form an $S(3,5,26)$. It is easy to establish that the mapping $(i,j) \rightarrow 13i + 2j\ (mod\ 26)$ is an isomorphism from the Hanani system to the one constructed in this paper.

6. Concluding Remarks

The obvious question to ask is whether the $S(3,5,26)$ dealt with in this paper is the unique Steiner system with these parameters. It would be particularly interesting if this were the case as all derived $S(2,4,25)$ subsystems of it are isomorphic. Since it is known that the number of non-isomorphic $S(2,4,25)$'s ≥ 6 [6] we would have an example of non-isomorphic Steiner systems on the same parameters which are not either all derived or all not derived. To the present authors' knowledge no occurrences of this are known.

It would also be of interest to construct MS-cyclic $S(3,5,v)$'s for other values of v, in particular $v=41,50,62$.

References

[1] P.J. Cameron, "Extremal results and configuration theorems for Steiner systems", in: *Topics on Steiner systems* (ed. C.C. Lindner and A. Rosa), *Ann. Discrete Math.* 7 (1980) 43-63.

[2] R.H.F. Denniston, "Some new 5-designs", *Bull. London Math. Soc.* 8 (1976) 263-267.

[3] M.J. Grannell and T.S. Griggs, "S-cyclic Steiner systems", *Ars Combinatoria* 16A (1983) 173-188.

[4] H. Hanani, "A class of 3-designs", *Combinatorial Mathematics* (Proc. Conf. Canberra 1977), *Lecture Notes Math. 686,* Springer, Berlin (1978) 34-46.

[5] H. Hanani, *Technion preprint* (earlier version of [4]).

[6] A. Ja. Petrenjuk, "O postroenii neizomorfnykh blok-skehm s pomoščiu izografičeskikh naborov blokov", *Kirovgrad Inst. S.-Kh. Masinostra.,* Kirovgrad (1983) 16 pp.

[7] E. Witt, "Über Steinersche systeme", *Abh. Math. Sem. Univ. Hamburg* 12 (1938) 265-275.

Appendix

Symmetric cyclic 5-element set orbits under C_{26}.

Difference quintuple		Difference triples <x,x,26-2x>		Other difference triples (together with their reverse triples)			
<1,1,22,1,1>	$a=b$						
<1,2,20,2,1>	$b=2a$						
<1,3,18,3,1>	1	1	4	<1,3,22>	<3,5,18>	<1,4,21>	<2,3,21>
<1,4,16,4,1>	2	1	5	<1,4,21>	<4,6,16>	<1,5,20>	<2,4,20>
<1,5,14,5,1>	3	1	6	<1,5,20>	<5,7,14>	<1,6,19>	<2,5,19>
<1,6,12,6,1>	4	1	7	<1,6,19>	<6,8,12>	<1,7,18>	<2,6,18>
<1,7,10,7,1>	5	1	8	<1,7,18>	<7,9,10>	<1,8,17>	<2,7,17>
<1,8,8,8,1>	$b=c$						
<1,9,6,9,1>	6	1	10	<1,9,16>	<6,9,11>	<1,10,15>	<2,9,15>
<1,10,4,10,1>	7	1	11	<1,10,15>	<4,10,12>	<1,11,14>	<2,10,14>
<1,11,2,11,1>	$c=2a$						
<2,1,20,1,2>	8	2	3	<1,2,23>	<1,5,20>	<2,3,21>	<1,4,21>
<2,2,18,2,2>	$a=b$						
<2,3,16,3,2>	9	2	5	<2,3,21>	<3,7,16>	<2,5,19>	<3,4,19>
<2,4,14,4,2>	$b=2a$						
<2,5,12,5,2>	10	2	7	<2,5,19>	<5,9,12>	<2,7,17>	<4,5,17>
<2,6,10,6,2>	$2a+b=c$						
<2,7,8,7,2>	11	2	9	<2,7,17>	<7,8,11>	<2,9,15>	<4,7,15>
<2,8,6,8,2>	12	2	10	<2,8,16>	<6,8,12>	<2,10,14>	<4,8,14>
<2,9,4,9,2>	$c=2a$						
<2,10,2,10,2>	$a=c$						
<3,1,18,1,3>	13	3	4	<1,3,22>	<1,7,18>	<3,4,19>	<1,6,19>
<3,2,16,2,3>	14	3	5	<2,3,21>	<2,8,16>	<3,5,18>	<2,6,18>
<3,3,14,3,3>	$a=b$						
<3,4,12,4,3>	15	3	7	<3,4,19>	<4,10,12>	<3,7,16>	<4,6,16>
<3,5,10,5,3>	16	3	8	<3,5,18>	<5,10,11>	<3,8,15>	<5,6,15>

<3,6,8,6,3>	$b=2a$						
<3,7,6,7,3>	$c=2a$						
<3,8,4,8,3>	17	3	11	<3,8,15>	<4,8,14>	<3,11,12>	<6,8,12>
<3,9,2,9,3>	18	3	12	<3,9,14>	<2,9,15>	<3,11,12>	<6,9,11>
<4,1,16,1,4>	19	4	5	<1,4,21>	<1,9,16>	<4,5,17>	<1,8,17>
<4,2,14,2,4>	20	4	6	<2,4,20>	<2,10,14>	<4,6,16>	<2,8,16>
<4,3,12,3,4>	21	4	7	<3,4,19>	<3,11,12>	<4,7,15>	<3,8,15>
<4,4,10,4,4>	$a=b$						
<4,5,8,5,4>	$c=2a$						
<4,6,6,6,4>	$b=c$						
<4,7,4,7,4>	$a=c$						
<4,8,2,8,4>	$b=2a$						
<5,1,14,1,5>	22	5	6	<1,5,20>	<1,11,14>	<5,6,15>	<1,10,15>
<5,2,12,2,5>	$2a+b=c$						
<5,3,10,3,5>	$c=2a$						
<5,4,8,4,5>	23	5	9	<4,5,17>	<4,8,14>	<5,9,12>	<4,10,12>
<5,5,6,5,5>	$a=b$						
<5,6,4,6,5>	$2a=b+c$						
<5,7,2,7,5>	24	5	12	<5,7,14>	<2,7,17>	<5,9,12>	<7,9,10>
<6,1,12,1,6>	$c=2a$						
<6,2,10,2,6>	$2a=b+c$						
<6,3,8,3,6>	25	6	9	<3,6,17>	<3,8,15>	<6,9,11>	<3,11,12>
<6,4,6,4,6>	$a=c$						
<6,5,4,5,6>	26	6	11	<5,6,15>	<4,5,17>	<6,9,11>	<5,9,12>
<6,6,2,6,6>	$a=b$						
<7,1,10,1,7>	27	7	8	<1,7,18>	<1,10,15>	<7,8,11>	<1,11,14>
<7,2,8,2,7>	28	7	9	<2,7,17>	<2,8,16>	<7,9,10>	<2,10,14>
<7,3,6,3,7>	29	7	10	<3,7,16>	<3,6,17>	<7,9,10>	<3,9,14>
<7,4,4,4,7>	$b=c$						
<7,5,2,5,7>	$a=b+c$						
<8,1,8,1,8>	$a=c$						
<8,2,6,2,8>	$a=b+c$						
<8,3,4,3,8>	30	8	11	<3,8,15>	<3,14,19>	<7,8,11>	<3,7,16>

<8,4,2,4,8>	31	8	12	<4,8,14>	<2,4,20>	<6,8,12>	<4,6,16>
<9,1,6,1,9>	32	9	10	<1,9,16>	<1,6,19>	<7,9,10>	<1,7,18>
<9,2,4,2,9>	33	9	11	<2,9,15>	<2,4,20>	<6,9,11>	<2,6,18>
<9,3,2,3,9>	34	9	12	<3,9,14>	<2,3,21>	<5,9,12>	<3,5,18>
<10,1,4,1,10>	35	10	11	<1,10,15>	<1,4,21>	<5,10,11>	<1,5,20>
<10,2,2,2,10>	$b=c$						
<11,1,2,1,11>	36	11	12	<1,11,14>	<1,2,23>	<3,11,12>	<1,3,22>

Annals of Discrete Mathematics 34 (1987) 207–224
© Elsevier Science Publishers B.V. (North-Holland)

Halving The Complete Design

Alan Hartman

IBM Israel Scientific Centre
Technion City
Haifa 32000
ISRAEL

TO ALEX ROSA ON HIS FIFTIETH BIRTHDAY

ABSTRACT

Let X be a finite set of cardinality v. We denote the set of all k-subsets of X by $\binom{X}{k}$. In this paper we consider the problem of partitioning $\binom{X}{k}$ into two parts of equal size, each of which is the block set of a $2-(v,k,\lambda)$ design. We determine necessary and sufficient conditions on v for the existence of such a partition when k=3 or 4. We also construct partitions for higher values of k and infinitely many values of v. The case where k=3 has been solved for all values of v by Dehon [5]. The case where v=2k has been solved for all values of k by Alltop [1]. The remaining results are new. The technique used is similar to that used by Denniston in his construction of a 4-(12,5,4) design without repeated blocks. We also prove an interesting corollary to Baranyai's theorem giving necessary and sufficient conditions for the existence of a partition of $\binom{X}{k}$ into $1-(v,k,\lambda)$ designs.

1. Introduction

Let X be a set of cardinality v. We denote the set of all k-subsets of X by $\binom{X}{k}$. For integers $v \geq k \geq t \geq 0$ and $\lambda \geq 1$, we define a $t-(v,k,\lambda)$ design to be a pair (X,B) where $B \subseteq \binom{X}{k}$ and every member of $\binom{X}{t}$ is contained in precisely λ members of B. The members of B are called blocks. Note that our definition excludes repeated blocks. The $k-(v,k,1)$ design $(X, \binom{X}{k})$ is referred to as the complete design with block size k.

In this paper we consider the problem of partitioning $\binom{X}{k}$ into two parts of equal size, each of which is the block set of a $2-(v,k,\lambda)$ design. Clearly we must have $\lambda = \frac{1}{2}\binom{v-2}{k-2}$. This problem is a special case of the problem of the existence of large sets of disjoint designs. A large set of disjoint $t-(v,k,\lambda)$ designs is a partition of $\binom{X}{k}$ into n parts, each of which is the block set of a $t-(v,k,\lambda)$ design where $\lambda = \frac{1}{n}\binom{v-t}{k-t}$. When $n=2$ we refer to the problem as the *halving* of $\binom{X}{k}$ into t-designs.

The usual approach to the problem of large sets of disjoint designs is to fix t, k, and λ and determine necessary and sufficient conditions on v for the existence of a large set. Note that if $n=ab$ then the existence of a large set of disjoint $t-(v,k,\lambda)$ designs implies the existence of a large set of disjoint $t-(v,k,a\lambda)$ designs.

The existence of large sets of 1-designs with $\lambda=1$ has been solved by Baranyai [2] for all values of k. Baranyai's theorem also implies the solution for all λ, and we shall prove this corollary later in the paper. The existence of large sets of 2-designs with $k=3$ and all λ has been effectively decided in the papers of Teirlinck [10,11] and Lu [8,9]. In [11] Teirlinck also gives large sets of disjoint $3-(v,4,\lambda)$ designs for all $v \equiv 0 (mod 3)$ and the smallest possible value for λ. In another remarkable paper, Teirlinck [12] constructs large sets of $t-(v,t+1,\lambda)$ designs for infinitely many values of v and all t when $\lambda=(t+1)!^{2t+1}$.

The halving problem has been investigated by Alltop [1] where he shows that the complete design with block size k on $2k$ points can be partitioned into two 2-designs if and only if k is not a power of 2. This result follows from his more general theorem that if t is odd and the complete design with block size k on $2k+1$ points can be partitioned into two t-designs, then the complete design with block size $k+1$ on $2k+2$ points can be partitioned into two $t+1$-designs.

A beautiful solution to the halving problem is given by Denniston [6] where he constructs a partition of $\binom{X}{5}$ into two 4-designs with $|X|=12$. In this paper we generalize Denniston's technique and apply it to the halving problem for 2-designs. Another solution to the halving problem has recently been found by Kreher and Radziszowski [7]. They partition $\binom{X}{7}$ into two 6-designs with $|X|=14$. A solution to the halving problem for 2-designs with $k=3$ can be found in Dehon [5], where he constructs $2-(v,3,\lambda)$ designs without repeated blocks. The only other example of a large set of disjoint designs of which we are aware is the partition into 2-(13,4,1) designs given in [4] by Chouinard.

In this paper we solve the halving problem for 2-designs when $k=3$ or 4. We also show how to partition $\binom{X}{k}$ into two 2-designs for higher values of k and infinitely many values of $|X|$. In Section 2 we define the concept of elegant matrices, deduce some of their properties, and show how they can be applied to the halving problem. In Section 3 we give constructions for families of elegant matrices, and in Section 4 we apply the constructions to the halving problem. Section 3 also contains the proof of the corollary to Baranyai's Theorem.

2. Halving With Elegant Matrices

Let I_n denote the set $\{0,1,2,...,n-1\}$, and let $M=(m_{I,J})$ be an $\binom{n}{i}\times\binom{n}{j}$ 0-1 matrix whose rows are indexed by the members of $\binom{I_n}{i}$ and whose columns are indexed by members of $\binom{I_n}{j}$. A matrix M with the above dimensions and row and column labels will be referred to as an (n,i,j)-matrix. For subsets X,Y of I_n we define $\sigma(M,X,Y)$ by

$$\sigma(M,X,Y) = \sum_{X\subseteq I,Y\subseteq J} m_{I,J};$$

that is, $\sigma(M,X,Y)$ is the number of 1's in the submatrix whose row indices contain X and whose column indices contain Y. By convention we take the empty sum to be zero.

The matrix M will be called *elegant* if it has the following properties:

$$\sigma(M,\{x\},\{y\}) = \sigma(M,\{y\},\{x\}) \quad \text{for all} \quad x,y \in I_n. \tag{E1}$$

$$\sigma(M,X,\emptyset) = \sigma(M,Y,\emptyset) \quad \text{for all} \quad X,Y\in\binom{I_n}{2}. \tag{E2}$$

$$\sigma(M,\emptyset,X) = \sigma(M,\emptyset,Y) \quad \text{for all} \quad X,Y\in\binom{I_n}{2}. \tag{E3}$$

The constant $\sigma(M,X,\emptyset)$ when $|X|=2$ will be denoted by $r_2(M)$, and we denote the constant $\sigma(M,\emptyset,X)$ by $c_2(M)$. These constants are not independent, since if we count the total number of 1's in two ways we obtain the following identity:

$$r_2(M)\frac{\binom{n}{2}}{\binom{i}{2}} = c_2(M)\frac{\binom{n}{2}}{\binom{j}{2}},$$

and hence

$$\binom{j}{2} r_2(M) = \binom{i}{2} c_2(M). \tag{2.1}$$

Note that elegant matrices are generalizations of designs since an elegant matrix M with $j=0$ is just the block incidence vector of a $2-(n,i,r_2(M))$ design.

Two operations which preserve elegance are the transpose and the complement. If M is an elegant (n,i,j)-matrix then M^T is an elegant (n,j,i)-matrix with $r_2(M^T) = c_2(M)$ and $c_2(M^T) = r_2(M)$. If J is the all 1's (n,i,j)-matrix then $J-M$ is also elegant with $r_2(J-M) = \binom{n}{j}\binom{n-2}{i-2} - r_2(M)$ and $c_2(J-M) = \binom{n}{i}\binom{n-2}{j-2} - c_2(M)$. Hence $(J-M)^T$ is elegant with

$$r_2((J-M)^T) = \binom{n}{i}\binom{n-2}{j-2} - c_2(M), \tag{2.2}$$

$$c_2((J-M)^T) = \binom{n}{j}\binom{n-2}{i-2} - r_2(M). \tag{2.3}$$

We now show how to use elegant matrices to solve the halving problem. Let $M^i = (m^i_{A,B})$, $i=0,1,2,\ldots,\lfloor (k-1)/2 \rfloor$ be a set of elegant $(n,k-i,i)$-matrices. (The notation $\lfloor x \rfloor$ denotes the greatest integer less than or equal to x). In the case when both k and $\binom{n}{k/2}$ are even, we define $M^{k/2}$ by

$$M^{k/2} = \begin{bmatrix} J0 \\ 0J \end{bmatrix},$$

where J and 0 are the all 1's and all 0's square matrices of side $\frac{1}{2}\binom{n}{k/2}$ respectively. It is a simple matter to check that $M^{k/2}$ is an elegant $(n,k/2,k/2)$-matrix. Now let $M^{k-i} = (J-M^i)^T$, for $i=0,1,2,\ldots,\lfloor (k-1)/2 \rfloor$ and we have a set of $k+1$ elegant matrices.

Theorem 2.1: If the matrices M^i defined above satisfy the equation

$$\sum_{i=0}^{k} r_2(M^i) = \sum_{i=0}^{k} c_2(M^i), \tag{2.4}$$

then there exists a partition of $\binom{I_n \times I_2}{k}$ into two $2-(2n,k,\frac{1}{2}\binom{2n-2}{k-2})$ designs.

Proof:

We construct the following set of blocks.

$$\beta = \{(A\times\{0\})\cup(B\times\{1\}) : m^i_{A,B} = 1,\ i=0,1,2,\ldots,k\}$$

It is sufficient to show that $(I_n \times I_2, \beta)$ is a $2-(2n,k,\frac{1}{2}\binom{2n-2}{k-2})$ design, since this implies that $\binom{I_n \times I_2}{k} - \beta$ is also the block set of a 2-design with the same parameters. We first count the blocks, by counting the 1's in M^i. Since $M^{k-i}=(J-M^i)^T$, the number of 1's in M^i plus the number of 1's in M^{k-i} is precisely $\binom{n}{k-i}\binom{n}{i}$ for $i=0,1,2,\ldots,\lfloor(k-1)/2\rfloor$. The number of 1's in $M^{k/2}$ is $\frac{1}{2}\binom{n}{k/2}\binom{n}{k/2}$. Hence

$$|\beta| = \frac{1}{2}\sum_{i=0}^{k}\binom{n}{k-i}\binom{n}{i} = \frac{1}{2}\binom{2n}{k}.$$

The number of blocks containing the pair of points $(x,0),(y,1)$ is given by

$$\sum_{i=0}^{k}\sigma(M^i,\{x\},\{y\}) = \sum_{i=0}^{\lfloor(k-1)/2\rfloor}\sigma(M^i,\{x\},\{y\}) + \binom{n-1}{i-1}\binom{n-1}{k-i-1} - \sigma(M^i,\{y\},\{x\})$$

when k is odd. When k is even we also have to add in

$$\sigma(M^{k/2},\{x\},\{y\}) = \binom{n-1}{\frac{k}{2}-1}\binom{n-1}{\frac{k}{2}-1}$$

In either case, using (E1), we see that

$$\sum_{i=0}^{k}\sigma(M^i,\{x\},\{y\}) = \frac{1}{2}\binom{2n-2}{k-2}.$$

The number of blocks containing the pair of points $(x,0),(y,0)$ is $\sum_{i=0}^{k} r_2(M^i)$ and the number of blocks containing the pair of points $(x,1),(y,1)$ is $\sum_{i=0}^{k} c_2(M^i)$. Hence to complete the proof, it is sufficient to prove that $\sum_{i=0}^{k} r_2(M^i)+c_2(M^i) = \binom{2n-2}{k-2}$. Using equations (2.2) and (2.3) we have for k odd:

$$\sum_{i=0}^{k} r_2(M^i) = \sum_{i=0}^{\lfloor(k-1)/2\rfloor} r_2(M^i) + \binom{n}{i}\binom{n-2}{k-i-2} - c_2(M^i)$$

$$\sum_{i=0}^{k} c_2(M^i) = \sum_{i=0}^{\lfloor(k-1)/2\rfloor} c_2(M^i) + \binom{n}{k-i}\binom{n-2}{i-2} - r_2(M^i)$$

When k is even, we must add

$$\frac{1}{2}\binom{n}{k/2}\binom{n-2}{\frac{k}{2}-2}$$

on the right hand side of both equations. Hence in either case

$$\sum_{i=0}^{k} r_2(M^i)+c_2(M^i) = \sum_{i=0}^{k} \binom{n}{i}\binom{n-2}{k-i-2} = \binom{2n-2}{k-2}.$$

Theorem 2.1 asserts that if we can construct elegant matrices M^i for $i=0,1,2,\ldots,\lfloor(k-1)/2\rfloor$ which satisfy equation (2.4) then we can halve the complete design on $2n$ points with block size k. Equation (2.4) contains only $\lfloor(k+1)/2\rfloor$ independent variables since M^{k-i} is constructed from M^i and equation (2.1) gives the relationship between $r_2(M^i)$ and $c_2(M^i)$. The most convenient form of equation (2.4) is the following

$$\sum_{i=0}^{\lfloor(k-1)/2\rfloor} \frac{k-2i}{(k-i)(k-i-1)} r_2(M^i) = \begin{cases} \frac{1}{k}\binom{n-1}{(k-2)/2}\binom{n-2}{(k-2)/2} & \text{if } k \text{ is even} \\ \frac{1}{k-1}\binom{n-1}{(k-1)/2}\binom{n-2}{(k-3)/2} & \text{if } k \text{ is odd.} \end{cases} \tag{2.5}$$

The proof that equations (2.4) and (2.5) are equivalent is tedious and is omitted. The interested reader can verify the truth of the assertion by using equations (2.1), (2.2), and (2.3) and standard combinatorial identities.

Before proceeding with general constructions for elegant matrices, we first give some examples of the construction described in Theorem 2.1.

Example 2.1: Let n be an odd integer, and let $k=3$. Let $L_{x,y}$ be a symmetric idempotent Latin square, for example $L_{x,y} = (x+y)/2 (mod\ n)$. (Division by 2 is well-defined, since n is odd.) Now consider the matrices $M^0 = J$, the all 1's $(n,3,0)$-vector, and define M^1 by

$$m^1_{\{x,y\},\{z\}} = 1 \quad \text{if and only if} \quad L_{x,y}=z.$$

It is easily verified that

$$\begin{array}{ll} r_2(M^0) = n-2 & c_2(M^0) = 0 \\ r_2(M^1) = 1 & c_2(M^1) = 0 \\ r_2(M^2) = 0 & c_2(M^2) = n-1 \\ r_2(M^3) = 0 & c_2(M^3) = 0 \end{array}$$

and to verify (E1) we note that

$$\sigma(M^1,\{x\},\{y\}) = \begin{cases} 1 & \text{if } x \neq y \\ 0 & \text{if } x=y \end{cases}$$

so in either case $\sigma(M^1,\{x\},\{y\})=\sigma(M^1,\{y\},\{x\})$. The block set of the design is then

$$\begin{aligned} \beta = {} & \{(x,0)(y,0)(z,0): x\neq y\neq z\neq x\} \cup \\ & \{(x,0)(y,0)(z,1): x\neq y; z=L_{x,y}\} \cup \\ & \{(x,0)(y,1)(z,1): y\neq z; x\neq L_{y,z}\}. \end{aligned}$$

Example 2.2: Here we give another construction for $k=3$. Let $n\equiv 1,3 \pmod 6$ and let (I_n,β) be a Steiner triple system of order n (i.e. a 2-$(n,3,1)$ design). Define M^0 and M^1 by

$$m^0_{B,\emptyset} = 1 \text{ if and only if } B \in \beta$$

$$m^1_{\{x,y\},\{z\}} = 1 \text{ if and only if } z \notin \{x,y\}$$

In this example we have

$$\begin{array}{ll} r_2(M^0) = 1 & c_2(M^0) = 0 \\ r_2(M^1) = n-2 & c_2(M^1) = 0 \\ r_2(M^2) = 0 & c_2(M^2) = 2 \\ r_2(M^3) = 0 & c_2(M^3) = n-3 \end{array}$$

and to verify (E1) we note that

$$\sigma(M^1,\{x\},\{y\}) = \begin{cases} n-2 & \text{if} x \neq y \\ 0 & \text{if} x = y \end{cases}$$

so in either case $\sigma(M^1,\{x\},\{y\}) = \sigma(M^1,\{y\},\{x\})$.

3. Constructing Elegant Matrices

Ideally one would like to determine necessary and sufficient conditions on the parameters n,i,j and $r_2(M)$ for the existence of an elegant (n,i,j)-matrix M. As noted in the previous section this is equivalent to the existence problem for $2-(n,i,r_2(M))$ designs without repeated blocks when $j=0$. This problem is very difficult and has only been solved in the case $i=3$ [5]. In this section we construct infinite families of elegant matrices and leave the general existence problem for further research. Because the operations of complementation and transposition preserve elegance it is sufficient to consider only the existence of elegant matrices with $n > i \geq j \geq 0$ and $0 \leq r_2(M) \leq \frac{1}{2}\binom{n}{j}\binom{i}{2}$. The rest of this section is devoted to constructing four families of elegant matrices.

The Design Family

If there exists a $t-(n,i,\lambda)$ design without repeated blocks, say (I_n,β), then we construct the matrix $Des(t,n,i,\lambda)$ to be the $(n,i,0)$-matrix defined by

$$Des(t,n,i,\lambda)_{I,\emptyset}=1 \quad \text{if and only if} \quad I \in \beta.$$

This family has parameters

$$r_2(Des(t,n,i,\lambda)) = \frac{\lambda\binom{n-2}{t-2}}{\binom{i-2}{t-2}}, \qquad c_2(Des(t,n,i,\lambda))=0.$$

The condition (E1) for elegant matrices is vacuous in the case when $j=0$. The matrix M^0 of Example 2.2 is an example of an elegant matrix from this family.

The Intersection Family

Let $n \geq i \geq j$ be non-negative integers and let X be a subset of $\{0,1,2, \cdots, j\}$. We define the (n,i,j)-matrix $Int(n,i,j,X)$ by

$$Int(n,i,j,X)_{I,J}=1 \quad \text{if and only if} \quad |I \cap J| \in X.$$

Note that when $X=\emptyset$ and when $X=I_{j+1}$ we obtain the all 0's and all 1's matrices respectively. In general the row sum of $Int(n,i,j,X)$ is given by the expression

$$\sigma(Int(n,i,j,X),I,\emptyset)=\sum_{x \in X} \binom{i}{x}\binom{n-i}{j-x}$$

and hence

$$r_2(Int(n,i,j,X))=\binom{n-2}{i-2}\sum_{x \in X} \binom{i}{x}\binom{n-i}{j-x}.$$

Similarly

$$c_2(Int(n,i,j,X))=\binom{n-2}{j-2}\sum_{x \in X} \binom{j}{x}\binom{n-j}{i-x}.$$

To verify condition (E1) we note that the number of pairs (I,J) such that $x \in I$, $y \in J$, and $|I \cap J| \in X$ depends only on whether $x=y$ or not. Hence condition (E1) holds for the Intersection Family, and thus they are elegant. The matrix M^1 of Example 2.2 is an example of an elegant matrix from this family.

The Average Family

Let $n>i>0$ be integers and let $0 \leq d \leq n$ be an integer. We define the symmetric set of d integers $Sym(d)$ by

$$Sym(d)=\begin{cases}\{\pm 1,\pm 2,\pm 3, \cdots \pm d/2\} & \text{if } d \text{ is even}\\ \{0,\pm 1,\pm 2,\pm 3, \cdots \pm (d-1)/2\} & \text{if } d \text{ is odd.}\end{cases}$$

Note that $|Sym(d)|=d$ and that $s \in Sym(d)$ if an only if $-s \in Sym(d)$.

If $gcd(n,i)=1$ we define the average $a(I)$, of an i-subset, I, of I_n to be the unique solution to the equation

$$i \;\; a(I) \equiv \sum_{x \in I} x \quad (mod \;\; n).$$

Now we define the average segment $A(I,d)$ of size d about an i-subset to be the set

$$A(I,d)=\{a(I)+s \quad (mod \;\; n) : s \in Sym(d)\}.$$

Finally we define the $(n,i,1)$-matrix $Avg(n,i,d)$ by

$$Avg(n,i,d)_{I,\{x\}}=1 \quad \text{if and only if} \quad x \in A(I,d).$$

Since $j=1$ we have $c_2(Avg(n,i,d))=0$. Since the row sum of $Avg(n,i,d)$ is d for all rows of the matrix we have

$$r_2(Avg(n,i,d))=\binom{n-2}{i-2}d.$$

To verify that condition (E1) holds for $Avg(n,i,d)$ we note that it is sufficient to verify the following proposition. Let $f(x,y,s)$ denote the number of i-subsets, I, such that $x \in I$ and $a(I)=y-s$.

Proposition $f(x,y,s)=f(y,x,-s)$

If the claim is true then we have

$$\sum_{s \in Sym(d)} f(x,y,s)= \sum_{s \in Sym(d)} f(y,x,-s)= \sum_{s \in Sym(d)} f(y,x,s)$$

since $s \in Sym(d)$ if an only if $-s \in Sym(d)$. Hence

$$\sigma(Avg(n,i,d),\{x\},\{y\})=\sigma(Avg(n,i,d),\{y\},\{x\}).$$

To prove the Proposition we note that $I'=(x+y)-I$ is a bijection between the objects being counted. That is, if $x \in I$ then $y \in (x+y)-I$, and if $y-s=a(I)$ then $x+s=a((x+y)-I)$. This completes the proof that $Avg(n,i,d)$ is an elegant matrix. The matrix M^1 of Example 2.1 is an example of an elegant matrix from this family.

We would also like to have a construction for elegant $(n,i,1)$-matrices when $gcd(n,i)\neq 1$, however a general method has thus far eluded us. Nevertheless we have the following constructions for $2 \leq i \leq 5$.

We say that an i-subset is averagable if its sum is congruent to zero modulo $g=gcd(n,i)$. If I is averagable then the equation defining $a(I)$ has g solutions of the form $a+kn/g$, with $0 \leq k < g$, and we now view $a(I)$ as the set of all solutions. We can then define $A(I,gd)$ by

$$A(I,gd)=\{a+s \pmod{n} : a \in a(I),\ s \in Sym(d)\}.$$

When I is not averagable, $|I|=2$ and n is even, then we define

$$A(I,2d)=\{a+2s \pmod{n} : a \in I,\ s \in Sym(d)\}.$$

The cardinality of $A(I,2d)$ is $2d$ since a non averagable set I contains two elements of different parity. We now have the following definition for $Avg(2n,2,2d)$

$$Avg(2n,2,2d)_{I,\{x\}}=1 \quad \text{if and only if} \quad x \in A(I,2d).$$

The argument showing that $Avg(2n,2,2d)$ is elegant is similar to that for $Avg(n,i,d)$ where $gcd(n,i)=1$. Note that the transformation $I'=(x+y)-I$ preserves averagability, and also serves as an appropriate bijection for the non averagable sets.

We move now to the construction of the matrices $Avg(3n,3,3d)$. In this case all the non averagable 3-subsets $\{x,y,z\}$ have the curious property that a unique member, x, of the set satisfies $x \not\equiv y \equiv z \pmod{3}$. We shall refer to x as the *odd-man-out* modulo 3. Now we construct the set $A(I,3d)$ for non averagable sets I by

$$A(I,3d)=\{x+s \pmod{3n} : s \in Sym(3d)\},$$

where x is the odd-man-out of I. This defines $Avg(3n,3,3d)$ in the standard way, and the proof of its elegance again uses the same bijection for all 3-subsets.

In general, when I contains an odd-man-out modulo m and m is a divisor of g we define $a(I)=x$ where x is the odd-man-out, and $A(I,gd)$ is defined by

$$A(I,gd)=\{a(I)+s \pmod{n} : s \in Sym(gd)\}.$$

The definition of $Avg(4n+2,4,2d)$ is similar. We note that a non averagable set contains a unique odd-man-out modulo 2, and we omit the details of the construction.

To define $Avg(4n,4,4d)$ we observe that the only 4-subsets which are neither averagable, nor have an odd-man-out modulo 2, are those whose sum is congruent to 2 modulo 4. These 4-subsets can be further subdivided into those which contain an odd-man-out modulo 4, and those which contain two even numbers and two odd numbers. Let $I=\{x,y,z,t\}$ be a 4-set whose sum is congruent to 2 modulo 4 and with $x \equiv y \not\equiv z \equiv t \pmod{2}$. We define $a(I)$ to be the set

$$a(I)=\{\frac{1}{2}(x+y),2n+\frac{1}{2}(x+y),\frac{1}{2}(z+t),2n+\frac{1}{2}(z+t)\},$$

and define $A(I,4d)$ by

$$A(I,4d)=\{a+2s \pmod{4n} : a \in a(I),\ s \in Sym(d)\}.$$

We have not been able to give a satisfactory definition of $Avg(5n,5,5d)$ for all vaues of n and d, however the following construction suffices when $n \not\equiv 0 \mod 6$.

Let I be a 5-subset of I_{5n} and let $P(I)=p_0+p_1+\cdots+p_4$ be the partition of the integer 5 where p_r is the number of members of I which are congruent to r modulo 5. If $P(I)=5$ or $P(I)=1+1+1+1+1$ then I is averagable. If $P(I)=1+4$ or $P(I)=1+2+2$ then I has an odd-man-out modulo 5. In the remaining cases $P(I)=2+3$, $P(I)=1+1+3$, and $P(I)=1+1+1+2$ we observe that I has both an odd-pair-out and an odd-triple-out. (An odd-pair-out is a unique part of size 2, or exactly two parts of size 1. An odd-triple-out is a unique part of size 3, exactly three parts of size 1, or the union of a unique part of size 1 with a unique part of size 2.)

If n is odd we define $a(I)$ to be the unique solution to the equation $2a \equiv x+y \mod 5n$ where $\{x,y\}$ is the odd-pair-out. If n is even but not divisible by 3 we define $a(I)$ to be the unique solution to the equation $3a \equiv x+y+z \mod 5n$ where $\{x,y,z\}$ is the odd-triple-out. We can then define $A(I,5d)$ and $Avg(5n,5,5d)$ in the standard way.

In all cases the cardinality of $A(I,d)$ is d and the transformation $I'=(x+y)-I$ is a bijection between the set of subsets I with $x \in I$ and $y \in A(I,d)$, and the set of subsets I' with $y \in I'$ and $x \in A(I',d)$. The transformation has the required property since if $P(I)$ is the partition of i induced by the set I reduced modulo m, then $P(I)=P(I')$ for any divisor m of n.

The Baranyai Family

We now begin the construction of a family of elegant matrices constructed using Baranyai's Theorem. A statement of this theorem, due to Cameron [3] is as follows: Let n be a non-negative integer. A *datum* on n is defined to be an r-tuple $(k_1,k_2,\cdots,k_r)$, where $r>0$ and $k_1,k_2,\cdots,k_r$ are integers satisfying $0 \le k_i \le n$ for all i, together with an $r \times s$ matrix $A=(a_{i,j})$ of non-negative integers satisfying

$$\sum_{j=1}^{s} a_{i,j} = \binom{n}{k_i}, \quad \text{for all } i,$$

$$\sum_{i=1}^{r} k_i \; a_{i,j} = n, \quad \text{for all } j.$$

(Since $\sum_{i=1}^{r}\sum_{j=1}^{s} k_i a_{i,j} = ns = \sum_{i=1}^{r} k_i \binom{n}{k_i}$, we have $s = \sum_{i=1}^{r} \binom{n-1}{k_i-1}$.)

Baranyai's Theorem: Given a datum $((k_i),(a_{i,j}))$ on n there exist sets $X_{i,j}$ $(1 \leq i \leq r,\ 1 \leq j \leq s)$ of subsets of I_n, $|X_{i,j}| = a_{i,j}$, having the properties

for fixed i, the sets $X_{i,j}$ form a partition of $\binom{I_n}{k_i}$;

for fixed j, the members of the sets $X_{i,j}$ form a partition of I_n.

We wish to prove the following

Corollary 3.1 Let n,λ be non-negative integers, and let k be an integer satisfying $0 \leq k \leq n$, $\lambda n \equiv 0 \pmod{k}$, and $\binom{n-1}{k-1} \equiv 0 \pmod{\lambda}$. Then there exist sets Y_j $(1 \leq j \leq \frac{1}{\lambda}\binom{n-1}{k-1})$ of subsets of I_n having the properties that

the sets Y_j form a partition of $\binom{I_n}{k}$;

for all j, the members of Y_j contain each member of I_n precisely λ times.

Proof: Write $\lambda n = (a_1 + a_2 + \cdots + a_p)k$ where $0 < a_j \leq n/k$ are integers. We wish to construct a $(p-\lambda) \times p$ matrix $B = (b_{i,j})$ such that

$$\sum_{i=1}^{p-\lambda} b_{i,j} = n - ka_j \quad \text{for all } 1 \leq j \leq p, \quad \text{and} \quad \sum_{j=1}^{p} b_{i,j} = n \text{ for all } 1 \leq i \leq p-\lambda.$$

Let $b_{1,1} = n - ka_1$ and $b_{i,1} = 0$ for $1 < i \leq p-\lambda$. We proceed to define the columns of B as follows. To define column j let i be the first row index such that $\sum_{1 \leq m < j} b_{i,m} < n$. Define $b_{m,j} = 0$ for all $m < i$, and let

$$b_{i,j} = min(n - ka_j, n - \sum_{m=1}^{j-1} b_{i,m}).$$

If $b_{i,j} = n - ka_j$ then set $b_{m,j} = 0$ for all $m > i$, otherwise set $b_{i+1,j} = n - ka_j - b_{i,j}$ and let $b_{m,j} = 0$ for all $m > i+1$. Note that in the second case the number of rows whose sum is n has been increased by 1.

By construction the sum of column j is $n - ka_j$ and hence

$$\sum_{j=1}^{p}\sum_{i=1}^{p-\lambda} b_{i,j} = pn - k\sum_{j=1}^{p} a_j = (p-\lambda)n.$$

But by construction, the sum of each row is at most n, and therefore the sum of each row is precisely n.

We now construct a datum on n with $(k_i)=(k,1,1,1,\cdots,1)$ and A is the $r\times s$ matrix with first row $a_1a_2\cdots a_p$ repeated $\frac{1}{\lambda}\binom{n-1}{k-1}$ times and blocks of the matrix B down the diagonal below the first row i.e.

$$\begin{vmatrix} a_1a_2\cdots a_p & a_1a_2\cdots a_p & & a_1a_2\cdots a_p \\ B & 0 & 0 & 0 \\ 0 & B & 0 & 0 \\ 0 & 0 & B & 0 \\ 0 & 0 & \cdots & B \end{vmatrix}$$

$$r=1+\frac{p-\lambda}{\lambda}\binom{n-1}{k-1}, \quad s=\frac{p}{\lambda}\binom{n-1}{k-1}.$$

The sum of the first row is

$$\frac{1}{\lambda}\binom{n-1}{k-1}\sum_{j=1}^{p}a_j=\frac{n}{k}\binom{n-1}{k-1}=\binom{n}{k}.$$

The sum of each of the remaining rows is $n=\binom{n}{1}$, and $\sum k_i a_{i,j}=n$ for all columns j. Hence by Baranyai's Theorem there exist sets $X_{i,j}$ of subsets of I_n with the properties described above. Let $Y_j=\bigcup_{m=1}^{p}X_{1,(j-1)p+m}$. Clearly Y_j is a partition of $\binom{I_n}{k}$. According to Baranyai's Theorem the union of any p columns of $X_{i,j}$ contains each member of I_n precisely p times. The union of $p-\lambda$ rows of $X_{i,j}$ with $i>1$ contains each member of I_n precisely $p-\lambda$ times. Hence each Y_j contains each member of I_n precisely $p-(p-\lambda)=\lambda$ times.

We now construct the Baranyai family of elegant matrices. Let $n\geq i\geq k\geq 2$ be integers and let λ,m be positive integers satisfying

$$\lambda n\equiv 0 \pmod{k}, \quad \binom{n-1}{k-1}\equiv 0 \pmod{\lambda},$$

$$0\leq m\leq\frac{1}{\lambda}\binom{n-1}{k-1}, \quad \text{and } m\binom{n}{i}=\frac{c}{\lambda}\binom{n-1}{k-1},$$

for some integer c. Let Y_j be the partition of the complete design with block size k constructed in the proof of Corollary 3.1, and let $f(I)$ be an arbitrary bijection from the set of i-subsets of I_n to the integers from 0 to $\binom{n}{i}-1$. Construct Y_I to be the set of k-subsets of I_n defined by

$$Y_I=\bigcup_{a=1}^{m}Y_{mf(I)+a}$$

reducing the subscripts modulo $\frac{1}{\lambda}\binom{n-1}{k-1}$. We now define $Bar(n,i,k,\lambda,m)$ to be the

(n,i,k)-matrix given by

$$Bar(n,i,k,\lambda,m)_{I,K}=1 \quad \text{if and only if} \quad K \in Y_I.$$

The basic idea behind the construction is to make each row of the matrix the incidence vector of m disjoint λ-factors of the complete design with block size k and ensure that the column sum is constant. The row sum of $Bar(n,i,k,\lambda,m)$ is $m\lambda n/k$ for all rows, and the column sum is c for all columns, and hence

$$r_2(Bar(n,i,k,\lambda,m))=\binom{n-2}{i-2}\frac{m\lambda n}{k},$$

$$c_2(Bar(n,i,k,\lambda,m))=\binom{n-2}{k-2}c=\frac{m\lambda(k-1)}{n-1}\binom{n}{i}.$$

To verify that $Bar(n,i,k,\lambda,m)$ satisfies (E1) we note that $\sigma(Bar(n,i,k,\lambda,m),I,\{x\})=m\lambda$ for all I and for all x. Hence

$$\sigma(Bar(n,i,k,\lambda,m),\{x\},\{y\})=\binom{n-1}{i-1}m\lambda=\sigma(Bar(n,i,k,\lambda,m),\{y\},\{x\}).$$

An example of an elegant Baranyai matrix with parameters $Bar(6,3,2,1,4)$ can be constructed as follows. Let $Y_1=\{e_1,e_2,e_3\}$, $Y_2=\{e_4,e_5,e_6\} \cdots Y_5=\{e_{13},e_{14},e_{15}\}$ be a 1-factorization of the complete graph on I_6. Label the columns of $Bar(6,3,2,1,4)$ by the edges $e_1,e_2, \cdots e_{15}$ in order. Then the construction defined above gives

$$Bar(6,3,2,1,4) = \begin{vmatrix} A \\ A \\ A \\ A \\ A \end{vmatrix} \quad \text{where } A = \begin{vmatrix} 111111111111000 \\ 111111111000111 \\ 111111000111111 \\ 111000111111111 \\ 000111111111111 \end{vmatrix}$$

4. Halving the Complete Design

In this section we apply the methods of the preceeding sections to construct partitions of the complete design with block size k into two 2-designs. We first derive necessary conditions for the existence of a partition into two t-designs.

Lemma 4.1: If there exists a partition of the complete design with block size k into two t-designs with parameters $t-(v,k,\frac{1}{2}\binom{v-t}{k-t})$ then $\binom{v-i}{k-i}$ is even for all $i=0,1,2, \cdots ,t$.

Proof:

The number $\frac{1}{2}\binom{v-i}{k-i}$ counts the number of blocks in each part which contain a fixed i-subset of I_v, and hence must be an integer.

In section 2 we gave a method for constructing such a partition when $t=2$ and v is even. The following lemma gives a method for the case when v is odd.

Lemma 4.2: If there exists a partition of the complete design with block size k into two designs with parameters $t-(v,k,\frac{1}{2}\binom{v-t}{k-t})$ and there exists a partition of the complete design with block size $k-1$ into two designs with parameters $t-(v,k-1,\frac{1}{2}\binom{v-t}{k-1-t})$ then there exists a partition of the complete design with block size k into two designs with parameters $t-(v+1,k,\frac{1}{2}\binom{v+1-t}{k-t})$.

Proof:

We construct the new block sets from the old ones, adding the new element (v) to each of the blocks of size $k-1$. The t-subsets of I_v occur precisely $\frac{1}{2}\binom{v-t}{k-t}+\frac{1}{2}\binom{v-t}{k-1-t}=\frac{1}{2}\binom{v+1-t}{k-t}$ times in each part of the new partition, and the t-subsets of I_{v+1} which contain the point v occur precisely $\frac{1}{2}\binom{v-(t-1)}{k-1-(t-1)}=\frac{1}{2}\binom{v+1-t}{k-t}$ times.

We now focus our attention on the cases where $t=2$ and k is small.

Theorem 4.3: (Dehon [5]) A partition of the complete design with block size 3 into two designs with parameters $2-(v,3,\frac{1}{2}(v-2))$ exists if and only if $v\equiv 2 \pmod 4$.

Proof:

Necessity follows from Lemma 4.1, and sufficiency is proved in Example 2.1.

Note that in Example 2.1 we constructed a $2-(2n,3,n-1)$ design containing a $2-(n,3,n-2)$ design for all odd values of $n>1$. Using the more general constructions of the previous section we obtain the following interesting embedding theorem.

Theorem 4.4: The block set of a $2-(n,3,\lambda)$ design can be embedded in the block set of a $2-(2n,3,n-1)$ design for all odd $n>1$.

Proof:

Using the construction given in Theorem 2.1, take $M^0=Des(2,n,3,\lambda)$ and $M^1=Avg(n,2,n-1-\lambda)$.

We turn now to the case where $k=4$.

Theorem 4.5: A partition of the complete design with block size 4 into two designs with parameters $2-(v,4,\frac{1}{4}(v-2)(v-3))$ exists if and only if $v\equiv 2$ or $3 \pmod 8$.

Proof:

Necessity follows from Lemma 4.1. By Lemma 4.2 and Theorem 4.3, it is sufficient to prove the existence of a partition when $v\equiv 2 \pmod 8$. Let $v=2n$ and use the construction of Theorem 2.1 with $M^0=0$ and $M^1=Avg(n,3,\frac{3}{4}(n-1))$. Since $n\equiv 1 \pmod 4$

the existence of these matrices is given in the previous section, both when $gcd(n,3)=1$ and when $gcd(n,3)=3$.

When $k>4$ we have only partial results, despite the fact that the number of integer solutions to Equation 2.5 increases rapidly. This is because the constructions for elegant matrices given in the previous section do not have a sufficiently wide range of values for the constant $r_2(M)$. When $k=5$ the necessary conditions of Lemma 4.1 are that $v\equiv 2,3$ or $4 \pmod 8$, but we are only able to find constructions for the following result. (Note that the case $v=10$ is covered by Alltop's Theorem D [1])

Theorem 4.6: There exists a partition of the complete design with block size 5 into two designs with parameters $2-(v,5,\frac{1}{2}\binom{v-2}{3})$ if $v\in\{10,11,12,18,19,20,34,35,36,58,59,60\}$ or $v\equiv 2,3$ or $4 \pmod{24}$, or $v\equiv 12,42,43,44,60,66,67$ or $68 \pmod{72}$.

Proof:

As before we rely on Lemma 4.2 and the previous theorem to reduce the problem to $v=10,18,34,58$ $v\equiv 2 \pmod{24}$, and $v\equiv 12,42,60$ or $66 \pmod{72}$. Let $v=2n$ and we take $M^0=0$, except when $v=34,58$. In these cases we let $M^0=Des(2,17,5,5)$ and $M^0=Des(2,29,5,45)$ respectively. The 2-(17,5,5) design is constructed by taking the orbit under the action of the affine group of the 5-set comprising 0 and the 4-th roots of unity in GF(17). The 2-(29,5,45) design is constructed similarly, adding two full affine orbits of 5-subsets to the orbit of 0 and the 4-th roots of unity in GF(29). The constructions for M^1 and M^2 are given in Figure 1.

n	M^1	M^2
5	0	$Int(5,3,2,\{1,2\})$
9	$Avg(9,4,6)$	$Int(9,3,2,\{0\})$
17	$Avg(17,4,9)$	$Bar(17,3,2,2,5)$
29	$Avg(29,4,23)$	$Bar(29,3,2,2,4)$
$12m+1$	$Avg(12m+1,4,8m)$	$Bar(12m+1,3,2,2,3m)$
$36m+6$	$Avg(36m+6,4,28m+4)$	$Bar(36m+6,3,2,1,12m+2)$
$36m+21$	$Avg(36m+21,4,32m+18)$	$Bar(36m+21,3,2,2,3m+2)$
$36m-6$	$Avg(36m-6,4,20m-4)$	$Bar(36m-6,3,2,1,24m-5)$
$36m-3$	$Avg(36m-3,4,16m-2)$	$Bar(36m-3,3,2,2,15m-2)$

Figure 1. Constructions for $k=5$

When $k=5$ and $v=12$ Denniston [6] has given the following construction. Let $Y_1,Y_2,\cdots,Y_5$ be a one-factorization of K_6, i.e. a partition of the 2-subsets of I_6 into five $1-(6,2,1)$ designs. For each 2-subset, E, of I_6 define $f(E)$ to be the unique index i such that $E\in Y_i$. Now define the (6,3,2)-matrix Den by

$$Den_{T,E}=1 \text{ if and only if } \begin{array}{l} |T\cap E|=0 \text{ or} \\ |T\cap E|=1 \text{ and } f(E)=f(T-E). \end{array}$$

Denniston showed that with $M^0=0$, $M^1=Int(6,4,1,\{1\})$ and $M^2=Den$ the design constructed by the technique of Theorem 2.1 is in fact a 4-(12,5,4) design. Generalizing this construction to give other partitions into 4-designs appears to be difficult. Another interesting problem is to generalize the construction of Denniston's matrix to give another family of elegant matrices.

Theorem 4.6 implies the existence of a 2-(11,5,42) design without repeated blocks. Applying Alltop's Theorem B [1] to this design yields a 3-(12,6,42) design. Thus we have also succeded in halving the complete design with block size 6 on 12 points into 3-designs. This result is not new since Kreher and Radziszowski's construction [7] implies a solution to the halving problem for 4-designs with $v=12$ and $k=6$.

Returning now to the halving problem for 2-designs we note that when $k=6$ the necessary conditions of Lemma 4.1 are that $v \equiv 2,3,4$ or $5 \pmod 8$, and we have the following constructions.

Theorem 4.7: There exists a partition of the complete design with block size 6 into two designs with parameters $2\text{–}(2n,6,\frac{1}{2}\binom{2n-2}{4})$ if $n \in \{5,10\}$, or $n \equiv 1 \pmod 8$, or $n \equiv 29 \pmod{40}$, or $n \equiv 14$ or $74 \pmod{80}$, or $n \equiv 2$ or $6 \pmod{16}$, but $n \not\equiv 150$ or $210 \pmod{240}$.

Proof:

When $n=5$ the result follows from Theorem 4.5, by taking the complement of each block in a partition of the complete design with block size 4 on 10 points. In the other cases we take $M^0=0$. The constructions for M^1 and M^2 are given in Figure 2. Note that a construction for for the matrices $Avg(5m,5,5d)$ with $m \equiv 0 \bmod 6$ would also give solutions for the cases where $n \equiv 150$ or $210 \pmod{240}$.

n	M^1	M^2
10	$Int(10,5,1,\{0\})$	$Int(10,4,2,\{1\})$
$8m+1$	$Avg(8m+1,5,5m)$	$Bar(8m+1,4,2,2,2m)$
$40m+29$	$Avg(40m+29,5,5m+3)$	$Bar(40m+29,4,2,2,18m+12)$
$80m+14$	$Avg(80m+14,5,15m+2)$	$Bar(80m+14,4,2,1,68m+10)$
$80m-6$	$Avg(80m-6,5,5m-1)$	$Bar(80m-6,4,2,1,76m-8)$
$16m+2$	$Avg(16m+2,5,5m)$	$Bar(16m+2,4,2,1,3m)$
$16m+6$	$Avg(16m+6,5,15m+5)$	$Bar(16m+6,4,2,1,4m+2)$

Figure 2. Constructions for $k=6$

Other partitions of the complete design with block size 6 are constructible using the results of Theorem 4.6 and Lemma 4.2. For $k=7$ the necessary conditions are $v \equiv 2,3,4,5$ or $6 \pmod 8$, and we have the constructions for all admissible $v=2n$ with $5 \leq n \leq 47$, $gcd(n,6)=1$, and $n \neq 11,25$. For $k=8$ the necessary conditions are $v \equiv 2,3,4,5,6$ or $7 \pmod{16}$, and we have the constructions for $v \in \{38,50,66\}$. All of these constructions have $M^0=0$, M^1 an Average matrix, and M^2,M^3 are Baranyai matrices.

5. Conclusions and Open Problems

We have given necessary and sufficient conditions for the halving of the complete design with block sizes 3 and 4 into 2-designs and infinitely many constructions for block sizes 5 and 6. We conjecture that the necessary conditions of Lemma 4.1 are sufficient for the existence of a halving of the complete design into t-designs. This appears to be a difficult conjecture, but when $t=2$ the methods of this paper appear to make the problem tractable. An important step in the proof of this conjecture is the construction of new families of elegant matrices.

Another important problem raised in this paper is the determination of necessary and sufficient conditions for the existence of elegant matrices. All the elegant matrices constructed here (with the exception of the Design family) have the property that they have constant row and column sums. This property is important when aiming for higher values of t, but may be unnecessarily restrictive when $t=2$.

Acknowledgements

I would like to express my gratitude to Alex Rosa and Earl Kramer for drawing my attention to the work of Kreher, Radziszowski, and Alltop on the subject. Thanks are also due to the referee for his suggestions on improving the readability of the paper.

References

[1] W. O. Alltop, "Extending t-designs", *J. Combinatorial Theory, Ser. A* 18 (1975) 177-186.

[2] Z. Baranyai, "On the factorization of the complete uniform hypergraph", *Finite and Infinite Sets, Colloq. Math. Soc. János Bolyai* 10 (1975) 91-108. North-Holland, Amsterdam.

[3] P. J. Cameron, *Parallelisms of Complete Designs* London Math. Soc. Lecture Note Series 23 (1976). Cambridge University Press, Cambridge UK.

[4] L. G. Chouinard II, "Partitions of the 4-subsets of a 13-set into disjoint projective planes", *Discrete Math.* 45 (1983) 297-300.

[5] M. Dehon, "On the existence of 2-designs $S_\lambda(2,3,v)$ without repeated blocks", *Discrete Math.* 43 (1983) 155-171.

[6] R. H. F. Denniston, "A small 4-design", *Annals of Discrete Math.* 18 (1983) 291-294.

[7] D. L. Kreher and S. P. Radziszowski, "The existence of simple 6-(14,7,4) designs", *J. Combinatorial Theory, Ser. A* 43 (1986) 237-243.

[8] J. X. Lu, "On large sets of disjoint Steiner triple systems, I, II, III", *J. Combinatorial Theory, Ser. A* 34 (1983) 140-146, 147-155, 156-182.

[9] J. X. Lu, "On large sets of disjoint Steiner triple systems, IV, V, VI", *J. Combinatorial Theory, Ser. A* 37 (1984) 136-163, 164-188, 189-192.

[10] L. Teirlinck, "On the maximum number of disjoint triple systems", *J. Geometry* 6 (1975) 93-96.

[11] L. Teirlinck, "On large sets of disjoint quadruple systems", *Ars Combinatoria* 17 (1984) 173-176.

[12] L. Teirlinck, "Non-trivial t-designs without repeated blocks exist for all t", to appear.

Annals of Discrete Mathematics 34 (1987) 225—242
© Elsevier Science Publishers B.V. (North-Holland)

Outlines of Latin Squares

A.J.W. Hilton

Department of Mathematics
University of Reading
Whiteknights
Reading RG6 2AX
UNITED KINGDOM

TO ALEX ROSA ON HIS FIFTIETH BIRTHDAY

ABSTRACT

A unified treatment of outline latin squares and of (p,q,r)-latin rectangles is presented, together with applications to embedding partial latin squares.

1. Introduction

The generalized latin square A of Figure 1 contains four symbols in each square, and each symbol is contained four times in each row and column. It can actually be obtained from the latin square B of Figure 1 by erasing the dotted cell boundaries and reducing the numbers modulo 3.

$A =$

1 1 2 2	1 1 2 3	2 3 3 3
1 1 2 3	2 2 3 3	1 1 2 3
2 3 3 3	1 1 2 3	1 1 2 2

$B =$

1	2	3	4	5	6
5	4	1	2	6	3
4	5	2	6	3	1
3	1	6	5	2	4
2	6	4	3	1	5
6	3	5	1	4	2

Figure 1

We show in this paper that a generalized latin square such as A can always be obtained from a latin square B by the kind of process described.

This interesting phenomenon has quite a number of applications, some of which are considered in this paper.

Some of the results described here are new, some of the proofs of known results are new, and the unified presentation is new. Most of the known results are contained in [1] and [2] by Andersen and Hilton, and the principal result, Theorem 1, is contained (in a slightly different form) in [10], by Hilton.

2. Outline latin squares

Let $S = (p_1,p_2, \cdots p_s)$, $T = (q_1,q_2, \cdots q_t)$ and $U = (r_1,r_2, \cdots r_u)$ be three compositions of n. (A *composition* of n is a vector with positive integer components which add up to n.) If L is an $n \times n$ latin square on symbols $1, \cdots n$ then we form the *(S,T,U)-amalgamated latin square* L^* of size $s \times t$ on u symbols $\sigma_1, ..., \sigma_u$ by placing σ_k in cell (i,j) of L^* each time that one of $r_1 + \cdots + r_{k-1}+1, \cdots r_1 + \cdots + r_k$ occurs in one of the cells (α,β), where

$$\alpha \in \{p_1 + \cdots + p_{i-1}+1, \cdots, p_1 + \cdots + p_i\}, \text{ and}$$
$$\beta \in \{q_1 + \cdots + q_{j-1}+1, \cdots, q_1 + \cdots + q_j\}.$$

If L is on some other set of symbols, say $\tau_1, ..., \tau_n$, then we may assume that $\tau_1, ..., \tau_n$ are replaced by $1,2, ..., n$ respectively, and apply the definition above. To illustrate this definition, let $n = 6$, $S = (1,3,2)$, $T = (2,2,2)$ and $U = (1,1,2,2)$, and let L be the Latin square C of Figure 2.

$C =$

1	2	3	4	5	6
4	6	1	2	3	5
3	4	5	1	6	2
2	3	6	5	1	4
5	1	2	6	4	3
6	5	4	3	2	1

Figure 2

Then the amalgamated latin square L^* on symbols $\{1,2,3,5\}$ is obtained by erasing the dotted lines, and replacing the 4's by 3's and the 6's by 5's (and keeping the existing 1's, 2's, 3's, and 5's). Then L^* is depicted as the generalized Latin square D of Figure 3. No significance is attached to the positions of symbols within cells. It is clear that the number of times each symbol appears in each row, column, or cell of D is determined by C and the compositions S, T, and U. Clearly in L^*

$D =$

<table>
<tr><td>1 2</td><td>3 3</td><td>5 5</td></tr>
<tr><td>2 3
3 3
3 5</td><td>1 1
2 5
5 5</td><td>1 2
3 3
5 5</td></tr>
<tr><td>1 5
5 5</td><td>2 3
3 5</td><td>1 2
3 3</td></tr>
</table>

Figure 3

(i) row i contains symbol σ_k $p_i r_k$ times,

(ii) column j contains symbol σ_k $q_j r_k$ times, and

(iii) cell (i,j) contains (counting repetitions) $p_i q_j$ symbols.

Let us call a generalized latin square L^{**} on symbols $\sigma_1, \cdots, \sigma_u$ in which n^2 symbols occur altogether (counting repetitions) and for which there are compositions $S = (p_1, \cdots, p_s)$, $T = (q_1, \cdots, q_t)$ and $U = (r_1, \cdots, r_u)$ of n such that (i), (ii), and (iii) above hold for L^{**}, an *(S,T,U)-outline latin square.* Thus we assume here that L^{**} satisfies various numerical constraints, but we *do not* assume that L^{**} is obtained from a latin square. Our main result is that, nonetheless, there is a latin square from which L^{**} can be obtained. Thus if we were presented with the outline latin square D of Figure 3 we would know that there is a latin square from which it could be obtained.

Theorem 1: Each outline latin square is an amalgamated latin square.

This theorem was originally proved in [10] (in a slightly different format). The special case of (S,T,U)-outline latin squares in which $s = t = n$, so that $S = T = I$ (the vector of all 1's), has been studied earlier under the name F-squares. The interest has mainly been in connection with orthogonality.

Before proceeding to prove Theorem 1 we point out that conditions (i), (ii) and (iii) are symmetrical. This symmetry can be seen more vividly if the $n \times n$ latin square is thought of as a triangulation (that is, a decomposition into K_3's) of the complete tripartite graph $K_{n,n,n}$; this triangulation may be written down in tabular form as an $n^2 \times 3$ orthogonal array. An amalgamated latin square then corresponds to the amalgamation (or identification) of various vertices of the $K_{n,n,n}$ (where vertices may be amalgamated only if they are in the same part); amalgamation can therefore be represented as an operation on an orthogonal array.

In order to prove Theorem 1 we need to give some graph theoretical definitions and to state a lemma. An *edge-colouring* of a multigraph G is a map $\phi: E(G) \rightarrow \mathcal{C}$, where $E(G)$ is the set of edges of G, and $\mathcal{C}$ is a set of colours (note that the edge-colouring is not intended necessarily to be "proper"). For $v \in V(G)$, the vertex set of G, and $k \in \mathcal{C}$, let $E_k(v)$ be the set of edges incident with v of colour k, and, for $u,v \in V(G)$, $u \neq v$, let $E_k(u,v)$ be the set of edges joining u and v of colour k. An edge colouring is called *equitable* if

$$||E_j(v)|-|E_k(v)|| \leq 1 \quad \text{for all } v \in V(G) \text{ and } j,k \in \mathcal{C}.$$

It is called *balanced* if, in addition,

$$||E_j(u,v)|-|E_k(u,v)|| \leq 1 \quad \text{for all } u,v \in V(G), u \neq v, \text{ and } j,k \in \mathcal{C}.$$

The lemma we need is due to D. de Werra (see [18,19]); another proof may be found in [3].

Lemma 2: Let $k \geq 1$ be an integer and let G be a bipartite multigraph. Then G has a balanced edge-colouring with k colours. □

There are two observations we would like to make about this lemma. The first is that it is quite easy to deduce that G has an equitable edge-colouring with k colours from the knowledge that the chromatic index of a bipartite graph equals the maximum degree; one need only split each vertex of G into a number of vertices each of degree k, plus possibly one of lower degree, sharing out edges in any suitable way. The difficulty lies in showing that the edge-colouring can be balanced. The second observation is that, in our proof of Theorem 1, we actually only need to know that G has an equitable edge-colouring with k colours. However, the full strength of Lemma 2 is needed in many extensions and analogues of Theorem 1.

We now turn to a proof of the theorem.

Proof of Theorem 1:

Let L^{**} be an (S,T,U)-outline latin square where $S = (p_1, \cdots, p_s)$, $T = (q_1, \cdots, q_t)$ and $U = (r_1, \cdots, r_u)$. If $S = T = U = (1,1, \cdots, 1)$ then L^{**} is a latin square and there is nothing to prove. So suppose that at least one of S, T, U is not equal to $(1,1, \cdots, 1)$. In view of the symmetry of the definition, there is no loss of generality in supposing that $S \neq (1,1, \cdots, 1)$. We may further suppose that $p_s \neq 1$. We shall show that row s of L^{**} can be split into p_s rows, giving an (S',T,U)-outline latin square on the same set of symbols, where $S' = (p_1, \cdots, p_{s-1}, 1, \cdots, 1)$. Repetition of this process on the rows, and the same process carried out on the columns and on the symbols, will yield an (S'',T'',U'')-outline latin square, where $S'' = T'' = U'' = (1,1, \cdots, 1)$, or, in other words, a latin square. It then follows that L^{**} is an (S,T,U)-amalgamated latin square.

Construct a bipartite graph B with vertex sets $\{c'_1, \cdots, c'_t\}$ and $\{\sigma'_1, \cdots, \sigma'_u\}$, representing columns and symbols respectively, and join c'_j to σ'_k by x edges if symbol σ_k appears x times in cell (s,j). Then c'_j has degree $p_s q_j$ and σ'_k has degree $p_s r_k$. Give the edges of B an equitable colouring with p_s colours $\kappa_1, \cdots, \kappa_{p_s}$. Then, for each $\nu \in \{1, \cdots, p_s\}$, vertex c'_j has q_j edges of colour κ_ν on it, and vertex σ'_k has r_k edges of colour κ_ν on it.

Now construct an (S',T,U)-outline latin square L' on the same set of symbols as follows. Let rows $1, \cdots, s-1$ of L' be the same as rows $1, \cdots, s-1$ of L^{**}. For $i = s-1+\nu$, where $\nu \in \{1, \cdots, p_s\}$, place symbol σ_k in cell (i,j) y times if there are y edges coloured κ_ν joining σ'_k to c'_j in B. We have to check that L' is in fact a (S',T,U)-outline latin square. Clearly L' satisfies (ii). For $1 \leq \nu \leq p_s$, row $s-1+\nu$ contains symbol σ_k r_k times: it now follows easily that L' satisfies (i). Finally, for $1 \leq \nu \leq p_s$, cell $(s-1+\nu, j)$ contains q_j symbols in it altogether, counting repetitions; it follows easily that L' satisfies (iii).

This proves Theorem 1. □

In the definition of an (S,T,U)-outline latin square, it was assumed that S, T and U were compositions of n; this is actually unnecessarily restrictive, for if we had supposed that S was a composition of n_1, T a composition of n_2 and U a composition of n_3, and that L^{**} was an (S,T,U)-outline latin rectangle if (i), (ii), and (iii) were obeyed, then it would follow that $n_1 = n_2 = n_3$. For the number of entries in row α is, by (i), $p_\alpha \sum r_k = p_\alpha n_3$, so the total number of entries is $\sum p_\alpha n_3 = n_1 n_3$. Similar arguments using (ii) and (iii) show that the total number of entries is also $n_2 n_3$ and $n_1 n_2$, so $n_1 = n_2 = n_3$.

3. Ryser's and Cruse's theorems

Ryser's theorem on embedding latin rectangles inside latin squares is an immediate consequence of Theorem 1. An $s\times t$ *partial filled*[*] *latin rectangle* R on symbols $1, \cdots, n$ is an $s\times t$ matrix in which each symbol from $1, \cdots, n$ occurs at most once in each row and at most once in each column (and each cell contains exactly one symbol). For $1 \le i \le n$, let $N(i)$ denote the number of times that i appears in R. Ryser's theorem [16] is as follows:

Theorem 3: Let R be an $s\times t$ partial filled latin rectangle on the symbols $1, \cdots, n$. Then R can be completed to an $n\times n$ latin square L if and only if for all $i \in \{1, \cdots, n\}$,

$$N(i) \ge s + t - n.$$

Proof:

Assume first that $s,t < n$. By Theorem 1, R can be completed to L if and only if there is an (S,T,U)-outline latin square as illustrated in Figure 4.

Here R is bordered by one row and one column of further cells, and $S = (1, \cdots, 1, n-s)$, $T = (1, \cdots, 1, n-t)$, and $U = (1, \cdots, 1)$. For $1 \le i \le s$, cell $(i, t+1)$ is filled with the symbols not occurring in the rest of row i, and, similarly, for $1 \le j \le t$, cell $(s+1, j)$ is filled with the symbols not occurring in the rest of column j. Cell $(s+1, t+1)$ is filled in such a way that each symbol occurs $n-t$ times altogether in the $(t+1)$-th column. (This requirement is equivalent to the requirement that each symbol occurs $n-s$ times altogether in the $(s+1)$-th row.) If a symbol occurs $N(i)$ times in R, then it occurs $s - N(i)$ times in the last column excluding cell $(s+1, t+1)$, and so it must occur $(n-t)-(s-N(i)) = N(i)-(s+t-n)$ times in cell $(s+1, t+1)$. Clearly L^{**} cannot exist if the construction requires any symbol to occur a negative number of times in cell $(s+1, t+1)$, but does exist otherwise. Hence, completion of R is possible if and only if, for $1 \le i \le n$, $N(i) \ge s+t-n$. The proof if $s = n$ or $t = n$ is similar (but easier). This proves Theorem 3. □

We remark that this technique is applicable to give a Ryser-type embedding theorem in a number of different contexts, and in fact it is applicable whenever we have a structure for which a theorem like Theorem 1 can be proved.

---

[*] The word 'filled' is used to indicate that each cell is filled. The word 'unfilled' will be used later to indicate that some cells may be empty, or, at least, not completely filled.

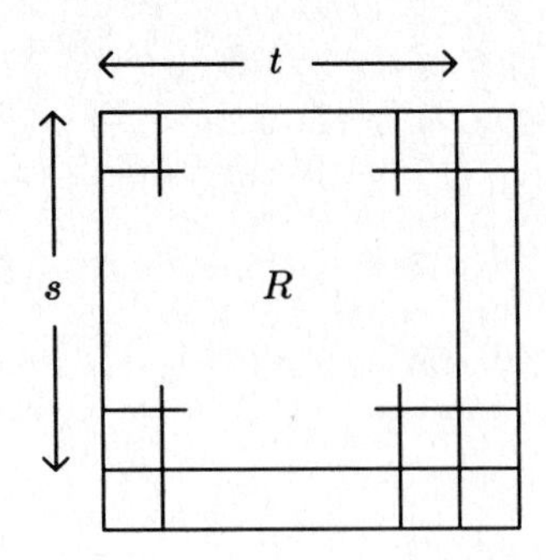

Figure 4

We can extend this technique to prove a theorem of Cruse [5] on embedding partial unfilled latin rectangles. An $s\times t$ *partial unfilled latin rectangle* R on symbols $1, \cdots, u$ is an $s\times t$ matrix in which each cell is either left empty or it is filled with an element from $\{1, \cdots, u\}$ in such a way that no element occurs twice in any row or column. Let N be the number of cells of R which are filled, let $N_\sigma(k)$ be the number of cells of R containing the symbol k, let $N_\rho(i)$ be the number of occupied cells of row i, and let $N_c(j)$ be the number of occupied cells of column j.

Theorem 4: Let R be an $s\times t$ partial unfilled latin rectangle on the symbols $1, \cdots, u$. Then R can be extended to an $n\times n$ latin square on symbols $1, \cdots, n$ in such a way that the empty cells of R are filled with the symbols $u+1, \cdots, n$ if and only if

(i) $N \leq (st+tu+us) - n(s+t+u-n)$, and

(ii) $N_\sigma(k) \geq s+t-n$, $(1 \leq k \leq u)$,
$N_\rho(i) \geq t+u-n$, $(1 \leq i \leq s)$, and
$N_c(j) \geq u+s-n$, $(1 \leq j \leq t)$.

Perhaps the most striking feature of these inequalities is that they are symmetric in s, t, and u. However, a moment's thought will convince the reader that this ought to be so, for the theorem can be reinterpreted as a theorem about extending a partial triangulation of $K_{n,n,n}$ in such a way that all the extra triangles have at most two vertices in the $K_{s,t,u}$, a situation which is symmetrical in s, t and u.

Proof: By Theorem 1, R can be embedded in a latin square on $1, \cdots, n$ with the empty cells of R filled with symbols $n-u+1, \cdots, n$ if and only if R can be embedded in the $(s+1)\times(t+1)$ (S,T,U)-outline latin square on symbols $1, \cdots, u, x$ as follows:

(i) Place x in the empty cells of R.

(ii) For $1 \leq i \leq s$, fill cells $(i,t+1)$ in such a way that $1, \cdots, u$ each occur once in row i and x occurs $n-u$ times in row i. (Note that (i) and (ii) can be carried out if and only if $(t - N_\rho(i))$, which is the number of empty cells in row i of R, satisfies $t - N_\rho(i) \leq n - u$, so that $N_\rho(i) \geq t+u-n$.)

(iii) For $1 \le j \le t$, fill cells $(s+1,j)$ in such a way that $1, \cdots ,u$ each occur once in column j and x occurs $n-u$ times. (Note that, similarly, (i) and (iii) can be carried out if and only if $N_c(j) \ge u+s-n$.)

(iv) Fill cell $(s+1,t+1)$ so that, in column $t+1$, each of $1, \cdots ,u$ occurs $n-t$ times and x occurs $(n-u)(n-t)$ times. (As in the proof of Theorem 3, (iv) can be carried out for symbols $1, \cdots ,u$ if and only if $N_\sigma(k) \ge s+t-n$ $(1 \le k \le u)$. The symbol x occurs $(st - N)$ times in R, and so it occurs $s(n-u) - (st-N)$ times in column $t+1$ without cell $(s+1,t+1)$. Therefore it occurs $(n-t)(n-u) - \{s(n-u)-(st-N)\} = (st+tu+us) - n(s+t+u-n) - N$ times altogether in cell $(s+1,t+1)$. Clearly this is possible if and only if $N \le (st+tu+us) - n(s+t+u-n)$.)

Here $S = (1, \cdots ,1,n-s)$, $T = (1, \cdots ,1,n-t)$ and $U = (1, \cdots ,1,n-u)$. This proves Theorem 4. □

Another proof of Cruse's theorem, one based directly on Ryser's theorem, was given in [7]. Actually, the expression "Cruse's theorem" is being used a little loosely here, since the inequalities involved are not the same in the different versions. However the differences are trivial. For example, let D be the number of unoccupied cells in R, $D_\rho(i)$ be the number of unoccupied cells in row i of R, and $D_c(j)$ be the number of unoccupied cells in column j of R. Then $D+N = st$, $D_\rho(i)+N_\rho(i) = t$, and $D_c(j)+N_c(j) = s$. Thus for example the inequalities $N_\rho(i) \ge t+u-n$ and $D_\rho(i) \le n-u$ are equivalent. Similarly, the inequality in [7] corresponding to (i) in the statement of Theorem 4 is

$$D \ge (s+t-n)(n-u).$$

This is not symmetrical in s, t, and u, but, substituting $D = st-N$, we obtain $st-N \ge (s+t-n)(n-u)$, which is a rearrangement of (i). Cruse's version of (i) was $N \le \frac{1}{n}\{stu-(n-s)(n-t)(n-u)\}$, but this reduces to our version easily.

4. (p,q,r)-latin rectangles

Some generalized latin rectangles have regular nicely behaved features which seem to make them worthy of study in their own right. For example, consider a generalized latin rectangle L in which each symbol occurs p times in each row, q times in each column, and in which there are r symbols in each cell (counting repetitions). We call such generalized latin rectangles, (p,q,r)-latin rectangles. These were considered by Andersen and Hilton in [1], [2], and [3]. If L has s rows, t columns, and f entries altogether (counting repetitions), then by counting three different ways we find that

$$f = pus = qtu = rst.$$

It is not quite obvious, but there is more symmetry in this definition than appears at first sight. For consider a tripartite graph G with vertex sets $(\rho'_1, \cdots ,\rho'_s)$, $(c'_1, \cdots ,c'_t)$, $(\sigma'_1, \cdots ,\sigma'_u)$, where, for $1 \le i \le s$, $1 \le j \le t$, and $1 \le k \le u$, ρ'_i is joined to c'_j by r edges, ρ'_i is joined to σ'_k by p edges and c'_j is joined to σ'_k by q edges. Then the (p,q,r)-latin rectangle L corresponds to a triangulation of G, with the occurrence of k in cell (i,j) corresponding to a triangle on vertices ρ'_i, c'_j, σ'_k in the triangulation of G. This triangulation could, of course, be represented in tabular form by a kind of generalized orthogonal array, in which the triangle on the vertices ρ'_i, c'_j, σ'_k corresponds to a row (i,j,k) of the generalized orthogonal array. Unfortunately the discussion of generalized latin rectangles by L.D. Andersen and A.J.W. Hilton in [1] obscured this symmetry. Note also that if there is no repetition of any element in any cell of L, then no two triangles of G are on the same set of vertices, and no two rows in the generalized orthogonal array are the same.

If we take the generalized orthogonal array corresponding to a (p,q,r)-latin rectangle L, and apply a permutation $\pi(1,2,3)$ to the columns of the orthogonal array, we can then write down a corresponding generalized latin rectangle $L_{\pi(1,2,3)}$. $L_{\pi(1,2,3)}$ is a conjugate of L, and there are six conjugates altogether, $L = L_{(1,2,3)}$, $L_{(2,1,3)}$, $L_{(3,2,1)}$, $L_{(2,3,1)}$, $L_{(1,3,2)}$ and $L_{(3,1,2)}$. Of course, these conjugates will not all be (p,q,r)-latin rectangles unless $p = q = r$. For example, $L_{(3,2,1)}$ is an (r,q,p)-latin rectangle with u rows, t columns, and s symbols. The process of obtaining $L_{(3,2,1)}$ can be described more succinctly as 'interchanging rows and symbols'; similarly for some of the other conjugates. We illustrate this in Figure 5, which shows a (2,3,5)-latin rectangle $L = L_{(1,2,3)}$ and two of its conjugates, $L_{(3,2,1)}$ and $L_{(1,3,2)}$. The other three conjugates are transposes of the three matrices shown. Figure 5 also shows the generalized orthogonal array corresponding to L.

The particular case of (p,q,r)-latin rectangles when $p = q = 1$, $r \neq 1$, has been studied under the name *semi-latin squares* in statistics (see [4] and [15]). The interest has mainly been in the statistical properties of various semi-latin squares of small size, for use in experimental design.

Under what circumstances do (p,q,r)-latin rectangles exist? This question is easily resolved using de Werra's theorem (Lemma 2).

Theorem 5: Let p,q,r,s,t,u be positive integers. A (p,q,r)-latin rectangle of size $s \times t$ on u symbols exists if and only if

$$\frac{s}{q} = \frac{t}{p} = \frac{u}{r}.$$

Proof: As remarked earlier, if a (p,q,r)-latin rectangle of size $s \times t$ on u symbols with f entries (counting repetitions) exists then

$$f = pus = qtu = rst,$$

so that $ps = qt$, $pu = rt$ and $qu = rs$, from which it follows that

$$\frac{s}{q} = \frac{t}{p} = \frac{u}{r}.$$

Conversely, let $\frac{s}{q} = \frac{t}{p} = \frac{u}{r}$. Construct a bipartite graph B with vertex sets $\{\rho_1, \cdots, \rho_s\}$ and $\{c_1, \cdots, c_t\}$, and, for $1 \leq i \leq s$ and $1 \leq j \leq t$, join ρ_i to c_j by r edges. Give B an equitable edge-colouring with u colours $\sigma_1, \cdots, \sigma_u$. Then ρ_i has degree tr, and so has $\frac{tr}{u} = p$ edges of each colour on it, and c_j has degree sr and so has $\frac{sr}{u} = q$ edges of each colour on it. From B construct the (p,q,r)-latin rectangle by placing symbol k in cell (i,j) whenever colour σ_k joins vertex ρ_i to vertex c_j. Since ρ_i has p edges of each colour on it, each symbol appears p times in each row; since c_j

$L = L_{(1,2,3)} =$

1 2 3 4 5	1 2 3 4 5
1 2 2 3 3	1 4 4 5 5
1 4 4 5 5	1 2 2 3 3

$L_{(3,2,1)} =$

1 2 3	1 2 3
1 2 2	1 3 3
1 2 2	1 3 3
1 3 3	1 2 2
1 3 3	1 2 2

$L_{(1,3,2)} =$

1 2	1 2	1 2	1 2	1 2
1 2	1 1	1 1	2 2	2 2
1 2	2 2	2 2	1 1	1 1

$$\begin{bmatrix} 1&1&1\\ 1&1&2\\ 1&1&3\\ 1&1&4\\ 1&1&5\\ 1&2&1\\ 1&2&2\\ 1&2&3\\ 1&2&4\\ 1&2&5\\ 2&1&1\\ 2&1&2\\ 2&1&2\\ 2&1&3\\ 2&1&3\\ 2&2&1\\ 2&2&4\\ 2&2&4\\ 2&2&5\\ 2&2&5\\ 3&1&1\\ 3&1&4\\ 3&1&4\\ 3&1&5\\ 3&1&5\\ 3&2&1\\ 3&2&2\\ 3&2&2\\ 3&2&3\\ 3&2&3 \end{bmatrix}$$

Figure 5: A (2,3,5)-latin rectangle, two of its conjugates, and the corresponding generalized orthogonal array.

has q edges of each colour on it, each symbol appears q times in each column; and since ρ_i and c_j are joined by r edges, cell (i,j) has r entries. This proves Theorem 5. □

We can also say exactly when a (p,q,r)-latin rectangle exists with no repetition in any cell.

Theorem 6: Let p,q,r,s,t,u be positive integers. A (p,q,r)-latin rectangle of size $s \times t$ on u symbols with no repetition in any cell exists if and only if

$$\frac{s}{q} = \frac{t}{p} = \frac{u}{r} \geq 1.$$

Proof: If a (p,q,r)-latin rectangle of size $s \times t$ on u symbols with no repetition in any cell exists, then u, the number of symbols, is at least as great as r, the number of entries in each cell. Therefore, by Theorem 5, $\frac{s}{q} = \frac{t}{p} = \frac{u}{r} \geq 1$.

Conversely, suppose that $\frac{s}{q} = \frac{t}{p} = \frac{u}{r} \geq 1$. Apply the argument of Theorem 5, but give the graph B a *balanced* edge-colouring with u colours. Since $u \geq r$, this means that the edges joining ρ_i and c_j all get different colours; this in turn yields a (p,q,r)-latin rectangle with no repetition in any cell. This proves Theorem 6. □

The next question one might ask about (p,q,r)-latin rectangles is which of them are (S,T,U)-amalgamations of latin squares, for appropriate compositions S, T and U? Again we can give a complete answer to this question.

Theorem 7: Let L be a (p,q,r)-latin rectangle and let

$$p = p_1^{\alpha_1} p_2^{\alpha_2} \cdots p_l^{\alpha_l},$$

$$q = p_1^{\beta_1} p_2^{\beta_2} \cdots p_l^{\beta_l}, \text{ and}$$

$$r = p_1^{\delta_1} p_2^{\delta_2} \cdots p_l^{\delta_l},$$

where $p_1, p_2, \cdots, p_l$ are distinct prime numbers and, for $1 \leq i \leq l$, $\alpha_i + \beta_i + \delta_i \geq 1$. Then L is the amalgamation of a latin square if and only if, for each $i \in \{1, \cdots, l\}$,

(i) $\alpha_i + \beta_i + \delta_i$ is even, and

(ii) $\lambda_i + \mu_i \geq \nu_i$

where $(\lambda_i, \mu_i, \nu_i)$ is a permutation of $(\alpha_i, \beta_i, \delta_i)$ with $\lambda_i \leq \mu_i \leq \nu_i$.

Proof: In view of Theorem 1, we need to show that a (p,q,r)-latin rectangle is an outline latin square if and only if, for each $i \in \{1, \cdots, l\}$, (i) and (ii) are satisfied.

Suppose first that (i) and (ii) in the statement of Theorem 7 are satisfied and that L is an $s \times t$ (p,q,r)-latin rectangle on u symbols. For $1 \leq i \leq l$, let

$$x_i = \tfrac{1}{2}(\alpha_i + \delta_i - \beta_i),$$

$$y_i = \tfrac{1}{2}(\delta_i + \beta_i - \alpha_i),$$

and

$$z_i = \tfrac{1}{2}(\beta_i + \alpha_i - \delta_i).$$

It is easy to see that, since (i) and (ii) are satisfied, all of x_i, y_i, and z_i are non-negative integers. Then

$$p = p_1^{\alpha_1} p_2^{\alpha_2} \cdots p_l^{\alpha_l} = \left(p_1^{x_1} \cdots p_l^{x_l}\right)\left(p_1^{z_1} \cdots p_l^{z_l}\right),$$

$$q = p_1^{\beta_1} p_2^{\beta_2} \cdots p_l^{\beta_l} = \left(p_1^{y_1} \cdots p_l^{y_l}\right)\left(p_1^{z_1} \cdots p_l^{z_l}\right),$$

and

$$r = p_1^{\delta_1} p_2^{\delta_2} \cdots p_l^{\delta_l} = \left(p_1^{x_1} \cdots p_l^{x_l}\right)\left(p_1^{y_1} \cdots p_l^{y_l}\right).$$

Let S be a composition with s components, each equal to $p_1^{x_1} \cdots p_l^{x_l}$, let T be a composition with t components, each equal to $p_1^{y_1} \cdots p_l^{y_l}$, and let U be a composition with u components, each equal to $p_1^{z_1} \cdots p_l^{z_l}$. This is possible since $\frac{s}{q} = \frac{t}{p} = \frac{u}{r} = x$, say, and so

$$sp_1^{x_1} \cdots p_l^{x_l} = tp_1^{y_1} \cdots p_l^{y_l} = up_1^{z_1} \cdots p_l^{z_l} =$$
$$x\left(p_1^{x_1} \cdots p_l^{x_l}\right)\left(p_1^{y_1} \cdots p_l^{y_l}\right)\left(p_1^{z_1} \cdots p_l^{z_l}\right).$$

Then (i), (ii) and (iii) of the definition of outline latin squares are satisfied, so, by definition, L is an (S,T,U)-outline latin square.

Conversely, if L is an (S,T,U)-outline latin square, then the reverse argument shows that (i) and (ii) of Theorem 7 are satisfied. This proves Theorem 7. □

Theorem 7 is not, however, the last word on the question of when a (p,q,r)-latin rectangle can be obtained from the amalgamation of a latin square, for we now show that, in one sense, all (p,q,r)-latin rectangles can be obtained from the amalgamation of latin squares.

Theorem 8: Let L be a (p,q,r)-latin rectangle. Then L is the top left-hand corner of an amalgamation of a latin square.

Proof: In view of Theorem 1, we need to show that L is the top left-hand corner of an outline latin square. Let L be an $s \times t$ rectangle on u symbols. Then, by Theorem 5, $\frac{s}{q} = \frac{t}{p} = \frac{u}{r}$.

We have a certain amount of choice in our construction of the containing outline latin square. We choose not to introduce any more symbols. Let $r = r_\rho r_c$ and let $m = \max\left(\lceil \frac{p}{r_\rho} \rceil, \lceil \frac{q}{r_c} \rceil\right)$. Suppose that L is not already the amalgamation of a latin square. We add one more row and one more column to L, giving an outline latin square L^*, as follows. For $1 \le i \le s$, in cell $(i,t+1)$ we place each symbol $mr_\rho - p$ times; for $1 \le j \le t$, in cell $(s+1,j)$ we place each symbol $mr_c - q$ times; finally in cell $(s+1,t+1)$ we place each symbol $m(um - tr_c) - s(mr_\rho - p)$ times.

Let S, T and U be compositions of um as follows:

$S = (r_\rho, \cdots, r_\rho, um - sr_\rho)$ (here s components equal r_ρ),

$T = (r_c, \cdots, r_c, um - tr_c)$ (here t components equal r_c), and

$U = (m, \cdots, m)$ (here u components equal m).

Then L^* is an (S,T,U)-outline latin square. To see this, we have to verify that (i), (ii), and (iii) of the definition of outline latin squares are satisfied:

Verification of (i):

For $1 \le i \le s$, row i contains each symbol $p + (mr_\rho - p) = r_\rho m$ times, and row $s+1$ contains each symbol

$$t(mr_c-q) + \{m(um-tr_c) - s(mr_\rho-p)\}$$
$$= m^2u - smr_\rho - tq + sp$$
$$= (um - sr_\rho)m$$

times, since $sp = tq$.

Verification of (ii):

For $1 \le j \le t$, column j contains each symbol $q + (mr_c-q) = r_c m$ times, and column $t+1$ contains each symbol

$$s(mr_\rho-p) + \{m(um-tr_c) - s(mr_\rho-p)\}$$
$$= (um-tr_c)m$$

times.

Verification of (iii):

For $1 \le i \le s$, $1 \le j \le t$, cell (i,j) contains $r = r_\rho r_c$ elements, cell $(s+1,i)$ contains

$$u(mr_c-q) = umr_c-uq = umr_c-sr = (um-sr_\rho)r_c$$

elements, cell $(j,t+1)$ contains

$$u(mr_\rho-p) = umr_\rho-up = umr_\rho-tr = r_\rho(um-tr_c)$$

elements, and finally cell $(s+1,t+1)$ contains

$$u\{m(um-tr_c) - s(mr_\rho-p)\}$$
$$= u^2m^2-umtr_c-umsr_\rho+usp$$
$$= u^2m^2-umtr_c-umsr_\rho+str$$
$$= (um-tr_c)(um-sr_\rho)$$

elements.

This proves Theorem 8. □

In the proof of Theorem 8, if we take m as small as possible (i.e. choose r_ρ and r_c so that $\max\left(\lceil \frac{p}{r_\rho} \rceil, \lceil \frac{q}{r_c} \rceil\right)$ is as small as possible) then the corresponding latin square whose amalgamation contains L is as small as possible; in particular, if $mr_\rho = p$ and $mr_c = q$ then $L = L^*$. Because of this, it is easy to see that Theorem 7 can actually be deduced from Theorem 8.

It must not be thought that in view of Theorems 7 and 8, there is no point in studying (p,q,r)-latin rectangles separately, because everything about them can be deduced from theorems on latin squares. The theorems in the rest of this paper in fact need separate treatment, and cannot easily be deduced from theorems on latin squares.

Moreover, even though a (p,q,r)-latin rectangle L is the top left-hand corner of an amalgamation of a latin square, it does not follow that the appropriate amalgamation of any latin square yields a (p,q,r)-latin rectangle. For example, consider a $(1,1,r)$-latin rectangle (a semi-latin square). Any $(1,1,r)$-latin rectangle may be formed from some

$nr \times nr$ latin square by amalgamating the first r columns, the second r columns, etc., and then erasing the bottom $(nr-n)$ rows. But if this procedure is applied to any $nr \times nr$ latin square, a $(1,1,r)$-latin rectangle may not be formed, because a symbol could occur up to r times in a column.

We can imitate the definitions of amalgamated latin squares and outline latin squares for (p,q,r)-latin rectangles. Let $V = \{v_1, \cdots, v_a\}$, $W = \{w_1, \cdots, w_b\}$ and $X = \{x_1, \cdots, x_c\}$ be compositions of s, t and u, respectively. If L is a (p,q,r)-latin rectangle of size $s \times t$ on the elements $1, \cdots, u$, then we form the *(V,W,X)-amalgamated (p,q,r)-latin rectangle L^** of size $a \times b$ on c symbols $\sigma_1, \cdots, \sigma_c$ by placing symbol σ_k in cell (i,j) of L^* each time that one of

$$x_1 + \cdots + x_{k-1} + 1, \quad \cdots \quad , x_1 + \cdots + x_k$$

appears in one of the cells (α,β), where

$$\alpha \in \{v_1 + \cdots + v_{i-1} + 1, \quad \cdots \quad , v_1 + \cdots + v_i\}, \text{ and}$$

$$\beta \in \{w_1 + \cdots + w_{j-1} + 1, \quad \cdots \quad , w_1 + \cdots + w_j\};$$

here $1 \le i \le a$, $1 \le j \le b$ and $1 \le k \le c$. If L is on some other set of symbols, say $\tau_1, \cdots, \tau_u$, then we may assume that $\tau_1, \cdots, \tau_u$ are replaced by $1, \cdots, u$ respectively, and apply the definition above. Clearly in L^*

(i') row i contains symbol σ_k $pv_i x_k$ times;

(ii') column j contains symbol σ_k $qw_j x_k$ times;

(iii') cell (i,j) contains $rv_i w_j$ symbols, counting repetitions.

Call a generalized latin rectangle L^{**} on symbols $\sigma_1, \cdots, \sigma_c$ for which there exist positive integers p,q,r,s,t,u and compositions $V = (v_1, \cdots, v_a)$, $W = (w_1, \cdots, w_b)$ and $X = (x_1, \cdots, x_c)$ of s, t and u such that (i'), (ii') and (iii') above hold, a *(V,W,X)-outline (p,q,r)-latin rectangle.*

Then we have the following analogue of Theorem 1.

Theorem 9: Each outline (p,q,r)-latin rectangle is an amalgamated (p,q,r)-latin rectangle.

Proof: This is obtained by imitating the proof of Theorem 1. □

If in a (p,q,r)-latin rectangle L there as no repetition of any symbol in any cell, then L is said to be *without repetition;* in that case, in the (V,W,X)-amalgamated (p,q,r)-latin rectangle we have:

(iv') in cell (i,j), no symbol is repeated more than $v_i w_j x_k$ times.

A (V,W,X)-outline (p,q,r)-latin rectangle satisfying (iv') is said to be a *(V,W,X)-outline (p,q,r)-latin rectangle without repetition.* We have the further analogue of Theorem 1.

Theorem 10: Each outline (V,W,X)-outline (p,q,r)-latin rectangle without repetition is an amalgamation of a (p,q,r)-latin rectangle without repetition.

Proof: This is similarly obtained by using the proof of Theorem 1, but using balanced instead of equitable edge-colourings. □

Using these two theorems, we can obtain the following two analogues of Cruse's theorem (Theorem 4) by imitating the proof of Cruse's theorem given here. Different proofs were given in [7] (depending on results in [1]).

A *partial unfilled (p,q,r)-latin rectangle* of size $s \times t$ on u symbols is an $s \times t$ matrix with up to r entries in each cell, counting repetitions, such that each element occurs at most p times in each row and at most q times in each column.

Theorem 11: Let R be an $s \times t$ partial unfilled (p,q,r)-latin rectangle on the symbols $1, \cdots, u$. Let l, m and n be positive integers with $\frac{l}{q} = \frac{m}{p} = \frac{n}{r}$. Then R can be extended to an $l \times m$ (p,q,r)-latin rectangle on the symbols $1, \cdots, n$ in such a way that the cells with fewer than r entries in R are filled up with the symbols $u+1, \cdots, n$ if and only if

(i) $N^* \leq r(st) + q(tu) + p(us) - s(pn) - t(rl) - u(qm) + qmn$,

(ii)
$$N_\sigma^*(k) \geq ps + qt - \left(\frac{pq}{r}\right) n, \quad (1 \leq k \leq u),$$
$$N_c^*(j) \geq qu + rs - \left(\frac{qr}{p}\right) m, \quad (1 \leq j \leq t), \text{ and}$$
$$N_\rho^*(i) \geq rt + pu - \left(\frac{rp}{q}\right) l, \quad (1 \leq i \leq s),$$

where $N_\sigma^*(k)$ is the number of occurrences of the symbol k, and $N_\rho^*(i) = \sum_{j=1}^{t} N^*(i,j)$, $N_c^*(j) = \sum_{i=1}^{s} N^*(i,j)$, $N^*(i,j)$ being the number of entries in cell (i,j). □

Note that these inequalities could be expressed differently, since $ln = qm$, $lr = qn$ and $rm = pn$.

If we wish the $l \times m$ (p,q,r)-latin rectangle into which R is embedded to be without repetition in any cell, then we have the following theorem.

Theorem 12: Let R be an $s \times t$ partial unfilled (p,q,r)-latin rectangle on the symbols $1, \cdots, u$ with no repeated entries in any cell. Let l, m, n be positive integers with $\frac{l}{q} = \frac{m}{p} = \frac{n}{r} \geq 1$. Then R can be extended to an $l \times m$ (p,q,r)-latin rectangle on the symbols $1, \cdots, n$ with no repetition in any cell in such a way that the cells in R with fewer than r entries are filled up with the symbols $u+1, \cdots, n$ if and only if

(i) $N^* \leq r(st) + q(tu) + p(us) - s(pn) - t(rl) - u(qm) + qmn$,

(ii) $N^* \geq r(st) + q(tu) + p(us) - s(pn) - t(rl) - u(qm) + qmn - (m-t)(l-s)(n-u)$,

(iii)
$$N_\sigma^*(k) \geq ps + qt - \left(\frac{pq}{r}\right) n, \quad (1 \leq k \leq u),$$
$$N_c^*(j) \geq qu + qt - \left(\frac{qr}{p}\right) m, \quad (1 \leq j \leq t),$$
$$N_\rho^*(i) \geq rt + pu - \left(\frac{rp}{q}\right) l, \quad (1 \leq i \leq s),$$

(iv) $$N_\sigma^*(k) \le ps + qt - \left\lceil \frac{pq}{r} \right\rceil n + (m-t)(l-s),\ (1 \le k \le u),$$

$$N_c^*(j) \le qu + rs - \left\lceil \frac{qr}{p} \right\rceil m + (l-s)(n-u),\ (1 \le j \le t),$$

$$N_\rho^*(i) \le rt + pu - \left\lceil \frac{rp}{q} \right\rceil l + (n-u)(m-t),\ (1 \le i \le s),$$

(v) $N_\rho^*(i,k) \ge p + t - m,\ (1 \le i \le s,\ 1 \le k \le u)$,
$N_c^*(j,k) \ge q + s - l,\ (1 \le j \le t,\ 1 \le k \le u)$, and
$N^*(i,j) \ge r + u - n,\ (1 \le i \le s,\ 1 \le j \le t)$,

where $N_\rho^*(i,k)$ is the number of times σ_k occurs in row i and $N_c^*(j,k)$ is the number of times σ_k appears in column j.

Note that N^*, $N_\sigma^*(k)$, $N_c^*(j)$, and $N_\rho^*(i)$ were defined in Theorem 11.

The demerging of an outline (p,q,r)-latin rectangle described in Theorem 9 can be carried a little further. For example, in Figure 6, there is a (2,2,2)-latin rectangle E which is the 'amalgamation' of the 'generalized latin rectangle' F of Figure 6. The process of going from E to F illustrated there is a general one; we now describe it more formally.

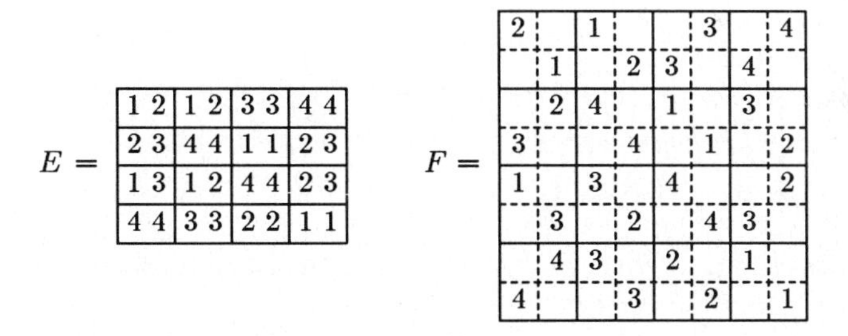

Figure 6

Let an $n \times n$ $\frac{u}{n}$-*stretched latin square* be an $n \times n$ matrix on u symbols in which each symbol appears once in each row, once in each column, and in which each cell contains either $\lfloor \frac{u}{n} \rfloor$ or $\lceil \frac{u}{n} \rceil$ symbols. We can again imitate the definition of amalgamated latin squares. Let $V = (v_1, \cdots, v_a)$ and $W = (w_1, \cdots, w_b)$ be two compositions of n. If L is an $n \times n$ $\frac{u}{n}$-stretched latin square on the elements $\sigma_1, \cdots, \sigma_u$, then we form the (V,W,I)-*amalgamation* L^* of L of size $a \times b$ by placing symbol σ_k in cell (i,j) of L^* each time σ_k occurs in one of the cells (α,β), where

$$\alpha \in \{v_1 + \cdots v_{i-1}+1, \cdots, v_1 + \cdots + v_i\}, \text{ and}$$
$$\beta \in \{w_1 + \cdots + w_{j-1}+1, \cdots, w_1 + \cdots + w_j\};$$

here $1 \leq i \leq a$ and $1 \leq j \leq b$.

Again by imitating the proof of Theorem 1 we obtain:

Theorem 13: A (p,q,r)-latin rectangle of size $s \times t$ on u symbols is an amalgamation of an $n \times n$ $\frac{u}{n}$-stretched latin square, where $n = ps$ $(= qt)$. □

It is easy to see that $r \leq pq$ if and only if $u \leq n$. If $r \leq pq$ we therefore have the case where each cell of the $\frac{u}{n}$-stretched latin square is either empty or contains one element. If $p = q = r$, as in E of Figure 6, then each original cell is replaced by a $p \times p$ matrix with one filled cell in each row and each column, as in F of Figure 6.

One problem closely related to Theorem 13 is the following. Given a (4,4,4)-latin rectangle of size $n \times n$, is it the amalgamation of a 'generalized latin rectangle' of size $2n \times 2n$ on the same set of symbols in which each of the original cells is split into four cells, one diagonal pair being empty, and the other having two symbols each, in such a way that each symbol occurs twice in each row and in each column? For example, in Figure 7, G is the amalgamation of H, which has the desired form.

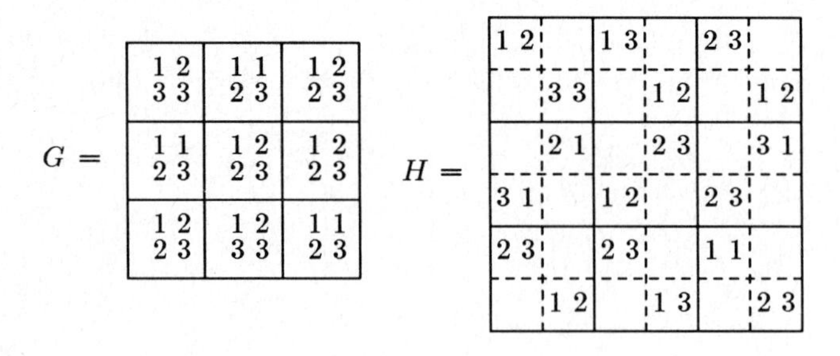

Figure 7

5. A final remark

One can obtain similar results when the matrices are all symmetric (see, for example, [1] and [2]), or when the conditions imposed are weaker (see [6], [8], and [17]), or both (see [12]). Analogous results also hold for Hamiltonian decompositions of complete graphs (see [9] and [11]) and under even more general conditions (see [13] and [14]).

Acknowledgement

I would like to thank the referee for pointing out a number of minor errors.

References

[1] L.D. Andersen and A.J.W. Hilton, "Generalized latin rectangles I: construction and decomposition", *Discrete Math.* 31 (1980) 125-152.

[2] L.D. Andersen and A.J.W. Hilton, "Generalized latin rectangles II: embedding", *Discrete Math.* 31 (1980) 235-260.

[3] L.D. Andersen and A.J.W. Hilton, "Generalized latin rectangles", *Graph Theory and Combinatorics* (Research Notes in Mathematics, Pitman, 1979), 1-17.

[4] R.A. Bailey, "Semi-latin squares", preprint.

[5] A.B. Cruse, "On extending incomplete latin rectangles", *Proc. Fifth Southeastern Conf. Combinatorics, Graph Theory, Computing,* Florida Atlantic University, Boca Raton, Florida (1974), 333-348.

[6] Deng Chai Ling and Lim Cheong Keang, "A result on generalized latin squares", Research Report 1/86, Department of Mathematics, University of Malaya.

[7] A.J.W. Hilton, "Embedding incomplete latin rectangles", *Annals of Discrete Math.* 13 (1982) 121-138.

[8] A.J.W. Hilton, "School timetables", *Studies on Graphs and Discrete Programming* (P. Hansen, ed.), North-Holland (1981), 177-188.

[9] A.J.W. Hilton, "Hamiltonian decompositions of complete graphs", *J. Comb. Theory* B36 (1984) 125-134.

[10] A.J.W. Hilton, "The reconstruction of latin squares with applications to school timetabling and to experimental design", *Math. Programming Study* 13 (1980) 68-77.

[11] A.J.W. Hilton and C.A. Rodger, "Hamiltonian decompositions of complete regular s-partite graphs", *Discrete Math.* 58 (1986) 63-78.

[12] A.J.W. Hilton and C.A. Rodger, "Matchtables", *Annals of Discrete Math.* 15 (1982) 239-251.

[13] C.St.J.A. Nash-Williams, "Detachments of graphs and generalized Euler trails", *Surveys in Combinatorics* (I. Anderson, ed.) (1986), 137-151.

[14] C.St.J.A. Nash-Williams, "Amalgamations of almost regular edge-colourings of simple graphs", *J. Comb. Theory* B, to appear.

[15] D.A. Preece and G.H. Freeman, "Semi-latin squares and related designs", *J. Roy. Statist. Soc.* B45 (1983) 267-277.

[16] H.J. Ryser, "A combinatorial theorem with an application to latin squares", *Proc. Amer. Math. Soc.* 2 (1951) 550-552.

[17] Tiong-Seng Tay, "Extending partial generalized latin rectangles with restraints", preprint.

[18] D. de Werra, "Balanced schedules", *INFOR* 9 (1971) 230-237.

[19] D. de Werra, "A few remarks on chromatic scheduling", *Combinatorial Programming: Methods and Applications* (B. Roy, ed.) Reidel, Dordrecht, Holland (1975) 337-342.

Annals of Discrete Mathematics 34 (1987) 243–248
© Elsevier Science Publishers B.V. (North-Holland)

The Flower Intersection Problem for Steiner Triple Systems

D.G. Hoffman and C.C. Lindner

Department of Mathematics
Auburn University
Auburn, Alabama 36849
U.S.A.

TO ALEX ROSA ON HIS FIFTIETH BIRTHDAY

ABSTRACT

We determine here those pairs (k,v) for which there exist a pair of Steiner triple systems on the same v-set, the triples in one system containing a particular point are the same as those in the other system containing that point, and the two systems otherwise have exactly k triples in common. The obvious necessary conditions are sufficient, with two exceptions $(k,v) = (1,9)$ and $(4,9)$.

1. Preliminaries

Alex Rosa loves intersection problems. In fact he and C.C. Lindner solved one of the first such problems. In [2], they determined those pairs (k,v) for which there exists a pair of Steiner triple systems (STS) on the same v-set having exactly k triples in common. In honour of Alex Rosa we solve here a variant of the problem: namely, the *flower intersection problem.*

The *flower* at a point x in a Steiner triple system (S,T) is the set of *all* triples containing x. The *flower intersection problem* for STSs is *the determination for each* $n = 2r+1 \equiv 1,3 \pmod 6$ of the set $I[r]$ of all k such that there exists a pair of triple systems (S,T_1) and (S,T_2) or order n having $k+r$ triples in common, r of them being the triples of a common flower. Note that $r \equiv 0,1 \pmod 3$; we call such a non-negative r *admissible.*

For each non-negative integer t, let $S(t)$ denote the set of non-negative integers less than or equal to t, with the exception of $t-1$, $t-2$, $t-3$, and $t-5$.

For each admissible r, let $J[r] = S\left(\frac{2r(r-1)}{3}\right)$. From [2], we see that $I[r] \subseteq J[r]$. Trivially, $I[0]=J[0]=I[1]=J[1]=\{0\}$, $I[3]=J[3]=\{0,4\}$. A bit of fiddling shows however that sadly,

$$I[4] = \{0,2,8\} = J[4] - \{1,4\}.$$

We proceed to prove that $I[r] = J[r]$ for all other admissible r.

In the next section, we will handle the special cases $r \in \{6,7,9,10,12,13\}$ by ad-hoc methods. In the last section we do the rest by induction using certain pairwise balanced designs.

2. Special Cases

Certainly if two Steiner triple systems on the same v-set have many triples in common, they must have at least one flower in common. To be more specific:

Lemma 1: Let $r \geq 6$ be admissible, let $k \in J[r]$ with $k \geq \frac{2r(r-3)}{3}$. Then $k \in I[r]$.

Proof:

From [2], there are two Steiner triple systems on the same $(2r+1)$-set which have exactly $k+r$ triples in common. Now if a point $x \in V$ is contained in a triple of the first system that is *not* a triple of the second, then obviously it must be contained in at least two such triples. A simple calculation now shows that there must be at least one point x for which the triples through x are the same in both systems. □

Lemma 2: $I[6] = J[6]$.

Proof:

This is not the most elegant proof in this paper. By Lemma 1, we need only show that $k \in I[6]$ for $0 \leq k \leq 11$. Consider the sets S_j, $1 \leq j \leq 6$, of triples in Table 1.

Let T be the set of triples $\{\{x,i,i+6\}: 1 \leq i \leq 6\}$ where x is a new symbol. Note that for $1 \leq j \leq 6$, $S_j \cup T$ is a Steiner triple system with flower T. Consider also the following permutations:

$$\pi_1 = (1, 3, 7, 9)\ (2, 11, 4)\ (5, 10, 8)\ (6, 12);$$
$$\pi_2 = (1, 6, 4, 9, 7, 12, 10, 3)\ (2, 5, 8, 11);$$
$$\pi_3 = (1, 3, 2, 12, 10, 7, 9, 8, 6, 4)\ (5, 11);$$
$$\pi_4 = (1, 6, 10)\ (3, 11)\ (4, 7, 12)\ (5, 9);$$
$$\pi_5 = (1, 7)\ (2, 10, 3, 12)\ (4, 9, 6, 8)\ (5, 11);$$
$$\pi_6 = (1, 5, 12, 7, 11, 6)\ (2, 10, 3, 8, 4, 9);$$
$$\pi_7 = (1, 11, 12, 10, 8, 9)\ (2, 3, 7, 5, 6, 4).$$

Note that $\pi_i(T) = T$ for $1 \leq i \leq 7$. Table 2 gives certain quadruples h,i,j,k for which $|S_h \cap \pi_i(S_j)| = k$, so $k \in I[6]$.

This completes the proof. □

Before starting on $I[7]$, we need some overtures.

Let us call a *symmetric* latin square of order 8 on symbols 0, 1, ..., 7 *special* if the four 2×2 diagonal blocks are $\begin{pmatrix} 0 & 1 \\ 1 & 0 \end{pmatrix}$, so it looks like Figure 1.

Denote by W the set of integers k for which there is a pair of special latin squares which agree in exactly k of the twenty-four cells above the 2×2 diagonal blocks.

Table 1: S_j, $1 \leq j \leq 6$					
S_1	S_2	S_3	S_4	S_5	S_6
3,6,11	3,8,11	6,7,10	1,5,9	1,9,11	1,4,11
4,8,11	1,3,6	2,10,11	3,7,12	2,4,9	10,11,12
8,9,12	8,9,12	5,7,8	4,5,12	1,4,12	3,7,12
5,6,9	5,6,9	3,5,6	7,9,11	5,7,12	1,2,10
1,4,6	1,4,8	6,9,11	3,6,11	7,8,10	3,5,10
4,7,12	4,7,12	8,11,12	2,3,4	4,6,8	6,8,10
5,10,12	5,10,12	3,4,8	4,8,9	3,4,5	8,9,11
1,2,5	1,2,5	1,9,12	6,9,10	5,9,10	4,5,9
3,7,10	3,7,10	4,7,12	1,2,6	6,10,11	1,5,6
2,10,11	2,10,11	1,6,8	7,8,10	1,3,6	6,7,11
3,4,5	3,4,5	3,7,11	1,10,12	8,9,12	2,3,11
1,11,12	1,11,12	8,9,10	1,4,11	4,7,11	1,3,8
5,7,8	5,7,8	2,3,12	4,6,7	1,5,8	7,9,10
1,9,10	1,9,10	4,5,9	2,9,12	2,5,6	5,8,12
2,6,7	2,6,7	1,4,11	5,6,8	3,10,12	2,6,9
2,3,12	2,3,12	1,2,5	2,10,11	6,7,9	3,4,6
1,3,8	4,6,11	1,3,10	1,3,8	3,8,11	2,4,12
7,9,11	7,9,11	2,7,9	2,5,7	2,11,12	4,7,8
6,8,10	6,8,10	5,10,12	8,11,12	1,2,10	1,9,12
2,4,9	2,4,9	2,4,6	3,5,10	2,3,7	2,5,7

Table 2: Constructions for $I[6]$			
h	i	j	k
3	1	6	0
1	1	6	1
4	1	6	2
2	1	6	3
1	2	4	4
4	2	4	5
4	3	5	6
6	5	1	7
5	1	6	8
4	4	5	9
4	6	3	10
1	7	1	11

Lemma 3: W contains (among other things) all non-negative integers $k \leq 14$.

Proof:

If k is such an integer, write $k = a+b+c$, where $a,c \in \{0,4\}$, and $0 \leq b \leq 9$, $b \neq 5$ or 7. Let B_1 and B_2 be two (easily found) latin squares of order 4 on symbols 4, 5, 6, and 7 which agree in exactly b cells. Let $A_1 = C_1 = \begin{pmatrix} 2 & 3 \\ 3 & 2 \end{pmatrix}$. If $a = 0$, let

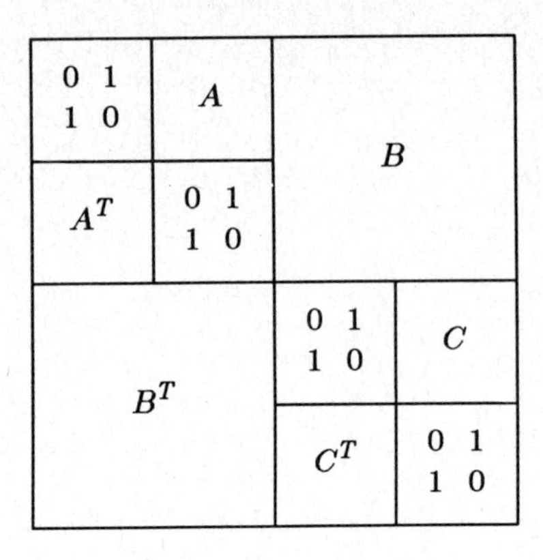

Figure 1: A special latin square

$A_2 = \begin{pmatrix} 3 & 2 \\ 2 & 3 \end{pmatrix}$; otherwise, let $A_2 = A_1$. Define C_2 similarly. For $i = 1,2$, replace regions A, B, C in Figure 1 by A_i, B_i, C_i respectively. □

Lemma 4: $I[7] = J[7]$.

Proof:

By Lemma 1, we need only show that $k \in I[7]$ for $0 \leq k \leq 18$. Write $k = a+b$, where $a \in \{0,4\}$ and $0 \leq b \leq 14$. By Lemma 3, there is a pair L_1, L_2 of special latin squares agreeing in exactly b cells above the diagonal blocks. We will construct a pair S_1, S_2 of Steiner triple systems on the set $\{\infty_i : 1 \leq i \leq 7\} \cup \{1,2,...,8\}$ which have exactly $k+7$ triples in common. Seven of these will constitute the flower at ∞_1.

For starters, on the ∞_i's, place a pair of systems of order 7 agreeing on a triples besides the three containing ∞_1. For the remaining triples of S_l, $l=1,2$, do the following: for each $1 \leq i < j \leq 8$, if symbol s appears in cell (i,j) of L_l, take the triple $\{\infty_s, i, j\}$. □

Before proceeding to $r=9$, we remark that H.L. Fu [1] proved that for $n \geq 5$, there is a pair of latin squares on the same n symbols agreeing in exactly k cells if and only if $k \in S(n^2)$.

Lemma 5: $I[9] = J[9]$.

Proof:

Let $k \in J[9]$, then we can write k as $k = a_1+a_2+a_3+b$, where a_1, a_2, $a_3 \in \{0,4\}$, and $b \in S(36)$. We define two systems S_1, S_2 on the set $(\{1,2,...,6\} \times \{1,2,3\}) \cup \{\infty\}$ which will agree on the flower at ∞. Let L_1 and L_2 be two latin squares on symbols $\{1,2,...,6\}$ agreeing in exactly b cells. For $t = 1,2,3$, let $S_{t,1}$, $S_{t,2}$ be a pair of systems on $\{\infty\} \cup (\{1,...,6\} \times \{t\})$ agreeing on the three triples containing ∞ and exactly a_t others. For $l = 1,2$, $S_{t,l}$ are some of the triples of S_l. For the rest, for each $1 \leq i, j \leq 6$, if the symbol s is in cell (i,j) of L_l, then the triple $\{(i,1),(j,2),(s,3)\}$ is in S_l. □

H.L. Fu also proved in [1] that for $n \geq 5$, there is a pair of latin squares on the same n symbols, with the same constant main diagonal, agreeing on exactly k cells off the main diagonal if and only if $k \in S(n^2-n)$.

Lemma 6: $I[10] = J[10]$.

Proof:

Let $k \in J[10]$, and write $k = a_1+a_2+b+c$, where $a_1,a_2 \in \{0,1,3,7\}$, $b \in \{0,4\}$, and $c \in S(7^2-7)$. We construct a pair of systems S_1, S_2 on the set $\{1,2,...,7\} \times \{1,2,3\}$. For $i = 1,2$, let $S_{i,1}$, $S_{i,2}$ be a pair of (easily found) systems on $\{1,2,...,7\} \times \{i\}$ with exactly a_i triples in common. Let $S_{3,1}$, $S_{3,2}$ be a pair of systems on $\{1,2,...,7\} \times \{3\}$ with the same three triples containing $(1,3)$, and b further triples in common. Let L_1, L_2 be a pair of latin squares on $\{1,...,7\}$ with the same constant main diagonal of 1's, and agreeing on c additional cells.

For $l = 1,2$, S_i contains the triples in $S_{s,l}$, $s = 1,2,3$ together with the triples $\{(i,1),(j,2),(s,3)\}$ for all $1 \leq i,j \leq 7$, where cell (i,j) of L_l contains the symbol s. Then S_1 and S_2 have $k+10$ triples in common, including the flower at $(1,3)$. □

Lemma 7: $I[12] = J[12]$. □

Lemma 8: $I[13] = J[13]$. □

We leave to the reader the easy tasks of mimicking the proofs of Lemmas 5 and 6 to establish Lemmas 7 and 8 respectively.

3. The Main Theorem

Recall that a *(pairwise-balanced) design* of *order* v (and *index* 1) is a collection of subsets, called *blocks,* of a v-set with the property that any two points of the set are contained together in exactly one block.

Lemma 9: Let $r \geq 15$ be admissible. Then there is a design of order r, each of whose blocks has size congruent to 0 or 1 (mod 3), and having at least one block of size t with $6 \leq t < r$.

Proof:

Let $t \neq 2$ or 6. Then it is well known that there is a design of order $4t$ with four disjoint blocks of size t, and the other blocks of size 4. (This is equivalent to a pair of orthogonal latin squares of order t.) Let $0 \leq s \leq t$, and delete $t-s$ points from one of the blocks of size t. This yields a design $D_{s,t}$ of order $3t+s$ with four disjoint blocks, three of size t and one of size s, and the remaining blocks of sizes 3 and 4. This construction also works if $t = 2$ or 6, and $s=0$, since then only a single latin square of order t is required. Further restricting s,t to be congruent to 0 or 1 (mod 3), and $t \geq 6$, covers all $r = 3t+s$ except $r = 15$, 16, and 19.

For $r = 15$, take any of the Steiner triple systems constructed in Lemma 4 and replace the subsystem on the ∞'s by a single block of size 7. For $r = 16$, adjoin a new point to the three disjoint blocks of size 5 in $D_{0,5}$. For $r = 19$, similarly adjoin a point to $D_{0,6}$. □

Main Theorem: For admissible r, $I[r] = J[r]$, except $1,4 \notin I[4]$.

Proof:

This has been proved for $r \leq 13$, so we assume that $r \geq 15$ and proceed by induction on r.

Let P be a design on the r-set R from Lemma 9. Suppose that $B_1, B_2, ..., B_s$ are the blocks of P, and suppose that $r_i = |B_i|$, $1 \leq i \leq s$, with $6 \leq r_1 \leq r$. Let $k \in J[r]$. We construct a pair S_1, S_2 of Steiner triple systems on $\{\infty\} \cup (R \times \{1,2\})$ as follows. Clearly, we can find for each $1 \leq i \leq s$, $a_i \in J[r_i]$ (and $a_i \neq 1$ or 4 if $r_i = 4$), with $\sum_{i=1}^{s} a_i = k$. Since $r_i < r$, we can find by induction a pair $S_{i,1}$, $S_{i,2}$ of systems on the set $\{\infty\} \cup (B_i \times \{1,2\})$ with exactly $a_i + r_i$ triples in common, r_i of them being the triples $\{\{\infty,(x,1),(x,2)\}: x \in B_i\}$. For $l = 1,2$, let the triples of S_l be the triples in the union of the $S_{i,l}$, $1 \leq i \leq s$. Then S_1 and S_2 have exactly $k + r$ triples in common, r of them being the flower $\{\{\infty,(x,1),(x,2): x \in R\}$. □

4. Concluding Remarks

Let (S,T) be a triple system and write $T = F_x \cup B$, where F_x is the flower at the point x. Set $X = S - \{x\}$ and $G = \{\{a,b\}: \{a,b,x\} \in F_x\}$. Then (X,G,B) is a group divisible design (GDD) with group size 2 and block size 3. When couched in these terms, the flower intersection problem becomes the intersection problem for GDDs with group size 2 and block size 3. A bit of reflection now reveals that the intersection problem for triple systems can be restated as the intersection problem for GDDs with group size 1 and block size 3. With the solution to these two problems in hand, the intersection problem for GDDs with all groups of the same size $k \geq 3$ and block size 3 is apparent. The authors have just recently solved this more general problem. However, the solution is just too long to be included here. Hence we content ourselves with an announcement of this result, with the details to appear elsewhere.

Happy birthday Alex!

Acknowledgement

Research of the second author is supported by NSF grant DMS-8401263.

References

[1] H.L. Fu, *On the construction of certain types of latin squares having prescribed intersections*, Ph.D. thesis, Auburn University, 1980.

[2] C.C. Lindner and A. Rosa, "Steiner triple systems with a prescribed number of triples in common", *Canad. J. Math.* 27 (1975) 1166-1175; corrigendum, *Canad. J. Math.* 30 (1978) 896.

Annals of Discrete Mathematics 34 (1987) 249–258
© Elsevier Science Publishers B.V. (North-Holland)

Embedding Totally Symmetric Quasigroups

D.G. Hoffman and C.A. Rodger

Department of Mathematics
Auburn University
Auburn, AL 36849
U.S.A.

TO ALEX ROSA ON HIS FIFTIETH BIRTHDAY

ABSTRACT

We show here that a totally symmetric quasigroup of order v can be embedded in one of order $w > v$ if and only if $w \geq 2v$, w is even if v is, and $(v,w) \neq (6k+5,12k+12)$.

1. Introduction

A set Q together with a binary operation $\circ$ on Q is a *quasigroup* if for every $a,b \in Q$, there exists unique $x,y \in Q$ satisfying the equations $a \circ x = b$ and $y \circ a = b$. If S is any subset of the set $\{x \circ x = x, x \circ y = y \circ x, y \circ (x \circ y) = x\}$, one can ask for which positive integers v and $w > v$ can a quasigroup on v elements, satisfying the identities in S, be embedded in one of order w, also satisfying the identities in S. (Note that the structure of the system of order v plays no role in the embedding.)

We list here some well known (and easily proved) necessary conditions:

i) $w \geq 2v$;

ii) if S contains the idempotent law $x \circ x = x$, then $w \geq 2v+1$, and neither v nor w can be two;

iii) if S contains the commutative law $x \circ y = y \circ x$, then we cannot have w odd if v is even;

iv) in the idempotent, commutative case, v and w must be odd;

v) if S contains both the idempotent law and Stein's third law $y \circ (x \circ y) = x$, then both v and w must be congruent to 0 or 1 (mod 3), and neither can be six. (These are equivalent to Mendelsohn triple systems [14].)

There are eight possible subsets S; in seven of the eight, the embedding problem has been solved by Ryser [15], Evans [6], Cruse [4], Doyen and Wilson [5], Hoffman and Lindner [11], and Hoffman [10]. See also [2], and for a survey, [13]. In these seven cases, the necessary conditions above are sufficient. In this paper, we deal with the remaining case, $S = \{xoy = yox, yo(xoy) = x\}$. Such quasigroups are called *totally symmetric.* We call one on v elements a $TSQG(v)$. Note that in a $TSQG$, any of the six equations

$$aob = c,\ aoc = b,\ boa = c,\ boc = a,\ coa = b,\ cob = a \qquad (*)$$

implies the other five.

In this case, the necessary conditions above are not sufficient. Here is our result:

Main Theorem Let v and w be positive integers, with $w > v$. Then a $TSQG(v)$ can be embedded in a $TSQG(w)$ if and only if

i) $w \geq 2v$;

ii) if v is even, then w is even; and

iii) $(v,w) \neq (6k+5, 12k+12)$, k a non-negative integer.

The problem naturally divides itself into the 36 cases for v and w (mod 6). The following table summarizes our approach:

v (mod 6) \ w (mod 6)	0	1	2	3	4	5
0	*D*	*I*	*D*	*I*	*S*	*I*
1	*D*	*D*	*D*	*D*	*V*	*D*
2	*D*	*I*	*D*	*I*	*D*	*I*
3	*D*	*D*	*D*	*D*	*V*	*S*
4	*S*	*I*	*D*	*I*	*D*	*I*
5	*V*	*D*	*V*	*S*	*V*	*D*

Table 1

The cases corresponding to the nine I's in Table 1 are impossible by condition ii) of the theorem. The eighteen D's are covered in section 3; the main tool there is the Doyen-Wilson theorem on embedding Steiner triple systems, (quasigroups for which S consists of all three identities). The four S's in Table 1 are done by difference methods in section 4; here the main tool is a theorem of Stern and Lenz [17]. The five V's in the table are handled in section 5 by an extension of Vizing's theorem. But first, we rephrase the problem in terms of graphs.

2. Graph Theoretic Considerations

For terms and notation not defined here, the reader is referred to our bible [3]. (none of our graphs here have multiple edges.)

An edge whose two ends are the same is a *loop*, an edge whose two ends are distinct is a *link*. A graph on two vertices consisting of exactly one loop and exactly one link is called a *lollipop*.

As usual, K_n denotes the complete graph on n vertices and $\binom{n}{2}$ links, while K_n^+ denotes K_n together with a loop at each vertex. K_3 is called a *triangle*.

Lemma 1 A $TSQG(v)$ is equivalent to a partition of the edges of K_v^+ into triangles, lollipops, and loops.

Sketch of Proof Consider the six equations in (*). If $a \neq b \neq c \neq a$, they correspond to a triangle in K_v^+ on the vertices a,b,c. If $a \neq b = c$, they reduce to three equations, and correspond to a lollipop on a,b with the loop at b. If $a = b = c$, they reduce to one equation, and correspond to a loop at a. □

A set M of edges of a graph is a *meeting* if every vertex of the graph is an end of exactly one edge in M. Thus a loopless meeting is what is usually called a *perfect matching*, or a *1-factor*.

Throughout the sequel, v,w, and d are positive integers with $d = w-v$.

Lemma 2. An embedding of a $TSQG(v)$ into a $TSQG(w)$ corresponds to a partition P of the edges of K_d^+ into triangles, lollipops, loops, and v meetings.

Sketch of proof. Given a 1-1 correspondence between the meetings and the elements of the $TSQG(v)$, if meeting M corresponds to element a, then a loop in M at vertex b corresponds to a lollipop on a,b with its loop at b, while a link in M on vertices b,c corresponds to a triangle on vertices a,b,c. □

3. The Doyen-Wilson Theorem.

A $TSQG(v)$ is idempotent (i.e. satisfies $x \circ x = x$) if and only if there are no lollipops in the corresponding partition of Lemma 1. For the embedding problem in Lemma 2, again there can be no lollipops, and further each of the meetings must be 1-factors. This problem was solved by Doyen and Wilson [5]; the necessary and sufficient conditions are that v and w must be congruent to 1 or 3 (mod 6), and $w \geq 2v+1$. This accounts for four of the eighteen D's in Table 1.

For the remainder of this section, let v and w satisfy the Doyen-Wilson conditions above. Thus we have a partition P of the links of K_d into triangles and v 1-factors.

Lemma 3. A $TSQG(v-1)$ can be embedded in a $TSQG(w-1)$.

Proof. Replace a 1-factor of P by $d/2$ loops and $d/2$ lollipops. □

Lemma 4. If $v \geq 3$, a $TSQG(v-2)$ can be embedded in a $TSQG(w-2)$.

Proof. Replace two of the 1-factors of P by d lollipops. □

Lemma 5. A $TSQG(v)$ can be embedded in a $TSQG(w-1)$.

Proof. Delete a point x from K_d. For every triangle in P on x,b,c, place a lollipop and a loop on b,c. In every 1-factor M of P, replace the link x,a by a loop on a to form a new meeting. □

Lemma 6. A $TSQG(v+1)$ can be embedded in a $TSQG(w+1)$.

Proof. Adjoin to P a $(v+1)^{st}$ meeting consisting of d loops. □

The four lemmas above give the other fourteen D's in Table 1.

4. The theorem of Stern and Lenz

Stern and Lenz [17] proved a beautiful theorem in order to give a shorter proof of the Doyen-Wilson theorem. We will need some definitions before stating this theorem.

If x is an integer, we define $|x|_d$ as follows: find the unique integer $y \equiv x \pmod{d}$ with $-d/2 < y \leq d/2$; then $|x|_d$ is defined to be the absolute value of y.

For the remainder of this section, we assume that the vertices of K_d^+ are the elements of the ring Z_d of integers modulo d. If the edge e has ends x and y, we define the *difference* of the edge e to be the non-negative integer $|x-y|_d \leq d/2$. So e is a loop if and only if its difference is 0. (By the way, the function $|x-y|_d$ makes Z_d into a metric space.)

Let $D = \{0,1, \cdots ,[d/2]\}$ be the set of differences of edges of K_d^+. For any $S \subseteq D$, we denote by $G_d(S)$ (or just $G(S)$) the graph on vertices Z_d with just the edges of K_d^+ whose differences lie in S. Thus $K_d^+ = G(D)$ and $K_d = G(D \backslash \{0\})$. If d is even, $G(\{d/2\})$ consists of a single 1-factor. If $x \in D \backslash \{d/2\}$, then $G(\{x\})$ is a disjoint union of g cycles, each of length d/g, where $g = \gcd(x,d)$.

If $x \in D$, we say x is a *good* difference (mod d) if d/g is even, where again $g = \gcd(x,d)$. Thus no difference is good if d is odd. If d is even, $d/2$ is good, as is any odd difference.

Finally, the theorem of Stern and Lenz:

Theorem 1. Let $\emptyset \neq S \subseteq D \backslash \{0\}$. Then the edges of $G(S)$ can be partitioned into 1-factors if and only if S contains at least one good difference. □

Before we can put this theorem to use, we need a preliminary lemma, and one more definition.

Lemma 7. Let d be even, let $a,b \in D \backslash \{0,d/2\}$, with a good, and either $b = 2a$, or $2a+b = d$. Then

i) the edges of $G_d(\{0,a,b\})$ can be partitioned into triangles, lollipops and loops, and

ii) the edges of $G_d(\{0,a,b\})$ can be partitioned into triangles, lollipops, loops, and one meeting.

Proof. Note first that $a \neq b$. Let $k = d/(2 \gcd(a,d))$, a positive integer since a is good. Let $G = G_{2k}(\{0,1,2\})$. Since each component of $G_d(\{0,a,b\})$ is isomorphic to G, we need only prove the lemma for G. Note that $G_{2k}(\{2\})$ consists of 2 disjoint k-cycles C_1 and C_2. First partition the edges of G into G_1 and G_2, where G_1 consists of the edges of difference 1 together with C_1, and G_2 the loops together with C_2. Obviously, G_1 is an edge-disjoint union of triangles, and the edges of G_2 can be partitioned into k lollipops and k loops, proving i). We leave to the reader the simple task of partitioning the edges of G_2 into lollipops, loops, and one meeting. (Hint - consider separately the two cases k (mod 2)). □

A three element set of integers is said to be a *sum-set* if one of the elements is the sum of the other two.

Lemma 8. Let k and ℓ be non-negative integers with $\ell \geq 2k+1$ and let ϵ be 0 or 1. Then a $TSQG(6k+3+\epsilon)$ can be embedded in a $TSQG(6l+5+\epsilon)$.

Proof. We consider the two cases $l-k$(mod 2).

Case 1. $\ell-k = 2s \geq k+1$.

Here $d = 12s+2$. We first partition $D\backslash\{0,4s,4s+1,5s+1,6s+1\}$ into $2s-1$ sum sets as follows:
for $1 \leq i \leq s$, $\{2i-1,3s-i,3s+i-1\}$ is a sum set, and
for $1 \leq i \leq s-1$, $\{2i,5s-i+1,5s+i+1\}$ is a sum set.

Select $2s-k-1$ of these sum sets. If $\{x,y,z\}$ is one of these, with say $x<y<z$, take the d triangles $\{i,x+i,x+y+i\}$, $i \in Z_d$, for a total of $d(2s-k-1)$ triangles. Let S be the set of $3k$ differences, not in the selected sum sets, together with the differences $5s+1$ and $6s+1$. Since $6s+1$ is a good difference, $G(S)$ can be partitioned into $6k+3$ 1-factors by the theorem of Stern and Lenz. Now apply Lemma 7 with $a = 4s+1, b = 4s$, using i) if $\epsilon = 0$ and ii) if $\epsilon = 1$.

Case 2. $\ell-k = 2s+1 \geq k+1$.

Here $d = 12s+8$. Now we partition $D\backslash\{0,3s+1,4s+2,4s+3,6s+4\}$ into $2s$ sum sets as follows:
for $1 \leq i \leq s$, $\{2i-1,5s-i+4,5s+i+3\}$ is a sum set, and
for $1 \leq i \leq s$, $\{2i,3s-i+1,3s+i+1\}$ is a sum set.

The rest of the proof is like Case 1, except we use Lemma 7 with $a = 4s+3$, $b = 4s+2$. □

Lemma 9. Let k,ℓ and ϵ be as in Lemma 8, except now $\ell \geq 2k+2$. Then a $TSQG(6k+5+\epsilon)$can be embedded in a $TSQG(6\ell+3+\epsilon)$.

Proof. This is just like the proof of Lemma 8, so we will only specify s,a,b and the sum sets in the two cases.

Case 1. $\ell-k = 2s$ (so $d = 12s-2$).

Here $a = 4s-1$, $b = 4s$, and the sum sets are
$\{2i-1,5s-i,5s+i-1\}$ for $1 \leq i \leq s-1$, and
$\{2i,3s-i-2,3s+i-2\}$ for $1 \leq i \leq s-1$.

Case 2. $\ell-k = 2s+1$ (so $d = 12s+4$).

Here $a = 4s+1$, $b = 4s+2$, and the sum sets are
$\{2i-1,3s-i,3s+i-1\}$ for $1 \leq i \leq s$, and
$\{2i,5s-i+2,5s+i+2\}$ for $1 \leq i \leq s-1$.□

Lemmas 8 and 9 together account for the four S's in Table 1.

5. An Extension of Vizing's Theorem

A *proper k edge-colouring* of a simple graph G is an assignment of one of k colours to each of the edges of G so that no vertex is incident with more than one edge of any given colour. We will abbreviate "proper k edge-colouring" to "k-colouring". Vizing's theorem [18], also proved by Gupta [9], states that if G has maximum degree Δ, then G has a $(\Delta+1)$-colouring. (By the way, Stern and Lenz used Vizing's theorem to prove their theorem.) The following extension of this theorem is implicit in the work of Vizing [19], was proved again by Fournier [7,8] and is also a corollary of a recent paper by the present authors [12]. So it surely must be true!

Theorem 2. Let G be a simple graph with maximum degree Δ. Suppose the subgraph of G induced by the vertices of degree Δ is a forest. Then G can be Δ-coloured. □

We need a few more preliminary results before we get down to work.

A *PSTS*(d) (*partial Steiner triple system* of order d) is a collection of edge-disjoint triangles in K_d. We denote by $s(d)$ the maximum number of triangles in a *PSTS*(d). Schonheim [16] determined $s(d)$. He showed that $s(d) = [(d/3)[(d-1)/2]]-\epsilon$, where $\epsilon = 0$ unless $d \equiv 5 \pmod 6$, in which case $\epsilon = 1$.

A *PSTS*(d) is said to be *equitable* if for all vertices x,y, $|t(x)-t(y)| \leq 1$, where $t(x)$ denotes the number of triangles containing the vertex x. The following theorem was established by L.D. Andersen, A.J.W. Hilton, and E. Mendelsohn [1]:

Theorem 3. Let d and k be positive integers. Then there exists an equitable *PSTS*(d) with k triangles if and only if $k \leq s(d)$. □

We must now examine more closely the structure of the partition of K_d^+ described in Lemma 2. So let P be such a partition. We define three simple graphs G_1, G_2 and G_3 on the d vertices as follows.

The edges of G_1 are the links in any of the v meetings of P.
The edges of G_2 are the links in the lollipops of P.
The edges of G_3 are the links in the triangles of P.

Obviously, the graphs G_1, G_2, and G_3 enjoy the following properties:

i) the edges of G_1, G_2, and G_3 partition the edges of K_d;

ii) the degree of each vertex in G_1 is either v or $v-1$, and G_1 can be v-coloured;

iii) if G_2^+ denotes the graph obtained by adding to G_2 a loop at each vertex which has degree v in G_1, then the edges of G_2^+ can be partitioned into lollipops and loops; and

iv) the edges of G_3 can be partitioned into triangles. In particular, every vertex has even degree in G_3, and the number of edges of G_3 is a multiple of three.

We leave to the reader the simple proof of the converse: if G_1, G_2, and G_3 are graphs on the same d vertices satisfying i) - iv) above, then there is a partition of K_d^+ satisfying the conditions of Lemma 2.

Now to work.

Lemma 10. Let v and w satisfy the conditions of the main theorem, with

$$(v,w) \equiv (1,4)(\text{mod } 6), \text{ or}$$
$$(v,w) \equiv (5,2)(\text{mod } 6), \text{ or}$$
$$(v,w) \equiv (5,4)(\text{mod } 6).$$

Then any $TSQG(v)$ can be embedded in a $TSQG(w)$.

Proof. Let T be an equitable $PSTS(d)$ with $(1/6)d(d-v)$ triangles and let G_3 be the graph whose edges are the edges of triangles of T. Then every vertex is in $(d-v)/6$ triangles of T, and hence has degree $d-v$ in G_3. Let G_2 be the graph with no edges, let G_1 be the graph consisting of the remaining edges of K_d. Then G_1 has every vertex of degree $v-1$, so it can be v-coloured by Vizing's theorem.□

Lemma 11. Let v and w satisfy the conditions of the main theorem, with $(v,w) \equiv (3,4)(\text{mod } 6)$. Then any $TSQG(v)$ van be embedded in a $TSQG(w)$.

Proof. Let T' be an equitable $PSTS(d)$ with $(1/6)(d(d-v)+2)$ triangles and let G_3' be the graph whose edges are the edges in triangles of T'. Then one vertex, say x, has degree $d-v+2$ in G_3', while the remaining vertices have degree $d-v$ in G_3'. Then T' must contain at least one triangle containing x; let a and b be the other two vertices in such a triangle t. Let T be the $PSTS(d)$ obtained from T' by removing triangle t, and let G_3 be the resulting graph. Let G_2 be the graph with one link on a,b; let G_1 be the graph consisting of the remaining edges of K_d. Then a and b have degree v in G_1, and the other vertices have degree $v-1$ in G_1, so G_1 can be v-coloured by Theorem 2. □

Lemma 12. Let v and w satisfy the conditions of the main theorem, with $(v,w) \equiv (5,0)(\text{mod } 6)$. Then any $TSQG(v)$ can be embedded in a $TSQG(w)$.

Proof. Let T' be an equitable $PSTS(d)$ with $(1/6)(d(d-v)-2)$ triangles, and let G_3' be the associated graph. Then one vertex, say x, has degree $d-v-2 \geq 6$ in G_3', the rest have degree $d-v$. Let a be any vertex adjacent to x. Since a has degree larger than x, there must be a vertex $c \neq x$ adjacent to a but not to x. Let t be the triangle containing a and c, let b be the third vertex of t.

Let T be the $PSTS(d)$ obtained from T' by removing t, and let G_3 be the associated graph. Let G_2 be the graph with exactly two links, one on a,b and the other on x,c. Let G_1 be the graph consisting of the remaining edges of K_d. Then a,b,c and x each have degree v in G_1, while the other vertices have degree $v-1$ in G_1. Moreover, the subgraph of G_1 induced by the vertices a,b,c and x is either a path of length three from a to x, or a path of length two from a to b together with the isolated vertex x. In any case, G_1 can be v-coloured by Theorem 2. □

We have accounted for the five V's in the table.

Note We would like to thank the referee for simplifying the proof of Lemma 12.

6. The Exception

Only one part of the main theorem remains.

Lemma 13. If k is a non-negative integer, then no $TSQG(6k+5)$ can be embedded in any $TSQG(12k+12)$.

Proof. Suppose there is such an embedding, let P be the associated partition of the edges of K_d^+ given by Lemma 2. Since $d = 6k+7$ is odd, each of the v meetings must contain exactly $2i+1$ loops for some i, $0 \leq i \leq 3k+3$. For each such i, let x_i be the number of meetings in P with exactly $2i+1$ loops. Let y be the number of lollipops in P, let z be the number of loops in P and let t be the number of triangles in P. Then

i) $\sum_{i=0}^{3k+3} x_i = 6k+5$ (counting meetings),

ii) $\sum_{i=0}^{3k+3} (2i+1)x_i + y + z = 6k+7$ (counting loops) and

iii) $\sum_{i=0}^{3k+3} (3k+3-i)x_i + y + 3t = (3k+3)(6k+7)$ (counting links).

Subtracting i) from ii), and subtracting $(3k+3)$ times i) from iii) shows that the only non-negative integer solution is $x_0 = 6k+5$, $x_i = 0$ for $1 \leq i \leq 3k+3$, $y = 0$, $z = 2$ and $t = 2k+2$.

Referring now to the partition of the edges of K_d into G_1, G_2, and G_3 given in section 5, we see that, since $z = 2$ and $y = 0$, there are two vertices of degree one in G_3, contradicting the fact that the edges of G_3 can be partitioned into triangles. □

References

[1] L.D. Andersen, A.J.W. Hilton, and E. Mendelsohn, "Embedding partial Steiner triple systems", *Proc. Lond. Math. Soc.*, 41 (1980), 557 - 576.

[2] F.E. Bennett and N.S. Mendelsohn, "On the existence of extended triple systems", *Utilitas Math.* 14 (1978), 249 - 267.

[3] A. Bondy and U.S.R. Murty, *Graph Theory with Applications,* American Elsevier, 1976.

[4] A. Cruse, "On embedding incomplete symmetric latin squares", *J. Combin. Theory Ser. A,* 16 (1974), 18 - 27.

[5] J. Doyen and R. Wilson, "Embeddings of Steiner triple systems", *Discrete Math.* 5 (1973), 409 - 416.

[6] T. Evans, "Embedding incomplete latin squares", *Amer. Math. Monthly,* 67 (1960), 958 -961.

[7] J.C. Fournier, "Colorations des arêtes d'un graphe", *Cahiers du C.E.R.O. (Bruxelles),* 15 (1973), 311 - 314.

[8] J.C. Fournier, "Méthode et théorème général de coloration des arêtes d'un multigraph", *J. Math. Pures Appl.,* 56 (1977), 437 - 453.

[9] R.P. Gupta, "The chromatic index and the degree of a graph", *Notices Amer. Math. Soc.* 13, abstract 66T-429.

[10] D.G. Hoffman, "Cyclic embeddings of semi-symmetric quasigroups and Mendelsohn triple systems", *Ars Combinatoria,* to appear.

[11] D.G. Hoffman and C.C. Lindner, "Embeddings of Mendelsohn triple systems", *Ars. Combin.* 2 (1981).

[12] D.G. Hoffman and C.A. Rodger, "Class one graphs", *J. Combin. Theory (B),* to appear.

[13] C.C. Lindner and T. Evans, *Finite embedding theorems for partial designs and algebras,* Les presses de l'Université de Montreal, 1977.

[14] N. Mendelsohn, "A natural generalization of Steiner triple systems", *Computers in Number Theory* Academic Press, New York, 1971, pp. 332 - 338.

[15] H.J. Ryser, "A combinatorial theorem with an application to latin rectangles", *Proc. Amer. Math. Soc.,* 2 (1951), 550 - 552.

[16] J. Schonheim, "On maximal systems of k-tuples", *Studia Sci. Math. Hungar.* 1 (1966), 363 - 368.

[17] G. Stern and H. Lenz, "Steiner triple systems with given subspaces : Another proof of the Doyen-Wilson theorem", *Boll. Un. Math. Ital. A(5)* 17 (1980), 109 - 114.

[18] V.G. Vizing, "On an estimate of the chromatic class of a p-graph"(Russian), *Diskret. Analiz.* 3 (1964), 25 - 30.

[19] V.G. Vizing, "The chromatic class of a multigraph", *Kibernetika (Kiev)* 3 (1965), 29 - 39; *Cybernetics* 3 (1965), 9 - 17. MR 34-17.

Annals of Discrete Mathematics 34 (1987) 259–272
© Elsevier Science Publishers B.V. (North-Holland)

Cyclic Perfect One Factorizations of K_{2n}

Edwin C. Ihrig

Department of Mathematics
Arizona State University
Tempe, Arizona
85287

TO ALEX ROSA ON HIS FIFTIETH BIRTHDAY

ABSTRACT

Recently Hartman and Rosa characterized those n for which K_{2n} admits a cyclic one factorization. We show that K_{2n} admits a cyclic perfect one factorization if and only if n is prime. We also show that if n is prime there is a one to one, onto correspondence between cyclic perfect one factorizations on K_{2n} and starter induced perfect one factorizations on K_{n+1}. Moreover the full symmetry group of the cyclic perfect one factorization is that of the corresponding starter induced perfect one factorization direct sum Z_2.

1. Introduction

In [6] Hartman and Rosa define a cyclic one factorization of K_{2n} as a one factorization of K_{2n} (the complete graph on $2n$ vertices) which has a cycle of length $2n$ in its symmetry group. They then show that a necessary and sufficient condition for K_{2n} to have a cyclic one factorization is that $n \neq 2^t$ for all $t \geq 2$. In this paper we study those cyclic one factorizations that are perfect.

The condition of being cyclic is a very natural one for perfect one factorizations. There are exactly three kinds of perfect one factorizations of K_{2n} which have a permutation consisting of of a single non trivial cycle in their symmetry groups. If this permutation has one fixed point then the perfect one factorization is called factor cyclic or vertex-1-rotational (see [11] p.46). The GK_{p+1} for p prime are examples of such perfect one factorizations. If the permutation has two fixed points then the perfect one factorization is called factor-1-rotational [11], bipyramidal [9], cyclic of type 2 [10] and generated by an even starter [1]. Again the GK_{p+1} are examples of this type of perfect one factorization. Both of these types of perfect one factorizations have been well studied, and a number of examples other than the GK_{p+1} are given in [4] and [13]. The only remaining possibility is that the permutation has no fixed points. These are the cyclic perfect one factorizations. GA_{2p} are examples of cyclic perfect one factorizations (see [11] p.49). However, despite the prominent role cyclic perfect one factorizations play, they do not seem to have been studied in their own right.

There are even more subtle reasons why cyclic perfect one factorizations are special. In [8] we have shown that cyclic perfect one factorizations constitute some of the few perfect one factorizations with symmetry groups that are not small. A perfect one factorization is said to have a small symmetry group if there is a vertex such that the only symmetry that fixes it is the identity. This means in particular that the symmetry group is in one to one correspondence with a subset of the vertex set. Then 4.1 of [8] says that any perfect one factorization which does not have a small symmetry group is either cyclic or starter induced. Starter induced perfect one factorizations are those perfect one factorizations that are generated by a starter on a not necessarily abelian group, and as such they constitute natural generalizations of 1-factor cyclic perfect one factorizations. With the help of [8] one can show that the condition of being starter induced is equivalent to being factor transitive. These one factorizations are described in more detail in 3.1 and 3.2. Since it appears that the new perfect one factorizations of K_{2n} which will be easiest to find will be the ones whose symmetry groups are not small, it is of interest to discover what perfect one factorizations are starter induced or cyclic. However not only have starter induced perfect one factorizations with starter group Z_{2n-1} been studied extensively, but also those with starter group $(Z_p)^m$ have been considered in some detail (see [2] and [12]). Thus the natural candidates for new perfect one factorizations are either starter induced perfect one factorizations with a more complicated starter group or cyclic perfect one factorizations.

Unfortunately our first result shows that no new orders have cyclic perfect one factorizations. We show that K_{2n} admits a cyclic perfect one factorization if and only if n is prime. This result is a modification of 4.2 of [8] to include the case when the cyclic perfect one factorization has a small symmetry group.

Our main result is slightly more positive in that it shows that there is a one to one onto correspondence between cyclic perfect one factorizations of K_{2n} and starter induced perfect one factorizations of K_{n+1} when n is prime. We state this more explicitly at the end of this section in 1.1. Since there are known starter induced perfect one factorizations on K_{p+1} which are different from GK_{p+1}, we now know of the existence of several new perfect one factorizations of K_{2p} which are different from GA_{2p}. In particular, using [2] (and 2.1 (m) for notation), we find there is a cyclic perfect one factorization on K_{22} with symmetry $[Z_{22}]Z_5$, one on K_{34} with symmetry Z_{34}, two on K_{38} with symmetry $[Z_{38}]Z_3$ and three on K_{38} with symmetry Z_{38}. Using [13] we find there are also new cyclic perfect one factorizations on K_{46} and K_{58}.

The main message of this paper is that the study of all perfect one factorizations with a non small symmetry group reduces to the study of starter induced perfect one factorizations. The part of this area which is as yet unexplored consists of starter induced perfect one factorizations with non abelian starter groups. The known constraints on such a starter group are given in 3.2.

If we wish to consider perfect one factorizations with small symmetry groups, the largest small symmetry groups consistent with our current knowledge of symmetry may also be of interest. Such a group would be strictly vertex transitive, and as such their perfect one factorizations would be a natural generalization of the cyclic perfect one factorizations. The group must be a semidirect product of Z_{n_1} with Z_{n_2} for n_1 and n_2 relatively prime with $n_1 n_2 = 2n$ (see [8] 3.13 and [5] p. 146, 9.4.3) and, of course, must not be Z_{2n}. Perhaps this symmetry group might prove a basis for generating new perfect one factorizations.

We now summarize the contents of the paper. In section 2 we give our notation and show that K_{2n} admits a cyclic perfect one factorization then n is prime. In section 3 we relate the cyclic perfect one factorizations of K_{2n} to the starter induced perfect one factorizations of K_{n-1}. Also in section 3 we summarize some basic facts about starter induced perfect one factorizations which are needed for the results about cyclic perfect one factorizations. These facts may have interest in their own right and are given in 3.2. For ease of reference we state the main result of the paper below.

Theorem 1.1 The following are true.

(a) There exists a cyclic perfect one factorization of K_{2n} if and only if n is prime.

(b) For each cyclic perfect one factorization P of K_{2n} one can construct a starter induced perfect one factorization $r(P)$ of K_{n+1} (use 3.8 and 3.4).

(c) Let n be prime. r defines a one to one onto correspondence between the isomorphism classes of cyclic perfect one factorizations of K_{2n} and the isomorphism classes of starter induced perfect one factorizations on K_{n+1}.

(d) We have (see 2.1 (g))

$$I_P^F = Z_2 \oplus I_{r(P)}^F.$$

Proof See 2.4, 3.15 and 3.16. □

2. Orders That Admit a Cyclic Perfect One Factorization

In this section we will show that K_{2n} admits a cyclic perfect one factorization if and only if n is prime. We will use the same notation as [7] which we briefly summarize in 2.1 and 2.2 below.

2.1 Notation

(a) $X = \{1,2, \cdots ,2n\}$

(b) If Y is a set then $S_Y = \{\tau \mid \tau$ is a one to one onto function from Y to $Y\}$. $\hat{1}$ is used to denote the identity function.

(c) Define

$$\iota : S_X \longrightarrow S_{S_X}$$

by

$$\iota(\tau)\sigma = \tau \sigma \tau^{-1}.$$

(d) We use σ to denote a one factor - an element of S_X that is conjugate to (1,2) (3,4) ... $(2n-1,2n)$.

(e) Let σ_1 and σ_2 be two one factors. $(\sigma_1,\sigma_2) \in H$ if and only if σ_1 and σ_2 form a Hamiltonian pair; that is, σ_1 together with σ_2 form a Hamiltonian circuit (see [7] 2.3 for equivalent conditions).

(f) We use P to denote a perfect one factorization of K_{2n} - P is a collection of $2n-1$ one factors any two of which form a Hamiltonian pair.

(g) I_P^F is the full symmetry group of P ($I_P^F = \{\tau \in S_X \mid \iota(\tau)P = P\}$). I_P is any subgroup of I_P^F. τ will be used to denote an element of I_P.

(h) Let $\tau \in S_Y$. Then $F_\tau = \{y \in Y \mid \tau(y) = y\}$ is the set of fixed points of τ.

(i) If $y \in Y$ and G is a subgroup of S_Y then

$$G_y = \{\tau \in G \mid \tau(y) = y\}$$

is the stabilizer (or isotropy) subgroup of y.

(j) $G \simeq G'$ if G is isomorphic to G'. However $G \subset G'$ will mean either G is a subgroup of G' or G is isomorphic to a subgroup of G' depending on the context.

(k) If $S \subset G$ a group then $<S>$ denotes the subgroup generated by S.

(l) $C(\tau) = \{\tau' \mid \tau\tau' = \tau'\tau\}$ is the centralizer of τ.

(m) $[G]_\psi G'$ denotes the semidirect product of G with G'. G is the normal subgroup and $\psi \in \mathrm{Hom}(G',\mathrm{Aut}(G))$. $[G]G'$ is used if ψ is clear from the context.

2.2 Definition

(a) A perfect one factorization is called non classical if it is neither GK_{p+1} nor GA_{2p}, the two known classes of perfect one factorizations on K_{p+1} and K_{2p} respectively where p is prime.

(b) P is called a cyclic perfect one factorization if I_P^F contains an element of order $2n$.

(c) An element τ of I_P^F is called reductive if $o(\tau) = 2$, F_τ is empty and $|F_{\iota(\tau)}| > 1$ (see 3.4 or [7] 3.12 for more information).

Note that 2.2 is apparently weaker than the Hartman Rosa definition. We will see in 2.6 that they are in fact equivalent. We start with some lemmas.

Lemma 2.3 Let $\tau \in I_P$ and $o(\tau) = 2n$. Then n is odd, $|F_{\iota(\tau^2)}| \geq n-1$ and τ^n is a reductive element.

Proof First we show that n is odd. τ^n is an element of I_P^F which has order 2. If τ^n had fixed points then let $I_P = <\tau>$ and use [7] 5.9 to find that $|I_P|$ divides $2n-2$ which contradicts $|I_P| = 2n$. Thus τ^n has no fixed points, and [7] 3.9 shows n is odd. Next we use [7] 3.15 to conclude that

$$\iota(\tau^2) = y_1 y_2 \cdots y_m$$

where y_i are disjoint n_i cycles and all the n_i are distinct. Since τ^n commutes with τ^2 we have that $\iota(\tau^n)^{-1} y_i \iota(\tau^n) = y_{j_i}$ for some j_i. Because the n_i are distinct j_i must be i. This means that $\iota(\tau^n)\{\sigma_1,\sigma_2, \cdots ,\sigma_{n_i}\} = \{\sigma_1,\sigma_2, \cdots ,\sigma_{n_i}\}$ where $\iota(y_i) = (\sigma_1,\sigma_2, \ldots ,\sigma_{n_i})$ and $\sigma_1,\sigma_2, \cdots ,\sigma_{n_i} \in P$. Now the order of τ^2 is n which is odd so that each n_i is odd. Since the order of τ^n is two there must be a $\sigma \in \{\sigma_1,\sigma_2, \cdots ,\sigma_{n_i}\}$ such that $\iota(\tau^n)\sigma = \sigma$. This together with $\iota(\tau^n)^{-1} y_i \iota(\tau^n) = y_i$

means that $\iota(\tau^n)\sigma = \sigma$ for all $\sigma \in \{\sigma_1,\sigma_2,...,\sigma_{n_i}\}$. Thus

$$P - F_{\iota(\tau^2)} \subset F_{\iota(\tau^n)}. \qquad *$$

It is now clear that τ^n is reductive since if not $|F_{\iota(\tau^2)}| = 1$ (see [7] 3.12). * would then say that $P = F_{\iota(\tau^2)}$ giving $\tau^2 = \hat{1}$ (use [7] 3.2) which is not possible since $o(\tau) = 2n$. Since τ^n is reductive [7] 3.13 gives us that $|F_{\iota(\tau^n)}| = n$ so that $|F_{\iota(\tau^2)}| \geq n-1$. This finishes the proof. □

We are now ready to prove the main result of this section.

Theorem 2.4 K_{2n} admits a cyclic perfect one factorization if and only if n is prime.

Proof If n is prime the standard perfect one factorization on K_{2n} is cyclic. This result is stated in [11] p. 49. It may be seen from the results of [3] which state that I_P^F has an element of order n and a central element of order 2. The product of these two elements is the desired element of order $2n$.

To finish we assume that K_{2n} admits a cyclic perfect one factorization. Let τ be an element of I_P^F which has order $2n$. 2.3 tells us that $\iota(\tau^2)$ has at least $n-1$ fixed points. Since $n > 2$ we have that $\tau^2 \in C(\sigma) \cap C(\sigma')$ for some σ, $\sigma' \in P$. [7] 2.9 says that all the disjoint cycles of τ^2 have the same order. This order must be $o(\tau^2) = n$. Thus there are two n cycles in the disjoint cycle decomposition of τ^2. After a relabelling of vertices we may assume that

$$\tau^2 = (1,3,5,...,2n-1)(2n,2n-2,2n-4, \cdots ,2).$$

Define τ_j by $\tau_j(i) = -i+2j+1 \bmod 2n$. One can verify that τ_j are the only elements of order two in S_X which commute with τ^2. 2.3 says that there are at least $n-1$ elements of P that commute with τ^2. Thus all of the τ_j except perhaps one must be in P. Observe that $\tau_k\tau_j(i) = i+2(k-j) \bmod 2n$, and thus the order of $\tau_k\tau_j$ is the order of $k-j \bmod n$. If τ_k and τ_j are both in P and $k \neq j \bmod n$ then [7] 2.3 says that $k-j$ must have order n in Z_n. If we fix a j so that τ_j is in P we find $\{k-j \mid \tau_k \in P \text{ and } k \neq j \bmod n\}$ has $n-2$ distinct elements mod n. Thus n must be prime since if n is composite there can not be $n-2$ distinct units mod n. □

Corollary 2.5 If $\tau \in I_P^F$ and $o(\tau) = 2n$ then τ is a $2n$ cycle.

Proof Use the same argument given in 2.4 to express τ^2 as

$$(1,3,5,...,2n-1)(2,4,6, \cdots ,2n).$$

Since τ^n is an element of order two that commutes with τ^2 we have that $\tau^n(1)$ must be even (if $\tau^n(1)$ were odd then $\tau^n\{1,3,...,2n-1\} = \{1,3,...,2n-1\}$ and thus τ^n would have to have a fixed point). Thus by cyclic relabelling of the even vertices we may insure that $\tau^n(1) = 2$. Since τ^n commute with τ^2 we then find that

$$\tau^n = (1,2)(3,4) \cdots (2n-1,2n).$$

We then find that (using that n is odd)

$$\tau^{n+2} = (1,4,5,8,9,12, \cdots ,2(n-1),2n-1,2,3,6,7, \cdots ,2n).$$

Thus τ^{n+2} is a $2n$ cycle which means that τ itself must be a $2n$ cycle since raising a cycle to a power can not increase its length. □

We have now established that cyclic perfect one factorizations can only arise when n is prime. In the next section we set up the correspondence with starter induced perfect one factorizations.

3. Cyclic Perfect One Factorizations of K_{2n} And Starter Perfect One Factorizations of K_{n+1}

In this section we will show that there is a one to one, onto correspondence between cyclic perfect one factorizations of K_{2n} and starter induced perfect one factorizations of K_{n+1}. We will also show that there is a simple relation between the symmetry groups of the two one factorizations. First we review the definition and basic properties of starter induced perfect one factorizations.

3.1 Definition

P is called a starter induced perfect one factorization of K_{2n} if there is a subgroup G of I_P^F such that the following are true.

(a) $|G| = 2n-1$.

(b) There is an $i \in X$ such that $G_i = \{\hat{1}\}$.

To relate this definition to a perfect one factorization generated by a starter as described in [11] we observe that condition 3.1 means that $|\{g(i) \mid g \in G\}| = |G| = 2n-1$. Thus X may be identified with $G \cup \{\infty\}$ where ∞ is identified with the only point not in $\{g(i) \mid g \in G\}$. Then the starter that generates P is the one factor that connects ∞ to $\hat{1}$. The following is the basic result concerning starter induced perfect one factorizations.

Theorem 3.2 If P is starter induced then the following are true.

(a) There is a unique $i_o \in X$ such that $\tau(i_o) = i_o$ for all $\tau \in I_P^F$. i_o is called the ideal point.

(b) There is only one subgroup G of I_P^F that satisfies 3.1(a) and (b). This subgroup is called the starter group.

(c) G is a normal subgroup of I_P^F, and for any $i \in X$ with $i \neq i_o$ we have

$$I_P^F = [G]_\psi(I_P^F)_i$$

where G is the starter group. $\psi(\tau)g = \tau g \tau^{-1}$ (see 2.1 (m)). $(I_P^F)_i$ acts on G by fixed point free automorphisms, and thus $|(I_P^F)_i|$ divides $2(n-1)$.

(d) If P is non classical (see 2.2 (a)) then $|(I_P^F)_i|$ is odd.

(e) $(I_P^F)_i$ is isomorphic to a subgroup of $\mathrm{Aut}(G)$ so that

$$I_P^F \subset [G]_I(\mathrm{Aut}(G))$$

where $I : \mathrm{Aut}(G) \longrightarrow \mathrm{Aut}(G)$ is the identity map.

Proof The proof of this result is essentially contained in [8]. If I_P^F is G or if P is classical (use [3]) all of the statements in 3.2 are trivial. If $I_P^F \neq G$ then I_P^F is not small since $|I_P^F| \geq 2|G| > 2n$. Then 4.1 of [8] gives the whole result except the uniqueness of G. The uniqueness of G comes from the following facts. [8] 3.1 says that I_P^F is solvable. Since $2n-1$ and $|I_P^F|/(2n-1)$ are relatively prime ([8] 3.4), the generalized Sylow theorem for solvable groups (see [5] 9.3.1 p.141) says that any subgroup of I_P^F of order $2n-1$ is conjugate to G. Since G is normal ([8] 4.1) this means that G is the unique subgroup of order $2n-1$ of I_P^F. □

We are now prepared to consider the relationship between cyclic perfect one factorizations and starter induced perfect one factorizations. Since it will be convenient to deal with vertices other than the integers $1, \cdots, 2n$, we start by expanding our notation to accomodate this need. We also make explicit the distinction between a perfect one factorization and an equivalence class of isomorphic perfect one factorizations. We need to do this since our results concern isomorphism classes while our constructions yield members of these classes.

3.3 Notation

(a) K_V is the complete graph on the vertex set V.

(b) $\mathcal{P} = (V,P)$ is a perfect one factorization of K_V if

  (1) $P \subset S_V$ and $|P| = |V|-1$

  (2) $\sigma \in P$ implies that $o(\sigma) = 2$ and $F_\sigma = \emptyset$

  (3) $\sigma_1, \sigma_2 \in H$; that is, $\sigma_1\sigma_2$ is the product of two disjoint n cycles.

(c) Let $\mathcal{P} = (V,P)$ and $\mathcal{P}' = (V',P')$ be perfect one factorizations. ρ is an isomorphism from $\mathcal{P}$ to $\mathcal{P}'$ if ρ is a set isomorphism from V to V' and $\rho P \rho^{-1} = \{\rho\sigma\rho^{-1} \mid \sigma \in P\} = P'$. $\mathcal{P} \simeq \mathcal{P}'$ means that $\mathcal{P}$ is isomorphic to $\mathcal{P}$.

(d) $[\mathcal{P}] = \{\mathcal{P}' \mid \mathcal{P}'$ is isomorphic to $\mathcal{P}\}$. $[\mathcal{P}]$ is said to be a perfect one factorization of $K_{|V|}$.

(e) $I_{\mathcal{P}}^F$ denotes the full symmetry group of $\mathcal{P}$. $I_{\mathcal{P}}$ denotes any subgroup of $I_{\mathcal{P}}^F$.

We phrase the reduction construction given in [7] in terms of this notation. The proof that the objects defined below are well defined and that $r_\tau(\mathcal{P})$ is a perfect one factorization is contained in [7] 3.13(c).

Definition-Lemma 3.4

Let $\mathcal{P} = (V,P)$ be a perfect one factorization, and let τ be a reductive element of $I_{\mathcal{P}}^F$ (see 2.2 (c)). Define a new perfect one factorization $r_\tau(\mathcal{P}) = (r_\tau(V), r_\tau(P))$ as follows:
(a) $r_\tau(V) = \{\infty\} \cup \{x_i \mid$ where $\tau = x_1 x_2 \cdots x_n$ is the disjoint cycle decomposition of $\tau\}$. Here ∞ represents an 'ideal' point which in this case may be any point which is not x_i for any i.
(b) $r_\tau(P) = \{r_\tau(\sigma) \mid \sigma \in C(\tau)\}$ where $r_\tau(\sigma)(x_i) = \sigma x_i \sigma^{-1}$ when $\sigma x_i \sigma^{-1} \neq x_i$ and $r_\tau(\sigma)(x_i) = \infty$ as well as $r_\tau(\sigma)(\infty) = x_i$ when $\sigma x_i \sigma^{-1} = x_i$.

Definition-Lemma 3.5 Let $\mathcal{P}$ be a perfect one factorization and let τ be a central reductive element of $I_{\mathcal{P}}^F$. Define

$$\pi_\tau : I_{\mathcal{P}}^F \to I_{r_\tau(\mathcal{P})}^F$$

by

$$\pi_\tau(\tau')(x_i) = \tau' x_i (\tau')^{-1}$$

$$\pi_\tau(\tau')(\infty) = \infty$$

π_τ is well defined and a group homomorphism.

Proof First observe that since τ is in the center of $I^F_{\mathcal{P}}$, $\tau' x_i (\tau')^{-1}$ is again a cycle in the disjoint cycle decomposition of τ. Thus $\pi_\tau(\tau')$ is a permutation on the set $r_\tau(V)$. It follows that $\pi_\tau(\tau')$ is a symmetry of $r_\tau(\mathcal{P})$ from the relation

$$\pi_\tau(\tau')^{-1} r_\tau(\sigma) \pi_\tau(\tau') = r_\tau((\tau')^{-1} \sigma \tau').$$

π_τ can be verified to be a group homomorphism by a straightforward calculation. □

3.6 Definition

(a) Let $CYC = \{\mathcal{P} \mid \mathcal{P}$ is a cyclic perfect one factorization $\}$, let $[CYC] = \{[\mathcal{P}] \mid \mathcal{P} \in CYC\}$, and let $[CYC]_{2n} = \{[\mathcal{P}] \mid \mathcal{P} \in CYC$ and $\mathcal{P} = (V,P)$ with $|V| = 2n\}$.

(b) Let $START = \{\mathcal{P} \mid \mathcal{P}$ is a starter induced perfect one factorization $\}$, let $[START] = \{[\mathcal{P}] \mid \mathcal{P} \in START\}$, and let $[START]_{2n} = \{[\mathcal{P}] \mid \mathcal{P} \in START$ and $\mathcal{P} = (V,P)$ with $|V| = 2n\}$.

We need the following lemma before we give the reduction construction that connects the cyclic perfect one factorizations to the starter induced perfect one factorizations.

Lemma 3.7 Let $\mathcal{P}$ be a non classical cyclic perfect one factorization. Then $I^F_{\mathcal{P}}$ has a unique element of order two which is thus central.

Proof If $I^F_{\mathcal{P}} = Z_{2n}$ then the result is clear. Otherwise $I^F_{\mathcal{P}}$ is not small and we may apply [7] 4.1(a) which says $I^F_{\mathcal{P}} = [Z_{2n}]H$ where H has odd order. Since n is also odd the Sylow 2 subgroup of $I^F_{\mathcal{P}}$ is isomorphic to Z_2. Let Z_2 be the unique subgroup of order two in Z_{2n}. Since Z_{2n} is normal in $I^F_{\mathcal{P}}$ any conjugate of Z_2 must be in Z_{2n} and so must be Z_2. This means Z_2 is normal and so the Sylow theorem says that it is the unique subgroup of order 2. □

Definition-Lemma 3.8 Let $|V|/2 = p$ be prime. If $\mathcal{P}$ is a perfect one factorization on K_V let $\tau \in I^F_{\mathcal{P}}$ be a reductive element of order two in $I^F_{\mathcal{P}}$. Define

$$r: [CYC] \to [START]$$

by

$$r([\mathcal{P}]) = [r_\tau(\mathcal{P})].$$

r is well defined.

Proof First consider the case when $\mathcal{P} = GA_{2p}$. Then $I^F_{\mathcal{P}}$ has $2p$ elements of order two, p of which are reductive (see [3]). One can verify by direct computation that $[r_\tau(\mathcal{P})] = GK_{p+1}$ when τ is any of the p reductive elements. Thus we need only consider the case when $\mathcal{P}$ is non classical. Then 2.3 and 3.7 gives us that $I^F_{\mathcal{P}}$ has a unique reductive element τ. Thus we must show two things to verify that r is well defined. We must show that $r_\tau(\mathcal{P}) \in START$ and that $r_\tau(\mathcal{P}) \simeq r_\tau(\mathcal{P}')$ if $\mathcal{P} \simeq \mathcal{P}'$. Let $\hat{\tau} \in I^F_{\mathcal{P}}$ and $o(\hat{\tau}) = 2n$. First we claim that $G = \pi_\tau \langle \hat{\tau}^2 \rangle$ is the starter group for $r_\tau(\mathcal{P})$. ∞ plays the role of i_o in 4.4(b). To verify that $\pi_\tau(\hat{\tau}^2)$ is an n cycle (and thus G satisfies the requirements of 3.1) one need only compute $(\hat{\tau})^2$ and x_i explicitly after performing a change of vertices so that $\hat{\tau}$ is in the form $(1,2,3,...,2n)$. Next we let ρ be an isomorphism from $\mathcal{P} = (V,P)$ to $\mathcal{P}' = (V',P')$. If τ and τ' are the elements of order two of $I^F_{\mathcal{P}'}$ respectively, then $\rho\tau\rho^{-1} = \tau'$ since $\rho\tau\rho^{-1}$ is an element of order two of $I^F_{\mathcal{P}'}$ and $I^F_{\mathcal{P}'}$ has only one element of order two. Thus $\rho x_i \rho^{-1}$ will be the disjoint cycles in the disjoint cycle decomposition of τ'. This enables us to define a map

$$\hat{\rho} : r_\tau(V) \rightarrow r_{\tau'}(V')$$

by

$$\hat{\rho}(x_i) = \rho x_i \rho^{-1} \text{ and } \hat{\rho}(\infty) = \infty$$

It can then be verified that $\hat{\rho} r_\tau(\sigma)\rho^{-1} = r_{\tau'}(\rho\sigma\rho^{-1})$, and thus $\hat{\rho} r_\tau(P)\rho^{-1} = P'$ as desired. □

Our next objective is to produce an inverse for r. We will call this function s, and we will use the function $\hat{s}$ defined below to construct s.

3.9 Definition

Let n be prime and let $\mathcal{P} \in START$ with $\mathcal{P} = (V,P)$. Let G be the starter group for $\mathcal{P}$ and i_o be the ideal point. Let i be any element of V with $i \neq i_o$. Define $\kappa_i : G \rightarrow V-\{i_o\}$ by $\kappa_i(g) = g(i)$. κ_i is one to one and onto. Now define

$$\hat{s}_i(\mathcal{P}) = (\hat{s}(V),\hat{s}_i(P))$$

where
(a) $\hat{s}(V) = G \oplus Z_2$ (here we use $Z_2 = \{-1,1\}$ with multiplication as the group operation) and
(b) $\hat{s}_i(P) = \{\sigma_g \mid g \in G \text{ and } g \neq \hat{1}\} \cup \{\hat{\sigma} \mid \sigma \in P\}$.
Here σ_g is defined by

$$\sigma_g((g',\epsilon)) = (g'g^\epsilon,-\epsilon)$$

for $g \in G$ and $\epsilon \in Z_2$.
$\hat{\sigma}$ is defined by

$$\hat{\sigma}((g,\epsilon)) = (\kappa_i^{-1}\sigma\kappa_i(g),\epsilon) \quad \text{if} \quad \sigma\kappa_i(g) \neq i_o$$
$$\hat{\sigma}((g,\epsilon)) = (g,-\epsilon) \quad \text{if} \quad \sigma\kappa_i(g) = i_o$$

The subscript i may be omitted if there is no ambiguity concerning which i is used to define κ_i.

Notice that the one factorization given in 3.9 does not depend on G or i_o because of 3.2. We know give a number of lemmas which will gather together enough properties of $\hat{s}$ to show that it may be used to define an inverse for r. The next lemma shows that i plays no essential role in the definition of $\hat{s}_i(\mathcal{P})$.

Lemma 3.10

$$\hat{s}_i(\mathcal{P}) \simeq \hat{s}_j(\mathcal{P}) \text{ for all } i \text{ and } j.$$

Proof Since $i,j \in X-\{i_o\}$ there is a $g \in G$ so that $g(i) = j$. Define

$$\rho\colon \hat{s}(V) \to \hat{s}(V) \quad \text{by} \quad \rho((g',\epsilon)) = (g'g^{-1},\epsilon).$$

Then a calculation yields

$$\rho\sigma_{g'}\rho^{-1} = \sigma_{gg'g^{-1}}.$$

Also if $\sigma \in P$, $\sigma_1 = \hat{\sigma} \in \hat{s}_i(P)$ and $\sigma_2 = \hat{\sigma} \in \hat{s}_j(P)$ then

$$\rho\sigma_1\rho^{-1} = \sigma_2.$$

This shows that $\rho(\hat{s}_i(P))\rho^{-1} = \hat{s}_j(P)$ as desired. □

Lemma 3.11

$\hat{s}(P)$ is a perfect one factorization.

Proof It is a straightforward computation to verify that σ_g and $\hat{\sigma}$ are one factors (that they are elements of order two with no fixed points). We must only consider which pairs of one factors are Hamiltonian since $|\hat{s}(P)| = |\hat{s}(V)|-1$. First observe that $G \simeq Z_n$ since n is prime. Because G is abelian we have

$$\sigma_g\sigma_{g'}(g'',\epsilon) = (g''(g'g^{-1})^{\epsilon},\epsilon).$$

Thus every cycle in the disjoint cycle decomposition of $\sigma_g\sigma_g{}'$ has order $o((g'g^{-1})^{\epsilon})$. If $g \neq g'$, this order is n since n is prime. Thus $\sigma_g\sigma_{g'}$ is a product of two disjoint n cycles and so $(\sigma_g,\sigma_{g'}) \in H$ for $g \neq g'$. Next we consider $\hat{\sigma}_1$ and $\hat{\sigma}_2$ for $\sigma_1,\sigma_2 \in P$ and $\sigma_1 \neq \sigma_2$. We have $\hat{\sigma}_1\hat{\sigma}_2(g,\epsilon) = (\kappa^{-1}\sigma_1\sigma_2\kappa(g),\epsilon)$ if $\sigma_2\kappa(g) \neq i$ and $\sigma_1\sigma_2\kappa(g) \neq i$. Using [7] 2.7 we may after a change of vertex labels assume that $i = 1$ and that $\sigma_1 = (1,2)(3,4)\cdots(2n-1,2n)$ and $\sigma_2 = (2,3)(4,5)\cdots(2n,1)$. Thus $\sigma_1\sigma_2 = (2n-1,2n-3,2n-5,\cdots,3,1)(2,4,6,\cdots,2n)$. If we label $(g,1)$ by $\kappa(g)$ and $(g,-1)$ by $\kappa(g)^-$ then we have

$$\hat{\sigma}_1\hat{\sigma}_2 = (2n-1,2n-3,\ldots,1,2^-,4^-,\ldots,2n^-)(2n-1^-,2n-3^-,\ldots,1^-,2,4,\ldots,2n).$$

Thus $(\hat{\sigma}_1,\hat{\sigma}_2) \in H$ since $\hat{\sigma}_1\hat{\sigma}_2$ is the product of two disjoint cycles. To finish we consider σ_g and $\hat{\sigma}$. We find that

$$(\sigma_g\hat{\sigma})^2(g',\epsilon) = (\kappa^{-1}(g^{-\epsilon}\sigma g^{\epsilon})\sigma\kappa(g'),\epsilon)$$

if $\sigma\kappa(g') \neq i$ and $\sigma g^{\epsilon}\sigma\kappa(g') \neq i$ (use $\kappa(g_1g_2) = g_1(\kappa(g_2))$). Now $g^{\epsilon} \neq \hat{1}$ so that $g^{-\epsilon}\sigma g^{\epsilon} \neq \sigma$. Thus $(\sigma,g^{-\epsilon}\sigma g^{\epsilon}) \in H$ and a similar argument to the one given above with σ_2 replaced by σ and σ_1 replaced by $g^{-\epsilon}\sigma g^{\epsilon}$ can be used to show that $(\sigma_g\hat{\sigma})^2$ is the product of two disjoint cycles. Thus $\sigma_g\hat{\sigma}$ is a product of two or fewer disjoint cycles. The [7] 2.3(c) shows that $(\sigma_g,\hat{\sigma}) \in H$ as desired. This completes the proof that $\hat{s}(P)$ is a perfect one factorization. □

Lemma 3.12

$\hat{s}(P) \in CYC.$

Proof Observe that $G \oplus Z_2 \simeq Z_{2n}$ since n is prime. Now $G \oplus Z_2$ acts on $\hat{s}(V) = G \oplus Z_2$ by left translation ($T_h h' = hh'$ for all $h,h' \in G \oplus Z_2$). We claim that $T_h \in I_P^F$ for all $h \in G \oplus Z_2$. This follows from the relations below:

(a) $T_{g_1}\sigma_{g_2}(T_{g_1})^{-1} = \sigma_{g_2}$ for all $g_1,g_2 \in G$

(b) $T_g\hat{\sigma}_1(T_g)sup-1 = \hat{\sigma}_2$ where $\sigma_2 = g\sigma_1 g^{-1}$ for all $g \in G$ and $\sigma_1 \in P$

(c) $T_{-1}\sigma_g(T_{-1})^{-1} = \sigma_{g^{-1}}$ for all $g \in G$ where -1 means $(\hat{1},-1) \in H$
(d) $T_{-1}\hat{\sigma}(T_{-1})^{-1} = \hat{\sigma}$ for all $\sigma \in P$.

Lemma 3.13

$$\text{If } [\mathcal{P}] = [\mathcal{P}'] \text{ then } [\hat{s}(\mathcal{P})] = [\hat{s}(\mathcal{P}')].$$

Proof Let ρ be an isomorphism from $\mathcal{P} = (V,P)$ to $\mathcal{P}' = (V',P')$. Let $\mathcal{P}$ (respectively $\mathcal{P}'$) have ideal point i_o (respectively i_o') and starter group G (respectively G'). Since ρ is an isomorphism $\rho(i_o)$ will be an ideal point for $\mathcal{P}'$, and $\rho G(\rho')^{-1}$ will be a starter group for $\mathcal{P}'$. Thus $\rho(i_o) = i_{o'}$ and $\rho G(\rho)^{-1} = G'$ by 3.2. We may assume that $\rho(i) = i'$ by 3.10. Define

$$\hat{\rho} : \hat{s}(V) \rightarrow \hat{s}(V')$$

by

$$\hat{\rho}((g,\epsilon)) = (\rho g \rho^{-1},\epsilon).$$

Observe by direct computation that

$$\hat{\rho}\sigma_g\hat{\rho}^{-1} = \sigma_{\rho g \rho^{-1}},$$

and that

$$\hat{\rho}\hat{\sigma}\hat{\rho}^{-1} = \hat{\sigma}', \text{ where } \sigma' = \rho\sigma\rho^{-1}.$$

This shows that $\hat{\rho}$ is the desired isomorphism. □

Now we are in the position to define the inverse for r.

Definition 3.14 Let $\mathcal{P}$ be a non classical perfect one factorization on K_V where $|V|-1$ is prime. Let $\mathcal{P} \in START$. Define

$$s:[START] \rightarrow [CYC]$$

by

$$s([\mathcal{P}]) = [\hat{s}(\mathcal{P})].$$

Now we come to the main result of the section.

Theorem 3.15 Let n be prime.

$$r : [CYC]_{2n} \rightarrow [START]_{(n+1)}$$

is a set isomorphism with s as its inverse.

Proof We must only verify that $rs = id$ and $sr = id$. First we show that $rs(\mathcal{P})$ is isomorphic to $\mathcal{P}$. Let $\mathcal{P} = (V,P)$, let i_o be the ideal point in V and let $i \neq i_o$ be in V. Define

$$\rho : r_\tau\hat{s}(V) \rightarrow V$$

by

$$\rho(((g,1),(g,-1))) = g(i)$$

and

$$\rho(\infty) = i_o.$$

Here τ is defined by $\tau((g,\epsilon)) = (g,-\epsilon)$. First observe that the only elements of $\hat{s}_i(P)$ that commute with τ are $\hat{\sigma}$ for $\sigma \in P$. One can verify by direct computation that

$$\rho r_\tau \hat{\sigma} \rho^{-1} = \sigma \text{ for every } \sigma \in P.$$

Thus ρ gives rise to the required isomorphism from $r_\tau \hat{s}_i(\mathcal{P})$ to $\mathcal{P}$. Next we must show that $\hat{s}_i r_\tau(\mathcal{P})$ is isomorphic to $\mathcal{P} = (V,P)$. Let $j \in V$ and let $\tau' \in I_P^F$ with $(\tau')^P = \tau$ (note that $o(\tau') = 2p$). Observe that the starter group G of $r_\tau(\mathcal{P})$ is $\{\pi_\tau(\tau')^{2k} \mid k \in Z\}$. We let $i = (j,\tau(j))$. Define

$$\rho : \hat{s}_i r_\tau(V) \rightarrow V$$

by

$$\rho((\pi_\tau(\tau')^{2k},\epsilon)) = (\tau')^{2k+(-\epsilon+1)p/2}(j).$$

One can show that $\rho\sigma_g\rho^{-1}$ commutes with $(\tau')^2$ and is not $(\tau')^n$ so that $\rho\sigma_g\rho^{-1}$ is in P for all $g \in G$. One can also verify that if $\sigma' = r_\tau(\sigma)$ then $\hat{\sigma}' = \rho\sigma\rho^{-1}$. This shows that $\rho P \rho^{-1} = \hat{s}_i r_\tau(P)$ and the theorem is now complete. □

We finish this section with a theorem that tells us the relation between the symmetry groups of the cyclic perfect one factorizations and those of the starter induced perfect one factorizations.

Theorem 3.16 Let $[\mathcal{P}] \in [CYC]$. Then

$$I_{[\mathcal{P}]}^F = I_{r([\mathcal{P}])}^F \oplus Z_2.$$

Proof First we establish some notation. Let $\tilde{\tau}_o$ be an element of order $2n$ of I_P^F. Let $\tilde{\tau} = (\tilde{\tau}_o)^n$, and $\tilde{\tau}_1 = (\tilde{\tau}_o)^2$. Let $\tilde{G}$ be the subgroup of I_P^F generated by $\tilde{\tau}_1$, and let $G = \pi(\tilde{G})$ where $\pi = \pi_{\tilde{\tau}} : I_P^F \rightarrow I_{r_\tau(P)}^F$ is the homomorphism defined in 3.5. Also use r to denote $r_{\tilde{\tau}}$. If v, $v' \in V$ we say $v \sim v'$ if v and v' are in the same $\tilde{G}$ orbit. Finally define f by

$$f : V \rightarrow r(V)$$

by

$$f(v) = (v,\tilde{\tau}(v)).$$

We now start the proof. First we show that $\ker(\pi) = \{\hat{1},\tilde{\tau}\} \simeq Z_2$. Suppose $\pi(\tau) = \hat{1}$. Then $\tau x_i \tau^{-1} = x_i$ for each i. Since x_i is a two cycle we have that τ^2 fixes the two points not fixed by x_i. As this is true for all i we have that τ^2 fixes every point in V, and so τ is either an element of order two or $\hat{1}$. 3.7 says $\tilde{\tau}$ is the only element of order two in I_P^F, and so $\ker(\pi) = \{\hat{1},\tilde{\tau}\}$ as claimed. To finish we need only produce a splitting α for π. Such a splitting is a homomorphism α from $I_{r(P)}^F$ to I_P^F in which $\pi\alpha$ is the identity on $I_{r(P)}^F$. Let $\tau \in I_{r(P)}^F$. By 2.2 we have $\tau(\infty) = \infty$. Thus $\tau(x_i) = x_j$ for some j. We now define $\alpha(\tau)$ as follows:

$$\alpha(\tau)(v) = v' \quad \text{if} \quad \tau(f(v)) = f(v') \text{ and } v \sim v'.$$

Using the specific representation of $\tilde{\tau}_o$ as the $2n$ cycle $(1,2,3,...,2n)$ we find that $f^{-1}(\tau(f(v)))$ is a set consisting of one odd number and one even number. Also one can verify that $v \sim v'$ if and only if v and v' have the same parity. Thus $\alpha(\tau)$ is well defined. Observe that $\pi\alpha(\tau)(x_i) = \alpha(\tau)x_i\alpha(\tau)^{-1}$. Let $x_i = f(v)$ for some $v \in V$. Then $x_i = (v, \tilde{\tau}(v))$ so that

$$\pi\alpha(\tau)(x_i) = (\alpha(\tau)(v), \alpha(\tau)(\tilde{\tau}(v))) = f(\alpha(\tau)(v)) = \tau(f(v)) = \tau(x_i).$$

This means that $\pi\alpha$ is the identity on $I^F_{r(P)}$. Next we show that α is a group homomorphism. We have $\alpha(\tau_2)(v) = v'$ if $\tau_2(f(v)) = f(v')$ with $v \sim v'$. Also $\alpha(\tau_1)(v') = v''$ if $\tau_1(f(v')) = f(v'')$ with $v' \sim v''$. Combining these two statements we find that $\tau_1\tau_2(f(v)) = f(v'')$ and $v \sim v''$ so that $\alpha(\tau_1\tau_2)(v) = v''$, and $\alpha(\tau_1\tau_2) = \alpha(\tau_1)\alpha(\tau_2)$ as desired. To finish we must show that $\alpha(\tau) \in I^F_P$ for each $\tau \in I^F_{r(P)}$. Let $C_2(\tilde{\tau}_1) = \{\sigma \mid o(\sigma) = 2 \text{ and } \sigma\tilde{\tau}_1\sigma^{-1} = \tilde{\tau}_1\}$. We have $|C_2(\tilde{\tau}_1)| = n$ from 2.3 and $P \cap C_2(\tilde{\tau}_1) = C_2(\tilde{\tau}_1) - \{\tilde{\tau}\}$. This second statement follows from the fact that $\iota(\tilde{\tau}_1)$ has order n (which is prime) and acts on P which has $2n-1$ elements so that $\iota(\tilde{\tau}_1)$ has $n-1$ fixed points. Thus $|P \cap C_2(\tilde{\tau}_1)| = n-1$. Now $\tilde{\tau} \in C_2(\tilde{\tau}_1)$ and $\tilde{\tau} \notin P$ showing that the second statement is correct. Let $P_1 = P \cap C_2(\tilde{\tau}_1)$ and let $P_2 = P \cap \{\sigma \mid \sigma\tilde{\tau}\sigma^{-1} = \tilde{\tau}\}$. Since $\tilde{\tau}$ is a reductive element of I^F_P we have that $|P_2| = n$. Thus there must be an element of P_2 which is not in P_1. Since P_2 is invariant under conjugation by $\tilde{\tau}_1$ we have $P_1 \cap P_2 = \emptyset$ and so $P = P_1 \cup P_2$. First let $\sigma \in P_1$. We will show that $\iota(\alpha(\tau))\sigma \in P$. Notice that $\iota(\alpha(\tau))\sigma$ is not $\tilde{\tau}$ for if so then

$$\sigma = \iota(\alpha(\tau))^{-1}\tilde{\tau} = \tilde{\tau} \notin P.$$

Thus we need only show that $\iota(\alpha(\tau))\sigma \in C_2(\tilde{\tau}_1)$. First we observe that

$$\tilde{\tau}_1\alpha(\tau)\tilde{\tau}_1^{-1} = \alpha(\pi(\tilde{\tau}_1)\tau\pi(\tilde{\tau}_1)^{-1})$$

Since α is a group homomorphism and $\alpha(\pi(\tilde{\tau}_1)) = \tilde{\tau}_1$. Let $\pi(\tilde{\tau}_1) = g$. We have

$$\tilde{\tau}_1\alpha(\tau)\sigma\alpha(\tau)^{-1}\tilde{\tau}_1^{-1} = \alpha(g\tau g^{-1})\sigma\alpha(g\tau^{-1}g^{-1})$$

since $\tilde{\tau}_1$ commutes with σ. Now since G is normal in $I^F_{r(P)}$ we have that $\tau^{-1}g\tau = g^{k+1}$ for some integer k. This gives us $g\tau g^{-1} = \tau g^k$ and

$$\iota(\tilde{\tau}_1)\iota(\alpha(\tau))(\sigma) = \alpha(\tau)\alpha(g)^k\sigma\alpha(g)^{-k}\alpha(\tau).$$

Again we use that $\alpha(g) = \tilde{\tau}_1$ and that $\tilde{\tau}_1$ commutes with σ so that the expression above becomes $\alpha(\tau)\sigma\alpha(\tau)^{-1}$ and so $\iota(\alpha(\tau))(\sigma)$ is invariant under conjugation by $\tilde{\tau}_1$. This means $\iota(\alpha(\tau))P_1 = P_1 \subset P$. To finish we must only consider the case in which $\sigma \in P_2$. In this case $r(\sigma) \in I^F_{r(P)}$ and so $\tau r(\sigma)\tau^{-1} = r(\sigma')$ for some $\sigma' \in P_2$. Notice that $(\sigma')^{-1}\alpha(\tau)\sigma\alpha(\tau)^{-1}$ commutes with $\hat{\tau}$ because each element in this product does. Thus we may apply π to this element to find that

$$\begin{aligned}\pi((\sigma')^{-1}\alpha(\tau)\sigma\alpha(\tau)^{-1}) &= \pi(\sigma')^{-1}\pi\alpha(\tau)\pi(\sigma)\pi\alpha(\tau)^{-1}\\ &= r(\sigma')^{-1}\tau r(\sigma)\tau^{-1}\\ &= \hat{1}.\end{aligned}$$

This means that in order to show that $(\sigma')^{-1}\alpha(\tau)\sigma\alpha(\tau)^{-1} = \hat{1}$ we must only show that $\sigma'(v)$ is in the same $\tilde{G}$ orbit as $\alpha(\tau)\sigma\alpha(\tau)^{-1}(v)$ for each $v \in V$. We claim that $\sigma'(v) \tilde{\sim} v$ unless $\sigma'(v) = \hat{\tau}(v)$. This is true since if v and $\sigma'(v)$ are in different orbits of $\tilde{G}$ then there is a $\sigma'' \in C_2(\tilde{\tau}_1)$ such that $\sigma''(v) = \sigma'(v)$ from 2.3. Thus $(\sigma'', \sigma') \notin H$ and so $\sigma'' = \hat{\tau}$ as desired. First assume that $\sigma'(v) \sim v$. Since $\alpha(\tau)$ preserves the G orbits, our conclusion is correct as long as $\sigma\alpha(\tau)^{-1}(v) \sim \alpha(\tau)^{-1}(v)$. If not then $\sigma\alpha(\tau)^{-1}(v) = \hat{\tau}\alpha(\tau)^{-1}(v) = \alpha(\tau)^{-1}(\hat{\tau}(v))$. This gives us that $\tau r(\sigma)\tau^{-1}$ fixes $f(v)$. Thus $r(\sigma')$ fixes $f(v)$ which implies that $\sigma'(v) = \hat{\tau}(v)$ since σ' has no fixed points. This contradicts our assumption that $\sigma'(v) \sim v$ since $\hat{\tau}$ interchanges the two orbits of $\tilde{G}$. The only other case we must consider is when $\sigma'(v) = \hat{\tau}(v)$. In this case the same calculation as given above shows that $\alpha(\tau)\sigma\alpha(\tau)^{-1}(v) = \hat{\tau}(v)$, and an analogous argument shows that $\alpha(\tau)\sigma\alpha(\tau)^{-1}(v) \sim \sigma'(v)$. This shows that $(\sigma')^{-1}\alpha(\tau)\sigma\alpha(\tau)^{-1} = \hat{1}$ and thus $\iota(\alpha(\tau))P_2 = P_2 \subset P$. This completes the proof that $\alpha(\tau) \in I_P^F$, and thus that α is a splitting. The theorem is now complete because this splitting shows that I_P^F is a semidirect product of Z_2 with $I_{r(P)}^F$. Since Z_2 is central in I_P^F this semidirect product must be a direct product. □

References

[1] B. Anderson. "Finite Topologies and Hamiltonian Paths", *J. Combin. Theory Ser. B,* 14 (1973), 87 - 93.

[2] B. Anderson. "Some Perfect 1-Factorizations", *Proc. of the 7th S.E. Conf. Combinatorics, Graph Theory, and Computing,* (1976) 79 -91.

[3] B. Anderson."Symmetry Groups of Some Perfect 1-Factorizations of Complete Graphs", *Discrete Math.,* 18, (1977), 227 - 234.

[4] B. Anderson and D. Morse. "Some Observations on Starters", *Proc. 5th S.E. Conf. Combinatorics, Graph Theory, and Computing,* (1974) 91 - 108.

[5] M. Hall. *The Theory of Groups,* Macmillan, New York, (1959).

[6] A. Hartman and A. Rosa. "Cyclic One Factorizations of the Complete Graph", *Europ. J. Combinat.,* 6, (1985) , 45 - 48.

[7] E. Ihrig. "Symmetry Groups Related to the Construction of Perfect One Factorizations of K_{2n}", *J. Combin. Theory Ser. B,* 40, (1986), 121 - 151.

[8] E. Ihrig. "The Structure of Symmetry Groups of Perfect One Factorizations of K_{2n} ", to appear.

[9] A. Kotzig. "Hamiltonian Graphs and Hamiltonian Circuits", in *Proc. Sympos. Smolenice 1963,* Nakl. CSAV, Praha, (1964), 63 - 82.

[10] N. Korovina. "Sistemy par cikliceskogo tipa", *Mat. Zametki* 28, (1980), 271- 278.

[11] E. Mendelsohn and A. Rosa. "One Factorizations of the Complete Graph - A Survey", *J. Graph Theory,* 9, (1985), 43 - 65.

[12] R. Mullin and E. Nemeth. "An Existence Theorem for Room Squares", *Canad. Math. Bull.* 12, (1969), 493 - 497.

[13] E. Seah and D. Stinson. "A Perfect One Factorization of K_{36} ", to appear.

Annals of Discrete Mathematics 34 (1987) 273–286
© Elsevier Science Publishers B.V. (North-Holland)

On edge but not vertex transitive regular graphs

A.V. Ivanov

Institute for System Studies
Academy of Sciences of the USSR
9 Prospect 60 Let Oktyabrya
117312 Moscow, USSR

TO ALEX ROSA ON HIS FIFTIETH BIRTHDAY

1. Main definitions and survey of results

Only nondirected graphs without loops and multiple edges are considered in this paper. A graph is called vertex (resp. edge) transitive if its automorphism group acts transitively on the vertices (resp. edges). A graph is called regular of valency d , if the valency of each vertex is d. Clearly, every vertex transitive graph is regular. Let us consider at greater length the relationship between these two sorts of symmetry.

It is easy to give many examples of vertex but not edge transitive graphs. The simplest infinite family of such graphs is described in [12]. Also it is not difficult to construct examples of edge but not vertex transitive graphs. For that we may take any complete bipartite graph $K_{m,n}$ $(m \neq n)$. But these graphs are not regular.

If requires much more effort for the construction of analogous examples of regular graphs. Examples of such graphs were first constructed by J. Folkman[7]. Using the terminology from [11], we shall call the graph G semisymmetric if G is regular and its automorphism group acts transitively on edges but intransitively on vertices.

Every semisymmetric graph is bipartite. This fact follows from a theorem proved by E. Dauber (see [8]). A number of other properties and some infinite families of semisymmetric graphs are described in [7]. We shall formulate sufficient conditions for the existence or nonexistence of semisymmetric graphs in the following.

Theorem 1 [7]: Let v be a positive integer. No semisymmetric graph with v vertices exists if v satisfies any of the following conditions:

1a) $v \equiv 1(\mathrm{mod}\ 2)$;
2a) $v \equiv 2p$ or $2p^2$, where p is a prime number;
3a) $v < 30$ and $v \equiv 0(\mathrm{mod}\ 4)$;
4a) $v < 20$.

A semisymmetric graph with v vertices exists if v satisfies any of the following conditions:

1b) $v \equiv 0(\mathrm{mod}\ 2p^3)$, where p is an odd prime number;
2b) $v \equiv 0(\mathrm{mod}\ 2pq)$, where p and q are odd prime numbers and $q \equiv 1(\mathrm{mod}\ p)$;
3b) $v \equiv 0(\mathrm{mod}\ 2pq^2)$, where p and q are prime numbers, q is odd and

$q \equiv -1 (\mathrm{mod}\ p)$;

4b) $v \geq 20$ and $v \equiv 0 (\mathrm{mod}\ 4)$.

J. Folkman[7] proposed a number of problems concerning semisymmetric graphs, some of which have been solved.

A semisymmetric graph with 54 vertices of valency 3 has been constructed in [3]. It provides an affirmative answer to problem (4.5) of [7] on the existence of semisymmetric graphs of valency d, where d is a prime number.

A semisymmetric graph with 70 vertices of valency 15 has been constructed in [4]. So problem (4.3) of [7] on the existence of semisymmetric graphs with $v = 2pq$ vertices, where p and q are odd prime numbers, $p < q$ and $q \equiv 1 (\mathrm{mod}\ p)$, also has an affirmative answer.

The answer to problem (4.4) of [7] on the existence of semisymmetric graphs of valency $d \geq \frac{v}{4}$ is contained in the papers [4] and [12]. The parameters of the such simplest graph are $(v,d) = (20,6)$.

The answer to problem (4.6) of [7] on the existence of semisymmetric graphs with v vertices of valency d, where d and $\frac{v}{2}$ are coprime, is contained in [11]. The first graph of the infinite family described in [11] has the parameters $(v,d) = (112,15)$.

Let us also note the procedure for construction of coverings (see [2]), allowing for every semisymmetric graph G of valency d to generate an infinite family of semisymmetric graphs of the same valency. This is the reason for the investigation of semisymmetric graphs with some supplementary properties. One of the results in this field is the complete classification of biprimitive cubic graphs [9].

All these results are very interesting, but for practical use complete lists of semisymmetric graphs are necessary. This work seems to be the first attempt to construct such lists. The semisymmetric graphs with v vertices ($v \leq 28$) are enumerated by computer. The nonexistence of semisymmetric graphs with 30 vertices is also proved by an exhaustive search. So a negative answer to Folkman's problem (4.2) of [7] is obtained. Let us note that 66 is now the smallest value of v for which the question of the existence of semisymmetric graphs is not settled.

2. Semisymmetric graphs as incidence systems

As was noted, every semisymmetric graph is bipartite with the same number of vertices in both parts. The bipartite graph G can be represented by (0,1)-matrices in a variety of ways. We shall interpret the bipartite graph G as the incidence system $S=S(E,B,I)$ with the set of elements E and the set of blocks B, corresponding to the vertices from different parts of the graph. Then the incidence $(e_i,b_j) \in I$ of an element $e_i \in E$ and a block $b_j \in B$ will correspond to the adjacency of the corresponding vertices of the graph. The automorphism group of the bipartite graph G is isomorphic to the automorphism group of the incidence system S. So, the construction of a semisymmetric graph G with v vertices of valency d is equivalent to the construction of some incidence system $S=S(E,B,I)$ with following properties:

1. $|E|=|B|=\frac{v}{2}=v_1$ and every element (resp. block) is incident to d blocks (resp. elements).
2. The automorphism group y=Aut S satisfies the following conditions :
   a) y acts transitively on elements and blocks;
   b) if y_e (resp. y_b) is the stabilizer of an element $e \in E$ (resp. block $b \in B$) in group y, then the blocks $b_1, \cdots, b_d$ (resp. the elements $e_1, \cdots, e_d$) incident to this element (resp. block) form the orbit of action of group y_e (resp. y_b).
3. The dual to S, the incidence system $S^*=S^*(B,E,I)$, is not isomorphic to S.

The construction of all mutually nonisomorphic semisymmetric graphs with v vertices was realized by the use of the technique of constructive enumeration of incidence systems [10]. The main difficulty was to choose the set of conditions corresponding to properties 1-3 which can be tested on the partially constructed incidence matrices.

In the beginning we intended to consider the incidence systems having the property 1 only. In this case it was possible to use the algorithm of constructive enumeration of regular bipartite graphs [1]. We intended to construct all such graphs and then to choose all graphs with automorphism groups acting transitively on both parts. And in the last stage of investigation we would select semisymmetric graphs. But this method would require very much processor time. For example, during the first two hours of the computer calculations more than 1500 graphs with 30 vertices of valency 3 were constructed and only one of them was transitive on the vertices of each part. Therefore it is desirable to construct solely the graphs which are transitive on the vertices of every part (i.e. incidence systems with properties 1 and 2a). One can not solve this problem in full measure. Instead, a set of the invariants, which approximate properties 1 and 2 for the graphs with $v \leq 30$, was used. The verification of property 3 was realized by some heuristic methods on the completely constructed incidence systems.

3. Invariants of the parts of semisymmetric graphs

Let d be the valency of every vertex of a semisymmetric graph G. The transitivity of the action of Aut S on the set of elements E implies the equality of every characteristic of every two vertices from the same part of G. Let us consider one of these invariants. Recall that the incidence of some element and some block corresponds to some edge of the semisymmetric graph. Taking this fact into account we shall use graph theoretic terminology from now on.

Let G be a bipartite graph with vertex set $V = V_1 \bigcup V_2$ and $u \in V_1$. Note by a_i^u the number of vertices belonging to V_1 and connecting with u exactly i paths of length 2. Clearly, the vector $a^u = (a_i^u)$ $0 \leq i \leq d$ is an invariant of the vertices in V_1. We shall denote this vector by a.

Let u and v be adjacent vertices, $u \in V_1, v \in V_2$. Denote by b_i^{uv} $(1 \leq i \leq d)$ the number of vertices adjacent to v and connecting with u by exactly i paths of length 2. From property 2 it follows that the vector $b^{uv}=(b_i^{uv})$ $1 \leq i \leq d$ does not depend on the choice of u or adjacent vertex v. Therefore later we shall omit the superscripts.

We shall prove the following relations in terms of incidence systems

$$\sum_{i=0}^{d} a_i = v_1-1$$

$$\sum_{i=1}^{d} ia_i = d(d-1) \qquad (1)$$

$$\sum_{i=1}^{d} b_i = d-1$$

$$db_i = ia_i \quad i = 1,2, \cdots ,d$$

a_i,b_i are nonnegative integers for i=1,2,...,d.

The first and third equations follow immediately from the definitions of a_i and b_i. To prove the second equation, count the incidences of blocks containing the element u with the other elements in two ways. To prove the last group of equations, count the incidences of blocks containing the element u with the elements corresponding to vertices of G connected to u by i paths of length 2, in two ways.

Let us rewrite the system (1) in the following form.

$$\sum_{i=0}^{d} a_i = v_1-1$$

$$\sum_{i=1}^{d} ia_i = d(d-1) \qquad (2)$$

$$ia_i \equiv 0 (\mathrm{mod}\ d) \quad i=1,2,...,d-1$$

a_i is nonnegative integer for i = 0,1,...,d.

Clearly, the systems (1) and (2) have the same solutions with respect to a_i, $i = 0,1, \cdots ,d$.

Now let us consider some other conditions for the solutions of (2) to be the invariants of a semisymmetric graph G.

Let $i^* = \max(0,2d-v_1)$. It is easy to see that $a_i = 0$ for $i < i^*$ when $2d > v_1$.

The condition $a_d \neq 0$ leads to the existence of $r = a_d+1$ vertices having the same set of neighbours. These vertices form a block of imprimitivity of size r for the transitive action of the group y= Aut S on the corresponding part of the graph G. Hence it is easy to show that

$$v \equiv 0(\mathrm{mod}\ r), \quad a_i \equiv 0(\mathrm{mod}\ r) \ \text{ for } i = 0,1,...,d$$

$$d \equiv 0(\mathrm{mod}\ r), \quad b_i \equiv 0(\mathrm{mod}\ r) \ \text{ for } i = 1,2,...,d \qquad (3)$$

One can show that $3 \leq d \leq v_1-3$ for semisymmetric graph G. From (2) we have $a_{d-1} = d$ or 0. If $a_{d-1} = d$, then $a_0 = v_1-d-1$, $a_i = 0$ for $1 \leq i \leq d-2$, $a_d = 0$. In this case it is easy to prove the transitivity of the action of the automorphism group y on the vertex set of the graph G from the transitivity of the action of y on one part of the graph. Therefore

$$a_{d-1} = 0 \tag{4}$$

The following inequalities can be proved in the same way:

$$a_d \neq d-1, \quad a_d \neq v_1-d-1 \quad \text{when } 2d > v_1 \tag{5}$$

The number of complete bipartite subgraphs $K_{2,i}$ $(1 \leq i \leq d)$ of the semisymmetric graph G, with two vertices from the first part having exactly i common neighbours in graph G, is equal to $v_1 \frac{a_i}{2}$. Therefore

$$a_i v_1 \equiv 0 (\text{mod } 2) \quad \text{for } i = 0,1, \cdots ,d \tag{6}$$

Counting the number of such groups with some fixed vertex gives $d\frac{b_i}{2}$. Therefore

$$b_i d \equiv 0 (\text{mod } 2) \quad \text{for} i = 1,2, \cdots ,d \tag{7}$$

Using (1) to rewrite (2) with restrictions (3) - (7) we have:

$$\begin{aligned}
&\sum_{i=0}^{d} a_i = v_1-1 \\
&\sum_{i=1}^{d} i a_i = d(d-1) \\
&v_1 \equiv 0 (\text{mod } (a_d+1)) \\
&d \equiv 0 (\text{mod } (a_d+1)) \\
&a_i \equiv 0 (\text{mod } (a_d+1)) \quad i = 0,1,...,d-1 \\
&i a_i \equiv 0 (\text{mod } (d(a_d+1))) \quad i = 0,1,...,d-1 \\
&v_1 a_i \equiv 0 (\text{mod } 2) \quad i = 0,1,...,d \\
&i a_i \equiv 0 (\text{mod } 2) \quad i = 0,1, \cdots ,d \\
&a_{d-1} = 0, \quad a_d \neq d-1, \quad a_d \neq v_1-d-1 \quad \text{if } 2d > v_1 \\
&a_i \geq 0, \quad i = 0,1,...,d
\end{aligned} \tag{8}$$

Denote by N_v^d the set of integer solutions of (8). Let a_i be the invariant described above of the vertices from part V_i $(i=1,2)$ of the semisymmetric graph G. By the reasoning given above we have proved

Theorem 2: $a^i \in N_v^d$ for $i = 1,2$.

The nonexistence of some semisymmetric graphs with parameters (v,d) immediately follows from the absence of solutions of (8). Examples of such parameters are (20,5), (20,7), (24, i) for $i \in \{7,8,9\}$, (28,i) for $i \in \{7,9,10,11\}$, (30,9), and (30,11).

Let a^* be the solution of system (8) and $a_0^* \neq 0$. Fix a pair of vertices from the same part V_* exactly i (resp. j) common neighbours with the first (resp. second) vertex of the fixed pair.

Theorem 3: For the vector $a^* \in N_v^d$ to be the above described invariant of the vertices of a bipartite graph transitive on each part, the system of relations (9) must have the solution in nonnegative integers.

$$\begin{aligned}
&\sum_{j=1}^{d} c_{ij} = a_i^* - \delta_i, \quad i = 0,1,\ldots,d \\
&\sum_{i=1}^{d} c_{ij} = a_j^* - \delta_j, \quad j = 0,1,\ldots,d \\
&c_{ij} \geq 0 \qquad\qquad\qquad\qquad\qquad\qquad\qquad\qquad\qquad\qquad\qquad (9) \\
&c_{ij} = 0 \text{ if } (i+j > d) \text{ or } (3d > v_1 \text{ and } i+j < 3d - v_1)
\end{aligned}$$

$$\text{where} \quad \delta_i = \begin{cases} 1 & \text{if } i = 0 \\ 0 & \text{if } i \neq 0 \end{cases}$$

Proof: The set of nonfixed vertices in part V_* of the bipartite graph G is divided into $l=(d+1)(d+1)$ classed by the possible numbers of common neighbours with the fixed vertices. The numbers of vertices c_{ij} in some of these classes must be equal to 0. Let

$$S_2 = \begin{pmatrix} 1\,1 & \cdots & 1\,1\,0\,0 & \cdots & 0\,0\,0\,0 & \cdots & 0\,0 \\ 0\,0 & \cdots & 0\,0\,1\,1 & \cdots & 1\,1\,0\,0 & \cdots & 0\,0 \end{pmatrix}$$

be the submatrix of the incidence matrix S corresponding to the fixed pair of vertices. Let us examine some other row $s_m = (s_{mk})$ of the matrix S corresponding to a vertex having exactly i (resp. j) of common neighbours with the first (resp. second) vertex of the fixed pair. Obviously,

$$\sum_{f=1}^{v_1} s_{mf} = \sum_{f=1}^{d} s_{mf} + \sum_{f=d+1}^{2d} s_{mf} + \sum_{f=2d+1}^{v_1} s_{mf} = d.$$

We have

$$\text{a)} \sum_{f=1}^{d} s_{mf} + \sum_{f=d+1}^{2d} s_{mf} = i+j = d - \sum_{f=2d+1}^{v_1} s_{mf}$$

$$\text{b)} \sum_{f=2d+1}^{v_1} s_{mf} = d - \sum_{f=1}^{d} s_{mf} - \sum_{f=d+1}^{2d} s_{mf} = d - i - j \leq v_1 - 2d$$

Clearly, the sum of the elements in row i, $i=1,2,\ldots,d$(resp. column j, $j=1,2,\ldots,d$) of the matrix $C = (c_{ij})$ must be equal to a_i^* (resp. a_j^*); the total number of vertices in V_* having exactly i (resp. j) common neighbours with the first (resp. second) vertex of the fixed pair. In the analogous equation for $i = 0$ (resp. $j = 0$) we must subtract 1 from a_0^* to take into account the second (resp. first) vertex of the fixed pair.

The absence of integer solutions of (9) proves the nonexistence of semisymmetric graphs for some invariants obtained from (8) for $v = 24,28,30$.

Until now we considered the relations for the invariants corresponding to one part of the graph G. From the semisymmetric graph G we can define two vectors (a^1,a^2) corresponding to the vertices from the different parts V_1,V_2 of the graph.

Theorem 4: Let N_v^d be the set of nonnegative solutions of (8). If (not obligatory different) vectors $a^1, a^2 \in N_v^d$ are the invariants of the vertices of the two parts of the semisymmetric graph G, then

$$\text{a)} \sum_{i=1}^{d} i^2(a_i^1 - a_i^2) = 0 \tag{10}$$

b) if $a_j^l \neq 0$ for some $0 \leq j \leq d$ and $l \in \{1,2\}$ then $a_i^{3-l}(1-2\frac{i}{d})+j \geq 0$ for $1 \leq i \leq d$.

Proof: By counting the number of subgraphs $K_{2,2}$ of the graph G in two different ways, we have:

$$(\frac{v_1}{2})\sum_{i=2}^{d} i(i-1)\frac{a_i^1}{2} = (\frac{v_1}{2})\sum_{i=2}^{d} i(i-1)\frac{a_i^2}{2}$$

Now with regard for the second equality from (1)

$$\sum_{i=1}^{d} i^2 a_i^1 = \sum_{i=1}^{d} i^2 a_i^2$$

So the relation (10a) has been proved.

Let us fix some edge q_1 of the semisymmetric graph G corresponding to the element $s_{mn} = 1$ of the incidence system S. Denote by $J_i(q_1) = s_f$ the set of rows of the matrix S satisfying the following conditions

$$s_f \in J_i(q_1) \Leftrightarrow \sum_{j=1}^{v_1} s_{fj} s_{mj} = i, \qquad s_{fn} = 1$$

Clearly, $J_i(q_1) = b_i^1$. Let us fix some edge q_2, corresponding to the element s_{mn^*} $(n^* \neq n)$ of the matrix S. For some j

$$a_j^2 \neq 0 \quad \text{and} \quad \sum_{f=1}^{v_1} s_{fn^*} s_{fn} = j$$

From the transitivity of the action of Aut S on the edges we have $g(q_2) = q_1$ for some automorphism $g \in \text{Aut } S$. Using the following notations

$$r_i = |J_i(q_1) \cap J_i(q_2)|, \quad p_i = |J_i(q_2)/J_i(q_1)|$$

from the equality

$$p_i + r_i = |J_i(q_2)| = b_i^1$$

and the obvious inequalities

$$p_i \leq a_i^1 - b_i^1, \qquad r_i \leq j - 1,$$

we have

$$a_i^1 - 2b_i^1 + j \geq 1$$

Now from (1) we have that the proposition (10b) is true for $l = 2$. Reasoning by the same way for S^T, we obtain (10b) for $l = 1$.

Apparently, considering more precisely the structures of the incidence systems corresponding to semisymmetric graphs one can add some other relations to the ones described above. In this case one can try to enumerate the semisymmetric graphs with $v > 30$.

Our experience of the construction of regular bipartite transitive on one part graphs show that including each new relation into consideration decreases the number of complete constructed graphs and the computing time of the enumeration. Sometimes if is not necessary to use the computer at all. For example, the nonexistence of semisymmetric graph G with $v = 30$, $d = 8, a^1 = (0,0,4,0,2,8,0,0,0)^T$ follows from (10). And the search for these graphs required more than four hours of processor time.

4. Algorithm of constructive enumeration of semisymmetric graphs

In [10] an algorithm for constructive enumeration of incidence systems is presented. By this algorithm all possibilities for the $n+1$st row of the incidence matrix when n rows are already constructed are generated and investigated. The computer program for the *ICL*–4/70 realizing this algorithm was used for an exhaustive search of semisymmetric graphs in the following manner.

In the beginning the system of relations (8) for $v = 20,24,28,30$ and $3 \leq d \leq \frac{v}{2} - 3$ was solved. Thus all possible values of the invariants described above for the vertices of semisymmetric graphs with parameters (v,d) were obtained. Some of these values were excluded by (9) and (10). Then all incidence systems which were transitive on the elements for every remaining value a^1 were constructed by the computer. In the process of construction the relations (9) and (10) were taken into account and verification of the requirement on subgroup y_e (property 2) was made. As a result, about 65% for $v \leq 28$ and and 10% for $v = 30$ of all constructed incidence systems actually satisfied property 2.

From the set of constructed incidence systems corresponding to edge transitive bipartite graphs we selected those satisfying property 3. For this the algorithm described in [6] would be used except for cases where $a^1 \neq a^2$.

5. Results of constructive enumeration of semisymmetric graphs with $v \leq 30$

The present paper is one of the first attempts to constructively enumerate combinatorial objects largely defined by the transitivity of their automorphism groups. Therefore information about all completely constructed incidence systems is presented later in the tables. This information allows one to estimate the degree of approximation of the transitivity properties by the described methods.

The invariants of the parts of semisymmetric graphs which are not excluded by (9) and (10) were obtained from (8) for different sets of parameters (v,d) and are presented in Tables 1a - 1d. In the same tables information on the numbers of constructed incidence systems is given.

Notations : d is the valency of the graph; N_d^o is the number of vectors (a_i) for given d; N is the number of complete constructed incidence systems (i.s); N_1 is the number of i.s. which are transitive on the elements; N_2 is the number of i.s. which are transitive on the elements and on the blocks; N_e is the number of i.s. corresponding to the edge transitive graphs; T is the processor time.

d	N_d^0	a_0	a_1	a_2	a_3	a_4	a_5	a_6	N	N_1	N_2	N_e	T(sec.)
3	1	3	6	0	0				3	1	1	1	6.7
4	1	4	0	4	0	1			1	1	1	1	0.5
	2	0	8	0	0	1			1	1	1	1	0.5
	3	3	0	6	0	0			1	1	1	1	1.2
	4	1	4	4	0	0			1	1	1	1	4.2
6	1	0	0	0	8	0	0	1	1	1	1	1	0.5
	2	0	0	3	0	6	0	0	1	1	1	1	1.8
	3	0	0	0	6	3	0	0	1	1	1	1	5.4
							Total:						
									10	8	8	8	20.8

Table 1a. $v=20$.

d	N_d^0	a_0	a_1	a_2	a_3	a_4	a_5	a_6	N	N_1	N_2	N_e	T(sec.)
3	1	5	6	0	0				1	1	1	1	30.5
4	1	6	0	4	0	1			2	2	2	2	1.1
	2	2	8	0	0	1			1	1	1	1	0.8
	3	5	0	6	0	0			1	1	1	1	3.3
	4	3	4	4	0	0			5	3	3	2	32.4
	5	1	8	2	0	0			2	1	1	1	6.0
5	1	1	0	10	0	0	0		1	1	1	1	4.7
6	1	3	0	0	6	0	0	2	1	1	1	1	0.5
	2	0	0	9	0	0	0	2	1	1	1	1	0.6
	3	2	0	0	8	0	0	1	1	1	1	1	0.7
	4	0	0	6	4	0	0	1	0	0	0	0	0.3
	5	1	0	3	4	3	0	0	0	0	0	0	83.6
	6	0	0	6	2	3	0	0	6	3	3	1	44.5
	7	1	0	0	10	0	0	0	0	0	0	0	4.3
	8	0	0	3	8	0	0	0	2	2	2	2	22.4
							Total:						
									24	18	18	15	3.3min.

Table 1b. $v=24$.

d	N_d^0	a_0	a_1	a_2	a_3	a_4	a_5	a_6	a_7	a_8	N	N_1	N_2	N_e	T(sec.)
3	1	7	6	0	0						4	2	2	1	10min.
4	1	8	0	4	0	1					2	1	1	1	1.2
	2	4	8	0	0	1					2	1	0	0	1.6
	3	7	0	6	0	0					3	1	1	1	13.4
	4	5	4	4	0	0					0	0	0	0	2min.
	5	3	8	2	0	0					0	0	0	0	1min.
	6	1	12	0	0	0					1	1	1	1	32.5
5	1	3	0	10	0	0	0				1	1	1	1	10.8
6	1	2	0	6	4	0	0	1			0	0	0	0	0.2
	2	0	0	12	0	0	0	1			3	2	2	2	2.1
	3	3	0	3	4	3	0	0			0	0	0	0	1.7min.
	4	2	0	6	2	3	0	0			0	0	0	0	1.0min.
	5	1	0	9	0	3	0	0			1	1	1	1	26.2
	6	1	0	6	6	0	0	0			0	0	0	0	3.8min.
	7	0	0	9	4	0	0	0			0	0	0	0	1.3min.
8	1	0	0	0	0	12	0	0	0	1	3	1	1	1	1.5
									Total:						
											20	11	10	9	22.5min.

Table 1c. $v=28$.

d	N_d^0	a_0	a_1	a_2	a_3	a_4	a_5	a_6	a_7	a_8	a_9	a_{10}	a_{11}	a_{12}	N	N_1	N_2	N_e	$T(sec.)$
3	1	8	6	0	0										57	2	2	1	2h.10m.
4	1	8	0	6	0	0									1	0	0	0	11.5
	2	6	4	4	0	0									4	2	2	1	18.1
	3	4	8	2	0	0									0	0	0	0	16.8min.
	4	2	12	0	0	0									4	3	3	1	13.8min.
5	1	4	0	10	0	0	0								0	0	0	0	5.0
6	1	6	0	0	6	0	0	2							1	1	1	1	0.9
	2	4	0	0	10	0	0	0							0	0	0	0	1.8
	3	0	6	0	8	0	0	0							1	1	1	1	41.0
	4	2	0	6	6	0	0	0							0	0	0	0	3.5min.
	5	0	0	12	2	0	0	0							0	0	0	0	1.2min.
7	1	0	0	0	14	0	0	0	0						1	1	1	1	25.2
8	1	0	0	0	8	2	0	4	0	0					7	2	2	1	3.0min.
	2	0	0	0	0	14	0	0	0	0					1	1	1	1	29.3
10	1	0	0	0	0	0	4	0	10	0	0	0			0	0	0	0	47.2
12	1	0	0	0	0	0	0	0	0	0	8	6	0	0	1	1	1	1	4.5min.
Total:															78	14	14	9	3h.11m.

Table 1d. $v=30$.

Table 2 contains the orders of automorphism groups of constructed incidence systems corresponding to the edge transitive graphs. Here N^0 is the number of i.s. S in succession; N_*^0 is the number of dual to S i.s. S^*. Clearly, the i.s. S corresponds to a semisymmetric graph iff $N^0 \neq N_*^0$.

N^0	v	d	N_d^0	Aut$S$	N_*^0
1	20	3	1	120	1
2		4	1	$2^{10}{\cdot}10$	2
3		4	2	$2^{5}{\cdot}120$	4
4		4	3	$2^{5}{\cdot}120$	3
5		4	4	40	5
6		6	1	$2^{5}{\cdot}120$	7
7		6	2	$2^{5}{\cdot}120$	6
8		6	3	120	8
9	24	3	1	72	9
10		4	1	$2^{12}{\cdot}72$	10
11		4	1	$2^{12}{\cdot}72$	11
12		4	2	$2^{6}{\cdot}48$	13
13		4	3	$2^{6}{\cdot}48$	12
14		4	4	48	14
15		4	4	48	15
16		4	5	384	16
17		5	1	240	17
18		6	1	$(3!)^4{\cdot}8$	18
19		6	2	$2^6(3!)^4{\cdot}24$	20
20		6	3	$2^6(3!)^4{\cdot}24$	19
21		6	6	144	21
22	24	6	8	576	22
23		6	8	576	23
24	28	3	1	56448	24
25		4	1	$2^{14}{\cdot}14$	25
26		4	3	56448	26
27		4	6	336	27
28		5	1	336	28
29		6	2	$2^{7}{\cdot}42$	31
30		6	2	$2^{14}{\cdot}168$	30
31		6	5	$2^{7}{\cdot}42$	29
32		8	1	$2^{14}{\cdot}168$	32
33	30	3	1	720	33
34		4	2	60	34
35		4	4	360	35
36		6	1	#$(3!)^{10}{\cdot}10$	36
37		6	3	720	37
38		7	1	20160	38
39		8	1	720	39
40		8	2	20160	40
41		12	1	720	41

Table 2

In summary, we constructed 5 pairs of nonselfdual incidence systems which were transitive on the elements and on the blocks. They correspond to 5 semisymmetric graphs with $20 \leq v \leq 28$. Note that all these graphs are known [7,4,12]. We proved only the completeness of the list of semisymmetric graphs for $v \leq 28$. Besides that we proved the nonexistence of semisymmetric graphs with 30 vertices, that is, we answered in the negative Folkman's problem (4.2) [7].

References

[1] A.M. Baraev and I.A. Faradžev, "Construction and investigation by the computer of regular bipartite graphs" (Russian), in *Algorithmic Research in Combinatorial Theory,* Nauka, Moscow, 1978, pp. 25ff.

[2] N.L. Biggs, *Algebraic Graph Theory,* Cambridge University Press, 1974.

[3] I.Z. Bouwer, "On edge but not vertex transitive cubic graphs", *Canad. Math. Bull.* 11 (1968) 533-535.

[4] I.Z. Bouwer, "On edge but not vertex transitive regular graphs", *Journal of Combinatorial Theory (B)* 12 (1972) 32-40.

[5] I.A. Faradžev, "Constructive enumeration of combinatorial objects" (Russian), in *Algorithmic Research in Combinatorial Theory,* Nauka, Moscow, 1978, pp. 3-11.

[6] I.A. Faradžev and V.A. Zaitchenko, "The algorithm for testing canonicity of incidence systems" (Russian), in *Algorithmic Research in Combinatorial Theory,* Nauka, Moscow, 1978, pp. 126-129.

[7] J. Folkman, "Regular line symmetric graphs", *Journal of Combinatorial Theory* 3 (1967) 215-232.

[8] F. Harary, *Graph Theory,* Addison-Wesley Publ. Co., 1969.

[9] M. E. Iofinova and A.A. Ivanov, "Biprimitive cubic graphs" (Russian), in *Investigation in Algebraic Theory of Combinatorial Objects* Proceedings of the seminar, Institute for System Studies, Moscow, 1985, pp. 124-134.

[10] A.V. Ivanov, "Constructive Enumeration of Incidence Systems", *Annals of Discrete Mathematics* 26 (1985) 227-246.

[11] M.H. Klin, "On edge but not vertex transitive regular graphs", in *Colloquia Mathematica Societatis Janos Bolyai, 25. Algebraic methods in graph theory, Szeged (Hungary), 1978* Budapest, 1981, pp. 399-403.

[12] V. K. Titov, "On symmetry in the graphs" (Russian), in *Voprocy Kibernetiki (15). Proceedings of the II All Union seminar on combinatorial mathematics, part 2,* Nauka, Moscow, 1975, pp. 76-109.

Annals of Discrete Mathematics 34 (1987) 287–296
© Elsevier Science Publishers B.V. (North-Holland)

A Product Theorem for Cyclic Graph Designs

Masakazu Jimbo

Institute for Socio-Economic Planning
The University of Tsukuba
Sakura, Ibaraki 305, JAPAN

Shinji Kuriki

Department of Applied Mathematics
Science University of Tokyo
1-3 Kagurazaka, Shinjuku-ku
Tokyo 162, JAPAN

TO ALEX ROSA ON HIS FIFTIETH BIRTHDAY

ABSTRACT

A graph design, which is a generalization of a path design, a circuit design or a bipartite design, has been studied by many authors (see, for example, Hell and Rosa [4] and Bermond and Sotteau [1]). In this paper, a product method for cyclic graph designs is given.

1. Introduction.

Balanced incomplete block designs were generalized to balanced graph designs by Hell and Rosa [4]. They also studied the effect of products upon the decomposition of graphs into various subgraphs. Many authors considered these topics (see, for example, [1], [3]-[7], [9] and [10]). In this paper, we consider a construction for cyclic graph designs by a product method. This extends known methods for cyclic block designs by Jimbo and Kuriki [8] and Colbourn and Colbourn [2].

Here, graphs are considered to be undirected unless otherwise stated. Edges are identified with pairs of endpoints but multiple edges and loops are allowed. $V(G)$ denotes the vertex set of a graph G, and $E(G)$ denotes the multiset of edges of G. The number of vertices (*order* of G) and the number of edges are denoted by $p(G)$ and $e(G)$, respectively. The addition of two graphs G and H, $G \oplus H$, is defined by a graph with the vertex set $V(G \oplus H)=V(G) \cup V(H)$ and the edge set $E(G \oplus H)$ which is a collection of edges of G and H. λG denotes a graph such that a graph G is added λ times. Let H and G be graphs. If there is a family $\mathcal{G} = \{G_1, G_2, \cdots\}$ of subgraphs of H, which are isomorphic to a graph G, such that each edge of H is in precisely one member of $\mathcal{G}$, then $\mathcal{G}$ is said to be a *G-decomposition* of H. Then $H = G_1 \oplus G_2 \oplus \cdots$ holds.

The complete graph on v vertices, in which each vertex is joined precisely once to each other vertex, is denoted by K_v. A decomposition of λK_v into subgraphs isomorphic to a graph G of order k is called a *graph design* and denoted by a (v,k,λ) *G-design.* If G is a path, a circuit or a bipartite graph, then a G-design is called a *path design*, a *circuit design* or a *bipartite design,* respectively. If each vertex occurs in precisely the same number of subgraphs in the decomposition, then the G-design is said to be *balanced.* If there is a balanced (v,k,λ) G-design, then

$$\lambda v(v-1)=2eb,\ vr=bk \text{ and } \lambda(v-1)\equiv 0\ (mod\ d)$$

hold, where $e=e(G)$, b is the number of subgraphs of the G-decomposition, r is the number of subgraphs containing a given vertex of λK_v and d is the *g.c.d.* of the degrees of the vertices of G. We can easily verify that if G is a *regular* graph, which is defined to be a graph with the same number of edges for each vertex, then a G-design is balanced. A K_k-decomposition of λK_v is simply a balanced incomplete block design.

A graph H is said to be *cyclic* if it has an automorphism of a single cycle of length $v=p(H)$. If H is cyclic, then without loss of generality, we can assume that $V(H)=Z_v$, the residues of modulo v. And if $\{a,b\}$ is an edge of H, then $\{a+c,b+c\}$ $(mod\ v)$ is also an edge of H. Let h_i be the number of edges $\{0,i\}$, in particular, h_0 is the number of loops $\{0,0\}$. Then obviously we have $h_i=h_{v-i}=h_{-i}$ for $i\in Z_v$. For a cyclic graph H, $H(x)=\sum_{i=0}^{v-1} h_i x^i$ is called a *characteristic polynomial* of H. The following polynomial is called a *reciprocal characteristic polynomial* of H:

$$\overline{H}(x)=\begin{cases}\displaystyle\sum_{i=-(v-1)/2}^{(v-1)/2} h_i x_i & \text{if } v \text{ is odd,}\\ \displaystyle\sum_{i=-v/2+1}^{v/2-1} h_i x^i+\frac{h_{v/2}}{2}(x^{v/2}+x^{-v/2}) & \text{if } v \text{ is even.}\end{cases}$$

Let H and F be cyclic graphs of order $v_1=p(H)$ and $v_2=p(F)$ with a reciprocal characteristic polynomial $\overline{H}(x)$ and a characteristic polynomial $F(y)$, respectively. A cyclic graph having a characteristic polynomial $\overline{H}(z)F(z^{v_1})$ with $z^{v_1 v_2}=1$ is called a *cyclic product* of H and F, denoted by $H*F$. Note that, in the case when v_1 is even, $H_{\frac{v_1}{2}}$ must be even in order that $\overline{H}(z)F(z^{v_1})$ is well-defined as a characteristic polynomial of a cyclic graph. Clearly, a cyclic product $*$ is not commutative. Note that for cyclic graphs H_1,H_2 and F, if $V(H_1)=V(H_2)$, then we have

$$(H_1\oplus H_2)*F=(H_1*F)\oplus(H_2*F),$$

and

$$F*(H_1\oplus H_2)=(F*H_1)\oplus(F*H_2).$$

Example 1. Let H and F be circuits of length 4 and 3, respectively. Then $\overline{H}(x)=x+x^{-1}$ and $F(y)=y+y^2$. Hence $H*F$ is a cyclic graph of order 12 with a characteristic polynomial

$$\overline{H}(z)F(z^4)=(z+z^{-1})(z^4+z^8)=z^3+z^5+z^7+z^9 \quad (z^{12}=1).$$

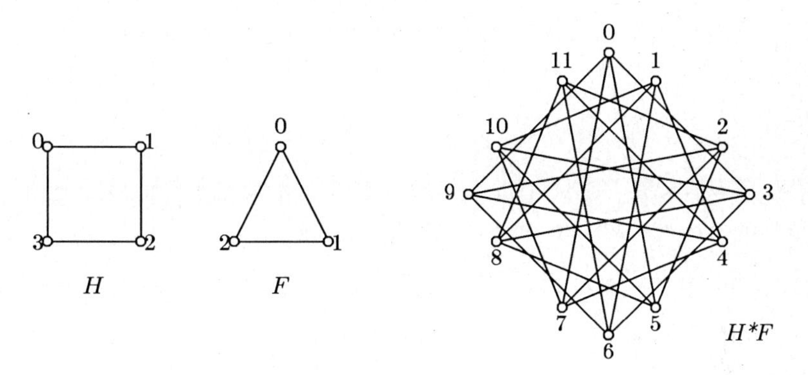

Lemma 1. Let $H=\lambda_1 K_{v_1}$ and $F=\lambda_2 K_{v_2}$, where λ_1 is even if v_1 is even. Then

$$\lambda_1\lambda_2 K_{v_1 v_2}=(H*F)\oplus(\lambda_2 H*I_{v_2})\oplus(\lambda_1 I_{v_1}*F) \tag{1}$$

holds, where I_{v_i} is a graph with v_i vertices and a loop for each vertex such that $V(I_{v_1})=V(H)$ and $V(I_{v_2})=V(F)$.

Proof:

Let $\overline{H}(x)$ be a reciprocal characteristic polynomial of H and let $H(x)$ be a characteristic polynomial of H. Then we have $H(x)=\lambda_1(x+x^2+\cdots+x^{v_1-1})$. Similarly, the characteristic polynomial of F is $F(y)=\lambda_2(y+y^2+\cdots+y^{v_2-1})$. Hence that of $F\oplus\lambda_2 I_{v_2}$ is $F'(y)=F(y)+\lambda_2=\lambda_2(1+y+y^2+\cdots+y^{v_2-1})$. Note that $z^{v_1}F'(z^{v_1})=F'(z^{v_1})$ and $z^{-1}=z^{v_1(v_2-1)+v_1-i}$ for any integer $0\leq i\leq v_1/2$, when $z^{v_1 v_2}=1$. Then we have $\overline{H}(z)F'(z^{v_1})=H(z)F'(z^{v_1})$ for $z^{v_1v_2}=1$. And the characteristic polynomial of $\lambda_1 I_{v_1}*F$ is $\lambda_1 F(z^{v_1})$. Thus the characteristic polynomial of the right hand side of (1) is

$$\begin{aligned}\overline{H}(z)F'(z^{v_1})+\lambda_1 F(z^{v_1})&=\{H(z)+\lambda_1\}F'(z^{v_1})-\lambda_1\lambda_2\\ &=\lambda_1\lambda_2(1+z+z^2+\cdots+z^{v_1-1})(1+z^{v_1}+z^{2v_1}+\cdots+z^{(v_2-1)v_1})-\lambda_1\lambda_2\\ &=\lambda_1\lambda_2(z+z^2+\cdots+z^{v_1v_2-1}),\end{aligned}$$

which is the characteristic polynomial of $\lambda_1\lambda_2 K_{v_1v_2}$. □

Let G be a subgraph of a cyclic graph H of order v. A graph $G+i$ is defined by $V(G+i)=\{g+i \ (mod\ v) \mid g\in V(G)\}$ and $E(G+i)=\{\{a+i,b+i\} \ (mod\ v) \mid \{a,b\}\in E(G)\}$. The set of graphs $\{G+i \mid i=0,1,2,\ldots\}$ is called an *orbit* of G. An orbit can be represented by any one of its graphs, which will be called a *base graph.* The smallest positive integer i such that $G+i=G$ is called the *length* of an orbit. If $i=v$ then an orbit is said to be *full* otherwise it is said to be *short.* For an orbit of length i with base graph G, let

$$<G>=G \oplus (G+1) \oplus (G+2) \oplus \cdots \oplus (G+i-1).$$

If a cyclic graph H of order v is represented as $H=<G_1>\oplus \cdots \oplus <G_\alpha>$ for G_ℓ's isomorphic to a graph G, then H is said to have a *cyclic G-decomposition.* In this paper, we consider only the case when any orbits corresponding to G_ℓ's are full. For each graph G_ℓ, let

$$\Delta G_\ell(x)=\sum_{\{i,j\}\in E(G_\ell)} (x^{i-j}+x^{j-i}).$$

Then $\overline{H}(x)=\Delta G_1(x)+\cdots+\Delta G_\alpha(x)$ holds. A cyclic G-decomposition of λK_v is called a *cyclic (v,k,λ) G-design,* where k is the order of G. Note that a cyclic G-design is always balanced.

Lemma 2. For an even integer v, if there exists a cyclic (v,k,λ) G-design with no short orbit, then λ must be even.

Proof:

Let $H=<G_1>\oplus\cdots\oplus<G_\alpha>$ be a cyclic G-decomposition of λK_v. If the edge $\{0, \frac{v}{2}\}$ is contained in a graph $G_\ell+j$ for a certain ℓ and j, then it is also contained in a graph $G_\ell+j+\frac{v}{2}$. Since the cyclic G-decomposition has no short orbit, $G_\ell+j$ and $G_\ell+j+\frac{v}{2}$ must be distinct. Thus the set of all graphs having the edge $\{0, \frac{v}{2}\}$ can be partitioned into pairs of graphs $(G_\ell+j, G_\ell+j+\frac{v}{2})$'s. Hence λ must be even. □

2. Difference arrays and Graph arrays.

Let F be a cyclic graph of order v with a characteristic polynomial $F(y)=\sum_{p=0}^{v-1} f_p y^p$, then $f=F(1)=\Sigma f_p$ is the degree of each vertex. Let G be a *simple* graph, which is defined to be a graph with no loops and no multiple edges, with vertex set $V(G)=\{0,1,\ldots,k-1\}$. A $k\times f$ array $D=(d_{ij})$ is called an *(F,G)-difference array* if $d_{ij}\in Z_v$ and every vertex p of F occurs exactly f_p times among the differences $\{d_{ij}-d_{i'j} \ (mod\ v) \mid j=1,2,\ldots,f\}$ for any two rows i and i' such that $\{i,i'\}\in E(G)$, that is,

$$\sum_{j=0}^{f} y^{d_{ij}-d_{i'j}} = F(y) \qquad (y^v=1) \tag{2}$$

holds for any $\{i,i'\}\in E(G)$.

Example 2. (i) Let F be a cyclic graph of order 5 with a characteristic polynomial $F(y)=1+y+y^4$ and G be a path of length 4. Then the following array is an (F,G)-difference array:

0	0	0
0	1	4
0	0	0
0	1	4

(ii) Let F be K_5 with a loop on each vertex and G be K_4, then the following array is an (F,G)-difference array:

0	0	0	0	0
0	1	2	3	4
0	2	4	1	3
0	3	1	4	2

In the case when a cyclic graph F is the complete graph K_v with a loop on each vertex, a $(\lambda F, K_k)$-difference array is called a (v,k,λ) *row difference scheme*, which is useful to construct cyclic *BIB* designs (see Jimbo and Kuriki [8]).

If there is a mapping ψ from $V(G)$ onto a set of n colors such that $\psi(a) \neq \psi(b)$ for any $\{a,b\} \in E(G)$, then a graph G is said to be n-colorable.

Lemma 3. If G is 2-colorable then an (F,G)-difference array exists for any cyclic graph F.

Proof:

We have the lemma by constructing an array such that the rows corresponding to the first color are all zero and in the rows corresponding to the second color, every vertex p of F adjacent to the vertex 0 occurs f_P times, where $F(y) = \Sigma f_p y^p$ is a characteristic polynomial of F. □

Let G be a simple graph of order $k = p(G)$ with $e = e(G)$ edges. Let G^* be a directed graph obtained from G by replacing each edge by two arcs, one directed each way, between the same vertices. A $k \times 2e$ matrix $A = (a_{ij})$ with elements from $V(G) = \{0, 1, \ldots, k-1\}$ is called a *graph balanced array* for G, if every arc of G^* occurs exactly once in the set of ordered pairs $\{(a_{ij}, a_{i'j}) \mid j = 1, 2, \ldots, 2e\}$, for any $\{i, i'\} \in E(G)$.

Lemma 4. A graph balanced array exists for any 2-colorable graph G.

An $m \times N$ matrix A with entries from a set of $s (\geq 2)$ elements is called an orthogonal array of size N, m constraints, s levels, strength t and index λ, if any $t \times N$ submatrix of A contains all possible $t \times 1$ column vectors with the same frequency λ.

Lemma 5. If there exists an orthogonal array of size k^2, $k+1$ constraints, k levels, strength 2 and index 1, then a K_k-array exists.

Proof:

Without loss of generality, we can assume that a $(k+1) \times k^2$ orthogonal array $A = (a_{ij})$ is standardized as follows:

$$a_{k+1, qk+r} = q \text{ for any } 0 \leq q \leq k-1 \text{ and } 1 \leq r \leq k,$$

and

$$a_{i,k(k-1)+j}=j-1 \text{ for any } 1\leq i\leq k \text{ and } 1\leq j\leq k.$$

Then the submatrix $\tilde{A}=(a_{ij})$ $(i=1,\ldots,k;j=1,\ldots,k(k-1))$ is a K_k-array. □

Example 3. The following are graph balanced arrays for graph G's which are not 2-colorable and $k\leq 5$ except K_4 and K_5:

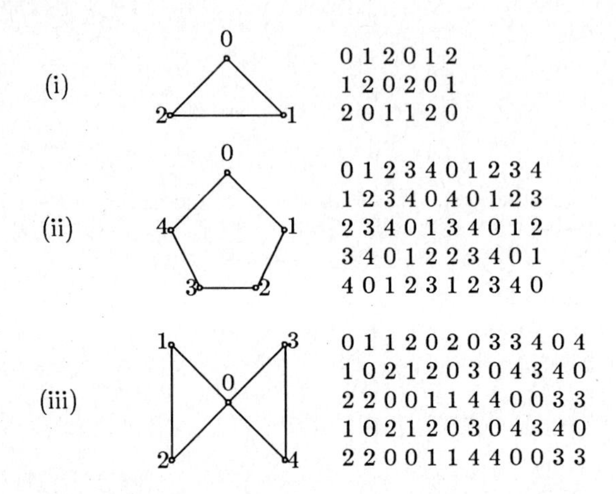

But the existence problem of graph balanced arrays is not solved, in general.

The following theorem shows a way to construct an (F,G)-difference array by using a graph balanced array.

Theorem 1: Let G be a simple graph with a graph balanced array. If a cyclic graph F has a cyclic G-decomposition with no short orbit, then there exists an (F,G)-difference array.

Proof:

Let $F=\langle G_1\rangle\oplus\cdots\oplus\langle G_\alpha\rangle$ be a cyclic G-decomposition, where G_ℓ's are isomorphic to a graph G with k vertices and e edges. Then we have $\bar{F}(y)=\triangle G_1(y)+\cdots+\triangle G_\alpha(y)$. Let $\phi_\ell:V(G)\rightarrow V(G_\ell)$ be isomorphisms. For a $k\times 2e$ G-array $A=(a_{ij})$, let $D_\ell=(\phi_\ell(a_{ij}))$ for $\ell=1,2,\ldots,\alpha$. Then every arc of G_ℓ^* occurs exactly once in the set of ordered pairs $\{(\phi_\ell(a_{ij}),\phi_\ell(a_{i'j}))\mid j=1,2,\ldots,2e\}$ for any $\{i,i'\}\in E(G)$, where G_ℓ^* is a directed graph corresponding to a graph G as before. Hence we have

$$\sum_{j=1}^{2e} y^{\phi_\ell(a_{ij})-\phi_\ell(a_{i'j})}=\triangle G_\ell(y).$$

Thus a $k\times 2e\alpha$ array $D=[D_1D_2\cdots D_\alpha]$ is an (F,G)-difference array. □

3. A product theorem for cyclic decompositions.

In this section we consider a composition method (or a product method) of two cyclic G-decompositions.

The following lemma is obvious:

Lemma 6. If cyclic graphs H and F with $V(H)=V(F)$ have cyclic G-decompositions, then $H \oplus F, \lambda H$ and $I_v * H$ have cyclic G-decompositions, where the graph I_v is a cyclic graph of order v with a loop for each vertex and no other edges.

Theorem 2. Let G be a simple graph. Let H and F be cyclic graphs. If

(i) H has a cyclic G-decomposition with no short orbit,

and if

(ii) there is an (F,G)-difference array,

then the cyclic product $H*F$ has a cyclic G-decomposition with no short orbit.

Proof:

Let a $k \times f$ array $D=(d_{ij})$ be an (F,G)-difference array. Let $H=\langle G_1\rangle \oplus \cdots \oplus \langle G_\alpha \rangle$ be a cyclic G-decomposition, where G_ℓ's are isomorphic to G. Then we have $\overline{H}(x) = \Delta G_1(x)+\cdots+\Delta G_\alpha(x)$. Let $\phi_\ell : V(G) = \{0,1,\ldots,k-1\} \rightarrow V(G_\ell)$ be an isomorphism. For each base graph G_ℓ of H, construct base graphs $G_{\ell j}$ $(j=1,2,\ldots,f)$ of $H*F$ which have vertex sets

$$V(G_{\ell j})=\{\phi_\ell(i)+v_1 d_{ij} \mid i \in V(G)\} \ (mod\ v_1 v_2)$$

and which is isomorphic to G, where v_1 and v_2 are the orders of H and F, respectively. Then we obtain by (2)

$$\begin{aligned}\sum_{j=1}^{f} \Delta G_{\ell j}(z) &= \sum_{j=1}^{f} \sum_{(i,i')\in E(G^*)} z^{\phi_\ell(i)-\phi_\ell(i')+v_1(d_{ij}-d_{i'j})} \\ &= \sum_{(i,i')\in E(G^*)} z^{\phi_\ell(i)-\phi_\ell(i')} \sum_{j=1}^{f} z^{v_1(d_{ij}-d_{i'j})} \\ &= \Delta G_\ell(z)\cdot F(z^{v_1}),\end{aligned}$$

where $z^{v_1 v_2}=1$. Hence

$$\begin{aligned}\overline{H}(z)\cdot F(z^{v_1}) &= \sum_{\ell=1}^{\alpha} \Delta G_\ell(z)\cdot F(z^{v_1}) \\ &= \sum_{\ell=1}^{\alpha}\sum_{j=1}^{f} \Delta G_{\ell j}(z),\end{aligned}$$

which implies that $H*F$ can be decomposed cyclically by base graphs $G_{\ell j}$'s. Hence the theorem is proved. □

By combining Theorems 1 and 2, we obtain the following corollary:

Corollary 1. If cyclic graphs H and F have cyclic G-decompositions with no short orbit, and if there is a graph balanced array for G, then a cyclic G-decomposition of $H*F$ exists.

The following corollary is the direct consequence of Theorem 2 and Lemma 3.

Corollary 2. Let a simple graph G be 2-colorable. If H has a cyclic G-decomposition with no short orbit, then for any cyclic graph F, H^*F has a cyclic G-decomposition with no short orbit.

Corollary 3. Under the same assumption as Theorem 2 $H^*(F\oplus\lambda I)$ has a cyclic G-decomposition with no short orbit, where $V(F)=V(I)$.

Proof:

Note that $H^*(F\oplus\lambda I)=(H^*F)\oplus(\lambda H^*I)$ and that $p(G)\times 1$ zero vector is an (I,G)-difference array. Hence H^*I has a cyclic G-decomposition with no short orbit by Theorem 2. Thus the corollary is proved by Lemma 6. □

Theorem 3. Let G be an n-colorable simple graph of order k. If there are

(i) a cyclic (v_1,k,λ_1) G-design with no short orbit,

(ii) a cyclic $(v_2,k,\lambda_1\lambda_2)$ G-design with no short orbit

and

(iii) a (v_2,n,λ_2) row difference scheme,

then there exists a cyclic $(v_1v_2,k,\lambda_1\lambda_2)$ G-design with no short orbit.

Proof:

We have only to show that the graph $\lambda_1\lambda_2K_{v_1v_2}$ has a cyclic G-decomposition. Let $A_1,\ldots,A_n$ be n color classes of G. Let $D=(d_{ij})$ be a (v_2,n,λ_2) row difference scheme. Construct a $k\times\lambda_2v_2$ array $\tilde{D}$ as follows. Let every vertex of G correspond to a row of an array $\tilde{D}$ arbitrarily. Make copies of the i-th row of D to the rows which correspond to vertices of A_i for every i. Then it is easily shown that the array $\tilde{D}$ is a $(\lambda_2(K_{v_2}\oplus I_{v_2}),G)$-difference array. From the assumption (i), $\lambda_1K_{v_1}$ has a cyclic G-decomposition with no short orbit. Hence by Theorem 2, the cyclic graph $\lambda_1K_{v_1}{}^*\lambda_2(K_{v_2}\oplus I_{v_2})$ has a cyclic G-decomposition with no short orbit. And $I_{v_1}{}^*\lambda_1\lambda_2K_{v_2}$ has a cyclic G-decomposition with no short orbit by Lemma 6. Finally, by (1) of Lemma 1

$$\lambda_1\lambda_2K_{v_1v_2}=\{\lambda_1K_{v_1}{}^*\lambda_2(K_{v_2}\oplus I_{v_2})\}\oplus(I_{v_1}{}^*\lambda_1\lambda_2K_{v_2}\},$$

we have the theorem. □

The following theorem can be shown from Corollary 1 by the similar argument to Theorem 3:

Theorem 4. Let G be a graph of order k with a graph balanced array. If there are

(i) a cyclic (v_1,k,λ_1) G-design with no short orbit

and

(ii) a cyclic (v_2,k,λ_2) G-design with no short orbit,

then there exists a cyclic $(v_1v_2,k,\lambda_1\lambda_2)$ G-design with no short orbit.

Acknowledgements

This research was supported in part by a Grant-in-Aid for Scientific Research of the Ministry of Education, Science, and Culture under Contract Number 321-6009-61530017, and by a Research Grant of Science University of Tokyo under Contract Number 86-1001.

References

[1] J.C. Bermond and D. Sotteau, "Graph decompositions and G-designs", *Proc. 5-th British Combinatorial Conf.* (1975) 53-72.

[2] M.J. Colbourn and C.J. Colbourn, "Recursive constructions for cyclic block designs", *J. Statist. Planning and Inference* 10 (1984) 97-103.

[3] H. Enomoto and K. Ushio, "C_k-factorizations of the complete bipartite graphs" (in Japanese), *Kyoto Univ. Inst. Math. Kokyuuroku* 587 (1986) 52-57.

[4] P. Hell and A. Rosa, "Graph decompositions, handcuffed prisoners and balanced P-designs", *Discrete Math.* 2 (1972) 229-252.

[5] J.D. Horton, "Resolvable path designs", *J. Combin. Theory (A)* 39 (1985) 117-131.

[6] C. Huang and A. Rosa, "On the existence of balanced bipartite designs", *Utilitas Math.* 4 (1973) 55-75.

[7] C. Huang, "Resolvable balanced bipartite designs", *Discrete Math.* 14 (1976) 319-335.

[8] M. Jimbo and S. Kuriki, "On a composition of cyclic 2-designs", *Discrete Math.* 46 (1983) 249-255.

[9] A. Rosa and C. Huang, "Another class of balanced graph designs: Balanced circuit designs", *Discrete Math.* 12 (1975) 269-293.

[10] S. Yamamoto, H. Ikeda, S. Shige-eda, K. Ushio and N. Hamada, "On claw-decomposition of complete graphs and complete bigraphs", *Hiroshima Math. J.* 5 (1975) 33-42.

Annals of Discrete Mathematics 34 (1987) 297—300
© Elsevier Science Publishers B.V. (North-Holland)

A New Class of Symmetric Divisible Designs

Dieter Jungnickel

Mathematisches Institut
Justus-Liebig-Universität Giessen
Arndtstrasse 2
D-6300 Giessen
Federal Republic of Germany

TO ALEX ROSA ON HIS FIFTIETH BIRTHDAY

ABSTRACT

Using affine designs, we give a new construction for symmetric divisible designs. This yields the existence of the class of such designs with parameters $m=q^d+\cdots+q+2$, $n=q^{d+1}$, $k=q^d(q^d+\cdots+q^2+q+1)+q^{d+1}$, $\lambda_1=q^d(q^{d-1}+\cdots+q+1)+q^{d+1}$, and $\lambda_2=q^d(q^{d-1}+\cdots+q+3)$ for all prime powers q and all positive integers d. We also construct relative difference sets with these parameters.

1. Introduction

Let A be an affine design $S_\lambda(2,k;v)$ (see Beth, Jungnickel, and Lenz [1] for background from design theory). In 1971, Wallis [6] proved that one can construct a symmetric design with parameters

(1) $v^* = (r+1)v$, $k^* = kr$ and $\lambda^* = k\lambda$ $(=(r-1)\mu)$;

In particular, using $AG_{d-1}(d,q)$ one gets the series of symmetric designs for

(2) $v^* = q^{d+1}(q^d+\cdots+q+2)$, $k^* = q^d(q^d+\cdots+q+1)$ and $\lambda^* = q^d(q^{d-1}+\cdots+q+1)$;

and using a Hadamard 3-design $S_{n-1}(3,2n;4n)$ one gets

(3) $v^* = 16n^2$, $k^* = 2n(4n-1)$, and $\lambda^* = 2n(2n-1)$.

Wallis used strongly regular graphs to prove this result; a considerably simpler construction using "auxiliary matrices" was given by Lenz and Jungnickel [4]. These authors also observed that their construction could be used to establish the existence of difference sets with parameters (2), a result due to McFarland [5]. In the present note, we use a simple modification of the Lenz-Jungnickel construction to produce a new series of symmetric divisible designs (from affine designs) and also a new series of relative difference sets.

2. New Symmetric Divisible Designs

In this section, we prove the analogue of Wallis's theorem (see e.g. Jungnickel [3] for divisible designs and the notation used here):

Theorem 1: Assume the existence of an affine design $S_\lambda(2,k;v)$. Then there also exists a symmetric divisible design with parameters

$$(4)\quad m=r+1,\ n=v,\ k'=kr+v,\ \lambda_1=k\lambda+v, \text{ and } \lambda_2=k(\lambda+2).$$

Proof:

Let A be the given affine $S_\lambda(2,k;v)$ and label the points and parallel classes of A as $p_1,...,p_v$ and $B_1,...,B_r$, respectively. Drake and Jungnickel [2] have constructed "auxiliary matrices" $M_1,...,M_r$ from A as follows. Put $M_h = (m_{ij}^h)$, $(h=1,...,r;\ i,j=1,...,v)$, where

$$m_{ij}^h = \begin{cases} 1 & \text{if } p_i \text{ and } p_j \text{ are on a block from } B_h \\ 0 & \text{otherwise.} \end{cases}$$

By the definition of an affine design, one then gets the following equations:

$$(5)\quad M_h M_{h'}^T = \begin{cases} \mu J & \text{if } h\neq h' \qquad (h,h'=1,...,r) \\ kM_h & \text{if } h=h' \end{cases}$$

and thus

$$(6)\quad M_1M_1^T + \cdots + M_rM_r^T = k(r-\lambda)I+k\lambda J.$$

Now let $M_0=J$ and replace in the group table of residues mod $r+1$ each entry i by M_i; that is, put

$$L = \begin{vmatrix} M_0 & M_1 & \cdots & M_r \\ M_1 & M_2 & \cdots & M_0 \\ & & \cdot & \\ & & \cdot & \\ & & \cdot & \\ M_r & M_0 & \cdots & M_{r-1} \end{vmatrix} = \begin{vmatrix} L_0 \\ L_1 \\ \cdot \\ \\ L_r \end{vmatrix}.$$

From (5) and (6), one obtains

$$L_hL_{h'}^T = \begin{cases} ((r-1)\mu+2k)J & \text{if } h\neq h' \\ k(r-\lambda)I+(k\lambda+v)J & \text{if } h=h'. \end{cases}$$

Thus L is the incidence matrix of the desired symmetric divisible design, since $(r-1)\mu = k\lambda$ for an affine design. □

We remark that replacing M_0 by the zero matrix in the above proof gives the Lenz-Jungnickel proof of Wallis's theorem (see also [1, II.8.16]). Using $AG_{d-1}(d,q)$ in Theorem 1 gives the following result:

Theorem 2: Let q be a prime power and d a positive integer. Then there exists a symmetric divisible design with parameters

(7) $m=q^d+\cdots+q+2,\ k=q^d(q^d+\cdots+q+1)+q^{d+1},$

$\lambda_1=q^d(q^{d-1}+\cdots+q+1)+q^{d+1}$ and $\lambda_2=q^d(q^{d-1}+\cdots+q+3)$. □

Note that using Hadamard 3-designs in Theorem 1 does not yield anything interesting. One obtains a symmetric design the complement of which has parameters (3) above. Similarly, the case $q=2$ of Theorem 2 results in the complement of a symmetric design with parameters (2). Thus Theorem 2 is interesting only for $q\neq 2$.

3. New relative difference sets

As in [4], one easily sees that the group $H = Z_m \oplus EA(q^{d+1})$ (where $EA(x)$ denotes the elementary abelian group of order x) acts regularly as an automorphism group of the symmetric divisible design $\mathcal{D}$ constructed in Theorem 2. Thus $\mathcal{D}$ corresponds to a relative difference set D with parameters (7) in G. (See [3] for relative difference sets and their relation to divisible designs). Using a modification of McFarland's construction [5] for difference sets with parameters (2), we can obtain relative difference sets with parameters (7) in other groups:

Theorem 3: Let q be a prime power and d a positive integer. Moreover, let G be any group of order $q^d+\cdots+q+2$. Then there exists a relative difference set with parameters (7) in $H = G \oplus EA(q^{d+1})$.

Proof:

Let $U_2,\ldots,U_m$ be the subgroups of $EA(q^{d+1})$ given by the hyperplanes of this group considered as the $(d+1)$-dimensional vector space over $GF(q)$. (Note that the number of hyperplanes is indeed $m-1$). Also, label the elements of G as $g_1,\ldots,g_m$. Now put $D = D_1 \cup D_2$ where $D_1=\{(g_i,u_i):i=2,\ldots,m;u_i\in U_i\}$ and $D_2=\{(g_1,x):x\in EA(q^{d+1})\}$. It is not difficult to check that D is the desired relative difference set. □

We remark that D_1 is a difference set with parameters (2) in H as constructed by McFarland [5] (see also [1, VI.7.3]).

References

[1] Th. Beth, D. Jungnickel, and H. Lenz, *Design Theory,* Bibliographisches Institut, Mannheim-Wien-Zürich, and Cambridge University Press, 1985.

[2] D.A. Drake and D. Jungnickel, "Klingenberg structures and partial designs II", *Pacific J. Math.* 77 (1978) 389-415.

[3] D. Jungnickel, "On automorphism groups of divisible designs", *Canadian J. Math.* 34 (1982) 257-297.

[4] H. Lenz and D. Jungnickel, "On a class of symmetric designs", *Arch. Math.* 33 (1979) 590-592.

[5] R.L. McFarland, "A family of difference sets in non-cyclic groups", *J. Comb. Th.* A15 (1973) 1-10.

[6] W.D. Wallis, "Construction of strongly regular graphs using affine designs", *Bull. Austral. Math. Soc.* 4 (1971) 41-49.

Annals of Discrete Mathematics 34 (1987) 301–306
© Elsevier Science Publishers B.V. (North-Holland)

2-(25,10,6) Designs Invariant under the Dihedral Group of Order Ten

Stojan Kapralov, Ivan Landgev, and Vladimir Tonchev

Institute of Mathematics
Sofia 1090
P.O. Box 373
BULGARIA

TO ALEX ROSA ON HIS FIFTIETH BIRTHDAY

In this paper we enumerate all 2-(25,10,6) designs which are invariant under the dihedral group of order 10, which splits the point set into 5 orbits of length 5. A few designs with these parameters were previously known; one design is mentioned in the survey paper of Mathon and Rosa [5], and others are constructed by Hanani [2] and Tran Van Trung [11].

The concepts and notations in this paper are in accordance with those in [7].

As a first step we construct all possible orbit matrices of a 2-(25,10,6) design with respect to an automorphism of order 5 without fixed points or blocks. If $M = (m_{ij})$ is such an orbit matrix, the following equalities hold:

$$\begin{aligned} \sum_{i=1}^{5} m_{ij} &= 10 \quad (1 \leq j \leq 8); \\ \sum_{j=1}^{8} m_{ij} &= 16 \quad (1 \leq i \leq 5); \\ \sum_{j=1}^{8} m_{ij}(m_{ij}-1) &= 24 \quad (1 \leq i \leq 5); \\ \sum_{j=1}^{8} m_{\alpha j} m_{\beta j} &= 30 \quad (\alpha \neq \beta). \end{aligned} \tag{1}$$

Using an algorithm described in [3], we found 50 integer matrices with nonnegative entries satisfying (1). Every such matrix is extended (if possible) to an incidence matrix of a 2-(25,10,6) design, by use of an algorithm from [4]. As a result we found 106 designs. To select the nonisomorphic ones we use some invariants introduced in [9].

Let D be a design. Let C_j^x be the number of pairs occurring together with a given point x in exactly j blocks. For every point x, we compute the vector

$$C^x = (C_0^x, C_1^x, \ldots, C_\lambda^x).$$

The set of 106 designs is divided into 43 classes by these invariants.

A more detailed investigation showed that designs belonging to the same class are isomorphic, i.e. 43 is the exact number of nonisomorphic 2-(25,10,6) designs invariant under D_{10}, dividing the point set into 5 orbits of length 5.

The 43 nonisomorphic 2-(25,10,6) designs are listed in Table 1, where we use the following notations:

$$A = U + U^4;\ B = U^2 + U^3;\ A' = I + A;$$
$$B' = I + B;\ H = A + B,$$

where U is the circulant with first row 0 1 0 0 0, and I is the 5×5 identity matrix; a zero matrix will be denoted simply by 0.

Table 1

No. 1

$$\begin{array}{cccccccc} I & A' & A' & B' & B' & I & I & I \\ 0 & I & B & B & A & B' & B' & A' \\ A' & A & 0 & A & A' & I & B & B' \\ B' & A & B & I & A & A' & A' & 0 \\ A' & B & B' & B & 0 & A & I & B' \end{array}$$

No. 2

$$\begin{array}{cccccccc} I & A' & A' & B' & B' & I & I & I \\ A' & A & B & I & A & H & I & I \\ A' & B & B' & B & 0 & I & B' & A \\ 0 & I & B & A & B & A' & B' & B' \\ B' & A & 0 & A & A' & I & B & A' \end{array}$$

No. 3

$$\begin{array}{cccccccc} I & A' & A' & B' & B' & I & I & I \\ 0 & I & B & B & A & B' & B' & A' \\ A' & A & 0 & A & A' & I & B & B' \\ B' & A & B & A' & 0 & A' & I & A \\ A' & B & B' & 0 & B & A & B' & I \end{array}$$

No. 4

$$\begin{array}{cccccccc} A & B & I & B' & H & I & I & B \\ A & B & H & A' & I & I & A & I \\ A' & A & A & I & B & H & I & I \\ B' & A' & I & I & I & I & A' & B' \\ 0 & I & B & A & B & A' & B' & B' \end{array}$$

No. 5

$$\begin{array}{cccccccc} 0 & I & B & B & B & B' & B' & B' \\ A & B & A & A & H & 0 & B & B \\ A & A & A' & B' & B & A' & 0 & I \\ A' & A & B & A' & 0 & I & B' & A \\ A' & A' & I & 0 & A & B' & A & B \end{array}$$

No. 6

$$\begin{array}{cccccccc} 0 & I & B & B & B & B' & B' & B' \\ A & B & A & A & H & 0 & B & B \\ A & A & B' & A' & B & A' & 0 & I \\ A' & A & A & B' & 0 & I & A' & B \\ A' & A' & I & 0 & A & B' & B & A \end{array}$$

No. 7

$$\begin{array}{cccccccc} 0 & I & A & B & B & A' & B' & B' \\ B & B & A & A & H & 0 & A & B \\ A & A & A' & A' & B & B' & 0 & I \\ A' & B & A & B' & 0 & I & A' & B \\ A' & B' & I & 0 & A & B' & B & B \end{array}$$

No. 8

$$\begin{array}{cccccccc} 0 & I & A & B & B & B' & A' & B' \\ A & B & B & A & H & 0 & A & B \\ B & A & B' & A' & B & B' & 0 & I \\ A' & A & A & A' & 0 & I & A' & A \\ A' & B' & I & 0 & B & B' & B & A \end{array}$$

No. 9

$$\begin{array}{cccccccc} 0 & I & A & B & B & B' & A' & B' \\ A & B & A & B & H & 0 & B & A \\ B & A & A' & B' & A & A' & 0 & I \\ B' & B & B & B' & 0 & I & A' & A \\ A' & B' & I & 0 & A & A' & A & B \end{array}$$

No. 10

$$\begin{array}{cccccccc} 0 & I & B & A & B & B' & A' & B' \\ A & B & B & A & H & 0 & B & A \\ B & A & B' & A' & A & A' & 0 & I \\ A' & A & A & A' & 0 & I & B' & B \\ B' & A' & I & 0 & B & B' & B & A \end{array}$$

No. 11

$$\begin{array}{cccccccc} 0 & I & A & B & A & A' & A' & B' \\ B & A & A & B & H & 0 & B & A \\ B & A & B' & A' & A & A' & 0 & I \\ B' & B & B & B' & 0 & I & B' & B \\ A' & B' & I & 0 & A & A' & B & A \end{array}$$

No. 12

$$\begin{array}{cccccccc} 0 & I & B & B & B & B' & B' & B' \\ A & B & A & A & H & 0 & B & B \\ A & A' & I & B' & A & A' & 0 & B \\ A' & A & A & I & B & B' & A' & 0 \\ A' & A & B' & A & 0 & I & B & A' \end{array}$$

No. 13

$$\begin{array}{cccccccc} 0 & I & B & B & B & B' & B' & B' \\ A & B & A & A & H & 0 & B & B \\ A & A' & I & A' & B & B' & 0 & A \\ A' & A & B & I & A & A' & B' & 0 \\ A' & A & A' & B & 0 & I & A & B' \end{array}$$

No. 14

$$\begin{array}{cccccccc} 0 & I & B & A & B & A' & B' & B' \\ B & B & A & A & H & 0 & A & B \\ A & A' & I & A' & A & B' & 0 & B \\ A' & B & A & I & A & A' & B' & 0 \\ B' & A & A' & B & 0 & I & B & B' \end{array}$$

No. 15

$$\begin{array}{cccccccc} 0 & I & B & B & A & A' & B' & B' \\ B & A & A & A & H & 0 & B & B \\ A & A' & I & A' & A & A' & 0 & A \\ A' & A & B & I & B & B' & B' & 0 \\ B' & B & B' & A & 0 & I & B & A' \end{array}$$

No. 16

$$\begin{array}{cccccccc} 0 & I & A & B & B & B' & A' & B' \\ A & B & A & B & H & 0 & B & A \\ B & B' & I & A' & B & B' & 0 & A \\ A' & A & B & I & B & B' & B' & 0 \\ A' & A & A' & A & 0 & I & A & A' \end{array}$$

No. 17

$$\begin{array}{cccccccc} 0 & I & B & A & B & B' & A' & B' \\ A & B & A & B & H & 0 & A & B \\ A & B' & I & A' & B & A' & 0 & A \\ A' & B & B & I & A & B' & B' & 0 \\ A' & A & B' & B & 0 & I & A & A' \end{array}$$

No. 18

$$\begin{array}{cccccccc} 0 & I & A & A & B & A' & A' & B' \\ B & B & A & B & H & 0 & A & A \\ B & B' & I & A' & A & B' & 0 & B \\ A' & B & A & I & B & B' & B' & 0 \\ B' & A & A' & A & 0 & I & B & A' \end{array}$$

No. 19

$$\begin{array}{cccccccc} 0 & I & B & B & B & B' & B' & B' \\ A & B & A & A & H & 0 & B & B \\ A & A' & I & B' & A & A' & 0 & B \\ A' & A & B & A' & 0 & I & B' & A \\ A' & A & A' & 0 & B & B' & A & I \end{array}$$

No. 20

$$\begin{array}{cccccccc} 0 & I & B & B & B & B' & B' & B' \\ A & B & A & A & H & 0 & B & B \\ A & A' & I & A' & B & B' & 0 & A \\ A' & A & A & B' & 0 & I & A' & B \\ A' & A & B' & 0 & A & A' & B & I \end{array}$$

No. 21

$$\begin{array}{cccccccc} 0 & I & A & B & B & A' & B' & B' \\ B & B & A & A & H & 0 & B & A \\ A & B' & I & A' & A & A' & 0 & B \\ A' & B & B & B' & 0 & I & B' & A \\ A' & A & A' & 0 & A & B' & B & I \end{array}$$

No. 22

$$\begin{array}{cccccccc} 0 & I & A & B & B & B' & A' & B' \\ A & B & B & A & H & 0 & A & B \\ B & B' & I & A' & B & B' & 0 & A \\ A' & A & A & A' & 0 & I & A' & A \\ A' & A & B' & 0 & B & B' & B & I \end{array}$$

No. 23

$$\begin{array}{cccccccc} 0 & I & B & A & B & B' & A' & B' \\ A & B & A & B & H & 0 & A & B \\ B & A' & I & B' & A & B' & 0 & B \\ A' & A & A & A' & 0 & I & B' & B \\ B' & A & A' & 0 & B & A' & A & I \end{array}$$

No. 24

$$\begin{array}{cccccccc} 0 & I & B & B & A & B' & A' & B' \\ B & B & B & A & H & 0 & A & A \\ B & A' & I & B' & B & B' & 0 & A \\ B' & A & A & A' & 0 & I & A' & B \\ A' & A & B' & 0 & B & A' & A & I \end{array}$$

No. 25

$$\begin{array}{cccccccc} 0 & I & A & B & A & A' & A' & B' \\ B & A & A & B & H & 0 & B & A \\ B & B' & I & A' & A & A' & 0 & A \\ B' & B & B & B' & 0 & I & B' & B \\ A' & A & B' & 0 & A & A' & B & I \end{array}$$

No. 26

$$\begin{array}{cccccccc} 0 & I & B & B & B & B' & B' & B' \\ A & B & A & A & H & 0 & B & B \\ A & A' & A' & B' & 0 & I & A & B \\ A' & A & I & A & B & B' & 0 & A' \\ A' & A & B & I & A & A' & B' & 0 \end{array}$$

No. 27

$$\begin{array}{cccccccc} 0 & I & A & B & B & A' & B' & B' \\ B & B & A & A & H & 0 & A & B \\ A & B' & A' & A' & 0 & I & B & A \\ A' & B & I & B & A & B' & 0 & B' \\ A' & A & A & I & B & B' & A' & 0 \end{array}$$

No. 28

$$\begin{array}{cccccccc} 0 & I & A & B & B & B' & A' & B' \\ B & A & A & A & H & 0 & B & B \\ A & A' & A' & A' & 0 & I & A & A \\ B' & B & I & A & B & B' & 0 & A' \\ A' & A & B & I & B & B' & B' & 0 \end{array}$$

No. 29

$$\begin{array}{cccccccc} 0 & I & B & A & B & B' & A' & B' \\ B & A & A & A & H & 0 & B & B \\ A & A' & A' & A' & 0 & I & A & A \\ B' & A & I & B & B & B' & 0 & A' \\ A' & B & A & I & B & B' & B' & 0 \end{array}$$

No. 30

$$\begin{array}{cccccccc} B & H & I & A & I & B & I & B' \\ A' & I & H & A & I & B & I & A \\ A' & B & I & B & A & B' & B' & 0 \\ 0 & I & A & B & B & B' & A' & B' \\ A & B & B & A & H & 0 & A & B \end{array}$$

No. 31

$$\begin{array}{cccccccc} A' & I & A' & B' & I & B' & I & I \\ B' & A' & 0 & B & I & A & B' & B \\ 0 & I & A & A & B & A' & A' & B' \\ A & A & B & A' & A' & 0 & I & B' \\ B & A' & B' & 0 & B' & B & A & I \end{array}$$

No. 32

$$\begin{array}{cccccccc} A' & I & B' & I & A' & I & B' & I \\ B' & B & B & B' & 0 & A' & A & I \\ 0 & I & A & B & A & B' & A' & A' \\ B & A' & 0 & B' & B' & I & B & A \\ A & B' & A' & I & B & A & 0 & A' \end{array}$$

No. 33

$$\begin{array}{cccccccc} B & B' & I & I & H & I & A & B \\ B & B' & I & H & I & B & I & A \\ B & B' & H & I & I & A & B & I \\ H & 0 & B & B & B & A & A & A \\ 0 & I & A & A & A & A' & A' & A' \end{array}$$

No. 34

$$\begin{array}{cccccccc} B & B' & I & I & H & I & A & B \\ B' & I & I & H & B & B & I & A \\ B' & I & H & I & B & A & B & I \\ 0 & I & A & A & A & A' & A' & A' \\ B & H & B & B & 0 & A & A & A \end{array}$$

No. 35

$$\begin{array}{cccccccc} A & I & I & B & B & B' & H & I \\ A & I & H & B & A' & A & I & I \\ B & H & A & B' & I & I & B & I \\ H & A & I & I & A & B & I & A' \\ 0 & B & A & B & A & B & A & H \end{array}$$

No. 36

$$\begin{array}{cccccccc} A & I & I & B & B & B' & H & I \\ A & B & A' & I & H & A & I & I \\ B & H & A & B' & I & I & B & I \\ H & I & A & A & I & B & I & A' \\ 0 & B & A & B & A & B & A & H \end{array}$$

No. 37

$$\begin{array}{cccccccc} I & I & I & I & A' & B' & A' & B' \\ I & B & A & H & I & I & A & A' \\ B & I & H & A & A & A' & I & I \\ B & H & I & A & B' & B & I & I \\ H & B & A & I & I & I & B' & B \end{array}$$

No. 38

$$\begin{array}{cccccccc} I & I & I & I & A' & A' & B' & B' \\ I & B & B & H & I & I & A & B' \\ A & I & H & A & B & A' & I & I \\ B & H & A & I & I & A & A' & I \\ H & A & I & B & B' & I & I & B \end{array}$$

No. 39

$$\begin{array}{cccccccc} I & I & I & I & A' & B' & A' & B' \\ I & B & A & H & I & I & A & A' \\ B & I & H & A & A & A' & I & I \\ B & H & A & I & I & B & B' & I \\ H & B & I & A & B' & I & I & B \end{array}$$

No. 40

$$\begin{array}{cccccccc} I & I & I & I & A' & A' & B' & B' \\ I & B & B & H & I & I & A & B' \\ A & I & H & A & B & A' & I & I \\ B & H & A & I & B' & I & I & B \\ H & A & I & B & I & A & A' & I \end{array}$$

No. 41

$$\begin{array}{cccccccc} I & I & I & I & A' & B' & A' & B' \\ I & A & B & H & I & I & A & A' \\ B & B & B' & A & I & H & I & I \\ B & A' & A & A & H & I & I & I \\ H & A & B & I & I & I & B' & B \end{array}$$

No. 42

$$\begin{array}{cccccccc} I & I & I & I & A' & A' & B' & B' \\ I & B & B & H & I & I & A & B' \\ A & B & A' & A & I & H & I & I \\ B & H & A & I & B' & I & I & B \\ H & I & A & B & A & I & A' & I \end{array}$$

No. 43

$$\begin{array}{cccccccc} I & I & I & I & A' & A' & B' & B' \\ A & B & B & B' & I & I & I & H \\ B & A & B' & B & I & I & H & I \\ A & A' & B & A & I & H & I & I \\ A' & A & A & B & H & I & I & I \end{array}$$

It is of interest to find out which of the above designs are embeddable into symmetric 2-(41,16,6) designs. The 2-(25,10,6) design given in [2] is non-embeddable, since it contains a pair of blocks intersecting in 7 points. Recently, Tran Van Trung [11] constructed a non-embeddable 2-(25,10,6) design invariant under D_{10} (isomorphic to number 20 in Table 1).

We checked the embeddability of our 43 quasi-residual 2-(25,10,6) designs with the help of an algorithm described in [6]. It turns out that the only embeddable designs are numbers 33 and 28. The first one extends to the symmetric 2-(41,16,6) design found by Bridges, Hall, and Hayden [1], and Tran Van Trung [11]; the second one extends to its dual [8].

The remaining designs are quasi-residual but not residual.

References

[1] W.G. Bridges, M. Hall Jr., and J.L. Hayden, "Codes and designs", *J. Comb. Theory* A31 (1981) 155-174.

[2] H. Hanani, "Balanced incomplete block designs and related designs", *Discrete Math.* 11 (1975) 255-369.

[3] S. Kapralov, "Algorithms for generating of orbit matrices", *Proceedings of the Conference; 100th anniversary of the birth of academician L. Tchakalov,* Samocov, 1986, proceedings to appear.

[4] I. Landgev, "On symmetric 2-(41,16,6) designs invariant under the Frobenius group of order 10", *Compt. Rend. de l'Acad. Bulg. des Sci.,* to appear.

[5] R. Mathon and A. Rosa, "Tables of parameters of BIBDs with $r \leq 41$ including existence, enumeration, and resolvability results", *Ann. Disc. Math.* 26 (1985) 275-308.

[6] V. Tonchev, "Quasi-residual designs, codes, and graphs", *Colloq. Math. Soc. Janós Bolyai* 37 (1981) 685-695.

[7] V. Tonchev, *Combinatorial Configurations: Designs, Codes, Graphs* Nauka i Iskustvo, Sofia, 1984.

[8] V. Tonchev, "The isomorphism of certain symmetric block designs", *Compt. Rend. de l'Acad. Bulg. des Sci.* 2 (1985) 161-164.

[9] V. Tonchev, "Hadamard matrices of order 28 with automorphisms of order 7", *J. Comb. Theory* A40 (1985) 62-81.

[10] Tran Van Trung, "The existence of symmetric block designs with parameters (41,16,6) and (66,26,10)", *J. Comb. Theory* A33 (1982) 201-204.

[11] Tran Van Trung, "Non-embeddable quasi-residual designs with $k<v/2$", *J. Comb. Theory* A, to appear.

Annals of Discrete Mathematics 34 (1987) 307—314
© Elsevier Science Publishers B.V. (North-Holland)

On the Steiner Systems S(2,4,25) invariant under a group of order 9

Earl S. Kramer

University of Nebraska,
Lincoln, Nebraska
68588-0323

Spyros S. Magliveras

University of Nebraska,
Lincoln, Nebraska
68588-0155

Vladimir D. Tonchev

Bulgarian Academy of Sciences

TO ALEX ROSA ON HIS FIFTIETH BIRTHDAY

ABSTRACT

We establish that there are exactly five non-isomorphic Steiner systems $S(2,4,25)$ invariant under a group of order 9. These designs and all other $S(2,4,25)$'s known to us are presented along with their full automorphism groups and block-graph invariants.

1. Introduction

We assume that the reader is familiar with the basic facts and notions from design theory (see [1], [5], [8]). We let G be the full automorphism group of a 2—(25,4,1) design, i.e. a Steiner system $S(2,4,25)$. In section 2 we show that if $p \,|\, |G|$ then p is 2,3,5 or 7. The following was shown by A.E. Brouwer (unpublished) and independently by the third author (also unpublished) :

Theorem 1.1 There are exactly 3 nonisomorphic Steiner systems $S(2,4,25)$ having an automorphism of order 7 and exactly one with an automorphism of order 5. The orders of G are 504, 63, 21 and 150 respectively.

The design with $|G|=150$ is the only one with a transitive automorphism group and was undoubtedly the first $S(2,4,25)$ discovered. It is a member of an infinite class of $S(2,4,v)$'s constructed by R. C. Bose in 1939 (cf [5], Ch.15,Th.15.3.5 or see [1],Th. VII.7.2).

Two non-isomorphic $S(2,4,25)$'s with $|G|=9$ were found by the third author [10], but A. E. Brouwer [2] pointed out that these two designs are in fact isomorphic with two designs constructed by L.P. and A.J. Petrenjuk [9]. These designs of the Petrenjuks were obtained by applying certain transformations to previously known designs. An $S(2,4,25)$ with a full automorphism group of order 6 was announced by H. Gropp [4]. In section 4 we present the only design with $|G|=6$ known to us, namely the one given by the Petrenjuks [9].

An invariant that has proven useful in establishing non-isomorphism of two designs is the *block-graph invariant*. The *block-graph* of a design **D** is the graph whose vertices are the blocks of **D**, and where two blocks B_1 and B_2 are adjacent whenever $B_1 \cap B_2 \neq \emptyset$. For a given vertex v, let n_i be the number of pairs of vertices $\dot{v},\ddot{v}$ different from v such that exactly i other vertices are simultaneously adjacent to $v,\dot{v},\ddot{v}$. The matrix of row vectors $(...,n_i,...)$ one for each representative v of a block orbit under the automorphism group **G** of **D** is the *block-graph invariant* of **D**.

2. Some facts about automorphisms of S(2,4,25)'s

We presently develop some facts about the structure of automorphisms of $S(2,4,25)$'s.

Lemma 2.1 Let α be an automorphism of an $S(2,4,25)$ where α has prime order $p \geq 3$. Then α fixes 0,1, or 4 points.

Proof: Let **B** be the 50 blocks of our $S(2,4,25)$ on the set $X = Y \cup Z$, with $Y \cap Z = \emptyset$, where Z is the set of fixed points of α. Let $\mathbf{B}_i = \{B \in \mathbf{B}: |B \cap Z| = i\}$, $0 \leq i \leq 4$, and let $b_i = |\mathbf{B}_i|$. Henceforth we assume that $z = |Z| \geq 2$. If $B \in \mathbf{B}$ covers a pair of points in Z and because $\lambda = 1$, it follows that $\alpha(B) = B$, so $B \subset Z$. Thus, $b_2 = b_3 = 0$ and $\mathbf{B}_4$ is the set of blocks of an $S(2,4,z)$ with point set Z. Now, the necessary conditions for an $S(2,4,z)$ with $z < 25$ require that $z = 4,13$ or 16. But to cover pairs from Y we need

$$6b_0 + 3b_1 = \binom{25-z}{2}, \quad \text{and} \quad b_0 + b_1 = 50 - b_4 \tag{1}$$

We easily conclude that $z = 4$, and we are done. □

Theorem 2.2 Let p be a prime dividing the order of the automorphism group G of an $S(2,4,25)$. Then, $p = 2,3,5$ or 7. Further, if $\alpha \in G$ has order p and:

(i) $p = 3$, then α fixes 1 or 4 points;
(ii) $p = 5$, then α fixes no points;
(iii) $p = 7$, then α fixes 4 points.

Proof: Let α be an automorphism of order p of an $S(2,4,25)$. If $p \in \{11,13,17,19,23\}$ then α fixes z points where $z \in \{2,3,6,8,12,14\}$. Our result then follows from the previous lemma □

Theorem 2.3 If $3^k \mid |G|$ then $k \leq 2$ and if H is a subgroup of G, with $|H| = 9$, then H is elementary abelian, i.e. H is isomorphic to $Z_3 \times Z_3$.

Proof: Assume that $k \geq 2$ so there is a subgroup H of order 9. If there were an automorphism of order 9 of the $S(2,4,25)$ then its cube would fix 7 or 16 points, contrary to Theorem 2.2., so H cannot be cyclic. Using the Cauchy-Frobenius-Burnside lemma, one may compute the number m of orbits of H on X by:

$$m = \frac{1}{9}\sum_{h\in H} \chi(h) \tag{2}$$

where $\chi(h)$ denotes the number of points fixed by h. Since the set of points fixed by an automorphism of order 3 forms a subsystem $S(2,4,z)$, any such automorphism can fix either 1 or 4 points of X. Moreover, the non-identity elements of H split into four subgroups of order 3 and any pair of elements of order 3 belonging to the same subgroup fix the same number of points. Thus, equation (2) can be rewritten as follows:

$$m = \frac{1}{9}\cdot(25+a.1+b.4) \tag{3}$$

where $a + b = 8$ and a,b are even. Under these conditions the only solution to (3) is $a = b = 4$. Hence, H partitions the point set X into 5 orbits, and the orbit lengths are clearly 1,3,3,9,9. Now, H cannot be contained in a subgroup $K \leq G$ of order 27, for otherwise the stabilizer in K of a point from an H-orbit of length 3 would be a subgroup of order 9 fixing at least 3 points, contradicting the above conclusion. This completes our proof □

3. The S(2,4,25)'s invariant under $G = Z_3 \times Z_3$

Having established the structure of a group of automorphisms of order 9, as a permutation group, we proceed to show that there are precisely five isomorphism classes of Steiner systems $S(2,4,25)$ admitting such a group H as a group of automorphisms.

From the orbit structure of H on the 25 points, we see that it is represented regularly on the two point-orbits of length 9, has two point-orbits of length 3, and fixes a point. The only question is whether the stabilizer in H of a point out of one orbit of length 3 is the same subgroup as the stabilizer of a point out of the other orbit of length 3. We quickly see, however, that this cannot happen because otherwise there would exist an element of order 3 fixing 7 points , contrary to Theorem 2.2. Without loss of generality we choose $H = \langle\alpha,\beta\rangle$, where:

$\alpha = (1\ 2\ 3)(4\ 5\ 6)(7\ 8\ 9)(10\ 11\ 12)(13\ 14\ 15)(16\ 17\ 18)(19)(20)(21)(22\ 23\ 24)(25)$

$\beta = (1\ 4\ 7)(2\ 5\ 8)(3\ 6\ 9)(10\ 13\ 16)(11\ 14\ 17)(12\ 15\ 18)(19\ 20\ 21)(22)(23)(24)(25)$

and proceed to generate the fused incidence matrix $\mathbf{A} = A_{2,4} = A_{2,4}(H) = (a_{i,j})$. For a definition of the matrices $A_{t,k}(G)$ and related results the reader is referred to [6] and [7]. H has 1422 orbits on the collection of all 4-subsets of $X = \{1,2, \cdots ,25\}$ and 36 orbits on the 2-subsets of X, so that $A_{2,4}$ is a 36×1422 matrix. An $S(2,4,25)$ exists with H as an automorphism group if and only if there is a collection of columns of A whose sum is the all 1's column vector $\mathbf{1} = (1,1, \cdots ,1)^T$, i.e. if and only if there is a binary vector $\mathbf{x} = (x_1,x_2, \cdots ,x_{1422})^T$ such that $\mathbf{Ax} = \mathbf{1}$. If a column of $\mathbf{A}$ has an entry $a_{i,j} > 1$ then obviously it can not be selected, so we delete all columns of $\mathbf{A}$ which have entries greater than 1 to arrive at a submatrix $\hat{\mathbf{A}}$ of size 36×739.

We can easily see that an $S(2,4,25)$ must be comprised of 2 orbits of length 1, and at least 4 orbits of length 3 on 4-sets. An analysis of the orbits of length 3 shows, however, that no more than 4 can be chosen. Therefore, a design consists of exactly 2 orbits of length 1, 4 orbits of length 3 and 4 orbits of length 9 of 4-sets.

There are exactly 2 H-orbits of length 1 on 4-sets, hence both must be chosen for an $S(2,4,25)$. There are exactly 20 orbits of length 3 on 4-subsets and exactly 4 must be chosen in a design. Using this information, we can further restrict the scope of the search for solutions. With the above restrictions, and using a new, recursive algorithm we find all binary solutions to

$$\hat{\mathbf{A}}\mathbf{x} = \mathbf{1} \tag{4}$$

with $\mathbf{x}$ of dimension 739.

There are a total of 1944 solutions to (4) which form 6 orbits, each of length 324, under the action of $N = N_{S_{25}}(H)$, the normalizer in S_{25} of H. Here the order of N is $9^3 \cdot 4$. Four of the N-orbits are distinguished among themselves and from the union of the remaining two by means of the block-graph invariant. The remaining two orbits have the same block-graph invariant and fuse under

$\pi = (1)(2\ 3)(4\ 7)(5\ 9)(6\ 8)(10\ 18)(11\ 17)(12\ 16)(13\ 15)(14)(19)(20\ 21)(22)(23\ 24)(25).$

Thus we have established:

Theorem 3.1 There are exactly 5 isomorphism classes of $S(2,4,25)$'s admitting a group of automorphisms of order 9. The orders of G are 504, 63, 9, 9 and 9 respectively.

As a consequence, the only remaining $S(2,4,25)$'s that can exist are those with automorphism group G, of order $|G| = 2^i 3^j,\ j \leq 1$.

4. The S(2,4,25)'s

In this section we present the $S(2,4,25)$'s known to us. For each design we give generators of the automorphism group G, representatives for each block-orbit, the orbit sizes, and the block-graph invariants. The permutations α and β appearing below have been given in section 3.

Design 1. $H = <\alpha,\beta> \leq G = <\alpha,\gamma_1>$, $|G| = 7 \cdot 8 \cdot 9 = 504$, where:

$\gamma_1 = (1\ 2\ 24\ 12\ 8\ 18\ 19\ 9\ 7\ 23\ 17\ 4\ 14\ 21\ 5\ 6\ 22\ 13\ 3\ 10\ 20)(11\ 16\ 15)(25)$

		BLOCK-GRAPH INVARIANT								
ORB.REP.	ORB.SIZE	n_4	n_5	n_6	n_7	n_8	n_9	n_{10}	n_{11}	n_{12}
1 2 3 19	42	1	0	30	204	411	372	158	0	0
1 5 9 25	8	0	0	63	126	462	378	147	0	0

Design 2. $H = <\alpha,\beta> \leq G = <\alpha,\gamma_2>$, $|G| = 7 \cdot 9 = 63$, where:

$\gamma_2 = (1\ 21\ 13\ 18\ 4\ 15\ 2\ 5\ 19\ 17\ 10\ 8\ 16\ 6\ 9\ 20\ 12\ 14\ 3\ 11\ 7)(22\ 23\ 24)(25)$

		BLOCK-GRAPH INVARIANT								
ORB.REP.	ORB.SIZE	n_4	n_5	n_6	n_7	n_8	n_9	n_{10}	n_{11}	n_{12}
1 2 3 19	21	0	30	208	396	414	110	18	0	0
1 4 7 22	21	0	33	202	399	411	116	15	0	0
19 20 21 25	7	3	27	207	393	414	120	12	0	0
22 23 24 25	1	0	63	126	462	378	147	0	0	0

Design 3. $G = <\alpha,\beta>$, $|G| = 9$,

		BLOCK-GRAPH INVARIANT								
ORB.REP.	ORB.SIZE	n_4	n_5	n_6	n_7	n_8	n_9	n_{10}	n_{11}	n_{12}
1 15 21 24	9	0	0	54	152	429	411	121	9	0
1 6 10 11	9	0	0	45	179	399	429	112	12	0
1 12 13 23	9	0	0	45	170	429	393	130	9	0
1 14 16 20	9	0	0	44	171	427	403	117	14	0
10 13 16 25	3	0	1	49	158	428	412	118	9	1
1 5 9 25	3	0	0	48	165	422	414	114	13	0
1 4 7 22	3	0	0	46	174	408	421	117	9	1
1 2 3 19	3	0	0	36	196	405	399	131	9	0
19 20 21 25	1	0	0	63	141	399	477	78	18	0
22 23 24 25	1	0	0	48	153	453	390	120	9	3

Design 4. $G = <\alpha,\beta>$, $|G| = 9$,

		BLOCK-GRAPH INVARIANT								
ORB.REP.	ORB.SIZE	n_4	n_5	n_6	n_7	n_8	n_9	n_{10}	n_{11}	n_{12}
1 11 13 21	9	0	0	41	185	409	403	130	8	0
1 16 18 23	9	0	0	39	191	405	399	136	6	0
1 6 10 14	9	0	0	39	187	415	393	134	8	0
1 15 20 24	9	0	0	39	179	432	390	121	15	0
10 13 16 25	3	0	1	31	203	401	403	127	9	1
1 2 3 19	3	0	0	54	142	462	372	140	6	0
1 4 7 22	3	0	0	43	183	405	406	135	3	1
1 5 9 25	3	0	0	36	192	419	381	141	7	0
19 20 21 25	1	0	0	63	141	399	477	78	18	0
22 23 24 25	1	0	0	39	198	372	453	102	9	3

Design 5. $G = <\alpha,\beta>$, $|G| = 9$,

ORB.REP.	ORB.SIZE	BLOCK-GRAPH INVARIANT								
		n_4	n_5	n_6	n_7	n_8	n_9	n_{10}	n_{11}	n_{12}
1 14 21 24	9	0	0	59	153	400	442	117	5	0
1 13 17 20	9	0	0	58	144	432	410	126	6	0
1 6 11 22	9	0	0	54	162	396	450	102	12	0
1 4 10 15	9	0	0	51	162	414	426	111	12	0
1 2 3 19	3	1	0	61	130	435	433	106	9	1
10 13 16 22	3	1	0	54	150	426	411	131	3	0
10 11 12 25	3	0	0	76	111	420	463	99	6	1
1 5 9 25	3	0	0	48	168	417	411	123	9	0
22 23 24 25	1	0	0	63	132	444	396	141	0	0
19 20 21 25	1	0	0	48	153	453	390	120	9	3

Design 6. $G = <\gamma_3,\gamma_4>$, $|G| = 2\cdot3\cdot25 = 150$, where:

$$\gamma_3 = (1\ 23)(2\ 24)(3\ 25)(4\ 21)(5\ 22)(6\ 17)(7\ 18)(8\ 19)(9\ 20)(10\ 16)(11)(12)(13)(14)(15)$$

$$\gamma_4 = (1\ 25\ 5)(2\ 19\ 10)(3\ 13\ 15)(4\ 7\ 20)(6\ 21\ 24)(8\ 14\ 9)(11\ 22\ 18)(12\ 16\ 23)(17)$$

ORB.REP.	ORB.SIZE	BLOCK-GRAPH INVARIANT								
		n_4	n_5	n_6	n_7	n_8	n_9	n_{10}	n_{11}	n_{12}
1 2 6 25	25	0	0	33	198	423	363	156	3	0
1 3 11 19	25	0	0	45	171	435	369	156	0	0

Design 7. $G = <\gamma_5,\gamma_6>$, $|G| = 21$, where:

$$\gamma_5 = (1\ 2\ 3\ 4\ 5\ 6\ 7)(8\ 9\ 10\ 11\ 12\ 13\ 14)(15\ 16\ 17\ 18\ 19\ 20\ 21)(22)(23)(24)(25)$$

$$\gamma_6 = (1\ 2\ 4)(3\ 6\ 5)(7)(8\ 9\ 11)(10\ 13\ 12)(14)(15\ 16\ 18)(17\ 20\ 19)(21)(22\ 23\ 24)(25)$$

ORB.REP.	ORB.SIZE	BLOCK-GRAPH INVARIANT								
		n_4	n_5	n_6	n_7	n_8	n_9	n_{10}	n_{11}	n_{12}
1 9 17 22	21	0	0	43	174	427	394	129	8	1
1 2 4 14	7	0	0	46	176	402	425	121	3	3
1 8 15 25	7	0	0	38	198	393	403	141	3	0
1 18 20 21	7	0	0	40	178	432	383	137	3	3
8 9 11 21	7	0	0	42	175	438	366	152	3	0
22 23 24 25	1	0	0	35	203	399	385	154	0	0

Design 8. $G = <\gamma_7,\gamma_8>$, $|G| = 6$, where:

$\gamma_7 = (1\ 20\ 16)(2\ 10\ 3)(5\ 17\ 18)(4\ 19\ 22)(6\ 9\ 25)(11\ 15\ 14)(7\ 21\ 12)(8\ 13\ 24)(23)$

$\gamma_8 = (1)(2\ 18)(3\ 5)(4\ 22)(6\ 15)(7)(8\ 24)(9\ 11)(10\ 17)(12\ 21)(13)(14\ 25)(16\ 20)(19)(23)$

		BLOCK-GRAPH INVARIANT								
ORB.REP.	ORB.SIZE	n_4	n_5	n_6	n_7	n_8	n_9	n_{10}	n_{11}	n_{12}
1 2 4 25	6	0	0	45	175	415	405	128	8	0
1 3 8 11	6	0	1	39	195	384	423	129	5	0
2 3 14 19	6	0	1	40	188	400	407	136	4	0
2 6 9 12	6	0	0	43	188	392	416	133	4	0
1 6 7 15	3	0	0	52	167	402	422	130	3	0
1 10 13 17	3	0	0	49	161	433	395	130	8	0
1 12 19 21	3	0	0	51	156	431	409	120	7	2
2 7 18 23	3	0	0	46	174	416	398	138	4	0
2 17 21 24	3	0	0	36	199	399	399	137	6	0
4 6 8 14	3	0	0	63	136	421	440	107	8	1
4 12 13 24	3	0	0	46	177	407	407	135	4	0
8 9 15 23	3	0	2	60	137	420	438	112	7	0
1 16 20 23	1	0	0	33	192	441	345	162	3	0
4 19 22 23	1	0	0	57	147	435	393	144	0	0

Added in Proof:

R. Mathon, E. Kramer, and S. Magliveras have found eight new S(2,4,25)'s admitting an automorphism of order 3. They also establish that there are no new S(2,4,25)'s with an automorphism of order 2. In view of the above, all S(2,4,25)'s with non-trivial automorphism group are now known.

References

[1] T. Beth, D. Jungnickel, H. Lenz, *Design Theory,* Bibliographisches Institut, Mannheim-Wien-Zurich, and Cambridge University Press, Cambridge, 1985.

[2] A. E. Brouwer , (personal communication)

[3] F. C. Bussemaker, R. Mathon, J.J. Seidel, "Tables of two-graphs", *Combinatorics and Graph Theory, Proc. Conf. Calcutta 1980,* Lecture Notes Math. 885, Springer 1981, pp. 70 - 112.

[4] H. Gropp (personal communication).

[5] M. Hall, *Combinatorial Theory,* 2nd Edition, Wiley, N.Y. 1986.

[6] E.S. Kramer, D.M. Mesner, "t-designs on hypergraphs", *Discrete Math.* 15 (1976) 263 - 296.

[7] E.S. Kramer, D. Leavitt, S.S. Magliveras, "Construction Procedures for t-designs and the existence of New Simple 6-designs", *Annals of Discrete Math.* 26 (1985) 247 - 274.

[8] R. Mathon, and A. Rosa, "Tables of Parameters of BIBDs with $r \leq 41$ including Existence, Enumeration, and Resolvability Results", *Annals of Discrete Math.* 26 (1985) 275 - 308.

[9] L.P. Petrenjuk and A.J. Petrenjuk, "List of 8 S(2,4,25)'s", (preprint).

[10] V.D. Tonchev, "Two new Steiner systems S(2,4,25)", *Compt. Rend. Acad. Bulg. Sci.* 39 (1986) 47 - 48.

Annals of Discrete Mathematics 34 (1987) 315—318
© Elsevier Science Publishers B.V. (North-Holland)

Simple 5–(28,6,λ) Designs from $PSL_2(27)$

Donald L. Kreher and Stanisław P. Radziszowski

School of Computer Science and Technology
Rochester Institute of Technology
Rochester, NY 14623
U.S.A.

TO ALEX ROSA ON HIS FIFTIETH BIRTHDAY

ABSTRACT

Simple 5–(28,6,λ) designs with $PSL_2(27)$ as an automorphism group are constructed for each λ, $2 \leq \lambda \leq 21$.

1. Introduction

A *t—design,* or t–(v,k,λ) *design* is a pair (X,B) with a v–set X of *points* and a family B of k–subsets of X called *blocks* such that any t points are contained in exactly λ blocks. A t–(v,k,λ) design (X,B) is *simple* if no block in B is repeated and is said to have $G \leq Sym(X)$ as an *automorphism group* if whenever K is a block $K^\alpha = \{x^\alpha : x \in K\}$ is also a block for all $\alpha \in G$.

In this paper we show that for each λ, $2 \leq \lambda \leq 21$, a 5–(28,6,λ) design exists with $G = PSL_2(27)$ as an automorphism group, leaving open only the existence of a 5–(28,6,1) design with some group other than $PSL_2(27)$. Incidently, Denniston showed that a 5–(28,7,1) design does indeed exist with $PSL_2(27)$, see [2].

2. Preliminaries

As with our construction of two disjoint nonisomorphic simple 6–(14,7,4) designs [6] as well as with the only other known examples of 6-designs with small λ [4,9], we start with the following observation of Kramer and Mesner [3]:

A t–(v,k,λ) design exists with $G \leq Sym(X)$ as an automorphism group if and only if there is a (0,1)–solution vector U to the matrix equation

$$A_{tk}U = \lambda J, \tag{1}$$

where

a. The rows of A_{tk} are indexed by the G–orbits of t–subsets of X;

b. The columns of A_{tk} are indexed by the G–orbits of k–subsets of X;

c. $A_{tk}[\Delta,\Gamma] = |\{K \in \Gamma : K \supset T_0\}|$ where $T_0 \in \Delta$ is any representative;

d. $J = [1,1,1,\ldots,1]^T$.

If we choose the group G to be $PSL_2(27)$ acting on the projective line $X = GF(3^3) \cup \infty$ then G has order 9828 and an isomorphic copy is generated by the permutations α, β, and γ on $\{0,1,2,...27\}$ given below:

$\alpha = (0,26,27)(1,2,10)(3,6,4)(5,15,19)(7,11,21)(8,17,14)(9,18,12)(13)(16,24,25)(20,23,22);$
$\beta = (0,13)(1,12)(2,11)(3,10)(4,9)(5,8)(6,7)(14,25)(15,24)(16,23)(17,22)(18,21)(19,20)(26,27);$
$\gamma = (0,23,22,10,11,1,27,26,25,15,16,4,3)(2,9,13,17,24,19,8,20,14,12,6,18,7)(5)(21).$

In this action G has exactly 10 orbits of 5–subsets and exactly 54 orbits of 6–subsets, see figure 1 and 2. Hence, the $A_{5,6}$ matrix belonging to G has 10 rows and 54 columns and is displayed in figure 3.

No.	Representative		
1	0 1 2 6 7	3	0 1 2 3 6
2	0 1 2 4 6	4	0 1 2 4 11
3	0 1 2 3 6	5	0 1 2 4 8
4	0 1 2 4 11	6	0 1 2 3 8
5	0 1 2 4 8	7	0 1 2 5 8

Figure 1. Orbit representatives of 5-subsets

No.	Representative				
1	0 1 2 6 7 16	19	0 1 2 6 7 14	37	0 1 2 3 8 17
2	0 1 2 6 7 12	20	0 1 2 3 6 14	38	0 1 2 3 6 20
3	0 1 2 6 7 8	21	0 1 2 6 7 13	39	0 1 2 6 7 19
4	0 1 2 4 6 7	22	0 1 2 4 11 13	40	0 1 2 4 8 18
5	0 1 2 4 5 6	23	0 1 2 4 6 14	41	0 1 2 4 6 20
6	0 1 2 3 4 6	24	0 1 2 4 6 15	42	0 1 2 6 7 21
7	0 1 2 3 6 7	25	0 1 2 3 6 15	43	0 1 2 4 6 21
8	0 1 2 5 6 7	26	0 1 2 4 11 14	44	0 1 2 3 6 21
9	0 1 2 4 6 8	27	0 1 2 4 11 15	45	0 1 2 4 6 22
10	0 1 2 6 7 10	28	0 1 2 6 7 15	46	0 1 2 4 11 21
11	0 1 2 4 6 9	29	0 1 2 3 6 16	47	0 1 2 6 7 25
12	0 1 2 3 6 9	30	0 1 2 6 7 22	48	0 1 2 4 6 23
13	0 1 2 6 7 9	31	0 1 2 6 7 20	49	0 1 2 3 6 23
14	0 1 2 6 7 11	32	0 1 2 6 7 18	50	0 1 2 4 11 22
15	0 1 2 4 7 11	33	0 1 2 6 7 17	51	0 1 2 4 6 25
16	0 1 2 5 8 10	34	0 1 2 3 8 16	52	0 1 2 4 6 26
17	0 1 2 4 8 11	35	0 1 2 3 12 16	53	0 1 2 3 6 26
18	0 1 2 3 8 11	36	0 1 2 4 6 18	54	0 1 2 6 7 27

Figure 2. Orbit representatives of 6-subsets

1 1 1 1 0 0 1 1 0 2 0 0 1 2 0 0 0 0 2 0 1 0 0 0 0 0 0 1 0 1 1 1 1 0 0 0 0 0 1 0 0 1 0 0 0 0 1 0 0 0 0 0 0 1

0 0 2 1 2 1 1 0 1 1 1 0 0 0 0 0 0 0 0 0 0 0 1 1 0 0 0 1 0 0 0 0 0 0 0 1 0 0 1 0 1 0 1 0 1 0 1 1 0 0 1 1 0 1

1 0 0 0 2 1 1 1 0 0 1 1 2 0 0 0 0 0 1 1 1 0 0 1 1 0 0 1 1 0 1 0 0 0 0 0 0 1 0 0 0 0 0 1 0 0 0 1 1 0 0 0 1 0

0 0 0 0 0 0 0 0 0 1 0 0 0 1 1 0 2 0 1 1 0 1 1 2 1 1 1 0 2 0 0 0 0 0 0 0 0 0 0 0 0 0 2 0 0 1 0 0 1 1 1 1 0 0

0 1 1 0 0 1 0 0 1 0 1 1 1 0 0 0 1 0 0 2 0 1 0 0 1 0 0 0 0 0 0 0 0 0 0 0 0 1 0 1 1 1 0 1 1 0 0 0 0 2 1 2 0 0

1 0 2 0 0 0 0 0 0 1 0 0 0 0 0 0 1 2 0 0 0 0 0 1 0 0 0 0 0 0 0 1 1 1 0 1 1 0 0 1 0 0 1 1 2 1 2 1 0 0 0 1 0 0

2 0 0 0 0 0 0 0 0 0 1 1 0 1 0 2 1 0 0 0 1 0 0 0 0 1 0 1 1 0 0 1 0 1 0 1 1 0 0 0 0 2 1 2 0 0 0 0 0 1 0 0 1 0

0 0 0 0 0 0 0 1 0 1 0 0 0 1 1 0 0 0 1 1 0 0 1 0 0 1 1 0 0 1 0 2 0 0 0 2 0 0 0 0 0 0 0 0 0 1 0 2 1 1 0 1 2 1

0 0 0 0 0 0 0 0 1 0 2 0 2 0 0 0 1 0 1 1 0 0 0 1 0 0 1 2 1 0 0 1 0 1 2 0 1 0 1 0 0 0 0 0 1 0 1 1 0 0 0 0 1 0

1 0 0 1 2 0 0 0 0 0 0 0 0 1 1 0 0 0 0 0 0 1 0 0 0 0 0 0 1 1 1 0 1 0 0 1 0 1 0 1 1 2 1 1 1 0 1 0 0 1 0 0 1 0

Figure 3. The $A_{5,6}$ matrix belonging to $PSL_2(27)$.

The task of solving the resulting system (1) of 10 linear Diophantine equations in 54 unknowns for a (0,1)–solution vector U was accomplished by our basis reduction algorithm presented at the 17-th Southeastern International Conference on Combinatorics Graph Theory and Computing, [5]. We remark that this algorithm is based in part on the $\mathbf{L}^3$ algorithm of A.K. Lenstra, H.W. Lenstra and L. Lovász [8], and was inspired by the work of J.C. Lagarias and A.M. Odlyzko [7].

3. Results

In figure 4 for each λ, $2 \leq \lambda \leq 11$ we give a list of orbits of 6-subsets by number whose union forms a 5–(28,6,λ) design. Noting that the complement of a 5–(28,6,λ) design is a 5–(28,6,23-λ) design, orbit number lists for $\lambda \geq 11$ were not given, and in the interest of brevity we give only one solution for each parameter situation. It is easy to see that these solutions satisfy equation (1). Furthermore, the industrious reader can check using an algorithm similar to the one found in [1] that these are indeed 5–(28,6,λ) designs.

λ	Orbit numbers
2	12 15 18 28 30 51
3	7 8 9 22 26 27 44 47
4	2 21 22 35 39 40 43 44 48 49 54
5	2 8 22 28 39 43 44 47 49 52 53
6	1 10 11 13 25 30 36 39 43 45 49 50
7	1 4 7 8 11 12 15 18 22 23 25 26 30 34 35 37 38 39 40 46 51 54
8	1 2 8 9 10 11 14 24 29 31 36 37 38 45 48 50 51
9	3 8 9 13 14 19 22 24 28 36 42 43 44 45 49 52 53
10	1 2 3 5 6 13 16 17 19 22 24 29 30 32 36 39 47 50 52 53
11	2 3 5 9 11 13 14 17 19 24 25 30 32 36 42 43 44 47 49 52 53

Figure 4. Designs from $PSL_2(27)$.

Finally, we state the following theorem.

Theorem 2.1 $PSL_2(27)$ *is an automorphism group of a simple* 5–(28,6,λ) *design for all* λ, $2 \leq \lambda \leq 21$ *and cannot be an automorphism group of a* 5–(28,6,λ) *design when* $\lambda=1$.

Proof: The first part of the theorem is established by figure 4. To show that $G=PSL_2(27)$ cannot be the automorphism of a 5–(28,6,1) design we consider the G-orbit lengths of 6-subsets. These are: 3276, 4914 and 9828. Thus if a 5–(28,6,1) design is to be constructed as a union of G-orbits then the number of blocks, 16380, must be written as a sum of these numbers. Congruence modulo 3 and the fact that there are only 4 orbits of length 3276, namely numbers 2, 16, 18, and 35, says that exactly two of them must be used. But, this is impossible since the row sum of any two of the corresponding columns of the $A_{5,6}$ matrix contains an entry greater than 1, contradicting $\lambda=1$.

Acknowledgements

Both authors' research was supported under NSF Grant number DCR-8606378.

References

[1] R.H.F. Denniston, "Some New 5–designs", *Bull. London Math. Soc.* 8 (1976) 263-267.

[2] R.H.F. Denniston, "On the problem of the Higher Values of t", *Annals of Discrete Math.* 7 (1980) 65-70.

[3] E.S. Kramer and D.M. Mesner, "t–Designs on Hypergraphs", *Discrete Mathematics* 15 (1976) 263–296.

[4] E.S. Kramer, D.W. Leavitt and S.S. Magliveras, "Construction Procedures for t-Designs and the existence of New Simple 6-Designs", *Annals of Discrete Mathematics* 26 (1985) 247-274.

[5] D.L. Kreher and S.P. Radziszowski, "Finding Simple t-Designs by Basis Reduction", *Proceedings of the 17-th Southeastern Conference on Combinatorics, Graph Theory and Computing,* Congressus Numerantium 55 (1986) 235-244.

[6] D.L. Kreher and S.P. Radziszowski, "The Existence of Simple 6-(14,7,4) designs", *Journal of Combinatorial Theory (A)* 43 (1986) 237-243.

[7] J.C. Lagarias and A.M. Odlyzko, "Solving Low–Density Subset Sum Problem", *Journal of the ACM* 32 (1985) 229–246.

[8] A.K. Lenstra, H.W. Lenstra and L. Lovász, "Factoring Polynomials with Rational Coefficients", *Mathematische Annalen* 261 (1982) 515–534.

[9] S.S. Magliveras and D.W. Leavitt, Simple 6-(33,8,36) Designs from $P\Gamma L_2(32)$, in *Computational Group Theory, Proceedings of the London Mathematical Society Symposium on Computational Group theory,* Academic Press (1984) 337-352.

Annals of Discrete Mathematics 34 (1987) 319–338
© Elsevier Science Publishers B.V. (North-Holland)

The existence of partitioned balanced tournament designs of side 4n + 3

E.R. Lamken

School of Mathematics
Georgia Institute of Technology
Atlanta, Georgia
U.S.A.

S.A. Vanstone

Department of Combinatorics and Optimization
University of Waterloo
Waterloo, Ontario, N2L 3G1
CANADA

TO ALEX ROSA ON HIS FIFTIETH BIRTHDAY

ABSTRACT

A balanced tournament design, $BTD(n)$, defined on a $2n$-set V is an arrangement of the $\binom{2n}{2}$ distinct unordered pairs of the elements of V into an $n \times 2n-1$ array such that (1) every element of V is contained in precisely one cell of each column and (2) every element of V is contained in at most two cells of each row. If we can partition the columns of a $BTD(n)$ into three sets C_1,C_2,C_3 of sizes 1, $n-1$, $n-1$ respectively so that the columns in $C_1 \cup C_2$ form an $H(n,2n)$ and the columns in $C_1 \cup C_3$ form an $H(n,2n)$, then the $BTD(n)$ is called partitionable. We denote a partitioned balanced tournament design of side n by $PBTD(n)$. In this paper, we prove the existence of $PBTD(n)$ for $n \equiv 3 \pmod 4$, $n \geq 7$ with three possible exceptions.

1. Introduction.

A balanced tournament design, $BTD(n)$, defined on a $2n$-set V is an arrangement of the $\binom{2n}{2}$ distinct unordered pairs of the elements of V into an $n \times 2n-1$ array such that

(1) every element of V is contained in precisely one cell of each column and

(2) every element of V is contained in at most two cells of each row.

An element which is contained only once in row i of a $BTD(n)$ is called a deficient element of row i. The two deficient elements of row i are referred to as the deficient pair of row i.

The existence of $BTD(n)$s was established in [11]. (A simpler proof of this result appears in [7].)

Theorem 1.1. *For n a positive integer, $n \neq 2$, there exists a $BTD(n)$.*

For the tournament scheduling aspects of these designs the reader is referred to [11].

Let V be a set of $2n$ elements. A Howell design of side s and order $2n$, or more briefly an $H(s,2n)$, is an $s \times s$ array in which each cell is either empty or contains an unordered pair of elements from V such that

(1) each row and each column is Latin (that is, every element of V is in precisely one cell of each row and column) and

(2) every unordered pair of elements of V is in at most one cell of the array.

It follows immediately from the definition of an $H(s,2n)$ that $n \leq s \leq 2n-1$.

If we can partition the columns of a $BTD(n)$ defined on V into three sets C_1, C_2, C_3 of sizes 1, $n-1$, $n-1$ respectively so that the columns in $C_1 \cup C_2$ form an $H(n,2n)$ and the columns in $C_1 \cup C_3$ form an $H(n,2n)$, then the $BTD(n)$ is called partitionable. We denote the design by $PBTD(n)$. An example of a $PBTD(5)$ is displayed in Figure 1. One motivation for studying partitioned designs is for the tournament scheduling problem itself. The reader is referred to [9] for details. Applications to other areas of design theory are described in the next paragraph.

Partitioned balanced tournament designs are related to Room squares. A Room square of side $2n+1$, $RS(2n+1)$, is an $H(2n+1,2n+2)$. Let R be a $RS(2n+1)$. Since each row or column of R contains n empty cells, a $t \times t$ subarray of empty cells in R must have $t \leq n$. If R contains an $n \times n$ subarray of empty cells, we say R contains a maximum empty subarray and denote R by $MESRS(2n+1)$. The connection between MESRSs and PBTDs was provided by D.R. Stinson in [13].

α 4	∞ 2	1 3	5 7	2 3	4 5	∞ 7	α 1	0 6
∞ 3	α 5	4 6	0 2	∞ 4	α 2	0 5	6 3	1 7
5 6	0 3	α 7	∞ 1	6 7	0 1	α 3	∞ 5	4 2
1 2	4 7	∞ 0	α 6	α 0	∞ 6	1 4	7 2	5 3
0 7	1 6	2 5	4 3	1 5	3 7	2 6	0 4	α ∞

C_2 (columns 2–3) C_3 (columns 6–7) C_1 (column 9)

Figure 1
A PBTD(5)

Theorem 1.2. *There exists a MESRS*$(2n+1)$ *if and only if there exists a PBTD*$(n+1)$.

There is an extensive literature on Room squares and it is known that Room squares of side $2n + 1$ exist if and only if $2n + 1 \neq 3$ or 5 (Mullin and Wallis [10]). In [13], D.R. Stinson established the nonexistence of a *MESRS*(7) and the existence of *MESRS*(m) for $m = 9$ and $m = 11$. He showed that two recursive constructions for Room squares could be used to provide infinite classes of MESRSs and conjectured that *MESRS*$(2n+1)$s exist for all $n \geq 4$.

New constructions for balanced tournament designs have led us to investigate the existence question for *MESRS*$(2n+1)$s by considering the existence of *PBTD*$(n+1)$s. We can now show that *PBTD*(n) exist for n a positive integer, $n \geq 5$, with 12 possible exceptions. In [8], we showed that *PBTD*(n) exist for $n \equiv 1 \pmod 4$, $n \geq 5$, with the possible exception of $n = 9$. In this paper, we will use this result and a frame construction to construct *PBTD*(n) for $n \equiv 3 \pmod 4$. In [9], we use the existence of *PBTD*(n) for $n \equiv 1 \pmod 2$ and a new recursive construction to construct *PBTD*(n) for $n \equiv 0 \pmod 2$.

In order to describe our constructions, we will need several definitions and preliminary results. These are contained in the next section. We include a summary of previous results on PBTDs in section 2. In section 3, we describe some new frame constructions. In order to use many of the constructions described in sections 2 and 3, we need the existence of *PBTD*(n) for some small values of n. These are provided in section 4. The frame constructions are used in section 5 to find *PBTD*(n) for $n \equiv 3$ (mod 4) with 3 possible exceptions.

2. Preliminary definitions and results.

In order to describe our constructions, we will need several definitions and results. These are collected in this section.

In [8] we described a direct construction for *PBTD*$(n+1)$s. This construction uses a starter for a Howell design of side $2n$ and order $2n + 2$, $H(2n,2n+2)$. (For definitions and results on starters and adders for $H(2n,2n+2)$, see [1].)

Theorem 2.1. [8] *Let* $n \equiv 1$ *(mod 2). Let* $S = \{\{x_1,y_1\},\{x_2,y_2\}, \ldots ,\{x_{n-1},y_{n-1}\},\{\alpha,z_1\},\{\infty,z_2\}\}$ *be a starter for an* $H(2n,2n+2)$ *defined on* $Z_{2n} \cup \{\alpha,\infty\}$. *Let* $S' = S - \{x_i,y_i\}$ *for some* i *where* $|x_i - y_i| \equiv 1$ *(mod 2),* $1 \leq i \leq n-1$. *Suppose there is an adder* $A = (a_1,a_2,\ldots,a_n)$ *of* n *distinct elements of*

Z_{2n} *such that*

(i) $a_i \in \{0,2,4,...,2n-2\}$ *for* $i = 1,2,...,n$ *and*

(ii) $S' + A = Z_{2n} - \{u_1,v_1\}$ *where* $|u_1 - v_1| = n$.

Then there is a PBTD$(n+1)$.

We use this result to find several small PBTDs of even side.

Lemma 2.2. [8,9] *There exist PBTD*$(n+1)$ *for* $n \in \{5,7,9,11,13,15,17,19,21\}$.

The main construction for finding $PBTD(n)$ where $n \equiv 1 \pmod 4$ is a frame construction. We will also use frames to construct $PBTD(n)$ for $n \equiv 3 \pmod 4$.

Let V be a set of v elements. Let $G_1,G_2,...,G_m$ be a partition of V into m sets. A $\{G_1,G_2,...,G_m\}$-frame F with block size k, index λ and latinicity μ is a square array of side v which satisfies the properties listed below. We index the rows and columns of F by the elements of V.

(1) Each cell is either empty or contains a k-subset of V.

(2) Let F_i be the subsquare of F indexed by the elements of G_i. F_i is empty for $i = 1,2,...,m$.

(3) Let $j \in G_i$. Row j of F contains each element of $V - G_i$ μ times and column j of F contains each element of $V - G_i$ μ times.

(4) The collection of blocks obtained from the nonempty cells of F is a $GDD(v;k;G_1,G_2,...,G_m;0,\lambda)$. (See [17] for the notation for group divisible designs (GDD).)

If there is a $\{G_1,G_2,...,G_m\}$-frame H with block size k, index λ and latinicity μ such that

(1) $H_i = F_i$ for $i = 1,2,...,m$ and

(2) H can be written in the empty cells of $F - \bigcup_{i=1}^{m} F_i$,

then H is called a complement of F and denoted $\bar{F}$. If a complement of F exists, we call F a complementary $\{G_1,G_2,...,G_m\}$-frame. A complementary $\{G_1,G_2,...,G_m\}$-frame F is said to be skew if at most one of the cells (i,j) and (j,i) $(i \neq j)$ is nonempty.

We will use the following notation for frames. If $|G_i| = h$ for $i = 1,2,...,m$, we call F a $(\mu,\lambda;k,m,h)$-frame. The type of a $\{G_1,G_2,...,G_m\}$-frame is the multiset $\{|G_1|,|G_2|,\ldots,|G_m|\}$. We will say that a frame has type $t_1^{u_1}t_2^{u_2}\cdots t_k^{u_k}$ if there are u_i G_j's of cardinality t_i, $1 \leq i \leq k$. For notational convenience, if $\mu = \lambda = 1$ and $k = 2$, we will denote a frame simply by its type.

The constructions which use frames also use sets of mutually orthogonal partitioned incomplete Latin squares (OPILS). Let $P = \{S_1,S_2,...,S_m\}$ be a partition of a set S $(m \geq 2)$. A partitioned incomplete Latin square, having partition P, is an $|S| \times |S|$ array L, indexed by the elements of S, satisfying the following properties.

(1) A cell of L either contains an element of S or is empty.

(2) The subarrays indexed by $S_i \times S_i$ are empty for $1 \leq i \leq m$.

(3) Let $j \in S_i$. Row j of L contains each element of $S - S_i$ precisely once and column j of L contains each element of $S - S_i$ precisely once.

The type of L is the multiset $\{|S_1|, |S_2|, \ldots, |S_m|\}$. If there are u_i S_j's of cardinality t_i, $1 \leq i \leq k$, we say L has type $t_1^{u_1} t_2^{u_2} \cdots t_k^{u_k}$.

Suppose L and M are a pair of partitioned incomplete Latin squares with partition P. L and M are called orthogonal if the array formed by the superposition of L and M, $L \circ M$, contains every ordered pair in $S \times S - \bigcup_{i=1}^{m} (S_i \times S_i)$ precisely once. A set of n partitioned incomplete Latin squares with partition P is called a set of n mutually orthogonal partitioned Latin squares of type $\{|S_1|, |S_2|, \ldots, |S_m|\}$ if each pair of distinct squares is orthogonal.

The following frame construction for PBTDs was proved in [8].

Theorem 2.3. *If there exist a complementary $\{G_1, G_2, \ldots, G_m\}$-frame ($m \geq 2$), a pair of orthogonal partitioned incomplete Latin squares of type $\{|G_1|, |G_2|, \ldots, |G_m|\}$ and PBTD$(|G_i|+1)$ for $i = 1,2,\ldots,m$, then there is a PBTD$((\sum_{i=1}^{m} |G_i|)+1)$.*

This construction combined with a direct product gives the next result.

Theorem 2.4. [8] *If there exist a complementary $\{G_1, G_2, \ldots, G_m\}$-frame ($m \geq 2$), a pair of orthogonal partitioned incomplete Latin squares of type $\{|G_1|, |G_2|, \ldots, |G_m|\}$, a pair of orthogonal Latin squares of side n and PBTD$(n|G_i|+1)$ for $i = 1,2,\ldots,m$, then there is a PBTD$(n\sum_{i'1}^{m} |G_i|+1)$.*

We used Theorem 2.3 and the existence of complementary 4^n frames and OPILs of type 4^n for $n \geq 4$, $n \neq 6$, to prove the following.

Theorem 2.5. [8] *Let n be a positive integer. There exists a PBTD$(4n+1)$ except possibly for $n = 2$.*

There are several other applications of the frame construction. One example is:

Theorem 2.6. [8] *Let $n \equiv 1 \pmod 2$, $n \geq 7$. If there is a PBTD$(m+1)$ ($m \neq 6$), then there is a PBTD$(mn+1)$.*

Other results using Theorems 2.3 and 2.4 can be found in [8].

Finally, we will also use two product constructions from [13].

Theorem 2.7. *Suppose there exist PBTD(m) and PBTD(n) and a pair of orthogonal Latin squares of order m. Then there exists a PBTD(mn).*

Theorem 2.8. *Suppose there exists a PBTD$(2m)$ and a pair of orthogonal Latin squares of side m. Then there exists a PBTD$(6m)$.*

3. Frame Constructions.

In this section, we describe some new frame constructions.

Theorem 3.1. *Let $n \equiv 0 \pmod 2$. If there exists a complementary 2^n frame and a pair of orthogonal partitioned incomplete Latin squares of type 2^n, then there is a PBTD$(2n+1)$.*

Proof. Let $V = \{u_i, v_i \mid i = 1,2,...,n\}$ and let $\overline{V} = \{\overline{u}_i, \overline{v}_i \mid i = 1,2,...,n\}$. Let $G_i = \{u_i, v_i\}$ and let $\overline{G}_i = \{\overline{u}_i, \overline{v}_i\}$ for $i = 1,2,...,n$.

Let F_1 be a complementary $\{G_1, G_2, ..., G_n\}$-frame defined on V. F_1 is of type 2^n. Let F_2 be a complement of F_1 defined on $\overline{V}$. F_2 will be a $\{\overline{G}_1, \overline{G}_2, \ldots, \overline{G}_n\}$-frame of type 2^n. F will denote the array of pairs formed by the superposition of F_1 and F_2, $F = F_1 \circ F_2$.

Let L_1 and L_2 be a pair of orthogonal partitioned incomplete Latin squares of type 2^n. Suppose L_1 is defined on V with partition $\{\{u_1,u_2\},\{v_1,v_2\}, \ldots, \{u_{n-1},u_n\},\{v_{n-1},v_n\}\}$ and suppose L_2 is defined on $\overline{V}$ with partition $\{\{\overline{u}_1,\overline{u}_2\},\{\overline{v}_1,\overline{v}_2\}, \ldots, \{\overline{u}_{n-1},\overline{u}_n\},\{\overline{v}_{n-1},\overline{v}_n\}\}$. L will be the array of pairs formed by the superposition of L_1 and L_2, $L = L_1 \circ L_2$.

Define the following arrays.

$$A_i = \begin{array}{|c|c|}\hline \alpha\overline{u}_i & \infty u_i \\ \hline \infty\overline{v}_i & \alpha v_i \\ \hline\end{array}\,, \qquad C_i = \begin{array}{|c|}\hline \overline{v}_i\ v_i \\ \hline \overline{u}_i\ u_i \\ \hline\end{array} \quad \text{and}$$

$$R_i = \begin{array}{|c|c|}\hline u_i\ v_i & \overline{u}_i\ \overline{v}_i \\ \hline\end{array} \quad \text{for } i = 1,2,...,n.$$

For $i = 1,3,...,n-1$ ($i \equiv 1 \bmod 2$),

$$B_i = \begin{array}{|c|c|}\hline \alpha u_i & \infty\overline{u}_i \\ \hline \infty\overline{u_{i+1}} & \alpha u_{i+1} \\ \hline\end{array}\,, \qquad C'_i = \begin{array}{|c|}\hline \overline{u}_{i+1}\ u_{i+1} \\ \hline u_i\ \overline{u}_i \\ \hline\end{array} \quad \text{and}$$

$$R'_i = \begin{array}{|c|c|}\hline \overline{u}_i\ u_{i+1} & u_i\ \overline{u}_{i+1} \\ \hline\end{array}\,.$$

For $i = 2,4,...,n$ ($i \equiv 0 \bmod 2$),

$$B_i = \begin{array}{|c|c|}\hline \alpha\overline{v}_{i-1} & \infty v_{i-1} \\ \hline \infty v_i & \alpha\overline{v}_i \\ \hline\end{array}\,, \qquad C'_i = \begin{array}{|c|}\hline v_i\ \overline{v}_i \\ \hline v_{i-1}\ \overline{v_{i-1}} \\ \hline\end{array} \quad \text{and}$$

$$R'_i = \begin{array}{|c|c|}\hline \overline{v}_i\ v_{i-1} & \overline{v_{i-1}}\ v_i \\ \hline\end{array}\,.$$

We construct two $2n+1 \times 2n+1$ arrays using these arrays and L and F.

$$H_1 = \left.\begin{array}{|ccc|c|}
\hline
A_1 & & & C_1 \\
 & A_2 \quad F & & C_2 \\
 & & & \vdots \\
 & & A_n & C_n \\
\hline
R_1 & R_2 & R_n & \alpha\infty \\
\hline
\end{array}\right.
\begin{array}{l}
\left.\begin{array}{c} \\ \\ \\ \\ \end{array}\right\} 2n \\
\left.\begin{array}{c} \\ \end{array}\right\} 1
\end{array}$$

$$H'_2 = \begin{array}{|ccc|c|}
\hline
B_1 & & & C'_1 \\
 & B_2 \quad L & & C'_2 \\
 & & & \vdots \\
 & & B_n & C'_n \\
\hline
R'_1 & R'_2 & R'_n & \alpha\infty \\
\hline
\end{array}$$

Every element in $V \cup \bar{V} \cup \{\alpha,\infty\}$ occurs once in each row and once in each column of H_1 and H'_2. Let C denote the $2n \times 1$ array $[C_1C_2 \cdots C_n]^T$. We permute the rows of H'_2 so that the last column of the resulting array H_2 is $[C_1C_2 \cdots C_n\{\alpha,\infty\}]^T$. We can now write H_1 and H_2 in the following forms.

$$H_1 = \begin{array}{|c|c|}
\hline
K_1 & C \\
\hline
R & \alpha\infty \\
\hline
\end{array}
\begin{array}{l} 2n \\ 1 \end{array}
\quad \text{and} \quad
H_2 = \begin{array}{|c|c|}
\hline
K_2 & C \\
\hline
R' & \alpha\infty \\
\hline
\end{array}
\begin{array}{l} 2n \\ 1 \end{array}$$

H_1 and H_2 are both $H(2n+1,4n+2)$s defined on $V \cup \bar{V} \cup \{\alpha,\infty\}$.

The array

$$B = \begin{array}{|c|c|c|}
\hline
K_1 & K_2 & C \\
\hline
R & R' & \alpha\infty \\
\hline
\end{array}$$

is a $PBTD(2n+1)$ defined on $V \cup \bar{V} \cup \{\alpha,\infty\}$. It is straightforward to verify that every distinct pair in $V \cup \bar{V} \cup \{\alpha,\infty\}$ occurs precisely once in B. Every element in $V \cup \bar{V} \cup \{\alpha,\infty\}$ occurs at most twice in each row of B and precisely once in each column of B. The partitioning of B is given by H_1 and H_2. □

Corollary 3.2. *There exists a PBTD*(13).

Proof. There exists a complementary 2^6 frame [8] and a pair of orthogonal partitioned incomplete Latin squares of type 2^6 [5]. □

This corollary provides a proof for the note added to the end of [8].

A $PBTD(n)$ is said to be missing a $PBTD(m)$ if it contains an $m \times (2m - 1)$ empty subarray which could be filled in with a $PBTD(m)$. The $PBTD(m)$ need not exist. We state without proof a result similar to Theorem 3.1 which provides $PBTD(n)$s which are missing a $PBTD(3)$.

Theorem 3.3. *Let $n \equiv 1 \pmod 2$. If there exists a complementary 2^n frame and a pair of orthogonal partitioned incomplete Latin squares of type 2^n, then there is a $PBTD(2n+1)$ which is missing a $PBTD(3)$.*

Corollary 3.4. *Let $n \equiv 1 \pmod 2$, $n \geq 5$. There is a $PBTD(2n+1)$ which is missing a $PBTD(3)$.*

Proof. There exist complementary 2^n frames [8] and a pair of *OPIL*s of type 2^n [5] for $n \equiv 1 \pmod 2$, $n > 5$. □

The idea of using different groups for the complementary frame and the pair of orthogonal partitioned incomplete Latin squares is also the basis of the following result.

Theorem 3.5. *Let $m \equiv 1 \pmod 2$, $m \geq 3$. If there exists a complementary 2^{mn} frame, a pair of orthogonal partitioned incomplete Latin squares of type m^{2n} and a pair of orthogonal Latin squares of side $m + 1$, then there is a $PBTD(2mn+1)$.*

Proof. Let $V = \{u_i^j, v_i^j \mid i = 1,2,\ldots,m,\ j = 1,2,\ldots,n\}$ and let $\overline{V} = \{\overline{u}_i^j, \overline{v}_i^j \mid i = 1,2,\ldots,m,\ j = 1,2,\ldots,n\}$. Let $G_i^j = \{u_i^j, v_i^j\}$ and let $\overline{G}_i^j = \{\overline{u}_i^j, \overline{v}_i^j\}$ for $i = 1,2,\ldots,m$ and $j = 1,2,\ldots,n$.

Let F_1 be a complementary $\{G_1^1,\ldots,G_m^1,G_1^2,\ldots,G_m^2,\ldots,G_1^n,\ldots,G_m^n\}$-frame defined on V. F_1 is of type 2^{mn}. Let F_2 be a complement of F_1 defined on $\overline{V}$. $\overline{F}_2$ will be a $\{\overline{G}_1^1,\ldots,\overline{G}_m^1,\overline{G}_1^2,\ldots,\overline{G}_m^2,\ldots,\overline{G}_1^n,\ldots,\overline{G}_m^n\}$-frame of type 2^{mn}. F will denote the array of pairs formed by the superposition of F_1 and F_2, $F_1 \circ F_2$.

Let L_1 and L_2 be a pair of orthogonal partitioned incomplete Latin squares of type m^{2n}. Suppose L_1 is defined on V with partition $\{U^1,\ldots,U^n,V^1,\ldots,V^n\}$ where $U^i = \{u_1^i,u_2^i,\ldots,u_m^i\}$ and $V^i = \{v_1^i,\ldots,v_m^i\}$. Suppose L_2 is defined on $\overline{V}$ with partition $\{\overline{U}^1,\ldots,\overline{U}^n,\overline{V}^1,\ldots,\overline{V}^n\}$ where $\overline{U}^i = \{\overline{u}_1^i,\ldots,\overline{u}_m^i\}$ and $\overline{V}^i = \{\overline{v}_1^i,\ldots,\overline{v}_m^i\}$. L will be the array of pairs formed by the superposition of L_1 and L_2, $L = L_1 \circ L_2$.

Let M_1 and M_2 be a pair of orthogonal Latin squares of side $m + 1$. Let M_i be the array of pairs formed by the superposition of M_1 and M_2 where M_1 is defined on the set $U^i \cup \{\alpha\}$ and M_2 is defined on the set $\overline{U}^i \cup \{\infty\}$. Let N_i be the array of pairs formed by the superposition of M_1 and M_2 where M_1 is defined on the set $V^i \cup \{\alpha\}$ and M_2 is defined on the set $\overline{V}_i \cup \{\alpha\}$. Furthermore, suppose M_i and N_i are partitioned in the following way.

$$M_i = \left[\begin{array}{c|c} & u_1^i\bar{u}_1^i \\ A_i & u_2^i\bar{u}_2^i \\ & \vdots \\ & u_m\bar{u}_m^i \\ \hline D_i & \alpha\infty \end{array}\right] \begin{array}{l} \left.\vphantom{\begin{array}{c}1\\1\\1\\1\end{array}}\right\} m \\ \left.\vphantom{1}\right\} 1 \end{array}$$

$$N_i = \left[\begin{array}{c|c} & v_1^i\bar{v}_1^i \\ B_i & v_2^i\bar{v}_2^i \\ & \vdots \\ & v_m^i\bar{v}_m^i \\ \hline E_i & \alpha\infty \end{array}\right] \begin{array}{l} \left.\vphantom{\begin{array}{c}1\\1\\1\\1\end{array}}\right\} m \\ \left.\vphantom{1}\right\} 1 \end{array}$$

Let $J_i = [\{u_1^i,\bar{u}_1^i\} \cdots \{u_m^i,\bar{u}_m^i\}]^T$ and let $K_i = [\{v_1^i,\bar{v}_1^i\} \cdots \{v_m^i,\bar{v}_m^i\}]^T$.

Define the following arrays for $i = 1,2,\ldots,m$ and $j = 1,2,\ldots,n$.

$$F_i^j = \begin{array}{|c|c|} \hline \alpha u_i^j & \infty\bar{u}_i^j \\ \hline \infty v_i^j & \alpha\bar{v}_i^j \\ \hline \end{array}, \quad C_i^j = \begin{array}{|c|} \hline v_i^j\bar{v}_i^j \\ \hline u_i^j\bar{u}_i^j \\ \hline \end{array} \quad \text{and}$$

$$R_i^j = \begin{array}{|c|c|} \hline \bar{u}_i^j\bar{v}_i^j & u_i^jv_i^j \\ \hline \end{array}.$$

We construct two $2mn + 1 \times 2mn + 1$ arrays as follows.

$$H_1 = \left[\begin{array}{ccccc|c} F_1^1 & & & & & C_1^1 \\ & & & F & & \vdots \\ & F_m^1 & & & & C_m^1 \\ & & F_1^2 & & & \vdots \\ & & & F_1^n & & C_1^n \\ & & & & F_m^n & C_m^n \\ \hline R_1^1 & R_m^1 & & R_1^n & R_m^n & \alpha\infty \end{array}\right]$$

$$H'_2 = \left[\begin{array}{cccccc|c} A_1 & & & & & & J_1 \\ & A_2 & & & L & & J_2 \\ & & & & & & \vdots \\ & & A_n & & & & J_n \\ & & & B_1 & & & K_1 \\ & & & & B_2 & & K_2 \\ & & & & & & \vdots \\ & & & & & B_n & K_n \\ \hline D_1 & D_2 & & E_1 & E_2 & E_n & \alpha\infty \end{array}\right].$$

Every element in $V \cup \overline{V} \cup \{\alpha,\infty\}$ occurs once in each row and once in each column of H_1 and H'_2. Let C denote the $2mn \times 1$ array $[C_1^1 \cdots C_m^1 \cdots C_1^n \cdots C_m^n]^T$. We permute the rows of H'_2 so that the last column of the resulting array H_2 is $[C_1^1 \cdots C_m^1 \cdots C_1^n \cdots C_m^n \{\alpha,\infty\}]^T$. We can now write H_1 and H_2 in the following forms.

$$H_1 = \left[\begin{array}{c|c} S_1 & C \\ \hline T_1 & \alpha\infty \end{array}\right] \quad \text{and} \quad H_2 = \left[\begin{array}{c|c} S_2 & C \\ \hline T_2 & \alpha\infty \end{array}\right].$$

H_1 and H_2 are $H(2mn+1,4mn+2)$s defined on $V \cup \overline{V} \cup \{\alpha,\infty\}$.

It is straightforward to verify that the array

$$B = \begin{array}{|c|c|c|} \hline S_1 & S_2 & C \\ \hline T_1 & T_2 & \alpha\infty \\ \hline \end{array}$$

is a $PBTD(2mn+1)$ defined on $V \cup \bar{V} \cup \{\alpha,\infty\}$. Every distinct pair in $V \cup \bar{V} \cup \{\alpha,\infty\}$ occurs at most twice in each row of B and every element occurs precisely once in each column of B. The partitioning of B is given by H_1 and H_2. □

Corollary 3.6. *Let $n \equiv 1$ (mod 2), $n \geq 3$. There is a $PBTD(6n+1)$.*

Proof. We use Theorem 3.3 with $m = 3$ and $n \equiv 1$ (mod 2), $n \geq 3$. Since there exists a complementary 2^{3n} frame [8], a pair of orthogonal partitioned incomplete Latin squares of type 3^{2n} [8,19] and a pair of orthogonal Latin squares of side 4, we can construct a $PBTD(6n+1)$. □

Corollary 3.7. *Let $N = \{71,99,155,183,111,131,171,191\}$. There exists a $PBTD(x)$ for $x \in N$.*

Proof. We use Theorem 3.3 with the values of m and n listed in the table below. The existence results for the complementary frame and pair of orthogonal partitioned Latin squares (OPILS) can be found in [8].

m	n	Complementary frame	OPILS	PBTD
7	5	2^{35}	7^{10}	71
7	7	2^{49}	7^{14}	99
7	11	2^{77}	7^{22}	155
7	13	2^{91}	7^{26}	183
11	5	2^{55}	11^{10}	111
13	5	2^{65}	13^{10}	131
17	5	2^{85}	17^{10}	171
19	5	2^{95}	19^{10}	191

□

The next results are generalizations of Theorem 3.5. We will use them to produce some small designs. Since the proofs are quite similar to the proof of Theorem 3.5, we omit them.

Theorem 3.8. *If there exists a complementary 2^n frame, a pair of orthogonal partitioned incomplete Latin squares of type $t_1^{u_1}t_2^{u_2} \cdots t_k^{u_k}$ where $\sum_{i=1}^{k} u_i t_i = 2n$ and $u_i \equiv 0$ (mod 2) for all i, and a pair of orthogonal Latin squares of order $t_i + 1$ for $i = 1,2,...,k$, then there is a $PBTD(2n+1)$.*

Corollary 3.9. *There exist $PBTD(\ell)$ for $\ell = 75,95$ and 167.*

Proof. (i) $\ell = 75$. There is a 2^{37} complementary frame [8]. We use a $GDD(74;\{7,8,9\};\{7,8\};0,1)$ to construct a pair of OPILs of type $7^6 8^4$. (This *GDD* is constructed in [3].)

(ii) $\ell = 95$. There is a 2^{47} complementary frame [8]. We use a $GDD(94;\{8,9,10\},\{3,11\};0,1)$ to construct a pair of OPILs of type $11^8 3^2$.

(iii) $\ell = 167$. There is a 2^{83} complementary frame. We use a $GDD(166;\{8,9,10\};\{19,7\};0,1)$ to construct a pair of OPILs of type $19^8 7^2$. □

Theorem 3.10. *Let $2n = tu + 6$ where $t \equiv 0 \pmod 2$. If there exists a complementary 2^n frame, a pair of orthogonal partitioned incomplete Latin squares of type $t^u 6$ and a pair of orthogonal Latin squares of order $t + 1$, then there is a PBTD$(2n+1)$.*

We note that this construction uses the special $H(7,14)$ constructed for a *PBTD*(7) in [12] in place of a pair of orthogonal Latin squares of order 7. (Both of the Howell designs D_2 and D_3 of Table 3 in [12] have the required property.)

Corollary 3.11. *There exist PBTD(ℓ) for $\ell = 39,47$ and 51.*

Proof. If there is a set of three mutually orthogonal Latin squares of side m where $m \geq 7$, then there is a $GDD(4m+6,\{4,5\},\{4,6^*\};0,1)$. This group divisible design can be used to construct a pair of *OPIL*s of type $4^m 6$.

(i) $\ell = 39$. Since there is a complementary 2^{19} frame [8] and a pair of *OPIL*s of type $4^8 6$, we can apply Theorem 3.10 with $t = 4$ and $u = 8$.

(ii) $\ell = 47$. There is a complementary 2^{23} frame [8] and a pair of *OPIL*s of type $8^5 6$. To construct the *OPIL*s of type $8^5 6$, we use the $GDD(23,\{4,5\},\{4,3^*\};0,1)$ from [19] as follows. Expand by a factor of 2, replacing each block with a pair of *OPIL*s of type 2^4 or 2^5. The resulting design is a pair of *OPIL*s of type $8^5 6$. We use Theorem 3.10 with $t = 8$ and $u = 5$ to construct a *PBTD*(47).

(iii) $\ell = 51$. Since there is a complementary 2^{25} frame [8] and a pair of *OPIL*s of type $4^{11} 6$, we can apply the theorem with $t = 4$ and $u = 11$. □

Let F be a complementary 2^n frame. F is called row complementary if we can construct a complement for F by interchanging rows i and $i + 1$ for $i = 1,3,\ldots,2n-1$. Suppose F is a $\{G_1,G_2,\ldots,G_n\}$-frame where $G_i = \{x_i,\bar{x}_i\}$. Let $V = \bigcup_{i=1}^{n} G_i$. Let F_1 be the $n \times n$ subarray of F indexed by rows $1,3,5,\ldots,2n-1$ and columns $1,3,5,\ldots,2n-1$. If F can be written so that row i and column i of F_1 contain every element of $V - G_i$ precisely once, we call F partitionable. In Figure 2, we display a row complementary partitioned 2^6 frame [4]. Row complementary partitioned 2^n frames can also be used to construct partitioned balanced tournament designs.

Theorem 3.12. *If there is a row complementary partitioned 2^n frame, a PBTD$(m+1)$ and a pair of orthogonal Latin squares of order m, then there is a PBTD$(mn+1)$.*

<table>
<tr><td></td><td></td><td>2 1
4 1</td><td></td><td>3 0
∞_2</td><td></td><td>1 0
2 0</td><td></td><td>1 1
∞_1</td><td></td><td>4 0
3 1</td><td></td></tr>
<tr><td></td><td></td><td></td><td>2 0
4 0</td><td></td><td>3 1
∞_2</td><td></td><td>1 1
2 1</td><td></td><td>1 0
∞_1</td><td></td><td>4 1
3 0</td></tr>
<tr><td>2 1
∞_1</td><td></td><td></td><td></td><td>3 1
0 1</td><td></td><td>4 0
∞_2</td><td></td><td>2 0
3 0</td><td></td><td>0 0
4 1</td><td></td></tr>
<tr><td></td><td>2 0
∞_1</td><td></td><td></td><td></td><td>3 0
0 0</td><td></td><td>4 1
∞_2</td><td></td><td>2 1
3 1</td><td></td><td>0 1
4 0</td></tr>
<tr><td>3 0
4 0</td><td></td><td>3 1
∞_1</td><td></td><td></td><td></td><td>4 1
1 1</td><td></td><td>0 0
∞_2</td><td></td><td>1 0
0 1</td><td></td></tr>
<tr><td></td><td>3 1
4 1</td><td></td><td>3 0
∞_1</td><td></td><td></td><td></td><td>4 0
1 0</td><td></td><td>0 1
∞_2</td><td></td><td>1 1
0 0</td></tr>
<tr><td>1 0
∞_2</td><td></td><td>4 0
0 0</td><td></td><td>4 1
∞_1</td><td></td><td></td><td></td><td>0 1
2 1</td><td></td><td>2 0
1 1</td><td></td></tr>
<tr><td></td><td>1 1
∞_2</td><td></td><td>4 1
0 1</td><td></td><td>4 0
∞_1</td><td></td><td></td><td></td><td>0 0
2 0</td><td></td><td>2 1
1 0</td></tr>
<tr><td>1 1
3 1</td><td></td><td>2 0
∞_2</td><td></td><td>0 0
1 0</td><td></td><td>0 1
∞_1</td><td></td><td></td><td></td><td>3 0
2 1</td><td></td></tr>
<tr><td></td><td>1 0
3 0</td><td></td><td>2 1
∞_2</td><td></td><td>0 1
1 1</td><td></td><td>0 0
∞_1</td><td></td><td></td><td></td><td>3 1
2 0</td></tr>
<tr><td>2 0
4 1</td><td></td><td>3 0
0 1</td><td></td><td>4 0
1 1</td><td></td><td>0 0
2 1</td><td></td><td>1 0
3 1</td><td></td><td></td><td></td></tr>
<tr><td></td><td>2 1
4 0</td><td></td><td>3 1
1 1</td><td></td><td>4 1
1 0</td><td></td><td>0 1
2 0</td><td></td><td>1 1
3 0</td><td></td><td></td></tr>
</table>

Figure 2

A row complementary partitioned 2^6 frame defined on $(Z_5 \times Z_2) \cup \{\infty_1,\infty_2\}$

Proof. Let $V = \{x_i, \bar{x}_i \mid i = 1,2,...,n\}$ and let $G_i = \{x_i, \bar{x}_i\}$. Let $M = \{1,2,...,m\}$.

Let F be a row complementary partitioned 2^n frame defined on V. F will be a $\{G_1, G_2, ..., G_n\}$-frame. Let R_i be the $1 \times 2n$ array formed by superimposing rows i and $i+1$ of F for $i = 1,3,5,...,2n-1$. Let R be the $n \times 2n$ array constructed as follows:

$$R = \boxed{\begin{array}{c} R_1 \\ R_3 \\ \cdot \\ \cdot \\ \cdot \\ R_{2n-1} \end{array}}\ .$$

Let N_1 and N_2 be a pair of orthogonal Latin squares of side m defined on M. Let N be the array of pairs formed by the superposition of N_1 and N_2, $N = N_1 \circ N_2$. N_{xy} will be the $m \times m$ array of pairs formed by replacing each pair (a,b) in M with the pair (a_x,b_y).

Let P_i be a $PBTD(m+1)$ defined on $(M \times G_i) \cup \{\alpha,\infty\}$. P_i can be written in the following form.

$$P_i = \begin{array}{|c|c|} \hline A_i & C_i \\ \hline B_i & \alpha\infty \\ \hline \end{array}\,.$$

We now use R to construct a $PBTD(mn+1)$ defined on $(M \times V) \cup \{\alpha,\infty\}$. Replace each pair (x,y) in R with the $m \times m$ array N_{xy}. Replace the 1×2 empty array in row i and columns $2i-1,2i$ $(i = 1,2,\ldots,n)$ with the $m \times 2m$ array A_i. We add a new row and a new column to the resulting $mn \times 2mn$ array, P'. P will be the following $(nm+1) \times (2nm+1)$ array.

$$P = \begin{array}{|c|c|} \hline & C_1 \\ P' & C_2 \\ & \vdots \\ & C_n \\ \hline B_1 \; B_2 \; \cdots \; B_n & \{\alpha,\infty\} \\ \hline \end{array}\,.$$

It is straightforward to verify that P is a $PBTD(nm+1)$ defined on $(M \times V) \cup \{\alpha,\infty\}$. The partitioning of P is given by columns $1,3,\ldots,2nm-1,2nm+1$ and columns $2,4,\ldots,2nm,2nm+1$. □

Corollary 3.13. *If there is a $PBTD(m+1)$ and a pair of orthogonal Latin squares of side m, then there is a $PBTD(6m+1)$.*

Proof. A row complementary partitioned 2^6 frame is displayed in Figure 2. □

4. Small designs.

In order to apply our constructions to the case $n \equiv 3 \pmod 4$, we will need the existence of several $PBTD(n)$s where n is small. We will use the following result which is proved in [9].

Theorem 4.1. *Let $n \equiv 1 \pmod 2$. If there exists a $PBTD(n+1)$ generated by a starter-adder pair on $Z_{2n} \cup \{\alpha,\infty\}$, a $PBTD(m)$, a $PBTD(m+k)$, a pair of orthogonal Latin squares of order m and an $IA(m+k,k,4)$, then there is a $PBTD((n+1)m+k)$.*

This theorem requires the existence of $IA(m+k,k,4)$s. These are provided by a recent result due to L. Zhu and K. Heinrich.

Theorem 4.2. [6] *An $IA(n,k,4)$ exists if and only if $n \geq 3k$ and $(n,k) \neq (6,1)$.*

Lemma 4.3. *Let $n \equiv 3 \pmod 4$. There exists a $PBTD(n)$ for $7 \leq n \leq 203$ except possibly for $n \in \{11,15,27\}$.*

Proof. Table 4.1 contains a list of constructions for $PBTD(n)$ for $n \equiv 3 \pmod 4$, $7 \leq n \leq 203$, and $n \notin \{11,15,27\}$.

Table 4.1

Constructions for $PBTD(n)$ for $n \equiv 3 \pmod 4$, $7 \leq n \leq 203$ and $n \notin \{11,15,27\}$.

n	Construction		n	Construction	
7	[12]		135	6.21+9	4.1
19	6.3+1	3.6			
23	4.4+7	2.3	139	6.23+1	3.6
31	6.5+1	3.6	143	8.17+7	4.1
35	5.7	2.7	147	8.18+3	4.1
39	4.8+7	3.11			
43	6.7+1	3.6	151	6.25+1	3.6
47	8.5+7	3.11			
51	4.11+7	3.11			
55	6.9+1	3.6	155	14.11+1	3.7
59	8.7+3	4.1	159	12.13+3	4.1
63	6.10+3	4.1	163	6.27+1	3.6
67	6.11+1	3.6	167		3.9
71	14.5+1	3.7	171	34.5+1	3.7
75		3.9	175	6.29+1	3.6
79	6.13+1	3.6	179	8.22+3	4.1
83	8.10+3	4.1	183	14.13+1	3.7
87	6.14+3	4.1	187	6.31+1	3.6
91	6.15+1	3.6	191	38.5+1	3.7
95		3.9	195	6.31+9	4.1 [9]
99	14.7+1	3.7	199	6.33+1	3.6
103	6.17+1	3.6	203	20.10+3	4.1
107	6.17+5	4.1			
111	22.5+1	3.7			
115	6.19+1	3.6			
119	8.14+7	4.1			
123	12.10+3	4.1			
127	6.21+1	3.6			
131	10.13+1	3.7			

5. *PBTD*(n) for $n \equiv 3$ (mod 4).

We will use the frame construction, Theorem 2.3, and the existence of $PBTD(n)$ for $n \equiv 1$ (mod 4) to construct $PBTD(n)$ for the case $n \equiv 3$ (mod 4). We require a complementary frame and a pair of orthogonal partitioned incomplete Latin squares of type $(4m)^4(2t)^1$. The complementary frame was constructed in [14].

Theorem 5.1. *Let m be a positive integer, $m \geq 4$, $m \neq 6,10$ and let t be a non-negative integer such that $0 \leq t \leq 3m$. Then there is a skew frame of type $(4m)^4(2t)^1$.*

We will use the same construction which was used to find the skew frame to provide the *OPIL*s of type $(4m)^4(2t)^1$. We need the existence of *OPIL*s of types 4^4, 4^5, $4^4 2$ and $4^4 6$.

Lemma 5.2. [8] *For $n = 4,5$ there is a pair of orthogonal partitioned incomplete Latin squares of type 4^n.*

Lemma 5.3. *There is a pair of orthogonal partitioned incomplete orthogonal Latin squares of type $4^4 2$.*

Proof. Let $G^1 = \{(0,0),(0,2),(2,0),(2,2)\}$, $G^2 = G^1 + (0,1)$, $G^3 = G^1 + (1,0)$ and $G^4 = G^1 + (1,1)$. Let $V = \bigcup_{i=1}^{\infty} G^i \cup \{\infty_1,\infty_2\}$. Let $G_i^j = G^j \times \{i\}$ for $i = 1,2$ and $j = 1,2,3,4$. Define $G_i^5 = \{(\infty_1,i),(\infty_2,i)\}$ for $i = 1,2$. Let $V_i = \bigcup_{j=1}^{5} G_i^j$ for $i = 1,2$.

Let F be a skew frame of type $4^4 2$ generated by an intransitive frame starter-adder (see [14]). $\overline{F}$ will denote the complement of F. We write F and $\overline{F}$ on the symbol set $V_1 \cup V_2$ so that $F \circ \overline{F}$ contains every pair in $V_1 \times V_2 - (\bigcup_{j=1}^{5} G_1^j \times G_2^j)$ precisely once. The starters and adders for the frames are listed below.

F:	Starter				Adders:
		S	11,1	∞_2,2	01
			∞_1,1	30,2	23
			12,1	01,2	31
			03,1	13,2	32
			10,1	23,2	11
			21,1	33,2	30
		C	31,1	32,2	
		R	10,1	33,2	

$\overline{F}$:	Starter	S	∞_2,1	12,2	Adders:	03
			13,1	∞_1,2		21
			23,1	11,2		10
			01,1	31,2		12
			30,1	21,2		33
			32,1	03,2		13
		C	33,1	10,2		
		R	32,1	31,2		

We can construct a pair of orthogonal partitioned incomplete orthogonal Latin squares of type $4^4 2$ from $F \circ \overline{F}$. □

Lemma 5.4. *There is a pair of orthogonal partitioned incomplete Latin squares of type* $4^4 6$.

Proof. Let G_i^j be defined as in the proof of Lemma 4.16 for $i = 1,2$ and $j = 1,2,3,4$. Define $G_i^5 = \{(\infty_1,i),(\infty_2,i),\ldots,(\infty_6,i)\}$ for $i = 1,2$. let $V_i = \bigcup_{j=1}^{5} G_i^j$ for $i = 1,2$.

Let F be a skew frame of type $4^4 6$ generated by an intransitive frame starter-adder (see [14]). $\overline{F}$ will denote the complement of F. We write F and $\overline{F}$ on the symbol set $V_1 \cup V_2$ so that $F \circ \overline{F}$ contains every pair in $V_1 \times V_2 - (\bigcup_{j=1}^{5} G_1^j \times G_2^j)$ precisely once. We list starters and adders for F and $\overline{F}$ below.

F:	S	30,1	∞_1,2	A:	23
		∞_2,1	03,2		32
		∞_3,1	12,2		31
		∞_4,1	11,2		01
		33,1	∞_5,2		30
		23,1	∞_6,2		11
	C	01,1	10,2		
		13,1	21,2		
		32,1	31,2		
	R	33,1	10,2		
		01,1	11,2		
		32,1	21,2		

$\bar{F}$:	S	∞_1,1	13,2	A:	21
		31,1	∞_2,2		12
		03,1	∞_3,2		13
		12,1	∞_4,2		03
		∞_5,1	23,2		10
		∞_6,1	30,2		33
	C	10,1	33,2		
		11,1	01,2		
		21,1	32,2		
	R	31,1	32,2		
		21,1	13,2		
		10,1	01,2		

We can construct a pair of orthogonal partitioned incomplete Latin squares of type $4^4 6$ from $F \circ \bar{F}$. □

Theorem 5.5. *Let $0 \leq t \leq 3m$. If there is a $TD(5,m)$, then there is a pair of orthogonal partitioned incomplete Latin squares of type $(4m)^4(2t)^1$.*

Proof. The construction for this proof is the same construction that was used to construct skew frames of type $(4m)^4(2t)^1$ in [14]. For completeness, we include the construction.

Let $(X,G,\mathbf{A})$ be a $TD(5,m)$ with $G = \{G_1,G_2,...,G_5\}$. Define a weighting $w: X \to \{0,2,4,6\}$ by setting $w(x) = 4$ if $x \in X - G_5$ and by defining $w(x)$ for $x \in G_5$ so that $\sum_{x \in G_5} w(x) = 2t$.

Apply Wilson's Fundamental Construction [18]. A block $A \in \mathbf{A}$ requires a pair of orthogonal partitioned incomplete Latin squares of one of the types 4^4, 4^5, $4^4 2$ or $4^4 6$. These designs are provided by Lemmas 5.2, 5.3 and 5.4. The result of this construction is a pair of orthogonal partitioned incomplete Latin squares of type $(4m)^4(2t)^1$. □

We can now apply Theorem 2.3 in the following form.

Theorem 5.6. *Let m be a positive integer, $m \geq 5$, $m \neq 6,10$ and let t be a non-negative integer such that $0 \leq t \leq 3m$. If there is a $PBTD(2t+1)$, then there is a $PBTD(16m+2t+1)$.*

We consider four cases $n \equiv 3,7,11$ and 15 (mod 16).

Lemma 5.7.

(i) *If $n \equiv 3$ (mod 16) and $n \geq 195$, then there exists a $PBTD(n)$.*

(ii) *If $n \equiv 7$ (mod 16) and $n \geq 183$, then there exists a $PBTD(n)$.*

(iii) *If $n \equiv 11$ (mod 16) and $n \geq 219$, then there exists a $PBTD(n)$.*

(iv) *If* $n \equiv 15$ *(mod 16) and* $n \geq 191$*, then there exists a PBTD*(n).

Proof. There exist *PBTD*(n) for $n = 19,7,43$ and 31 (Lemma 4.3). We apply Theorem 5.6 with

(i) $t = 9$ and $m \geq 11$, (ii) $t = 3$ and $m \geq 11$,

(iii) $t = 21$ and $m \geq 11$ and (iv) $t = 15$ and $m \geq 11$. □

Combining the results of Lemma 4.3 and Lemma 5.7, we have the following.

Theorem 5.8. *Let* $n \equiv 3$ *(mod 4). There exists a PBTD*(n) *except possibly for* $n \in \{11,15,27\}$.

Added In Proof

There exists a *PBTD*(27). Apply Lemma 4.2 in [19] to construct a pair of *OPILS* of type $4^5 6$. Use Theorem 3.10 of this paper to construct a *PBTD*(27).

References.

[1] Anderson, B.A., *Howell designs of type H(p−1,p+1),* J. Combinatorial Theory (A) 24 (1978), 131-140.

[2] Bose, R.C., Parker, E.T. and Shrikhande, S., *Further results on the construction of mutually orthogonal Latin squares and the falsity of Euler's conjecture,* Canadian J. Math. 12 (1960), 189-203.

[3] Colbourn, C.J., Manson, K.E. and Wallis, W.D., *Frames for twofold triple systems,* Ars Combinatoria 17 (1984), 69-78.

[4] Dinitz, J.H. and Stinson, D.R., *Further results on frames,* Ars Combinatoria 11 (1981), 275-288.

[5] Dinitz, J.H. and Stinson, D.R., *MOLS with holes,* Discrete Math. (1983), 145-154.

[6] Heinrich, K. and Zhu, L., *Existence of orthogonal Latin squares with aligned subsquares,* Discrete Math. 59 (1986) 69-78.

[7] Lamken, E.R. and Vanstone, S.A., *The existence of factored balanced tournament designs,* Ars Combinatoria 19 (1985) 157-160.

[8] Lamken, E.R. and Vanstone, S.A., *Partitioned balanced tournament designs of side 4n+1,* Ars Combinatoria 20 (1985) 29-44.

[9] Lamken, E.R. and Vanstone, S.A., *The existence of partitioned balanced tournament designs,* Annals of Discrete Math. (elsewhere in this volume).

[10] Mullin, R.C. and Wallis, W.D., *The existence of Room squares,* Aequationes Math. 13 (1975) 1-7.

[11] Schellenberg, P.J., van Rees, G.H.J. and Vanstone, S.A., *The existence of balanced tournament designs,* Ars Combinatoria 3 (1977) 303-318.

[12] Seah, E. and Stinson, D.R., *An assortment of new Howell designs,* Utilitas Math. (to appear).

[13] Stinson, D.R., *Room squares with maximum empty subarrays,* Ars Combinatoria 20 (1985) 159-166.

[14] Stinson, D.R., *Some Classes of Frames and the Spectrum of Skew Room Squares and Howell Designs,* Ph.D. Thesis, University of Waterloo, 1981.

[15] Stinson, D.R. and Wallis, W.D., *An even side analogue of Room squares,* Aequationes Math. 27 (1984) 201-213.

[16] Todorov, D.T., *Three mutually orthogonal Latin squares of order* 14, Ars Combinatoria 20 (1985) 45-48.

[17] Vanstone, S.A., *Doubly resolvable designs,* Discrete Math. 29 (1980) 77-86.

[18] Wilson, R.M., *Constructions and uses of pairwise balanced designs,* Math. Centre Tracts 55 (1974) 18-41.

[19] Stinson, D.R., and Zhu, L. *On the existence of MOLS with equal-sized holes,* Aequationes Math. (to appear).

Annals of Discrete Mathematics 34 (1987) 339–352
© Elsevier Science Publishers B.V. (North-Holland)

The existence of partitioned balanced tournament designs

E.R. Lamken

School of Mathematics
Georgia Institute of Technology
Atlanta, Georgia
U.S.A.

S.A. Vanstone

Department of Combinatorics and Optimization
University of Waterloo
Waterloo, Ontario, N2L 3G1
CANADA

TO ALEX ROSA ON HIS FIFTIETH BIRTHDAY

ABSTRACT

A balanced tournament design, $BTD(n)$, defined on a $2n$-set V is an arrangement of the $\binom{2n}{2}$ distinct unordered pairs of the elements of V into an $n \times 2n-1$ array such that (1) every element of V is contained in precisely one cell of each column and (2) every element of V is contained in at most two cells of each row. If we can partition the columns of a $BTD(n)$ into three sets C_1, C_2, C_3 of sizes $1, n-1, n-1$ respectively so that the columns in $C_1 \cup C_2$ form an $H(n,2n)$ and the columns in $C_1 \cup C_3$ form an $H(n,2n)$, then the $BTD(n)$ is called partitionable. We denote a partitioned balanced tournament design of side n by $PBTD(n)$. In this paper, we prove the existence of $PBTD(n)$ for $n \geq 5$ with 12 possible exceptions.

1. Introduction.

A balanced tournament design, $BTD(n)$, defined on a $2n$-set V is an arrangement of the $\binom{2n}{2}$ distinct unordered pairs of the elements of V into an $n \times 2n-1$ array such that

(1) every element of V is contained in precisely one cell of each column and

(2) every element of V is contained in at most two cells of each row.

An element which is contained only once in row i of a $BTD(n)$ is called a deficient element of row i. The two deficient elements of row i are referred to as the deficient pair of row i.

The existence of $BTD(n)$s was established in [9]. (A simpler proof of this result appears in [4].)

Theorem 1.1. *For n a positive integer, $n \neq 2$, there exists a $BTD(n)$.*

Balanced tournament designs can be used to represent round robin tennis tournaments. In terms of scheduling a tennis tournament we think of the rows of the design as representing courts, the columns as rounds of play and the elements as players or teams. The properties of the array give us a schedule of play which has each player playing each other player precisely once, each player plays each round and no player plays more than twice on any court. We would now like to consider $BTD(n)$s with additional properties. In terms of tournaments, suppose that we also require that each player plays precisely once on each court in the first n rounds of the tournament. A $BTD(n)$ with this property is called a factor balanced $BTD(n)$ and is denoted by $FBBTD(n)$. In graphical terminology, the first n pairs of any row in the design form a 1-factor of K_{2n}. In addition to this, if each player plays precisely once on each court during the final n rounds, the design is said to be partitioned and is denoted by $PBTD(n)$. Clearly, the existence of a $PBTD(n)$ implies (by definition) that there exists a $FBBTD(n)$. The converse is false as the example in Figure 1 illustrates a $FBBTD(6)$ which is not a $PBTD(6)$. It is a relatively easy task to determine the spectrum of $FBBTD(n)$s. ([7]) In the sequel we study the more difficult problem of the existence of $PBTD(n)$. What make the problem difficult is the fact that these designs are not *pairwise balanced design closed* (*PBD*-closed). We begin by giving a more formal definition.

$0\ \bar{1}$	$\alpha\ \bar{4}$	$3\ 4$	$\bar{2}\ \bar{3}$	$\infty\ 2$	$\bar{0}\ 1$	$1\ \bar{1}$	$\infty\ \bar{4}$	$2\ 4$	$\bar{0}\ \bar{2}$	$\alpha\ 3$
$\infty\ 3$	$1\ \bar{2}$	$\alpha\ \bar{0}$	$4\ 0$	$\bar{3}\ \bar{4}$	$\bar{1}\ 2$	$\alpha\ 4$	$2\bar{2}$	$\infty\ \bar{0}$	30	$\bar{1}\ \bar{3}$
$\bar{4}\ \bar{0}$	$\infty\ 4$	$2\ \bar{3}$	$\alpha\ \bar{1}$	$0\ 1$	$\bar{2}\ 3$	$\bar{2}\ \bar{4}$	$\alpha\ 0$	$3\ \bar{3}$	$\infty\ \bar{1}$	$4\ 1$
$1\ 2$	$\bar{0}\ \bar{1}$	$\infty\ 0$	$3\ \bar{4}$	$\alpha\ \bar{2}$	$\bar{3}\ 4$	$0\ 2$	$\bar{3}\ \bar{0}$	$\alpha\ 1$	$4\ \bar{4}$	$\infty\ \bar{2}$
$\alpha\ \bar{3}$	$2\ 3$	$\bar{1}\ \bar{2}$	$\infty\ 1$	$4\ \bar{0}$	$\bar{4}\ 0$	$\infty\ \bar{3}$	$1\ 3$	$\bar{4}\ \bar{1}$	$\alpha\ 2$	$0\ \bar{0}$
$\bar{2}\ 4$	$\bar{3}\ 0$	$\bar{4}\ 1$	$\bar{0}\ 2$	$\bar{1}\ 3$	$\alpha\ \infty$	$\bar{0}\ 3$	$\bar{1}\ 4$	$\bar{2}\ 0$	$\bar{3}\ 1$	$\bar{4}\ 2$

Figure 1
FBBTD(6)

Let V be a set of $2n$ elements. A Howell design of side s and order $2n$, or more briefly an $H(s,2n)$, is an $s \times s$ array in which each cell is either empty or contains an unordered pair of elements from V such that

(1) each row and each column is Latin (that is, every element of V is in precisely one cell of each row and column) and

(2) every unordered pair of elements of V is in at most one cell of the array.

It follows immediately from the definition of an $H(s,2n)$ that $n \leq s \leq 2n-1$.

If we can partition the columns of a $BTD(n)$ defined on V into three sets C_1,C_2,C_3 of sizes $1,n-1,n-1$ respectively so that the columns in $C_1 \cup C_2$ form an $H(n,2n)$ and the columns in $C_1 \cup C_3$ form an $H(n,2n)$, then the $BTD(n)$ is called partitionable. We denote the design by $PBTD(n)$. An example of a $PBTD(5)$ is displayed in Figure 2.

Partitioned balanced tournament designs are related to Room squares. A Room square of side $2n+1$, $RS(2n+1)$, is an $H(2n+1,2n+2)$. Let R be a $RS(2n+1)$. Since each row or column of R contains n empty cells, a $t \times t$ subarray of empty cells in R must have $t \leq n$. If R contains an $n \times n$ subarray of empty cells, we say R contains a maximum empty subarray and denote R by $MESRS(2n+1)$. The connection between *MESRS*s and *PBTD*s was provided by D.R. Stinson in [10].

α 4	∞ 2	1 3	5 7	2 3	4 5	∞ 7	α 1	0 6
∞ 3	α 5	4 6	0 2	∞ 4	α 2	0 5	6 3	1 7
5 6	0 3	α 7	∞ 1	6 7	0 1	α 3	∞ 5	4 2
1 2	4 7	∞ 0	α 6	α 0	∞ 6	1 4	7 2	5 3
0 7	1 6	2 5	4 3	1 5	3 7	2 6	0 4	α ∞
C_2				C_3				C_1

Figure 2
A $PBTD(5)$

Theorem 1.2. *There exists a $MESRS(2n+1)$ if and only if there exists a $PBTD(n+1)$.*

There is an extensive literature on Room squares and it is known that Room squares of side $2n+1$ exist if and only if $2n+1 \neq 3$ or 5 (Mullin and Wallis [8]). In [10], D.R. Stinson established the nonexistence of a $MESRS(7)$ and the existence of $MESRS(m)$ for $m = 9$ and $m = 11$. He showed that two recursive constructions for Room squares could be used to provide infinite classes of *MESRS*s and conjectured that $MESRS(2n+1)$s exist for all $n \geq 4$. This paper settles the conjecture in the affirmative with the possible exception of 12 values of n.

We have investigated the existence of $MESRS(2n+1)$s by considering the existence of $PBTD(n+1)$s. In [5] and [6], we showed that $PBTD(n)$ exist for $n \equiv 1 \pmod 2$, $n \geq 5$, with four possible exceptions. In this paper, we describe a new recursive construction for $PBTD(n)$s. This construction along with the existence of $PBTD(n)$s for $n \equiv 1 \pmod 2$ is used to establish the existence of $PBTD(n)$s for $n \equiv 0 \pmod 2$. We describe this construction in section 3. In the next section, we collect several results which we will need from previous papers. In section 4, we apply the construction and provide the small designs needed to complete the case $n \equiv 0 \pmod 2$ with eight possible exceptions.

2. Preliminary results.

In this section, we collect for future reference some of the previous constructions and results on partitioned balanced tournament designs.

In [5], we described a direct construction for $PBTD(n+1)$s. This construction uses a starter for a Howell design of side $2n$ and order $2n+2$, $H(2n,2n+2)$. (For definitions and results on starters and adders for $H(2n,2n+2)$, see [1].)

Theorem 2.1. [5] *Let* $n \equiv 1 \pmod 2$. *Let* $S = \{\{x_1,y_1\},\{x_2,y_2\}, \ldots ,\{x_{n-1},y_{n-1}\},\{\alpha,z_1\},\{\infty,z_2\}\}$ *be a starter for an* $H(2n,2n+2)$ *defined on* $Z_{2n} \cup \{\alpha,\infty\}$. *Let* $S' = S - \{x_i,y_i\}$ *for some* i *where* $|x_i - y_i| \equiv 1 \pmod 2$, $1 \le i \le n-1$. *Suppose there is an adder* $A = (a_1,a_2,\ldots,a_n)$ *of* n *distinct elements of* Z_{2n} *such that*

(i) $a_i \in \{0,2,4,\ldots,2n-2\}$ *for* $i = 1,2,\ldots,n$ *and*

(ii) $S' + A = Z_{2n} - \{u_1,v_1\}$ *where* $|u_1 - v_1| = n$

Then there is a $PBTD(n+1)$.

We use this result to find several small $PBTD$s of even side.

Lemma 2.2. *There exist* $PBTD(n+1)$ *for* $n \in \{5,7,9,11,13,15,17,19,21\}$.

Proof. Starters and adders for $PBTD(n+1)$ for $n = 5,7,9$ and 11 appear in [5]. We list starters and adders for $PBTD(n+1)$ for $n = 13,15,17,19$ and 21 in Table 2.1. □

The main constructions for finding $PBTD(n)$ when $n \equiv 1 \pmod 2$ are frame constructions. The details of these constructions can be found in [5] and [6]. An example of this type of construction is the following.

Theorem 2.3. [5] *If there exists a complementary* $\{G_1,G_2,\ldots,G_m\}$*-frame* ($m \ge 2$), *a pair of orthogonal partitioned incomplete Latin squares with partition* $\{G_1,G_2,\ldots,G_m\}$ *and* $PBTD(|G_i|+1)$ *for* $i = 1,2,\ldots,m$, *then there is a* $PBTD((\sum_{i=1}^{m} |G_i|)+1)$.

We used Theorem 2.3 together with existence results for frames and orthogonal partitioned incomplete Latin squares to prove the existence of $PBTD(n)$ for $n \equiv 1 \pmod 2$.

Theorem 2.4. [5] *Let* $n \equiv 1 \pmod 4$, $n \ge 5$. *There exists a* $PBTD(n)$ *except possibly for* $n = 9$.

Theorem 2.5. [6] *Let* $n \equiv 3 \pmod 4$, $n \ge 7$. *There exists a* $PBTD(n)$ *except possibly for* $n \in \{11,15,27\}$.

Several other applications of the frame constructions can be found in [5]. We will use the following result in this paper.

Theorem 2.6. [5] *Let* $n \equiv 1 \pmod 2$, $n \geq 7$. *If there is a* $PBTD(m+1)$, $m \neq 6$, *then there is a* $PBTD(mn+1)$.

Table 2.1

Starter-adder pairs for $PBTD(n+1)$ for n = 13,14,17,19,21.

n = 13

S	2,4	3,6	5,9	7,12	10,16	14,21	17,25	15,24	13,23
A	0	12	18	4	24	8	14	2	22
	11,22	8,20	α,18	∞,19					
	10	16	20	6					

$\{x_i, y_i\}$: 0,1

$\{u_1, v_1\}$: 7,20

n = 15

S	2,4	3,6	5,9	7,12	8,14	15,22	17,25	20,29	18,28
A	0	2	10	16	12	26	22	4	24
	16,27	11,23	13,26	10,24	α,19	∞,21			
	29	20	14	6	18	8			

$\{x_i, y_i\}$: 0,1

$\{u_1, v_1\}$: 6,21

n = 17

S	2,4	3,6	5,9	7,12	8,14	11,18	17,25	19,28	21,31
A	0	2	16	22	12	6	32	18	14
	22,33	20,32	16,29	13,27	15,20	10,26	α,23	∞,24	
	10	30	24	20	26	4	8	28	

$\{x_i, y_i\}$: 0,1

$\{u_1, v_1\}$: 10,27

n = 19

S	2,4	3,6	5,9	7,12	8,14	10,17	15,23	18,27	21,31	
A	0	2	4	14	22	12	10	26	6	
	24,35	25,37	20,33	22,36	19,34	16,32	13,30	11,29	α,26	∞,28
	34	24	8	16	36	18	32	28	30	20

$\{x_i, y_i\}$: 0,1

$\{u_1, v_1\}$: 16,35

$n = 21$

S	2,4	3,6	5,9	7,12	8,14	10,17	11,19	20,29	21,31
A	0	2	4	14	24	10	20	8	12
	25,37	28,41	26,40	24,39	22,38	15,32	18,36	16,35	13,33
	40	26	16	6	38	34	22	36	28
	α,27	∞,30							
	32	18							

$\{x_i,y_i\}$: 0,1

$\{u_1,v_1\}$: 15,36

In section 4, we will also use two product constructions from [10].

Theorem 2.6. *Suppose there exist PBTD(m) and PBTD(n) and a pair of orthogonal Latin squares of order m. Then there exists a PBTD(mn).*

Theorem 2.7. *Suppose there exists a PBTD$(2m)$ and a pair of orthogonal Latin squares of side m. Then there exists a PBTD$(6m)$.*

3. Recursive Constructions.

In this section, we prove the main recursive construction of the paper. This construction uses incomplete orthogonal arrays, $IA(n,k,s)$'s.

Let V be a finite set of size n. Let K be a subset of V of size k. An incomplete orthogonal array $IA(n,k,s)$ is an $(n^2 - k^2) \times s$ array written on the symbol set V such that every ordered pair of $V \times V - (K \times K)$ occurs in any ordered pair of columns from the array. An $IA(n,k,s)$ is equivalent to a set of $s - 2$ mutually orthogonal Latin squares of order n which are missing a subsquare of order k. We need not be able to fill in the $k \times k$ missing subsquares with Latin squares of order k.

Theorem 3.1. *Let $n \equiv 1 \pmod 2$. If there exists a PBTD$(n+1)$ generated by a starter-adder pair on $Z_{2n} \cup \{\alpha,\infty\}$, a PBTD$(m)$, a PBTD$(m+k)$, a pair of orthogonal Latin squares of order m and an $IA(m+k,k,4)$, then there is a PBTD$((n+1)m+k)$.*

Proof. Let B be a PBTD$(n+1)$ generated by a starter-adder pair on $Z_{2n} \cup \{\alpha,\infty\}$. Let C_i denote column i of B for $i = 1,2,\ldots,2n+1$. Suppose row i of C_{2n+1} contains the pair $\{u_i,v_i\}$ where $|u_i - v_i| = n$ for $i = 1,2,\ldots,n$ and row $n+1$ of C_{2n+1} contains the pair $\{\alpha,\infty\}$. The partitioning of B into $H(n+1,2n+2)$s will be $C_1 \cup C_3 \cup \cdots \cup C_{2n-1} \cup C_{2n+1}$ and $C_2 \cup C_4 \cup \cdots \cup C_{2n} \cup C_{2n+1}$.

Let $\{s_1,t_1\}$ be a pair in column 1, row j, $1 \le j \le n$, of B such that $|s_1 - t_1| \equiv 1 \pmod 2$. Let $s_i = s_1 + i - 1 \pmod{2n}$ and let $t_i = t_1 + i - 1 \pmod{2n}$ for $i = 1,2,\ldots,2n$. Then $\{s_i,t_i\}$ occurs in column i of B. Since $|s_1 - t_1| \equiv 1 \pmod 2$, $\bigcup_{j=0}^{n-1}\{s_{2j+1},t_{2j+1}\} = Z_{2n}$ and $\bigcup_{j=1}^{n}\{s_{2j},t_{2j}\} = Z_{2n}$. Every element of Z_{2n} occurs once in $\{s_1,s_2,\ldots,s_{2n}\}$ and once in $\{t_1,t_2,\ldots,t_{2n}\}$. If $\{s_i,t_i\}$ occurs in row j of B, then $\{s_{n+i},t_{n+i}\}$

also occurs in row j. Note that if i is odd, then $n + i$ is even.

Let $M = \{1,2,...,m\}$. Let L_1 and L_2 be a pair of orthogonal Latin squares of side m defined on M. L will be the array of pairs formed by the superposition of L_1 and L_2. L_{xy} will denote the array of pairs formed by replacing each pair (a,b) in L with the pair (a_x,b_y).

We use an $IA(m+k,k,4)$ to construct a pair of orthogonal Latin squares of order $m + k$, N_{s_i} and N_{t_i}, which are missing subsquares of order k. Let $\beta = \{\beta_1,\beta_2, \ldots ,\beta_k\}$ and let $\gamma = \{\gamma_1,\gamma_2, \ldots ,\gamma_k\}$. Suppose N_{s_i} is defined on $(M \times \{s_i\}) \cup \beta$ and is missing a subsquare of order k defined on β. Suppose N_{t_i} is defined on $(M \times \{t_i\}) \cup \gamma$ and is missing a subsquare of order k defined on γ. Let $N_{s_it_i}$ be the array of pairs formed by the superposition of N_{s_i} and N_{t_i}. $N_{s_it_i}$ can be written in the following form.

$$N_{s_it_i} = \begin{array}{|c|c|} \hline A_i & E_i \\ \hline D_i & O \\ \hline \end{array} \quad \text{where } O \text{ is an empty array}$$

of size $k \times k$. A_i is an $m \times m$ array, E_i is $m \times k$ and D_i is $k \times m$.

Let B_i be a $PBTD(m)$ defined on $M \times \{u_i,v_i\}$ for $i = 1,2,...,n$. The partitioning of B_i is as follows.

$$B_i = \begin{array}{|c|c|c|} \hline F_i & G_i & H_i \\ \hline \end{array}$$
$$\quad m-1 \quad m-1 \quad 1$$

$F_i \cup H_i$ is an $H(m,2m)$ and $G_i \cup H_i$ is an $H(m,2m)$. Let B_{n+1} be a $PBTD(m+k)$ defined on $(M \times \{\alpha,\infty\}) \cup (\beta \cup \gamma)$. The partitioning of B_{n+1} is as follows.

$$B_{n+1} = \begin{array}{|c|c|c|c|c|} \hline K_1 & K_2 & K_3 & K_4 & K_5 \\ \hline \end{array}$$
$$\quad m-1 \quad m-1 \quad 1 \quad k \quad k$$

$K_1 \cup K_4 \cup K_3$ is an $H(m+k,2m+2k)$ and $K_2 \cup K_5 \cup K_3$ is an $H(m+k,2m+2k)$.

Let $V = \{0,1,2,...,2n-1\} \cup \{\alpha,\infty\}$ and let $W = \beta \cup \gamma$. We can now construct a $PBTD((n+1)m+k)$ defined on $(M \times V) \cup W$ as follows.

Replace each pair $\{x,y\}$ in B which does not belong to the set $\{\{u_1,v_1\}, \ldots ,\{u_n,v_n\},\{s_1,t_1\}, \ldots ,\{s_{2n},t_{2n}\},\{\alpha,\infty\}\}$ with the $m \times m$ array L_{xy}. Replace each pair $\{s_i,t_i\}$ with the $m \times m$ array A_i for $i = 1,2,...,2n$. Replace each pair $\{u_i,v_i\}$ with the $m \times 2m - 1$ array B_i for $i = 1,2,...,n$. Add k new rows to the $(n+1)m \times (2mn+2m-1)$ array: $[D_1D_2...D_{2n}O_{2m+2k-1}]$ where O_{2m+k-1} is a $k \times 2m + 2k - 1$ empty array. Add $2k$ new columns to this array. The new columns are of the following form.

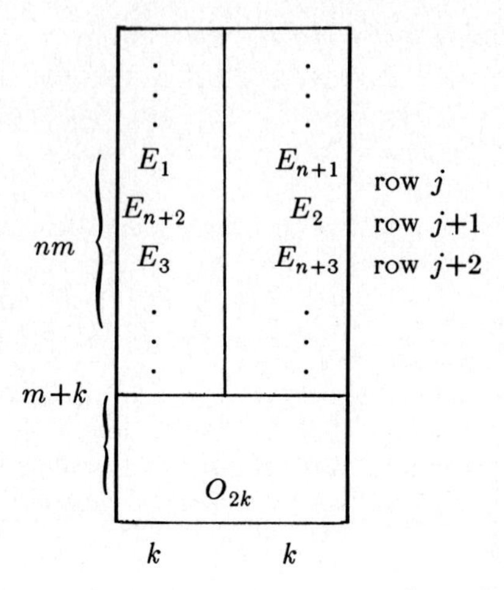

where O_{2k} is an empty array of size $(m+k) \times 2k$. The resulting array is of size $(n+1)m + k \times (2mn+2m-1+2k)$ and is missing a subarray of size $(m+k) \times (2(m+k)-1)$ in the lower right hand corner. Fill this array in with the array B_{n+1}. Let P denote the resulting array. The form of P is indicated below. The rows and columns of B have been permuted so that $\{s_1,t_1\}$ occurs in cell (1,1) of B.

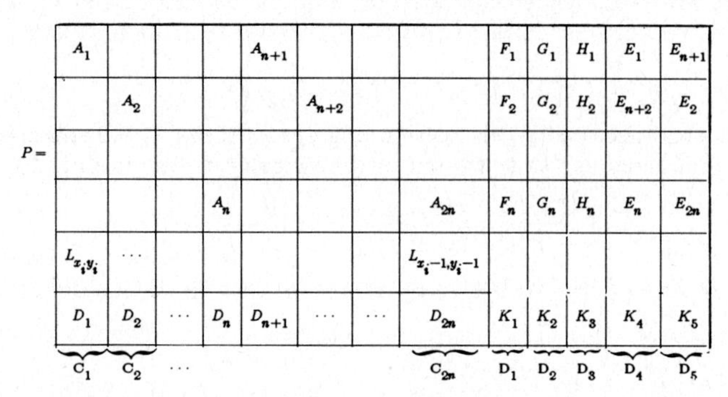

P is a $PBTD((n+1)m+k)$ defined on $\mathbf{V} = (M \times V) \cup W$. Every element of $\mathbf{V}$ occurs once in each column and at most twice in each row of P. Every distinct unordered pair in $\mathbf{V}$ occurs once in P. We use the collections of columns indicated in the diagram above to construct a partitioning of P.

$$\mathbf{C}_1 \cup \mathbf{C}_3 \cup \cdots \cup \mathbf{C}_{2n-1} \cup \mathbf{D}_1 \cup \mathbf{D}_4 \cup \mathbf{D}_3$$

is an $H((n+1)m + k, 2(n+1)m+2k)$ and

$$\mathbf{C}_2 \cup \mathbf{C}_4 \cup \cdots \cup \mathbf{C}_{2n} \cup \mathbf{D}_2 \cup \mathbf{D}_5 \cup \mathbf{D}_3$$

is an $H((n+1)m + k, 2(n+1)m + 2k)$. □

This construction can be used with any $PBTD(n+1)$ which has a partitionable transversal. Let B be a $PBTD(n+1)$ defined on $V \cup \{\alpha,\infty\}$. We write B in the following form.

$$B = \begin{array}{|cc|c|} \hline A & & C \\ & & \\ \hline R & & \alpha\infty \\ \hline \end{array}$$
$$\underbrace{}_{\mathbf{C}_1}\ \underbrace{}_{\mathbf{C}_2}\ \underbrace{}_{\mathbf{C}_3}$$

The partitioning of B is given by $\mathbf{C}_1 \cup \mathbf{C}_3$ and $\mathbf{C}_2 \cup \mathbf{C}_3$. We consider the pairs in B as ordered pairs. Suppose there is a collection of $2n$ pairs T in A, one in each column and two in each row such that each element of V occurs once as a first coordinate and once as a second coordinate in T, (i.e. a transversal of A). Let T_i denote the pairs of T in $\mathbf{C}_i$ for $i = 1,2$. If the pairs in T_i occur in n distinct rows and n distinct columns and every element of V occurs precisely once in T_i for $i = 1$ and $i = 2$, then we say B has a partitionable transversal T.

Theorem 3.2. *If there exists a $PBTD(n+1)$ with a partitionable transversal, a $PBTD(m)$, a $PBTD(m+k)$, a pair of orthogonal Latin squares of order m and an $IA(m+k,k,4)$, then there is a $PBTD((n+1)m+k)$.*

We omit the proof of Theorem 3.2 since it is quite similar to that of Theorem 3.1.

Theorem 3.1 is the main construction used to find $PBTD(n)$ for $n \equiv 0 \pmod 2$. In order to use this construction, we need the following recent result on $IA(n,k,4)$s.

Theorem 3.3. *[3] An $IA(n,k,4)$ exists if and only if $n \geq 3k$.*

The last construction in this section is a special case of the singular indirect product. The proof is similar to the proof of the singular direct product for *MESRS*s [10] and is omitted for that reason.

Theorem 3.4. *If there exists a $PBTD(n)$, a $PBTD(m)$ which is missing as a subarray a $PBTD(k)$ (the $PBTD(k)$ need not exist), a $PBTD(kn)$ and an $IA(m,k,4)$, then there is a $PBTD(nm)$.*

Corollary 3.5. *There is a $PBTD(152)$.*

Proof. Let $n = 8$, $m = 19$ and $k = 3$ in Theorem 4.3. There exists a $PBTD(8)$, a $PBTD(19)$ which is missing a $PBTD(3)$ (Corollary 3.4), a $PBTD(24)$ and an $IA(19,3,4)$ [3]. □

4. *PBTD*(n) for $n \equiv 0 \pmod 2$.

In order to apply our construction to the case $n \equiv 0 \pmod 2$, we will need the existence of *PBTD*(n) for $n \equiv 0 \pmod 2$ and $6 \le n \le 364$.

Lemma 4.1. *Let $n \equiv 0 \pmod 2$. There exists a PBTD(n) for $6 \le n \le 230$ except possibly for $n \in \{26,28,32,34,38,44,58,94\}$.*

Proof. In Table 4.1, we list constructions for *PBTD*(n) for $n \equiv 0 \pmod 2$, $6 \le n \le 230$, and $n \notin \{26,28,32,34,38,44,58,94\}$. □

Lemma 4.2. *Let $n \equiv 0 \pmod 2$. There exists a PBTD(n) for $232 \le n \le 364$.*

Proof. The constructions for *PBTD*(n) where $n \equiv 0 \pmod 2$ and $232 \le n \le 364$ are listed in Table 4.2. □

We can now prove the existence of *PBTD*(n) where $n \equiv 0 \pmod 2$, $n \ge 5$, with 8 possible exceptions.

Theorem 4.3. *Let $n \equiv 0 \pmod 2$, $n \ge 6$. There exists a PBTD(n) except possibly for $n \in \{26,28,32,34,38,44,58,94\}$.*

Proof. There exists a *PBTD*(n) for $n \equiv 0 \pmod 2$, $6 \le n \le 364$, except possibly for $n \in \{26,28,32,34,38,44,58,94\}$ (Lemmas 4.1 and 4.2).

Let $n \equiv 0 \pmod 2$, $n \ge 366$. We write $n = 6(4m+1) + k$ where $k \in \{0,2,4,...,20,22\}$. For $m \ge 15$ and $k \in \{0,2,...,22\}$ there exist *PBTD*$(4m+1)$ (Theorem 2.4) and *PBTD*$(4m+1+k)$ (Theorem 2.5). Since there exists a pair of orthogonal Latin squares of order $4m + 1$ ([2]) and an *IA*$(4m+1+k,k,4)$ (Theorem 3.3), we can apply Theorem 3.1 to construct *PBTD*(n) for $n \equiv 0 \pmod 2$ and $n \ge 366$. □

5. Conclusions.

We have settled the existence question for *PBTD*(n)s with twelve possible exceptions. Theorems 2.4, 2.5 and 4.3 can be summarized as follows.

Theorem 5.1. *Let n be a positive integer, $n \ge 5$. There exists a PBTD(n) except possibly for*

$$n \in \{9,11,15,26,27,28,32,34,38,44,58,94\}.$$

It should be noted that the recursive constructions together with the existence of *PBTD*(n) for small n could be used to reduce the number of exceptional cases. The next result illustrates this.

Table 4.1.

Constructions for $PBTD(n)$ for $n \equiv 0 \pmod 2$, $6 \leq n \leq 230$ and $n \notin \{26,28,32,35,38,44,58,94\}$.

n	Construction		n	Construction		n	Construction	
6	Starter-Adder	2.2	90	6.15	2.7	162	10.16+2	3.1
8	Starter-Adder	2.2	92	7.13+1	2.5	164	10.16+4	3.1
10	Starter-Adder	2.2	96	5.19+1	2.5	166	10.16+6	3.1
12	Starter-Adder	2.2	98	8.12+2	3.1	168	10.16+8	3.1
14	Starter-Adder	2.2	100	11.9+1	2.5	170	17.10	2.6
16	Starter-Adder	2.2	102	6.17	2.6	172	8.21+4	3.1
18	Starter-Adder	2.2	104	10.10+4	3.1	174	10.17+4	3.1
20	Starter-Adder	2.2	106	21.5+1	2.5	176	8.21+8	3.1
22	Starter-Adder	2.2	108	6.18	2.6	178	8.22+2	3.1
24	4.6	2.7	110	6.17+8	3.1	180	10.18	2.6
30	5.6	2.6	112	8.14	2.6	182	10.18+2	3.1
36	7.5+1	2.5	114	8.14+2	3.1	184	10.18+4	3.1
40	5.8	2.6	116	23.5+1	2.5	186	10.18+6	3.1
42	6.7	2.7	118	9.13+1	2.5	188	11.17+1	2.5
46	9.5+1	2.5	120	6.20	2.6	190	9.21+1	2.5
48	6.8	2.6	122	10.12+2	3.1	192	8.24	2.6
50	7.7+1	2.5	124	10.12+4	3.1	194	8.24+2	3.1
52	6.8+4	3.1	126	10.12+6	3.1	196	12.16+4	3.1
54	6.9	2.7	128	8.16	2.6	198	16.12+6	3.1
56	11.5+1	2.5	130	13.10	2.6	200	10.20	2.6
60	6.10	2.6	132	16.8+4	3.1	202	20.10+2	3.1
62	6.10+2	3.1	134	8.16+6	3.1	204	20.10+4	3.1
64	8.8	2.6	136	8.16+8	3.1	206	8.25+6	3.1
66	13.5+1	2.5	138	6.23	2.7	208	8.25+8	3.1
68	8.8+4	3.1	140	10.14	2.6	210	10.21	2.6
70	5.14	2.6	142	10.14+2	3.1	212	8.25+12	3.1
72	6.12	2.6	144	10.14+4	3.1	214	10.21+4	3.1
74	6.12+2	3.1	146	10.14+6	3.1	216	6.36	2.6
76	15.5+1	2.5	148	12.12+4	3.1	218	10.21+8	3.1
78	7.11+1	2.5	150	12.12+6	3.1	220	10.22	2.6
80	10.8	2.6	152	19.8	3.5	222	22.10+2	3.1
82	9.9+1	2.5	154	6.25+4	3.1	224	22.10+4	3.1
84	10.8+4	3.1	156	6.24+12	3.1	226	14.16+2	3.1
86	17.5+1	2.5	158	6.25+8	3.1	228	16.14+4	3.1
88	6.14+4	3.1	160	10.16	2.6	230	16.14+6	3.1

Table 4.2.

Constructions for $PBTD(n)$ for $n \equiv 0 \pmod 2$, $232 \leq n \leq 364$.

n	Construction		n	Construction	
232	14.16+8	3.1	300	10.30	2.6
234	6.39	2.7	302	6.50+2	3.1
236	8.29+4	3.1	304	6.50+4	3.1
238	14.17	2.6	306	6.50+6	3.1
240	10.24	2.6	308	14.22	2.6
242	12.20+2	3.1	310	22.14+2	3.1
244	12.20+4	3.1	312	22.14+4	3.1
246	10.24+6	3.1	314	22.14+6	3.1
248	8.31	2.6	316	14.22+8	3.1
250	10.25	2.6	318	6.53	2.6
252	10.24+12	3.1	320	6.53+2	3.1
254	10.25+4	3.1	322	6.53+4	3.1
256	12.21+4	3.1	324	8.41	2.6
258	10.25+8	3.1	326	8.41+2	3.1
260	20.13	2.6	328	8.41+4	3.1
262	12.21+10	3.1	330	6.54+6	3.1
264	16.16+8	3.1	332	6.54+8	3.1
266	12.22+2	3.1	334	6.54+10	3.1
268	6.43+10	3.1	336	6.54+12	3.1
270	6.45	2.7	338	6.54+14	3.1
272	12.22+8	3.1	340	6.54+16	3.1
274	7.39+1	2.5	342	6.54+18	3.1
276	6.46	2.7	344	6.54+20	3.1
278	6.46+2	3.1	346	6.54+22	3.1
280	6.46+4	3.1	348	6.54+24	3.1
282	6.46+6	3.1	350	6.54+26	3.1
284	6.46+8	3.1	352	6.56+16	3.1
286	6.46+10	3.1	354	6.56+18	3.1
288	6.48	2.6	356	6.56+20	3.1
290	6.48+2	3.1	358	6.56+22	3.1
292	6.48+4	3.1	360	6.56+24	3.1
294	6.48+6	3.1	362	6.60+2	3.1
296	6.48+8	3.1	364	6.60+4	3.1
298	14.21+4	3.1			

Lemma 5.2. *If there exists a PBTD(9), then there exist PBTD(n) for* $n \in \{44,58,94\}$.

Proof. The following table lists the constructions to be used.

n	Construction	
44	6.7+2	3.1
58	6.9+4	3.1
94	10.9+4	3.1

□

It is interesting that the existence of partitioned balanced tournament designs implies the existence of several other types of designs. We have already noted the equivalence between *MESRS*s and *PBTD*s. *PBTD*s can also be used to prove the existence of an even sided analogue for Room Squares.

Let n be a non-negative integer and let S be a set of size $2n + 2$. Let F be a partition of S into unordered pairs. A house of order $n + 1$ is a $2n+2 \times 2n+2$ array H which satisfies the following:

(1) every cell of H either is empty or contains an unordered pair of elements of S,

(2) every element of S occurs in precisely one cell of each row and column of H,

(3) the pairs in F each occur in precisely two cells of H, whereas every other pair of elements occurs in exactly one cell of H,

(4) the pairs in the first and second rows of H are precisely those in F, and

(5) every column of H contains one pair from F.

A house of order $n + 1$ is an approximation to a Room square of side $2n + 2$. It is known that houses of order n exist for n a positive integer and $n \neq 2$, [11].

If we can permute the columns $1,...,2n+2$ and rows $3,...,2n+2$ of a house of order $n + 1$ H so that the $n+1 \times n+1$ subarray H_1 indexed by rows and columns $2j + 1$ for $j = 0,1,...,n$ is an $H(n+1,2n+2)$ and the $n+1 \times n+1$ subarray H_2 indexed by rows and columns $2j$ for $j = 1,2,...,n+1$ is an $H(n+1,2n+2)$, then we call H a partitionable house of order $n + 1$. We note that H_1 and H_2 will form a pair of almost disjoint $H(n+1,2n+2)$s. (They will have the same first row.)

Therefore, a partitionable house of order $n + 1$ can be used to construct a *PBTD*$(n+1)$. Conversely, it is easy to construct a partitionable house of order $n + 1$ from a *PBTD*$(n+1)$.

The following theorem lists the designs which are equivalent to *PBTD*s.

Theorem 5.3. *For n a positive integer, the following designs are equivalent.*

(i) *a PBTD*$(n+1)$

(ii) *a MESRS*$(2n+1)$ *[10]*

(iii) *a pair of almost disjoint* $H(n+1,2n+2)$*s and [10]*

(iv) *a partitionable house of order* $n + 1$.

Added In Proof

Recently, $PBTD(n)$'s for $n = 27$, 32, 38, 58, and 94 have been constructed. This leaves seven possible exceptions to the existence of *PBTD*s. For the case $n = 27$, the reader is referred to [6], and for $n = 32$, 38, 58 and 94 to E.R. Lamken, "A note on partitioned balanced tournament designs", *Ars Combinatoria* 24 (1987).

References

[1] Anderson, B.A., *Howell designs of type* $H(p-1,p+1)$, J. Combinatorial Theory (A) 24 (1978), 131-140.

[2] Bose, R.C., Parker, E.T. and Shrikhande, S., *Further results on the construction of mutually orthogonal Latin squares and the falsity of Euler's conjecture,* Canadian J. Math. 12 (1960), 189-203.

[3] Heinrich, K. and Zhu, L., *Existence of orthogonal Latin squares with aligned subsquares,* Discrete Math. 59 (1986) 69-78.

[4] Lamken, E.R. and Vanstone, S.A., *The existence of factored balanced tournament designs,* Ars Combinatoria 19 (1985) 157-160.

[5] Lamken, E.R. and Vanstone, S.A., *Partitioned balanced tournament designs of side* $4n+1$, Ars Combinatoria 20 (1985) 29-44.

[6] Lamken, E.R. and Vanstone, S.A., *The existence of partitioned balanced tournament designs of side* $4n+3$, Annals of Discrete Math. (elsewhere in this volume).

[7] Lamken, E.R. and Vanstone, S.A., *Balanced tournament designs and related topics,* Preprint.

[8] Mullin, R.C. and Wallis, W.D., *The existence of Room squares,* Aequationes Math. 13 (1975) 1-7.

[9] Schellenberg, P.J., van Rees, G.H.J. and Vanstone, S.A., *The existence of balanced tournament designs,* Ars Combinatoria 3 (1977) 303-318.

[10] Stinson, D.R., *Room squares with maximum empty subarrays,* Ars Combinatoria 20 (1985) 159-166.

[11] Stinson, D.R. and Wallis, W.D., *An even side analogue of Room squares,* Aequationes Math. 27 (1984) 201-213.

[12] Todorov, D.T., *Three mutually orthogonal Latin squares of order* 14, Ars Combinatoria 20 (1985) 45-48.

Annals of Discrete Mathematics 34 (1987) 353—362
© Elsevier Science Publishers B.V. (North-Holland)

Constructions for Cyclic Steiner 2-designs

Rudolf Mathon

Department of Computer Science
University of Toronto
Toronto, Ontario, Canada M5S 1A4

TO ALEX ROSA ON HIS FIFTIETH BIRTHDAY

ABSTRACT

This paper surveys direct and recursive constructions for cyclic Steiner 2-designs. A new method is presented for cyclic designs with blocks having a prime number of elements. Several new constructions are given for designs with block size 4 which are based on perfect systems of difference sets and additive sequences of permutations.

1. Introduction

A *balanced incomplete block design* (briefly BIBD) with parameters (v,k,λ) is a pair (V,B) where V is a v-set and B is a collection of k-subsets of V (called *blocks*) such that every 2-subset of V is contained in exactly λ blocks. A *Steiner 2-design* is a (v,k,λ) BIBD with $\lambda = 1$. An *automorphism* of a BIBD (V,B) is a bijection $\phi: V \rightarrow V$ such that the induced mapping $\Phi: B \rightarrow B$ is also a bijection. The set of all such mappings forms a group under composition called the automorphism group of the design.

A (v,k,λ) BIBD is *cyclic* if it has an automorphism consisting of a single cycle of length v. Cyclic (v,k,λ) BIBD's will be denoted by $C(v,k,\lambda)$. A (v,k,λ) *difference family* (briefly DF) is a collection of k-subsets $D_1, \ldots, D_t$ of the integers Z_v modulo v such that for each nonzero $x \in Z_v$ the congruence $d_i - d_j \equiv x (mod\ v)$ has exactly λ solution pairs (d_i,d_j) with $d_i,d_j \in D_l$, for some l. A (v,k,λ) DF is called *simple* if $\lambda = 1$. It is easily verified that a necessary condition for the existence of a (v,k,λ) DF is $\lambda(v-1) \equiv 0\ mod\ k(k-1)$. In particular, if a simple DF exists then $v \equiv 1$ $mod\ k(k-1)$. A (v,k,λ) DF generates a cyclic BIBD $C(v,k,\lambda)$ with $V = Z_v$ and $B = \{\sigma^i D_l \mid 0 \leq i < v,\ 1 \leq l \leq t\}$, where $\sigma: V \rightarrow V$, $\sigma(x) = x + 1\ mod\ v$ and $t = \lambda(v-1)/(k(k-1))$. The t blocks $D_1, \ldots, D_t$ are called *starter* or *base blocks* of the design (V,B) (they are representatives of the orbits of B under σ). An orbit analysis of a cyclic Steiner 2-design $C(v,k,1)$ yields the following necessary existence condition:

$$v \equiv 1,\ k \ mod\ k(k-1). \tag{1}$$

The case $v = k(k-1)t + 1$ corresponds to a simple DF. If $v = k(k-1)t + k$ then there are $t + 1$ starter blocks $D_0, D_1, \ldots, D_t$, where $D_0 = \{0, m, 2m, \ldots, (k-1)m\}$, $m = (k-1)t + 1$ generates an m-orbit and $D_1, \ldots, D_t$ generate t v-orbits under σ. It is clear, that the differences in $D_1, \ldots, D_t$ cover the elements $Z_v \backslash D_0$ exactly once.

Two difference families $\mathbf{D} = \{D_1, \ldots, D_t\}$ and $\mathbf{D}' = \{D_1', \ldots, D_t'\}$ are said to be *equivalent* if for some integers $r, s_1, \ldots, s_t$

$$\{D_1', \ldots, D_t'\} = \{rD_1 + s_1, \ldots, rD_t + s_t\} \ mod\ v. \tag{2}$$

If $\mathbf{D}$ is equivalent with itself, then the corresponding r is called a *multiplier* of $\mathbf{D}$ and $\tau: x \rightarrow rx,\ x \in Z_v$ is an automorphism of the cyclic design.

Cyclic designs have a nice structure and interesting algebraic properties. Their concise representation makes them attractive in applications and for testing purposes. Cyclic BIBD's and difference systems have been studied by many authors [3], [7], [10], [13]. Results concerning cyclic Steiner 2-designs are surveyed in [5] which also contains a fairly extensive bibliography.

The present paper addresses the problem of existence of cyclic Steiner 2-designs $C(v,k,1)$. In the next two sections we discuss direct and recursive constructions for general block sizes k. In addition to known techniques, several new constructions are presented for $k = 4$ and 5. We conclude with a list of open problems. The paper significantly extends the existence results given in [5] for cyclic Steiner 2-designs with block sizes $k > 3$.

2. Direct Constructions

The majority of direct methods for constructing cyclic designs are based on finite fields. In this section we survey those constructions which apply to Steiner 2-designs and apply them to generate some new designs with blocks of prime size.

We begin with two general constructions of Wilson for $(v,k,1)$ difference families [13].

Theorem 1 Let $p = k(k-1)t + 1$ be a prime and α a primitive root of Z_p. Let H^m be the multiplicative subgroup of $Z_p \backslash \{0\}$ generated by α^m and let $\omega = \alpha^{2mt}$.

(i) If $k = 2m + 1$ is odd and $\{\omega - 1, \omega^2 - 1, \ldots, \omega^m - 1\}$ is a system of representatives for the cosets $\alpha^i H^m$, $i = 0,1,\ldots,m-1$, then the blocks $D_{i+1} = \{\alpha^{mi}, \omega\alpha^{mi}, \ldots, \omega^{2m}\alpha^{mi}\}$, $i = 0,1,\ldots,t-1$ form a $(p,k,1)$ DF.

(ii) If $k = 2m$ is even and $\{1, \omega - 1, \ldots, \omega^{m-1} - 1\}$ is a system of representatives for the cosets $\alpha^i H^m$, $i = 0,1,\ldots,m-1$, then the blocks $D_{i+1} = \{0, \alpha^{mi}, \omega\alpha^{mi}, \ldots, \omega^{2m-2}\alpha^{mi}\}$, $i = 0,1,\ldots,t-1$ form a $(p,k,1)$ DF in Z_p.

Theorem 2 Let $p = k(k-1)t + 1$ be a prime and α a primitive root of Z_p. If there exists a set $B = \{b_1, \ldots, b_k\}$ *subset* Z_p such that $\{b_j - b_i \mid 1 \leq i < j \leq k\}$ is a system of representatives for the cosets $\alpha^i H^m$, $i = 0,1,\ldots,m-1$, where $m = k(k-1)/2$ and H^m is the subgroup of $Z_p \backslash \{0\}$ generated by α^m, then $D_{i+1} = \alpha^{2mi} B$, $i = 0,1,\ldots,t-1$ is a $(p,k,1)$ DF in Z_p.

Our next result concerns the case $v \equiv k \bmod k(k-1)$.

Theorem 3 Let $k = 2m + 1$ and $p = 2mt + 1$, $t \geq 2$ be two odd primes and let α be a primitive root of Z_p. Define $m - 1$ numbers r_i by the equations $\alpha^{r_i} = \alpha^{ti} - 1$, $i = 1, \ldots, m - 1$. If there exists a $\beta \in Z_k$ such that the $2m$ elements ± 1, $\pm(\beta^{ti} - 1)\beta^{-r_i}$, $i = 1, \ldots, m - 1$ are all distinct in Z_k, then the blocks

$$D_0 = \{0_0, 0_1, \ldots, 0_{2m}\}$$
$$D_{i+1} = \{0_0, \alpha^i{}_{\beta^i}, \alpha^{t+i}{}_{\beta^{t+i}}, \ldots, \alpha^{2mt-t+i}{}_{\beta^{2mt-t+i}}\}, \quad i = 0,1,\ldots,t-1 \tag{3}$$

form a $(kp,k,1)$ DF in Z_{kp}.

Proof We note, that since in the family of blocks $\mathbf{B} = \{B_1, \ldots, B_t\}$, $B_{i+1} = \{0, \alpha^i, \ldots, \alpha^{2mt-t+i}\}$ each nonzero difference appears exactly $k - 2m + 1$ times, $\mathbf{B}$ forms a (p,k,k) DF in Z_p. To complete the proof, it suffices to show that for any fixed difference in $\mathbf{B}$ the corresponding subscript differences cover every non-zero element of Z_k exactly once. Since for each i, $(\alpha^{ti} - 1)\,\alpha^{-r_i} = 1$ this is equivalent to the assumption that ± 1, $\pm(\beta^{ti} - 1)\beta^{-r_i}$, $i = 1, \ldots, m - 1$ are distinct in Z_k. Finally, since k and p are distinct primes the design is cyclic in Z_{kp}. □

We will apply Theorem 3 to blocksize $k = 7$. Then $m = 3$ and p is a prime of the form $p = 6t + 1$, $t \geq 2$. If α is a primitive root of Z_p, then $\alpha^{3t} = -1$ and since

$$(\alpha^t + 1)\alpha^{2t} = \alpha^{2t} - 1 = (\alpha^t + 1)(\alpha^t - 1)$$

we have $\alpha^t - 1 = \alpha^{2t}$. Let r be the solution of $\alpha^r = \alpha^{2t} - 1$. We require that for some $\beta \in Z_7$ the 6 numbers

$$\pm\beta^{2t}, \quad \pm(\beta^t - 1), \quad \pm\beta^{2t-r}(\beta^{2t} - 1) \tag{4}$$

cover the non-zero elements of Z_7. Since β^{2t} cannot be congruent to 1 modulo 7, we see that $t \equiv 1$ or 2 mod 3. If $t \equiv 1 \bmod 3$, then (4) are distinct if either $\beta = 2$ and $r \equiv 0 \bmod 3$, or $\beta = 4$ and $r \equiv 2 \bmod 3$. If $t \equiv 2 \bmod 3$, then we need either $\beta = 2$ and $r \equiv 1 \bmod 3$, or $\beta = 4$ and $r \equiv 0 \bmod 3$. Combining all these conditions we obtain the following result.

Corollary 4 Let $p = 6t + 1$ be a prime, $t \geq 2$, $t \not\equiv 0 \bmod 3$, and let α be a primitive root in Z_p. Then the blocks (3) form a $(7p,7,1)$ DF for some $\beta \in Z_7$ if and only if $t \not\equiv r \bmod 3$, where r satisfies $\alpha^r = \alpha^{2t} - 1$.

We note, that for some values of t we obtain two non-isomorphic cyclic designs. If $t \equiv 4 \bmod 6$, then (4) are distinct also if either $\beta = 3$ and $r \equiv 2 \bmod 3$, or $\beta = 5$ and $r \equiv 0 \bmod 3$. If $t \equiv 2 \bmod 6$, then (4) are distinct also if either $\beta = 3$ and $r \equiv 0 \bmod 3$, or $\beta = 5$ and $r \equiv 1 \bmod 3$.

For $k = 7$ solutions exist when t = 2*, 5, 7, 13, 16*, 26*, 35, 37, 38*, 40*, 46*, 47, etc. The base blocks for t = 2*, 5 and 7 are

$$\begin{array}{ccccccc} 0_0 & 1_1 & 4_4 & 3_2 & 12_1 & 9_4 & 10_2 \\ 0_0 & 2_2 & 8_1 & 6_4 & 11_2 & 5_1 & 7_4 \end{array}$$

$$\begin{array}{ccccccc} 0_0 & 1_1 & 4_4 & 3_2 & 12_1 & 9_4 & 10_2 \\ 0_0 & 2_5 & 8_6 & 6_3 & 11_5 & 5_6 & 7_3 \end{array}$$

$$\begin{array}{ccccccc} 0_0 & 1_1 & 26_4 & 25_2 & 30_1 & 5_4 & 6_2 \\ 0_0 & 3_2 & 16_1 & 13_4 & 28_2 & 15_1 & 18_4 \\ 0_0 & 9_4 & 17_2 & 8_1 & 22_4 & 14_2 & 23_1 \\ 0_0 & 27_1 & 20_4 & 24_2 & 4_1 & 11_4 & 7_2 \\ 0_0 & 19_2 & 29_1 & 10_4 & 12_2 & 2_1 & 21_4 \end{array}$$

$$\begin{array}{ccccccc} 0_0 & 1_1 & 37_2 & 36_4 & 42_1 & 6_2 & 7_4 \\ 0_0 & 3_2 & 25_4 & 22_1 & 40_2 & 18_4 & 21_1 \\ 0_0 & 9_4 & 32_1 & 23_2 & 34_4 & 11_1 & 20_2 \\ 0_0 & 27_1 & 10_2 & 26_4 & 16_1 & 33_2 & 17_4 \\ 0_0 & 38_2 & 30_4 & 35_1 & 5_2 & 13_4 & 8_1 \\ 0_0 & 28_4 & 4_1 & 19_2 & 15_4 & 39_1 & 24_2 \\ 0_0 & 41_1 & 12_2 & 14_4 & 2_1 & 31_2 & 29_4 \end{array}$$

The solutions for $t = 5$ and 7 are first examples of BIBD's with the parameters (217,7,1) and (301,7,1), respectively. For $k = 11$ solutions exist when $t = 33$, 54*, 57, 91, 94*, etc. and for $k = 13$, $t = 13$, 19, 59, etc. (* indicates 2 solutions).

We conclude this section with a well-known result in finite geometries [6].

Theorem 5 Let q be a prime power. Then the lines in the projective geometry $PG(n,q)$, $n \geq 2$ form a cyclic design with parameters $((q^{n+1} - 1)/(q - 1),\ q + 1,\ 1)$.

3. Recursive Constructions

Given two difference families it is sometimes possible to combine them to construct a new one. Several such constructions are known for general cyclic BIBD's [4] [8] [14]. To apply them, various conditions on the block sizes are usually required.

We begin with a construction by C.J. Colbourn and M.J. Colbourn [4].

Theorem 6 Let $A^t_i = \{0,a^i_1, \ldots ,a^i_{k-1}\}$, $i = 1, \ldots ,t$ be a $(v,k,1)$ DF in Z_v and let $B^s_j = \{0,b^j_1, \ldots ,b^j_{k-1}\}$, $j = 1, \ldots ,s$ be a $(w,k,1)$ DF in Z_w.

(i) If $v = k(k - 1)t + 1$ and w is relatively prime to $(k - 1)!$, then for $i = 1, \ldots ,t$, $j = 1, \ldots ,s$ and $l = 0,1, \ldots ,w - 1$

$$\left.\begin{array}{l} \{0,a^i_1 + lv,a^i_2 + 2lv, \ldots ,a^i_{k-1} + (k-1)lv\} \\ \{0,vb^j_1,vb^j_i, \ldots ,vb^j_{k-1}\} \end{array}\right\} \qquad (5)$$

is a $(vw,k,1)$ DF in Z_{vw}.

(ii) If $v = k\alpha$, $w = k\beta$ and β is relatively prime to $(k - 1)!$, then for $i = 1, \ldots ,t$, $j = 1, \ldots ,s$ and $l = 0,1, \ldots ,w - 1$

$$\left.\begin{array}{l} \{0,a^i_1 + lv,a^i_2 + 2lv, \ldots ,a^i_{k-1} + (k-1)lv\} \\ \{0,\alpha b^j_1,\alpha b^j_2, \ldots ,\alpha b^j_{k-1}\} \\ \{0,\alpha\beta,2\alpha\beta, \ldots ,(k-1)\alpha\beta\} \end{array}\right\} \qquad (6)$$

is a $(k\alpha\beta,k,1)$ DF in $Z_{k\alpha\beta}$. Here $\alpha = (k - 1)t + 1$, $\beta = (k - 1)s + 1$, and only full orbit base blocks A^t_i, B^s_j are considered.

We note that the construction can be used if either w or β are prime. Then the existence of a $(w,k,1)$ DF implies the existence of a $(w^n,k,1)$ DF for every $n \geq 1$. Similarly, from a $(k\beta,k,1)$ DF we obtain a $(k\beta^n,k,1)$ DF. Also, if a $(v,k,1)$ DF exists with $v \equiv 1 \bmod k(k-1)$ and prime k then there exists a $(vk,k,1)$ DF.

In [8] M. Jimbo and S. Kuriki have introduced a more general construction for cyclic BIBD's which is based on orthogonal arrays. Applying it to Steiner 2-designs we obtain the following typical result.

Theorem 7 Suppose there exists a $C(v,k,1)$ and a $C(w,k,1)$, where $v \equiv 1 \bmod k(k-1)$ and k is an odd prime. Then there exists a $C(vw,k,1)$. If, in addition, $w \equiv 1 \bmod k(k-1)$, then the conclusion holds for k a prime power.

So, for example, if k is an odd prime not dividing v, then the existence of a $C(v,k,1)$ implies the existence of both $C(v^n,k,1)$ and $C(kv^n,k,1)$ for any $n \geq 1$.

The next construction employs cyclic pairwise balanced designs. A *pairwise balanced design* (briefly PBD) is a pair (V,B) where V is a v-set and B is a collection of subsets of V (blocks) such that every 2-subset of V is contained in exactly one block. A PBD will be denoted by $(v,K,1)$, where $K = \{k_1, \ldots, k_n\}$ is the set of block sizes.

Theorem 8 Suppose there exists a cyclic $(v,K,1)$ PBD with $K = \{k_1, \ldots, k_n\}$ and that for each k_i there exists a $(k_i,k,1)$ Steiner 2-design. Then there exists a $C(v,k,1)$.

Proof Replace each base block in the PBD by the blocks of the corresponding Steiner 2-design to obtain the base blocks of the final $C(v,k,1)$. □

In the next section we shall give some other recursive constructions for cyclic designs with blocks of size 4 and 5 which are based on the concepts of perfect systems of difference sets and additive sequences of permutations.

4. Special Constructions

The existence question for cyclic Steiner triple systems has been completely settled by Peltesohn [10], who constructed $C(v,3,1)$ for all $v \equiv 1,3 \bmod 6$, $v \neq 9$.

For block sizes $k > 3$ the existence problem for $C(v,k,1)$ remains unsolved. The state of affairs is most promising for the cases $k = 4$ and 5.

In order to present additional recursive constructions we require a few more definitions.

A collection of t k-subsets $D_i = \{d^i_0, d^i_1, \ldots, d^i_{k-1}\}$, $0 = d^i_0 < d^i_1 < \cdots < d^i_{k-1}$, $i = 1, \ldots, t$ is said to be a *perfect difference family* (PDF) in Z_v, $v = k(k-1)t + 1$, if the $tk(k-1)/2$ differences $d^i_l - d^i_j$, $0 \leq j < l < k$ cover the set $\{1,2, \ldots, tk(k-1)/2\}$. PDF's are equivalent to regular perfect systems of difference sets starting with 1, which have been studied by many authors (see [1] for a recent survey). It has been shown [2] that PDF's can exist only when k is 3,4 or 5. For $k = 3$ the existence of a PDF is related to Skolem's partitioning problem [1].

Let X^1 be the m-vector $(-r,-r+1,\ldots,-1,0,1,\ldots,r-1,r)$, $m = 2r + 1$ and let $X^2,\ldots,X^n$ be permutations of X^1. Then $X^1,\ldots,X^n$ is an *additive sequence of permutations* (ASP) of order m and length n if the vector sum of every subsequence of consecutive permutations is again a permutation of X^1. ASP play an important role in recursive constructions for PDF's and vice versa [1] [11] [12].

Block size 4

We begin with two direct constructions.

Theorem 9 let $p = 12t + 1$, $t \geq 1$ be a prime and let α be a primitive root of Z_p.

(i) ([3] [13]) If $p \neq x^2 + 36y^2$ for any integers x and y then

$$\{0,\alpha^{2i},\alpha^{4t+2i},\alpha^{8t+2i}\} \quad i = 0,1,\ldots,t-1 \tag{7}$$

is a $(p,4,1)$ DF in Z_p.

(ii) ([5]) If $\alpha \equiv 3 \bmod 4$ (and such an α always exists in Z_p) then

$$\left.\begin{array}{l} \{0,\alpha^{4i},\alpha^{4i+3},\alpha^{4i+6}\} \quad i = 0,\ldots,3t-1\} \\ \{0,\alpha^{4j+1},\alpha^{4t+4j+1},\alpha^{8t+4j+1}\} \quad j = 0,\ldots,t-1\} \\ \{0,p,2p,3p\} \end{array}\right\} \tag{8}$$

form a $(4p,4,1)$ DF in Z_{4p}.

The next two constructions will exhibit the relationship between PDF's and ASP.

Theorem 10 ([3] [13]) Let $D_i = \{0,a_i,b_i,c_i\}$, $i = 1,\ldots,t$ be a PDF in Z_{12t+1} and let X^1, X^2, X^3 be an ASP of order $m = 2r + 1$, $r \geq 2$ and length 3. Then

(i) For $i = 1,\ldots,t$ and $j = 1,\ldots,m$ the $6tm$ positive differences in the family

$$\Delta_{mi-m+j} = \{0,ma_i + \alpha_j,mb_i + \beta_j,mc_i + \gamma_j\} \tag{9}$$

cover the set $\{r+1,r+2,\ldots,r+6tm\}$. Here α,β and γ are the m-vectors X^1, $X^1 + X^2$, $X^1 + X^2 + X^3$, respectively.

(ii) For $i = 1,\ldots,t$

$$\left.\begin{array}{l} X_i^1 = (-c,a-c,-b,b-c,a-b,-a,a,b-a,c-b,b,c-a,c)_i \\ X_i^2 = (c-b,c,b-a,c-a,b-c,a-c,-b,a,b,-c,a-b,-a)_i \\ X_i^3 = (b-a,-b,a-c,a,c,c-b,b-c,-c,-a,c-a,b,a-b)_i \end{array}\right\} \tag{10}$$

the $(12t+1)$-vectors $X^j = (0,X^j_1,\ldots,X^j_t)$, $j = 1,2,3$ form an ASP of order $12t + 1$ and length 3.

In order to utilize products of the form (9) for constructing new difference families we need to find additional base blocks with differences covering the set $\{1,\ldots,r\}$ and possibly $\{r + 6tm + 1,\ldots,6x\}$ for some $x \geq 1$.

We list now the known recursive constructions for $1 \leq m \leq 25$.

Theorem 11 Let $D(t) = \{D_1, \ldots, D_t\}$ be a PDF and let $\Delta(mt) = \{\Delta_1, \ldots, \Delta_{mt}\}$ be defined by (9), where $m = 2r + 1$ and $\alpha = (-r, -r+1, \cdots, -1, 0, 1, \ldots, r-1, r)$. Then

1. For $r = 2$

$\beta = (-2,0,2,-1,1), \quad \gamma = (0,-2,1,-1,2)$

$$D(5t+1) = \Delta(5t) \cup \{0,1,30t+4,30t+6\}$$

is a PDF in Z_{60t+13}.

2. For $r = 3$

$\beta = (-1,-2,-3,3,2,1,0), \quad \gamma = (-2,1,-3,0,3,-1,2)$

$$D(7t+1) = \Delta(7t) \cup \{0,2,3,42t+7\}$$

is a DF in Z_{84t+13}.

3. For $r = 6$

$\beta = (-4,-5,-1,-2,3,-6,6,5,1,-3,0,2,4)$

$\gamma = (-1,-5,-6,3,-3,-4,4,2,5,-2,6,1,0)$

$$D(13t+1) = \Delta(13t) \cup \{0,1,4,6\}$$

is a PDF in $Z_{156t+13}$.

4. For $r = 9$

$\beta = (-6,-7,1,-2,-3,5,3,-9,-4,7,-8,0,-5,9,-1,6,2,4,8)$

$\gamma = (-2,1,-8,-9,3,-1,-5,-6,5,2,-7,7,-3,8,-4,6,0,9,4)$

$$D(19t+4) = \Delta(19t) \cup \{0,1,7,x+23\} \cup \{0,2,x+14,x+19\}$$
$$\cup \{0,3,x+13,x+21\} \cup \{0,4,x+15,x+24\}, \quad x = 114t$$

is a PDF in $Z_{228t+49}$.

5. For $r = 11$

$\beta = (0,-2,1,-7,2,-6,-1,-5,4,-10,-11,-9,6,-4,-8,-3,11,8,10,3,5,7,9)$

$\gamma = (9,5,4,-10,0,-7,10,-9,8,-4,-3,1,-5,-11,-8,2,6,-2,11,-6,7,-1,3)$

$$D(23t+5) = \Delta(23t) \cup \{0,1,8,x+28\} \cup \{0,2,x+14,x+24\} \cup$$
$$\{0,3,x+18,x+29\} \cup \{0,4,x+17,x+23\} \cup \{0,5,x+21,x+30\}, \quad x = 138t$$

is a PDF in $Z_{276t+61}$.

6. For $r = 12$
using α, β, γ and $\Delta(5t)$ from 1 to obtain $\Delta(25t)$

$$D(25t+5) = \Delta(25t) \cup \{0,1,x+18,x+29\} \cup \{0,4,x+20,x+26\}$$

$$\cup \{0,3,8,x+27\} \cup \{0,7,x+21,x+30\} \cup \{0,10,12,x+25\}, \quad x = 150t$$

is a PDF in $Z_{300t+61}$, and

$$D(25t+6) = \Delta(25t) \cup \{0,1,x+34,x+36\} \cup \{0,3,x+18,x+29\}$$
$$\cup \{0,4,x+20,x+28\} \cup \{0,5,x+22,x+32\}$$
$$\cup \{0,6,x+19,x+31\} \cup \{0,7,x+21,x+30\}, \quad x = 150t$$

is a PDF in $Z_{300t+79}$.

Proof Use (9) to check that the required sets are covered by all differences from the base blocks. □

We note that the constructions 2,5,6b are new and that 1,3,4,6a have been known [11]. The ASP with $r = 11$ has been found by P.J. Laufer.

If we apply all methods listed in Sections 2, 3 and 4 and add the computer generated DF's from [5] we obtain the following results for $1 \leq t \leq 50$:

$(12t+1,4,1)$ PDF

$t = 1,4\text{-}8,14,21,23,26,28,30\text{-}31,36,41$

$(12t+1,4,1)$ DF

$t = 1,3\text{-}10,14\text{-}15,19\text{-}21,23,26,28\text{-}31,34\text{-}36,38,40\text{-}41,43,45,50$

$(12t+4,4,1)$ DF

$t = 3\text{-}6,12,20,24,30,32,36,43.$

Block size 5

As before, two direct constructions are known.

Theorem 12 Let $p = 20t + 1$, $t \geq 1$ be a prime and let α be a primitive root of Z_p.
(i) ([3] [13]) If $p \neq x^2 + 100y^2$ for any integers x and y then

$$\{\alpha^{2i},\alpha^{4t+2i},\alpha^{8t+2i},\alpha^{12t+2i},\alpha^{16t+2i}\} \quad i = 0,1,\ldots,t-1 \tag{11}$$

is a $(p,5,1)$ DF in Z_p.
(ii) ([5]) If $\alpha^r + 1 = \alpha^s(\alpha^r - 1)$ for some odd integers r and s then

$$\{0,\alpha^{2i},\alpha^{2i+r},\alpha^{2t+2i},\alpha^{2t+2i+r}\} \quad i = 0,1,\ldots,t-1 \tag{12}$$
$$\{0,p,2p,3p,4p\}$$

form a $(5p,5,1)$ DF in Z_{5p}.

Concerning PDF's with blocks of size 5 and ASP of length 4, results can be proved which are similar to those stated in Theorem 10 [1]. They can be used to derive the following construction.

Theorem 13 Let $D(t) = \{D_1, \ldots, D_t\}$ be a PDF in C_{20t+1} and let $D(s) = \{D_1, \ldots, D_s\}$ be a DF in C_{20s+1}. Then a DF $D(r)$ exists in C_{20r+1}, $r = 20st+s+t$ and $D(r)$ is perfect whenever $D(s)$ is perfect.

Proof Use $D(t)$ to construct an ASP of length 4 and order $m = 20t+1$ [1]. With help of this ASP construct the blocks $\Delta(ms)$ in a similar way as in (9). Then $D(r) = \Delta(ms) \cup D(s)$. □

PDF's with $k = 5$ can exist only if t is even and $t \geq 6$ [1]. They have been enumerated for $t = 6$ [9] and examples are known for $t = 8$, 10, 732, 974, etc.

Difference families are known for the following values of t, $1 \leq t \leq 50$:

(20t+1) PDF

t = 6, 8, 10

(20t+1) DF

t = 1-3,6,8,10,12,14,21-22,30,32-33,35,41,43-44

(20t+5) DF

t = 3-5,7,9-10,13,15,18,22,24-25,27-28,30,34,37,39-40,42-43,45,48-50.

Open Problems

1. Does there exist a $(12t+1,4,1)$ DF for every $t \geq 3$? Can all of these DF's be perfect if $t \geq 4$?
2. Does there exist a $C(v,4,1)$ for every $v \neq 16$, 25 and 28?
3. Does there exist an ASP of length 3 for every order $m \geq 5$, $m \neq 9,10$?
4. Do there exist $C(v,5,1)$ for $v = 81$ and 85?
5. Construct examples of PDF's $D(t)$ for $k = 5$ and even $t \geq 12$.
6. Construct examples of ASP of length 4 for orders $m \geq 7$.

Acknowledgement

Research supported by NSERC Grant A8651.

References

[1] J. Abrham, "Perfect systems of difference sets - a survey", *Ars Combinatoria* 17A (1984) 5-36.

[2] J.-C. Bermond, A. Kotzig, J. Turgeon, "On a combinatorial problem of antennas in radioastronomy", *Proc. 18th Hungarian Combinatorial Colloquium*, North Holland, 1976, 135-149.

[3] R.C. Bose, "On the construction of balanced incomplete block designs", *Ann. Eugenics* 9 (1939) 353-399.

[4] M.J. Colbourn, C.J. Colbourn, "On cyclic block designs", *Math. Report of Canadian Academy of Science* 2 (1980) 21-26.

[5] M.J. Colbourn, R.A. Mathon, "On cyclic Steiner 2-designs", *Ann. Discrete Math.* 7 (1980) 215-253.

[6] M. Hall, Jr., *Combinatorial Theory*, Blaisdell, Waldham, Mass. (1967).

[7] H. Hanani, "Balanced incomplete block designs and related designs", *Discrete Math.* 11 (1975) 255-369.

[8] M. Jimbo, S. Kuriki, "On a composition of cyclic 2-designs", *Discrete Math.* 46 (1983) 249-255.

[9] P.J. Laufer, "Regular perfect systems of difference sets of size 4 and extremal systems of size 3", *Ann. Discrete Math.* 12 (1982) 193-201.

[10] R. Peltesohn, "Eine Lösung der beiden Heffterschen Differenzenprobleme", *Compositio Math.* 6 (1939) 251-257.

[11] D.G. Rogers, "Addition theorems for perfect systems of difference sets", *J. Lond. Math. Soc.* (2) 23 (1981) 385-395.

[12] J.M. Turgeon, "Construction of additive sequences of permutations of arbitrary lengths", *Ann. Discrete Math.* 12 (1982) 239-242.

[13] R.M. Wilson, "Cyclotomy and difference families in elementary abelian groups", *J. Number Theory* 4 (1972) 17-47.

[14] R.M. Wilson, "Constructions and uses of pairwise balanced designs", *Combinatorics* (eds. M. Hall, Jr. and J.H. van Lint), Mathematical Centre, Amsterdam, 1975, 19-42.

Annals of Discrete Mathematics 34 (1987) 363–370
© Elsevier Science Publishers B.V. (North-Holland)

On the Spectrum of Imbrical Designs

E. Mendelsohn and A. Assaf

Department of Mathematics
University of Toronto
Toronto, Ontario, M5S 1A1
CANADA

TO ALEX ROSA ON THE ELEVENTH ANNIVERSARY OF HIS THIRTY-NINTH BIRTHDAY

ABSTRACT

An imbrical design is a minimal, but not necessarily minimum, covering. Thus, a pair (V,D) is an ID(v,k,λ,b) if and only if

(a) $|V| = v$;

(b) D is a collection of b k-subsets of V called blocks;

(c) every pair of elements of V is in at least λ blocks;

(d) for every $b \in D$ there is a pair $\{x,y\} \subset b$ so that xy is in exactly λ blocks ($\{x,y\}$ is called *essential* for b);

(d') equivalently, if $D' \subset D$, (V,D') is not an imbrical design.

Given v, k, and λ, the spectrum $Spec(v,k,\lambda) = \{b$: there exists ID$(v,k,\lambda,b)\}$. In this paper we determine $Spec(v,3,1)$ for all v and $Spec(v,4,1)$ except for (v,b), where $v \in \{25,28,37,40,85\}$, $b = \frac{1}{12}v(v-1) + 1$. The results are that min $Spec(v,3,1) = \lceil \frac{v}{3} \lceil \frac{v-1}{2} \rceil \rceil$, max $Spec(v,3,1) = \binom{v-1}{2}$, and that $Spec(v,3,1)$ is an interval except that $v=3$, $b=2$ and $v=7$, $b=8$ do not exist. For $k=4$ we have similar results: min $Spec(v,4,1) = \lceil \frac{v}{4} \lceil \frac{v-1}{3} \rceil \rceil$, max $Spec(v,4,1) = \binom{v-2}{2}$ and $Spec(v,4,1)$ is an interval except possibly for the noted values, and $v=4$, $b=2$, $v=13$, $b=14$ do not exist.

1. Introduction

The idea of a maximal packing which is not maximum in the case $k=3$, $\lambda=1$ is called a maximal partial triple system, and many papers have been written about them (see [1,2,5,7], for example). The concept has not been extended to $k=4$, and seems to be quite difficult. In this paper, we dualize the results to coverings and obtain results both for $k=3$, $\lambda=1$, and for $k=4$, $\lambda=1$. It is quite likely that if the covering of pairs by quintuples were solved, we could extend our methods to them as well.

We introduce the new term *imbrical design* for a covering which is not necessarily the smallest, to avoid confusion. This is the same reason that the term maximal partial triple system replaced the term maximal non-maximum packing. A pair (V,D) is an $\mathrm{ID}(v,k,\lambda,b)$ if and only if

(a) $|V| = v$;

(b) D is a collection of b k-subsets of V called blocks;

(c) every pair of elements of V is in at least λ blocks;

(d) for every $b \in D$ there is a pair $\{x,y\} \subset b$ so that xy is in exactly λ blocks ($\{x,y\}$ is called *essential* for b);

(d') equivalently, if $D' \subset D$, (V,D') is not an imbrical design.

Given v, k, and λ, the spectrum $Spec(v,k,\lambda) = \{b : \text{there exists } \mathrm{ID}(v,k,\lambda,b)\}$. In this paper we determine $Spec(v,3,1)$ for all v and $Spec(v,4,1)$ except for (v,b), where $v \in \{25,28,37,40,85\}$, $b = \frac{1}{12}v(v-1) + 1$. The results are that min $Spec(v,3,1) = \lceil \frac{v}{3} \lceil \frac{v-1}{2} \rceil \rceil$, max $Spec(v,3,1) \doteq \binom{v-1}{2}$, and that $Spec(v,3,1)$ is an interval except that $v=3$, $b=2$ and $v=7$, $b=8$ do not exist. For $k=4$ we have similar results: min $Spec(v,4,1) = \lceil \frac{v}{4} \lceil \frac{v-1}{3} \rceil \rceil$, max $Spec(v,4,1) = \binom{v-2}{2}$ and $Spec(v,4,1)$ is an interval except possibly for the noted values, and $v=4$, $b=2$, $v=13$, $b=14$ do not exist.

2. Augmentations and interval designs

In this section we determine max $Spec(v,k,1)$ for $k=3$ and $k=4$, and find a condition on an imbrical design with b blocks so that if such a design exists then $[b, \max Spec(v,k,1)] \subseteq Spec(v,k,1)$. These are called *interval designs*.

The remainder of this section is devoted to showing that

(a) If $k=3$, $v \not\equiv 1,3 \bmod 6$, then the minimal covering is an interval design.

(b) If $k=3$, $v \equiv 1,3 \bmod 6$, there exists an interval design with $\frac{v(v-1)}{6} + 2$ blocks.

(c) If $k=4$, $v \not\equiv 1,4 \bmod 12$, the minimal covering is an interval design.

(d) If $k=4$, $v \equiv 1,4 \bmod 12$, an interval design exists with $\frac{v(v-1)}{12} + 2$ blocks.

We now give a basic augmentation procedure. Let (V,B) be an $\mathrm{ID}(v,k,1,b)$. Let $\infty = \{\infty_0,\infty_1, \cdots \infty_{k-1}\} \in B$ be fixed. Let a be a block of B, $a = \{a_0,a_1, \cdots a_{k-1}\}$, with $\{a_0,a_{k-1}\}$ essential (i.e., in no other block). Thus not both $a_0,a_{k-1} \in \infty$. Assume without loss of generality that if one of them is in ∞, it is a_0, and $a_0 = \infty_{k-1}$. The basic augmentation procedure is

(1) Delete from B the block $\{a_0, \cdots a_{k-1}\}$.

(2) Add to B the block $\{\infty_i, a_0, a_1, \cdots a_{k-2}\}$ where ∞_i has the least i such that $\infty_i \in \infty - a$, provided there is an essential edge in this new block.

(3) Add to B the block $\{\infty_i, a_1, a_2, \cdots a_{k-1}\}$ where ∞_i has the least i such that $\infty_i \in \infty - a$, provided there is an essential edge in this new block.

(4) Add the block $\{\infty_0, \infty_1, \cdots \infty_{k-3}, a_0, a_{k-1}\}$.

This process yields $A_\infty(V,B,a)$.

Lemma 1: $A_\infty(V,B,a)$ is an ID$(v,k,1,b+i)$ where $i \in \{0,1,2\}$, and increases the number of blocks of the form $\{\infty_0, \infty_1, \cdots \infty_{k-3}, u, v\}$.

Proof: As $\{a_0, a_k\}$ is an essential edge of a, the block added in (4) must be present if the block a is deleted. □

Theorem 1: Max $Spec(v,k,1) = \binom{v-k+2}{2}$.

Proof: Let $V_m = \{\infty_0, \cdots \infty_{k-3}\} \cup V'$, and $B_m = \{\{\infty_0, \cdots \infty_{k-3}, u, v\}: u, v \in V'\}$. It is clear that every pair is covered at least once, and that $\{u,v\}$ is essential [property (d)] for $\{\infty_0, \cdots \infty_{k-3}, u, v\}$. Thus (V_m, B_m) is an imbrical design with $b = \binom{v-k+2}{2}$.

Suppose that we have an imbrical design with a fixed block $\infty = \{\infty_0, \infty_1, \cdots \infty_{k-1}\}$, and a block a so that $a \cap \infty \neq \{\infty_0, \infty_1, \cdots \infty_{k-3}\}$; then form $A_\infty(V,B,a)$. This does not decrease the number of blocks, and increases the number of blocks containing $\{\infty_0, \cdots \infty_{k-3}\}$. This process only terminates when we achieve (V_m, B_m). □

We call an ID an interval design if there exists a block a^* so that $A_\infty(V,B,a^*)$ is an ID$(v,k,1,b+1)$, and further that no augmentation $A_\infty(V,B,a')$ changes this property of augmenting by one.

Theorem 2: If an ID$(v,k,1,b)$ is an interval design then $\left[b, \binom{v-k+2}{2}\right] \subseteq Spec(v,k,1)$. Further, such designs exist for $k=3$ and $k=4$, $b = \min Spec(v,k,1)$ (or $\min Spec(v,k,1) + 2$ for the cases where the BIBD exists).

Proof: Any augmentation that increases the number of blocks increases it by one or by two. Since $b < b' \leq \binom{v-k+2}{2}$, by augmenting using blocks other than a^*, we can get an ID$(v,k,1,b')$ or an ID$(v,k,1,b'-1)$; in the latter case, we can augment by a^* as well.

Let $k = 3$. Suppose that we have a block $\{a,b,c\}$ all distinct from ∞_0, with ab repeated, and ac and bc both essential. Then the augmentation to ∞ac, ∞bc increases the size by one, and no other augmentation by ∞ affects this. It is clear that every minimal cover which is not a BIBD has this. If $v \equiv 1,3 \bmod 6$, perform any augmentation to any block; say abc is replaced by dac, dab, dbc; now let $\infty_0\infty_1\infty_2$ be dab.

Now consider $k = 4$. Suppose that we have blocks $\infty_0\infty_1\infty_2\infty_3$, $d\infty_1 ab$, and $abXY$. If da and db are essential, the augmentation of the second block yields $\infty_0 d\infty_1 a$ and $\infty_0\infty_1 db$. As da and db are essential, no other augmentation can prevent this one from increasing the number of blocks by exactly one. It is clear that the minimal cover for $v \not\equiv 1,4 \bmod 12$ has such a configuration [7,8]. For $v \equiv 1,4 \bmod 12$, the same trick of an extra augmentation creates the desired configuration. □

Summarizing, we see that the only imbrical designs left in doubt for $k = 3,4$ are those with **one more block** than a BIBD of the same size.

3. Failed designs

We define a *failed design* FD(v,k) to be a pair (V,B) where B is a collection of subsets of V of size k, so that every pair of subsets of V is in at least one block of B, and $|B| = \dfrac{v(k-1)}{2k} + 1$, and B has no repeated blocks. We note that for $k = 3$ and $k = 4$, the failed designs are exactly what is missing from the spectrum.

Lemma 2: There does not exist

(a) ID(7,3,1,8), i.e. an FD(7,3)

(b) ID(13,4,1,14), i.e. an FD(13,4).

Proof:

(a) There must be three repeated pairs and the graph of the repeated pairs must form a triangle, say $\{ab,ac,bc\}$. Now we must have blocks through a, say $ab1$, $ab2$, $ac3$, $ac4$ since the four elements are distinct (and are not b or c). However, we have two blocks bcx, bcy; $x \neq 1,2$ since $b1$, $b2$ are not repeated, and $x \neq 3,4$ since $c3$, $c4$ are not repeated.

(b) We have for $k=4$ the following possibilities for repeated pairs.

(i) One pair repeated 7 times, but this implies $v > 14+2$.

(ii) A set of three points a,b,c with ab repeated four times and bc repeated four times. If there is a block containing a,b,c, then we must have $abc1$, $ab23$, $ab45$, $ab67$, $bc89$, $bc1011$, $bc1213$, for a total of at least sixteen points. If there is no block containing a,b,c, it is even worse.

(iii) Two disjoint pairs ab, cd each appearing 4 times. In this case, we must have for $ab12$, $ab34$, $ab56$, $ab78$, $ac910$, $ad1112$, a total of at least sixteen points. Thus we must either have a pair of blocks $abc1$, $cda2$, or a block $abcd$. In the first case, we must have $abc1$, $acd2$, $ab34$, $ab56$, $ab78$, $a91011$, for $v \geq 15$. If $abcd$ is a block, then $ab12$, $ab34$, $ab56$, $a789$, $b71011$ is forced, for $v \geq 15$.

(iv) Finally, we have the possibility of six pairs ab, ac, ad, bc, bd, cd each appearing twice. Then we have $ab12$, $ab34$, $ac56$, $ad910$, $ad1112$, forcing $v \geq 16$. □

Lemma 3: For fixed k, if a failed design with v points exists, and a BIBD of order w with a subdesign of order v exists, then a failed design of order w exists.

Proof: Simply replace the blocks on the subdesign by the blocks of a failed design on the same set. □

Theorem 3: (k=3, λ=1) For all $v \neq 3,7$, $v \equiv 1,3$ mod 6, a failed design FD(v,3) exists.

The cases v = 9, 13, and 15 are done in the next section. For $v \geq 19$, the theorem of Doyen and Wilson [4] provides us with a (v,3,1)-design containing a (9,3,1)-subdesign.

The analogous result for k=4 is more difficult. The next sequence of lemmata is to provide us with the fact that failed designs exist for all v>13, except possibly for $v \in \{25,37,40,85\}$. We use *pairwise balanced designs* to establish this; a PBD(v,$\exists k$,K) is a pair (V,B) which is a PBD(v,$K \cup \{k\}$) containing at least one block of size k.

Lemma 4: If there is a PBD(v,$\exists 16$,K) where every $k \in K$ satisfies $k \equiv 1,4$ mod 12, then there is an FD(v,4).

Proof: Replace one of the blocks of size 16 by an FD(16,4) (constructed in the next section), and the remaining blocks by BIBD(k,4)'s. □

Lemma 5: [7] If there exists a PBD(v,$\exists 5$,K) where $k \in K$ implies $k \equiv 0,1$ mod 4, then there exists a PBD($3v$+1,$\exists 16$,4).

Proof: Let (X,B) be a PBD(v,$\exists 5$,K). Let $Y = X \times \{0,1,2\} \cup \{\infty\}$. Let us define B' as follows:

(a) $\{\infty, x_0, x_1, x_2\} \in B'$;

(b) Given any pair $\{x_i, y_i\}$, where $x \neq y$, there is a unique $b \in B$; since $|b| \equiv 0,1$ mod 4, $k' = |b \times \{0,1,2\} \cup \{\infty\}| \equiv 1,4$ mod 12. Thus there is a BIBD(k',4) on this set, say (K',B''), and further all blocks with ∞ are of the form of part (a). So let B' also contain all blocks of B'' not containing ∞. □

Lemma 6: [6] There exists a PBD(v+q,$\exists 5$,$\{4,q\}$) for $v \equiv 4$ mod 12, $1 \leq q \leq \frac{v-1}{3}$.

Proof: Since a resolvable BIBD(v,4) exists [7] when $v \equiv 4$ mod 12, we can complete q parallel classes by adding ∞_i to the blocks of the ith parallel class for $1 \leq i \leq q$, and finally adding the block $\{\infty_1, \cdots \infty_q\}$. □

Lemma 7: For $v \equiv 1,4$ mod 12, $v \geq 16$, $v \notin \{25,28,37,40,85\}$, a PBD($v$,$\exists 16$,4) exists.

Proof: Apply Lemma 6.

(a) $v \equiv 1$ mod 36, $v = 36t+37 = 3[(12t+4)+8]+1$
(b) $v \equiv 4$ mod 36, $v = 36t+40 = 3[(12t+4)+9]+1$
(c) $v \equiv 13$ mod 36, $v = 36t+49 = 3[(12t+4)+12]+1$
(d) $v \equiv 16$ mod 36, $v = 36t+16 = 3[(12t+4)+1]+1$
(e) $v \equiv 25$ mod 36, $v = 36t+25 = 3[(12t+4)+4]+1$
(f) $v \equiv 28$ mod 36, $v = 36t+28 = 3[(12t+4)+5]+1$

The exceptions to this construction are (a) 37,73; (b) 40,76,112; (c) 49,85,121,157; (d) 16; (e) 25; and (f) 28.

Now v = 16 will be done later; 49 = 3(16)+1, so we may use the $v \to 3v+1$ construction of BIBD(v,4,1).

For v = 73, 73 = 3(24)+1. Remove one point from the affine plane on 24 points to get a PBD(24,$\exists 5$,$\{4,5\}$).

For $v = 76$, $76 = 3(25)+1$; the affine plane on 25 points is a PBD(25,∃5,5).

For $v = 112$, $112 = 3(37)+1$; take a transversal design on 40 points coming from 4 orthogonal Latin squares of side 8. Drop three points from one group to get a PBD(37,∃5,{8,4,5}).

For $v = 121$, $121 = 3(40)+1$; take the TD design on 40 points, and we have a PBD(40,∃5,{8,5}).

For $v = 157$, $157 = 3(52)+1$. Take the TD design coming from four orthogonal Latin squares of order 12, and truncate one group to size 4. This gives a PBD(52,∃5,{4,5,12}) [3]. □

These lemmata give

Theorem 4: For $k=4$, $\lambda=1$, $v \neq 4,13$, $v \equiv 1,4 \bmod 12$, there exists an FD(v,4) except possibly for $v \in \{25,28,37,40,85\}$. □

4. Four small failed designs

(a) $v=9$, $k=3$, $V = \{0,1,2,3,4,5,6,7,8\}$, $B = \{345, 678, 258, 048, 237, 156, 246, 057, 013, 026, 128, 368\}$.

(b) $v=13$, $k=3$, $V = Z_{13}$, $B = \{(0,1,4), (0,2,8) \bmod 13\} - \{(0,1,4), (1,3,9), (0,3,12)\} \cup \{(0,1,3), (0,3,12), (1,4,9), (3,9,12)\}$.

(c) $v=15$, $k=3$, $V = Z_{15}$, $B = \{(0,1,4), (0,6,8), (0,5,10) \bmod 15\} - \{(0,1,4), (1,2,5), (0,2,9)\} \cup \{(1,4,5), (2,5,9), (0,4,9), (0,1,2)\}$.

(d) $v=16$, $k=4$, $V = Z_{14} \cup \{a,b\}$, $B = \{(a,b,0,7), (0,1,4,6) \bmod 14\}$.

5. Conclusion

Imbrical designs may soon become as important as maximal partial designs. There are many questions. We close with some conjectures.

An *affine failed design* AFD(p) is a FD(p^2,p), and a *projective failed design* is a FD(p^2+p+1,$p+1$). It is conjectured that projective failed designs never exist, and affine failed designs exist when the corresponding plane does.

Acknowledgements

The authors would like to thank A. Hartman for valuable computer mail on the subject.

References

[1] C.J. Colbourn, E. Mendelsohn, and A. Rosa, "Extending the concept of decomposability for triple systems", *Annals of Discrete Math.*, to appear.

[2] C.J. Colbourn and A. Rosa, "Quadratic leaves of maximal partial triple systems", *Graphs and Combinatorics* 2 (1986) 317-337.

[3] A.L. Dulmage, D.M. Johnson and N.S. Mendelsohn, "Orthomorphisms of Groups and Orthogonal Latin Squares", *Can. J. Math.* 13 (1961) 356-372.

[4] J. Doyen and R.M. Wilson, "Embeddings of Steiner triple systems", *Discrete Math.* 5 (1973) 229-239.

[5] J. Novák, "On certain minimal systems of triples", *Colloques Internationaux CNRS #260,* Orsay, 1976.

[6] H. Hanani, "Balanced incomplete block designs and related designs", *Discrete Math.* 11 (1975) 255-369.

[7] H. Hanani, D. Ray Chaudhuri, and R.M. Wilson, "On resolvable designs", *Discrete Math.* 3 (1972) 343-357.

[8] W.H. Mills, "On the covering of pairs by quadruples I", *J. Comb. Theory* A13 (1972) 55-78.

[9] W.H. Mills, "On the covering of pairs by quadruples II", *J. Comb. Theory* A15 (1973) 138-166.

[10] E. Severn, *On the spectrum of maximal partial triple systems,* Ph.D. thesis, University of Toronto, 1978.

Annals of Discrete Mathematics 34 (1987) 371–378
© Elsevier Science Publishers B.V. (North-Holland)

Some Remarks on n-Clusters on Cubic Curves

N.S. Mendelsohn, R. Padmanabhan and Barry Wolk

Department of Mathematics,
University of Manitoba
Winnipeg, Manitoba, R3T 2N2
CANADA

TO ALEX ROSA ON HIS FIFTIETH BIRTHDAY

1. Introduction

A pair $(\mathcal{P},\mathcal{L})$ is an *n-cluster* on a cubic curve if $\mathcal{P}$ is a set of n points on the curve and if the line joining two points of $\mathcal{P}$ meets the cubic only in points of $\mathcal{P}$ and the tangent at a point of $\mathcal{P}$ also meets the cubic only in points of $\mathcal{P}$. The set $\mathcal{L}$ is the set of all lines joining two points of $\mathcal{P}$ together with the lines which are tangent to the cubic at some point of $\mathcal{P}$. Obviously $|\mathcal{L}|$ is finite. The points of $\mathcal{P}$ are referred to as the vertices of the cluster. In the case where n is a prime number and the curve is a non-singular complex cubic curve the structure of the cluster has been determined in [4]. We note particularly, that for a given n it is possible to have non-isomorphic clusters even when the group structure of the vertex set $\mathcal{P}$ is the same. In what follows this will be illustrated with examples. We also note that birational transformations which map cubic curves of genus 1 onto cubic curves usually destroy the incidence structure of the lines of $\mathcal{L}$. Hence we only consider cubic curves to be equivalent if they can be mapped onto each other by a projective transformation.

2. The Group Structure of the Vertex set $\mathcal{P}$

For any non-singular cubic curve over a field $\mathcal{F}$ one can define uniquely a medial extended triple system as follows. The elements of the system are the points of the cubic curve and a binary operator $\bullet$ is defined by $a \bullet b = c$ if the line joining a to b meets the cubic in c. In case the line joining a to b is tangent to the cubic at a we have $a \bullet b = a$. If a is a point of inflection of the cubic then $a \bullet a = a$. The algebra satisfies the following identities:

1) $a \bullet b = b \bullet a$

2) $a \bullet (a \bullet b) = b$

3) $(a \bullet b) \bullet (c \bullet d) = (a \bullet c) \bullet (b \bullet d)$

Equation 3) is the only one which is not trivially obvious. It is based on a configuration theorem valid for all non-singular cubic curves over any field. For details see [3].

From any such algebra we can construct an Abelian group as follows. We take a fixed element e of the algebra and let $f = e \bullet e$. We now define a group operation $+$ by $a + b = (a \bullet b) \bullet e$, and $-a = a \bullet f$. It can be proved that the system under the operation $+$ is an Abelian group with e the identity element (cf. pages 328-329 of [5]).

The $\bullet$-algebra of the vertex set P of an n-cluster is a subalgebra of the algebra of the whole cubic curve. Also, if the identity element e of the group of the algebra is a vertex of the n-cluster, then so if f and the vertices of the n-cluster are a subgroup of the whole group of the cubic.

In the case where the cluster contains no point of inflection then if the group of the cubic is defined using a point of inflexion as the identity element of the group, it follows that the group operation applied to points of the cluster will yield points outside the cluster and as a result will in general not be useful in studying the cluster. As will be seen in what follows, even when a cluster has a point of inflexion which is the identity element of the group, it may happen that there is a sub-cluster of the given cluster whose vertices are not a subgroup of the group.

Example 1

We illustrate some of these concepts by means of an example of a cubic curve of genus 1, over the rational field Q . Consider the curve whose equation is given by $f(x,y,z) = -3x^2y+xy^2+3x^2z-2xz^2-2yz^2-2y^2z+5xyz = 0$. A straightforward calculation shows that this curve is of genus 1. Using the following notation: $[x]$ is a point on the curve; $[x]\rightarrow(a,b,c)$ means that (a,b,c) are homogeneous coordinates for the point $[x]$; $[x][y][z]$ means that the points $[x],[y],[z]$ are collinear; and $[x][x][y]$ means that the tangent to the curve at $[x]$ meets the curve at $[y]$, the following table completely describes a 9-cluster on the curve.

coordinates	collinearities	tangencies
[1] → (0,1,0)	[8][3][2]	[8][8][6]
[2] → (2,2,1)	{3][4][6]	[3][3][7]
[3] → (1,3,0)	[4][2][7]	[4][4][5]
[4] → (1,1,1)	[2][6][5]	[2][2][0]
[5] → (0,1,-1)	[6][7][0]	[6][6][1]
[6] → (2,0,3)	[7][5][1]	[7][7][8]
[7] → (0,0,1)	[5][0][8]	[5][5][3]
[8] → (-1,1,-1)	[0][1][3]	[1][1][2]
[0] → (1,0,0)	[1][8][4]	[0][0][4]

All of these relationships are readily verified. It is immediately clear that the points and lines form a 9-cluster. From the tangency table it is clear that none of the vertices are points of inflexion. Geometrically the lines may be described as follows. The collinearities form a 9_3 configuration; that is a configuration of nine points and nine lines such that every line is incident with three points and every point is on three lines. The tangencies form an inscribed-circumscribed 9-gon. The name derives from the fact that if we look at the vertices in the order [0],[4],[5],[3],[7],[8],[6],[1],[2] then each vertex is on the cubic and each edge is tangent to the cubic. Figure 1 illustrates the cluster. It is easily verified that if one of the vertices is taken as the identity element of the group of the cubic then the vertices of the cluster are a cyclic subgroup of the group. If any other point of the cubic is taken as the identity of the group, then

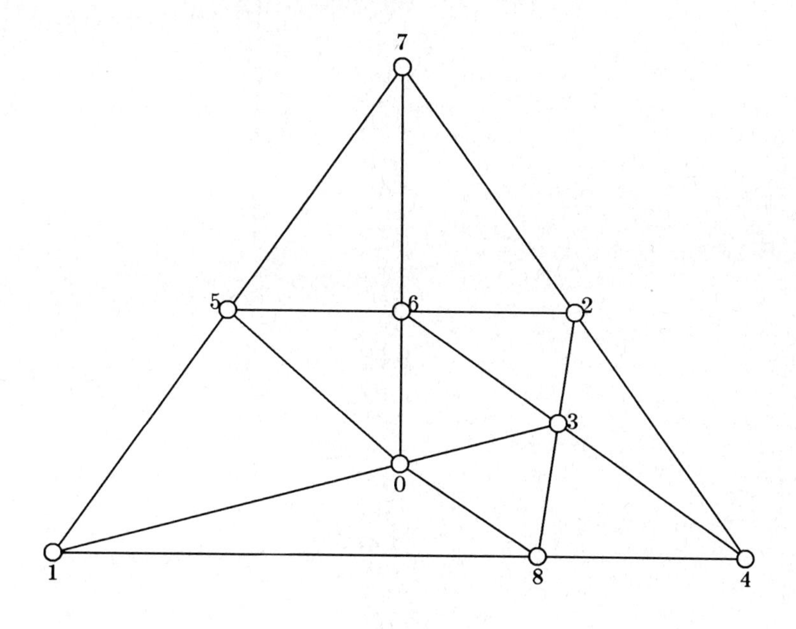

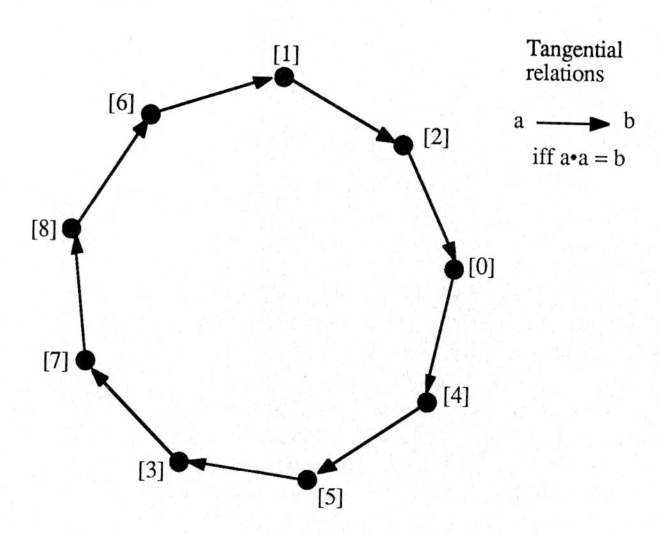

Figure 1

the vertices of the cluster do not form a subgroup and it is difficult to describe the relationship of the vertices of the cluster to the group.

In the next section we will give examples of clusters whose vertices have isomorphic groups but which are not isomorphic as a whole.

3. n-Clusters in Cubic Curves over the Complex Field

In this section we confine our study to the case of a curve of genus 1 whose coefficients lie in the field of complex numbers. The group of the curve is taken in standard form, that is we take as the identity element one of the nine points of inflexion. We denote this point by 0. The group in this form is well known (e.g. [3]). For our purposes here it is sufficient to know that every cyclic group is a subgroup. The relation between the group and the extended triple system algebra is given by the equations: $a+b=(a\bullet b)\bullet 0$; $-a=a\bullet 0$; $a\bullet b=-a-b$. If we take an element (say 1) of the group which is not of finite order, the subgroup it generates is the infinite cyclic group C_∞. In the extended triple system the element 1 generates a proper subset of this group (e.g. it does not generate 0). The distinction is of great importance in the study of clusters. We now describe this subset. Let S be the subset generated by 1 under the operation $\bullet$. If $a,b\in S$ then $-a-b\in S$. In particular, $-2a\in S$. Also, if $a\equiv 1$ mod 3 and $b\equiv 1$ mod 3 then $-a-b\equiv 1$ mod 3. Hence S contains only integers which are $\equiv 1$ mod 3. We show that S consists of all integers $\equiv 1$ mod 3. From the fact that $1\in S$ and $a\in S$ implies $-2a\in S$ it follows that 1,-2,4,-8, ... all are in S. Again since S contains 1 and 4, it follows that S contains -5,10, -20,... Inductively, suppose $b\equiv 1$ mod 3 and if $a\equiv 1$ mod 3 and $|a|<|b|$ then $a\in S$. We distinguish two cases.

Case 1 $b>0$; $-b+2\equiv 1$ mod 3; $|-b+2|<b$; $b=-((-b+2)-2)\in S$.

Case 2 $b<0$; $-b-1\equiv 1$ mod 3; $|-b-1|<|b|$; $b=-(1+(-b-1))\in S$.

In the same way, we can verify that the element 2 generates an algebra under the operation $\bullet$ which consists of all integers congruent to 2 mod 3. It follows that if we look at the elements of the infinite cyclic group as an extended triple system algebra under the definition $a\bullet b=-a-b$ then the algebra contains three disjoint subalgebras, namely the three congruence classes mod 3. Next we look at the case where the element 1 is of finite order n. In this case we represent the cyclic group C_n by the residue classes mod n. If $n\equiv 1$ mod 3 then if we take all the integers congruent to 1 mod 3, and reduce these mod n we obtain all residues mod n. Hence the algebra generated by 1 contains all the elements of C_n. The case $n\equiv 2$ mod 3 yields the same result that is the algebra generated by 1 consists of all the elements of the group. The case $n\equiv 0$ mod 3 is more interesting. In this case the group contains a subgroup of index 3, and each coset of this subgroup contains the elements congruent to one of the residues mod 3. The elements of each coset form a subalgebra of the algebra whose elements are the elements of the group. Each of the subalgebras are 1-generated. As the next example will show, the clusters corresponding to the different subalgebras need not be isomorphic.

Example 2

Since the group of a complex cubic curve of genus 1 contains the cyclic group C_{27} we consider the 27-cluster whose vertices are the elements of C_{27}. We represent the elements of this group by the residue classes mod 27. The lines of the cluster consist of unordered triples of vertices (a,b,c) where $a+b+c \equiv 0$ mod 27. The residue class 3 generates a subgroup $\mathcal{H}_0$ of order 9 and we list the cosets as follows:

$\mathcal{H}_0$	0	3	6	9	12	15	18	21	24
$\mathcal{H}_1$	1	4	7	10	13	16	19	22	25
$\mathcal{H}_2$	2	5	8	11	14	17	20	23	26

The lines of the 27-cluster consist of those of the three 9-subclusters together with 81 lines of the form (a,b,c) where $a \in \mathcal{H}_0, b \in \mathcal{H}_1, c \in \mathcal{H}_2$, and $a + b + c \equiv 0$ mod 27. On deleting these 81 lines what remains is three disjoint 9-clusters. The following tables list the lines of these three clusters.

$\mathcal{H}_0$

Collinearities			Tangencies		
0	3	24	0	0	0
0	6	21	9	9	9
0	9	18	18	18	18
0	12	15	3	3	21
3	6	18	21	21	12
3	9	15	12	12	3
6	9	12	6	6	15
9	21	24	15	15	24
12	18	24	24	24	6
15	18	21			

$\mathcal{H}_1$

Collinearities			Tangencies		
1	4	22	1	1	25
1	7	19	25	25	4
1	10	16	4	4	19
4	7	16	19	19	16
4	10	13	16	16	22
7	22	25	22	22	10
10	19	25	10	10	7
13	16	25	7	7	13
13	19	22	13	13	1

$\mathcal{H}_2$

Collinearities			Tangencies		
2	5	20	2	2	23
2	8	17	23	23	8
2	11	14	8	8	11
5	8	14	11	11	5
5	23	29	5	5	17
8	20	29	17	17	20
11	17	29	20	20	14
11	20	26	14	14	26
14	17	26	26	26	2

It is clear that the cluster whose vertex set is $\mathcal{H}_0$ is not isomorphic to the remaining two. The lines of this cluster consist of three inflexion tangents, two inscribed-circumscribed triangles and a configuration consisting of 9 points, 10 lines with each line containing 3 points and each point of inflection is on 4 lines and each of the remaining points is on 3 lines. This cluster is illustrated in Figure 2. In order to illustrate the symmetries more succinctly the three inflexion points were drawn on a circle rather than on a line. The remaining two clusters are isomorphic to the one given in Figure 1. There is an interesting way of looking at the configuration given in Figure 2. If we remove the line joining the three points of inflexion what remains is the Pappus configuration. This configuration can be drawn on a real cubic curve as illustrated in Figure 3. We conclude that a Pappus configuration can be drawn on a real cubic curve where three of the points which are not joined in the Pappus configuration are points of inflexion of the cubic. In fact, if we take a real cubic start with the three inflexion points and take any other point on the cubic, then the Pappus configuration can be uniquely completed. Hence there is a one parameter family of Pappus configurations which can be drawn on a real cubic curve with three of the points of the configuration being the three points of inflection of the cubic. Of course the one parameter class is repeated six times under the various permutations of the points of inflexion.

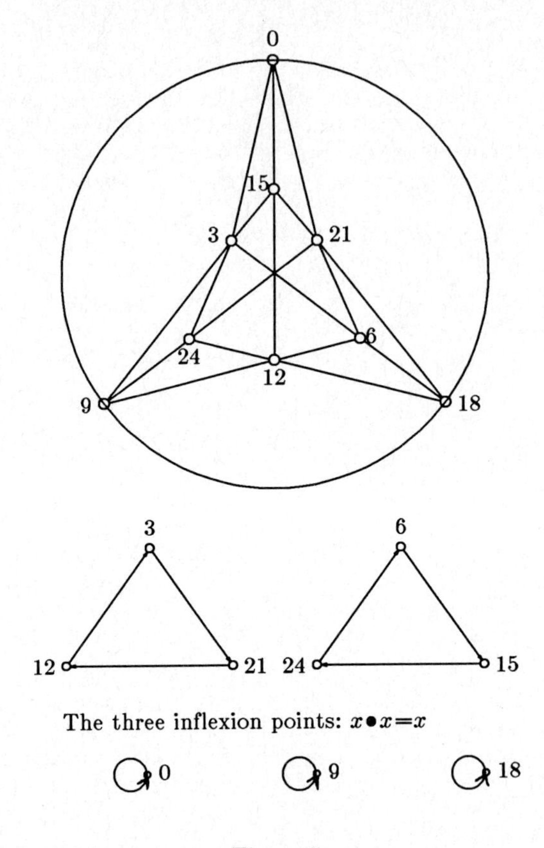

Figure 2

This example generalizes immediately, as follows. Let n be an integer such that $n \equiv 0 \bmod 3$. In an occurrence of the group $\mathcal{C}_{3n}$ on the cubic, the corresponding points are the vertices of a $3n$-cluster. Let $\mathcal{H}_0$ be the unique subgroup of order n of the group $\mathcal{C}_{3n}$, and let $\mathcal{H}_1$ and $\mathcal{H}_2$ be the two other cosets. The group $\mathcal{H}_0$ has three elements of order 3 (including the identity element). This means that the corresponding n-cluster has three tangents at inflexion points. The remaining cosets contain no elements of order 3 and hence the corresponding n-clusters have no vertices which are points of inflexion of the cubic curve. We summarize these results in the following theorem.

Theorem 1 Let $n \equiv 0 \bmod 3$. The complex cubic curve of genus 1 contains at least two isomorphism classes of n-clusters. The clusters of one of the classes contain three vertices which are points of inflexion of the curve while those of a second class contain no points of inflexion.

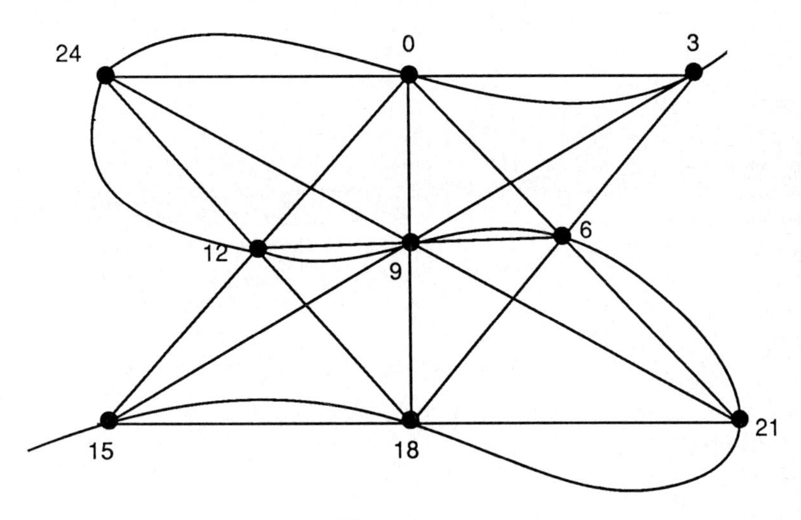

Figure 3

In the language of extended triple system algebra these results can be stated as follows. Let $n \equiv 0 \bmod 3$. There are at least two isomorphism classes of medial extended triple systems of order n. In one of these classes the algebras each have three idempotent elements. In the second class the algebras have no idempotents.

Remark Other types of n-cluster exist when $n \equiv 0 \bmod 3$. For instance, if $n \equiv 0 \bmod 9$ the complex cubic curve contains the group $C_n \times C_n$. Using essentially the same argument we obtain an n-cluster with 9 inflexion points among its vertices.

In the cases where $n \equiv 1 \bmod 3$ or $n \equiv 2 \bmod 3$ the argument is identical. In these cases each of the cosets $\mathcal{H}_0\ \mathcal{H}_1\ \mathcal{H}_2$ contain one element of order 3 and hence the corresponding n-clusters may be isomorphic. In any case we have the following theorem.

Theorem 2 Let $n \equiv 1 \bmod 3$ or $n \equiv 2 \bmod 3$. The complex cubic curves contain three disjoint n-clusters. Each of these clusters contains exactly one point of inflexion of the curve as one of its vertices.

In the language of medial extended triple systems this theorem implies that for n not a multiple of 3 the theorem establishes the existence of a medial extended triple system of order n with exactly one idempotent. Of course this result could be obtained directly.

4. A further application of n-clusters

A problem of some interest to number theorists is whether a rational cubic curve contains a specific cyclic group. A number of results are known along these lines. For instance Billing and Mahler [1] have shown that a rational cubic curve of genus 1 cannot contain the group C_{11}. The methods used by us enable us to find a small (perhaps minimal) extension. The basic idea which appears in [2] and [3] is that the lines of a cluster contain subconfigurations of the type k_3 with a collineation which is cyclic on its points and lines. In embedding such configurations in a cubic curve, the coefficients of the cubic must satisfy a number of polynomial equations with rational coefficients. The extension of the rationals by the roots of these equations yield a field in which the cubic contains the given group. In the case of the group C_{11} it turns out that by using a 10_3 configuration that there is a cubic with the group C_{11} where the coefficients of the curve lie in the field $Q(\sqrt{2})$. Without giving the calculations the example we obtained is the following. The equation of the cubic is

$$x^2y+(-2-\sqrt{2})xy^2-2x^2z+(2+\sqrt{2})xz^2+(2+\sqrt{2})y^2z-(4+3\sqrt{2})+3xyz=0$$

The vertices of the 11-cluster are: (1,0,0),(0,1,0), (0,0,1), $(2+\sqrt{2},1,0)$, $(0,2+\sqrt{2},2)$, $(2+\sqrt{2},1,1)$, (1,2,1), $(\sqrt{2},\sqrt{2},1)$, $(1+\sqrt{2},\sqrt{2},1)$, $(1+\sqrt{2},0,1)$, (1,1,1). Here, the point of inflexion is (1,1,1).

References

[1] G. Billing and K. Mahler, "On Exceptional points on Cubic Curves", *J. London Math. Society* 15 (1940) 32-43.

[2] N.S. Mendelsohn, R. Padmanabhan and Barry Wolk, "Planar Projective Configurations", to appear in *Note di Matematica.*

[3] N.S. Mendelsohn, R. Padmanabhan and Barry Wolk, "Designs Embeddable in a Plane Cubic Curve", to appear in *Note di Matematica.*

[4] E. Mendelsohn, N.S. Mendelsohn, R. Padmanabhan and Barry Wolk, "Decomposition of Prime Clusters on a Complex Cubic", *Congressus Numerantium* 57 (1987) 55-61.

[5] R. Padmanabhan, "Logic of equality in geometry", *Annals of Discrete Mathematics* 15 (1982) 313-331.

Annals of Discrete Mathematics 34 (1987) 379–384
© Elsevier Science Publishers B.V. (North-Holland)

A Few More BIBD's with $k = 6$ and $\lambda = 1$

R.C. Mullin

Department of Combinatorics and Optimization
University of Waterloo
Waterloo, Ontario, N2L 3G1
CANADA

D.G. Hoffman and C.C. Lindner

Department of Mathematics
Auburn University
Auburn, Alabama
U.S.A.

TO ALEX ROSA ON HIS FIFTIETH BIRTHDAY

ABSTRACT

W.H. Mills has shown that there exists a BIBD $(v,6,1)$ for all $v \equiv 1$ or 6 (modulo 15), $v > 36$, with 165 possible exceptions. We show that such designs exist for 35 of these values.

1. Introduction

A *balanced incomplete block design* (BIBD) with parameters (v,k,λ) is a pair (V,B) where V is a v-set and B is a family of k-subsets (where $2 < k < v$) called blocks, in which B has the property that every pair of distinct points of V occurs in precisely λ of its members.

It is well known that

(i) $\lambda(v-1) \equiv 0 \ (mod\ k-1)$

and

(ii) $\lambda v(v-1) \equiv 0 \ (mod\ k(k-1))$

are necessary conditions for the existence of a BIBD (v,k,λ). For $v = 6$ and $\lambda = 1$, these reduce to the condition that v be congruent to 1 or 6 (mod 15). Let $B(6) = \{v : \exists$ a BIBD $(v,6,1)\}$. Since there is a projective plane of order 5, then 31 lies in $B(6)$, and it is easily shown that 16 and 21 do not lie in $B(6)$. Since an affine plane of order 6 does not exist, there is no BIBD(36,6,1).

In a remarkable series of papers (see references), W.H. Mills has shown that there exists a BIBD(v,6,1) for all $v > 36$ which satisfy $v \equiv 1$ or 6 (modulo 15) with at most 165 possible exceptions.

It is our purpose here to reduce this number of possible exceptions to 130.

To construct designs to eliminate these exceptions, we require other combinatorial configurations. For the definition of pairwise balanced design, transversal design, and group divisible design, see [12]. For the definition of incomplete transversal design, see [2]. We also adopt the notation of these references. A *resolution class* of a BIBD, say D, is a set of blocks which contains each point of D exactly once. Further, D is *resolvable* if the blocks of D can be partitioned into resolution classes. Since each point of D is contained in precisely one class, the number of resolution classes is r, the frequency with which each point occurs in D. Let $RN(k) = \{r : \exists$ a resolvable BIBD $(v,k,1)$ with precisely r resolution classes$\}$. As a notational convenience, we use the symbol PBD(v,$\{6,f^*\}$,1) to denote a pairwise balanced design of order v and index 1 in which there is precisely one block of size f, and all other blocks have size k.

Theorem 1.1. *Suppose that*

(i) there exists a PBD(v,$\{6,f^*\}$,1);

(ii) there exists a TD($6,v-f+a$)$-TD(6,a)$; and

(iii) $f+5a \in B(6)$, where $f \geq a$.

Then there exists a BIBD(w,6,1), where $w = 6v-5f+5a$.

Proof. The proof follows by applying a slight generalization of the indirect product for BIBD's (see [10, p.96, construction 4.1). □

Clearly, to apply the above theorem, certain ingredients are required. A standard method for constructing pairwise balanced designs employs resolvable block designs, as follows.

Lemma 1.2. *If there is a resolvable BIBD with block size 5 and r resolution classes, then there exists a PBD*($5r+1$,$\{6,r^*\}$,1).

Proof. The pairwise balanced design is created by adjoining r new points $\infty_1,\infty_2,\ldots,\infty_r$ to the balanced incomplete design, adjoining ∞_i to each block of the ith resolution class, and adjoining a block consisting of the set $\{\infty_1,\infty_2,\ldots,\infty_r\}$ as a block. □

The set of replication numbers $RN(k)$ is PBD-closed (see [11]). Since the affine plane of order 5 is a resolvable BIBD with block size 5 which has 6 resolution classes, it follows that 6 is in $RN(5)$, hence if v belongs to $B(6)$, then v also belongs to $RN(5)$. Thus the designs with block size 6 and $\lambda = 1$ created by Mills yield resolvable designs with block size 5 and $\lambda = 1$.

Lemma 1.3. *The set* $\{16,21,51,81\}$ *is contained in* $RN(5)$.

Proof. As noted in [3], there are resolvable BIBD's with $\lambda = 1$ and $v = 65$ and 85. Therefore 16 and 21 belong to $RN(5)$. It is shown in [11] that there is a resolvable BIBD with $\lambda = 1$ and $v = 205$, hence 51 belongs to $RN(5)$.

By Lemma 1.2 there exists a PBD(81,{6,16},1). Since $RN(5)$ is PBD closed, the lemma follows. □

The following lemma is a good source of incomplete transversal designs.

Lemma 1.4. *Suppose that* a, t, *and* m *are integers satisfying* $0 \le t \le m$ *and* $0 \le a \le m$. *If there exist a* $TD(8,m)$ *and a* $TD(6,t)$, *then there exists a* $TD(6,7m+t+a)-TD(6,a)$.

Proof. The proof consists of an appropriate application of Wilson's construction for transversal designs (see [12]).

Result 1.5. *There exists a* $BIBD(v,6,1)$ *for* $v \in \{586, 1306, 1336, 1341, 1386, 1411, 2061\}$.

Proof. We apply Theorem 1.1 with the parameters given in table I. To interpret the following table, w is the order of the design constructed. A PBD(v,{6,f*},1) is created by using a resolvable BIBD with f resolution classes, and an incomplete transversal design $TD(6,v-f+a)-TD(6,a)$ can be created by employing the listed values of m, t and a in Lemma 1.4. The existence of the ingredient transversal design is shown in [1], except for the existence of TD(6,24) which has been established by R. Roth, Emory University.

Table I

w	f	v	m	t	a
586	21	106	11	8	11
1306	51	256	27	16	5
1336	51	256	27	16	11
1341	51	256	27	16	12
1386	51	256	27	16	21
1411	51	256	27	16	26
2061	81	406	43	24	6

2. Other constructions

In this section we construct more BIBDs with $k = 6$ and $\lambda = 1$.

Theorem 2.1. *Suppose that* m *and* t *are integers satisfying* $0 \le t \le m$. *If there exists a* $TD(14,m)$, *and if* $\{5m+1,15t+1\} \subset B(6)$, *then* $65m+15t+1 \in B(6)$.

Proof. Since $16 \in RN(5)$, there exists a PBD(81,{6,16*},1). By deleting a point which occurs in the block size of 16, one obtains a group divisible design GDD($5^{13}15^1$,{6},1). Since $66 \in B(6)$ there exists a GDD(5^{13},{6},1). By applying Wilson's fundamental construction [see 12], the result follows. □

The lemma is easily extended to the following.

Theorem 2.1a. *Suppose that m and t are integers satisfying $0 \leq t \leq m$. If there exists a TD(14,m), and if there is a BIBD($5m+u$,6,1) and a BIBD($15t+u$,6,1), both of which contain a flat of order u, then $65m+15t+u \in B(6)$.*

Result 2.2. *There exists a BIBD(v,6,1) for $v \in$* {921, 951, 966, 981, 996, 1026, 1216, 1311, 1591, 1606, 1701, 1746, 1761, 1846, 1966, 2121, 2151, 2871, 4821, 5301, 5991, 6801, 8001, 11151}.

Proof. The proof consists of an application of Theorem 2.1 and Theorem 2.1a with $u = 6$ (see Table II and Table III). □

Table II is based on Theorem 2.1.

Table II

v	m	t	v	m	t
921	13	5	1846	27	6
951	13	7	1966	27	14
966	13	8	2121	31	7
981	13	9	2151	31	9
996	13	10	2871	43	5
1026	13	12	4821	61	57
1311	19	5	5301	73	37
1701	25	5	5991	79	57
1746	25	8	6801	103	7
1761	25	9	8001	103	87

v	m	t
11151	151	89

Table III is based on Theorem 2.1a with $u = 6$.

Table III

v	m	t
1216	17	7
1591	23	6
1606	23	7

Lemma 2.3. *There exists a BIBD*$(v,6,1)$ *for* $v \in \{6016, 6166, 8151, 10851\}$.

Proof. The proof follows from another application of Theorem 1.1. It is well known that if there exists a BIBD$(f,6,1)$ and if there exists a TD$(6,f)$, then there exists a PBD$(6f,\{6,f^*\},1)$. Using this fact, and Lemma 1.4, this lemma is established in Table IV. □

Table IV

w	f	m	t	a
6016	181	128	9	81
6166	186	131	13	80
8151	241	171	8	136
10851	331	233	24	118

3. Conclusion

It has been shown that {586, 921, 951, 966, 981, 996, 1026, 1216, 1306, 1311, 1336, 1341, 1386, 1411, 1591, 1606, 1701, 1746, 1761, 1846, 1966, 2061, 2121, 2151, 2871, 4821, 5301, 5991, 6016, 6166, 6801, 8001, 8151, 10851, 11151} $\subset B(6)$. Updating Mills' list, this reduces the number of cases for which the existence of a BIBD$(v,k,1)$ is in doubt to 130. These 130 values are listed in Table V.

Table V

46	51	61	81	141	166	171	196	201
226	231	246	256	261	276	286	291	316
321	336	346	351	376	406	411	436	441
466	471	486	496	501	526	561	591	616
621	646	651	676	706	711	736	741	766
771	796	801	831	871	886	891	916	946
1006	1011	1036	1041	1066	1071	1096	1101	1131
1141	1156	1161	1176	1186	1191	1221	1246	1251
1276	1396	1401	1456	1461	1476	1486	1491	1516
1521	1546	1551	1566	1611	1636	1641	1666	1671
1686	1696	1726	1771	1816	1821	1851	1881	1941
1971	2031	2241	2331	2571	2601	3201	3471	3501
3621	3771	3861	4191	4221	4246	4251	4971	5241
5391	5481	5866	5901	6021	6046	6051	6141	6921
7101	8241	8271	8541					

Acknowledgement

Research supported in part by the Natural Sciences and Engineering Research Council Canada under Grant A3071.

References

[1] A.E. Brouwer, The number of mutually orthogonal latin squares - a table up to order 10000, *Mathematisch Centrum,* report ZW123/79, Amsterdam, June 1979.

[2] A.E. Brouwer and G.H.J. van Rees, More mutually orthogonal latin squares, *Discrete Math.* 39 (1982) 263-281.

[3] R. Mathon and A. Rosa, Tables for parameters of BIBDs with $r \leq 41$ including existence, enumeration, and resolvability results, *Annals of Discrete Math.* 26 (1985) 275-308.

[4] W.H. Mills, A new block design, Proceedings of the Southeastern Conference on Combinatorics, Graph Theory and Computing, Florida Atlantic University, Boca Raton, Florida (1975) 461-465.

[5] W.H. Mills, Two new block designs, *Utilitas Math.* 7 (1975) 73-75.

[6] W.H. Mills, The construction of balanced incomplete block designs with $\lambda = 1$, Proceedings of the Seventh Manitoba Conference on Numerical Mathematics and Computing, University of Manitoba, Winnipeg, Manitoba (1977) 131-148.

[7] W.H. Mills, The construction of BIBDs using non-abelian groups, Proc. Ninth Southeastern Conf. on Combinatorics, Graph Theory and Computing, Florida Atlantic University, Boca Raton, Florida (1978) 519-526.

[8] W.H. Mills, The construction of balanced incomplete block designs, Proc. Tenth Southeastern Conf. on Combinatorics, Graph Theory and Computing (1979), 73-86.

[9] W.H. Mills, Balanced incomplete block designs with $k = 6$ and $\lambda = 1$, in *Enumeration and Design,* D.M. Jackson and S.A. Vanstone, eds. Academic Press, New York (1984), 239-244.

[10] R.C. Mullin, On the extension of the Moore-Sade construction for combinatorial configurations, *Cong. Num.* 23 (1979) 87-102.

[11] D.K. Ray-Chaudhuri and R.M. Wilson, *The Existence of Resolvable Block Designs, A Survey of Combinatorial Theory,* North Holland (1973) 361-375.

[12] R.M. Wilson, Construction and uses of pairwise balanced designs, *Math. Centre Tracts* 55 Amsterdam (1974) 18-41.

Annals of Discrete Mathematics 34 (1987) 385—392
© Elsevier Science Publishers B.V. (North-Holland)

Isomorphism Problems for Cyclic Block Designs

*K.T. Phelps**

School of Mathematics
Georgia Institute of Technology
Atlanta, Georgia, 30332
U.S.A.

TO ALEX ROSA ON HIS FIFTIETH BIRTHDAY

ABSTRACT

We establish that for all $\lambda > 0$, $k \geq 3$, there exists cyclic block designs (v,k,λ) which are isomorphic but not multiplier equivalent, for infinitely many values of v. The constructions are based on previous work of the author [14], [13] and N. Brand [3], [4]. In particular we establish this result when $v = pq$ and $v = p^t$, $t \geq 2$, for primes $p,q;p,q$ sufficiently large.

1. Introduction

A block design with parameters (n,k,λ) is a pair (S,B) where S is an $n-$ set and B is a collection of $k-$element subsets of S with the property that every pair of S is in exactly λ blocks of B. A block design is **cyclic** if the automorphism group contains a regular cyclic subgroup, in which case we identify S with $Z_n = \{0,1,2,...,n-1\}$ and the regular cyclic subgroup with $\langle Z_n,+\rangle$, the integers under addition modulo n.

Two isomorphic cyclic block designs (Z_n,B_1) and (Z_n,B_2) are said to be multiplier equivalent if in fact there is a multiplier isomorphism between them, (i.e. , $\beta: x \rightarrow ax (\text{mod } n)$, a relatively prime to n). In this paper, we will give constructions for cyclic block designs with parameters (n,k,λ) which have isomorphic cyclic mates but which are not multiplier equivalent. Before proceeding let us review the background of this problem.

* Current Address:
Department of Algebra, Combinatorics and Analysis
Auburn University
Auburn, Alabama 36849-3501
U.S.A.

2. Background

In general, the spectrum for cyclic block designs is unknown. The case $k=3,\lambda=1$ (i.e. Steiner triple systems) has been extensively studied. Peltesohn [12] established that cyclic Steiner triple systems exist if and only if $n\equiv1$ or 3 mod 6 and $n\neq9$. Prior to this, S. Bays and others enumerated cyclic $(n,3,1)$ for $n\leq43$. As part of any enumeration study, one must face the question of isomorphism. S. Bays [2] was able to prove that if n is a prime then any two isomorphic cyclic $(n,3,1)$ must be multiplier equivalent. S. Bays naturally conjectured that this is true for all cyclic $(n,3,1)$; the results of the enumeration studies supported this conjecture. In fact, one can say that the conjecture is valid for all $n\leq45$.

S. Bays' proof that isomorphic cyclic Steiner triple systems of prime order are multiplier isomorphic was purely group-theoretic and thus applied to any cyclic combinatorial structure. He established that any two isomorphic cyclic structures will be multiplier equivalent if and only if the corresponding regular cyclic subgroup are conjugate in the full automorphism group.

Colbourn and Mathon have an excellent survey [6] article on cyclic designs and the interested reader is referred to that article for more information. However, one important fact that Colbourn and Mathon fail to mention is that Lambossy [9] constructed isomorphic cyclic combinatorial structures on 16 and on 27 points, which were not multiplier equivalent. Moreover, Lambossy gave necessary and sufficient conditions for the affine group, in the ring of integers mod n, to have non-conjugate n-cycles. (The affine group, also called the metacyclic group is the group of all transformations of the form $\alpha{:}x\rightarrow bx+c(\text{mod } n)$ where $(b,n)=1$.) He further proved that when n is a prime power (i.e. $n=p^t, t>1$) then $\alpha{:}x\rightarrow bx+c(\text{mod } p^t)$ is a p^t-cycle if and only if $b\equiv1$ mod p and $c\equiv0(\text{mod } p)$ for primes $p>2$. (For $p=2$, the condition becomes $b\equiv1$ mod $4, c\neq=0$ mod 2.) Lambossy also established that these regular cyclic subgroups are not conjugate in the affine group — but this is a trivial exercise.

The above results seem to have been ignored by most, if not all, researchers and it was not until recently that the Bays-Lambossy conjecture was disproved by N. Brand [3].

3. Cyclic Designs of Order p^t

To construct isomorphic cyclic designs which are not multiplier equivalent one must first be able to construct cyclic designs which is, in general, a difficult and unresolved issue. However, R. M. Wilson [14] established that for primes $p, p\equiv1(\text{mod } k(k-1))$, which are sufficiently large and satisfy the obvious necessary conditions, there exist cyclic $(p,k,1)$ designs (and hence cyclic (p,k,λ) designs).

Lemma 3.1: If there exists a cyclic (p,k,λ) then there exists cyclic (p^t,k,λ) designs for all $t>0$ which have multiplier automorphism $\beta{:}x\rightarrow bx(\text{mod } p^t)$ for $b\equiv1$ mod p.

Proof: Let p be an odd prime, (i.e., assume $k>2$). The group of units in the ring of integers $(\text{mod } p^t)$ has order $p^{t-1}(p-1)$ and is cyclic. Let a be the generator of this multiplicative group. Note that $b\equiv a^{(p-1)}(\text{mod } p^t)$ is a generator for the subgroup of units $m, m\equiv1$ mod p. Moreover, $a\equiv\underline{a}(\text{mod } p)$ where $\underline{a}$ is a generator in the multiplicative group of integers $(\text{mod } p)$.

Let $A_i = \{x \mid x \equiv i \bmod p\}$ and let B_t denote the set of orbit representations in a cyclic (p^t,k,λ) design. For each block $l=\{i_1,i_2, \cdots ,i_k\}\in B_1$ choose a block $l'=\{x_1,x_2, \cdots ,x_k\}$ such that $x_j \equiv i_j \bmod p$, for $j=1,2,...,k$. Define

$$U_t = \{b^i l' = \{b^i x_1, b^i x_2, \cdots ,b^i x_k\} \mid i=0,1, \cdots ,p^{t-1}-1\}$$

and

$$pB_{t-1} = \{pl = \{px_1,px_2, \cdots ,px_k\} \mid l \in B_{t-1}\}$$

and

$$B_t = U_t \bigcup pB_{t-1}$$

where the computations are all mod p^t and b is the generator of the multiplicative subgroup of units $m, m \equiv 1 (\bmod\ p)$.

The proof that B_t is a set of orbit representatives for a cyclic (p^t,k,λ) design follows inductively from the existence of B_1 and is straightforward.

Note that B_t will always have some multiplier automorphism $m \equiv 1 \bmod p$, regardless of the choice of B_{t-1}. Also note that there are many different choices of l' for each $l \in B$, (essentially $(p^{t-1})^{k-1}$ choices).

From the previous discussion of P. Lambossy's work we can conclude that the designs constructed above will have regular cyclic subgroups and these will have isomorphic mates. N. Brand [4] shows that the isomorphism must be of a certain form and hence we can conclude that the fixed points of the isomorphism are all $x \in Z_{p^t}$ such that $x \equiv 0$ (mod p). This also follows from Lambossy's characterization of the p^t-cycles, $(\beta : x \rightarrow bx+c, b \equiv 1 \bmod p)$. We conclude that both the original design and its isomorphic mate will have the same identical cyclic subdesign on the set $A_0 = \{x \mid x=0 \bmod p\}$. Hence any multiplier isomorphism between these designs must be congruent (modulo p^{t-1}) to a multiplier automorphism of the cyclic subdesign. If we construct a cyclic design for which this is impossible then we have a design with an isomorphic mate which is not multiplier equivalent.

Lemma 3.2: If m is a multiplier automorphism of a cyclic (p,k,λ) design (Z_p,B) and m fixes one orbit, then m fixes a block in that orbit.

Proof: Assume for $l=\{x_1,x_2, \cdots ,x_k\}\in B, ml=l+i$ (i.e. , $ml=\{mx \mid x \in l\}$ and $l+i=\{x+i \mid x \in l\}$) where all computations are reduced modulo p. Then $m^r l = l+(m^{r-1}+m^{r-2}+ \cdots +1)i = l$ for some r. But each orbit under the additive group is full having length p. Hence $m^{r-1}+ \cdots +1 \equiv (m^r-1)/(m-1) \equiv 0$ (mod p). Thus $m^r \equiv 1 \bmod p$ or $i \equiv 0 \bmod p$, but r divides $p-1$ and since every sub-orbit under multiplication by m has length r or 1, then $kr+n \cdot 1 = p$ and thus there is a block l', with $ml'=l'$.

Theorem 3.3: If there exists a cyclic (p,k,λ) design with multiplier automorphism m then there exists a cyclic (p^t,k,λ) design with all multiplier automorphisms $m' \equiv m$ (mod p).

Proof: We can assume that m generates the multiplicative automorphism group for the cyclic (p,k,λ) design (Z_p,B) and that we can choose orbit representatives $m^i l \in B$ such that $l, ml, \cdots ,m^r l$ are in different orbits and $m^{r+1} l = l$ whether or not

$m^{r+1} \equiv 1 \bmod p$. Let B_1 be the set of these orbit representatives and let a be the generator of the multiplicative group in $Z_p t, t > 1$. Then $a \equiv \underline{a} \bmod p$ where $\underline{a}$ is a generator for the multiplicative group in Z_p. Let $l = \{\underline{a}^{\overline{r_1}}, \underline{a}^{r_2}, \cdots, \underline{a}^{r_k}\} \in \overline{B}$ and $m \equiv \underline{a}^s (\text{mod } p)$. Let $c = a^{p^{t-1}}$, and $m' \equiv c^s (\text{mod } p^t)$. Define $l' = \{c^{\overline{r_1}}, c^{r_2}, \cdots, c^{\overline{r_k}}\} \in U_t$ for each $l \in B_1$ (as in Lemma 3.1). Since $c \equiv \underline{a} \bmod p$ then $c^s \equiv \underline{a}^s \equiv m \bmod p$ and for each $c^r \in l', \underline{a}^r \in l, c^r \equiv \underline{a}^r (\text{mod } p)$. Moreover, $\overline{c^s l'}$ corresponds to $ml \in B_1$. Proceed as in Lemma 3.1 so that $b_i l' \in U_t$ for each $l' \in U_t$ and $B_t = pB_{t-1} \bigcup U_t$. Obviously, $m' \equiv m \bmod p$ will always be a multiplier automorphism for U_t since $m' \equiv c^s b^i \bmod p^t$ for some i. The result now follows by induction on t assuming that every $m' \equiv m \bmod p$ is a multiplier automorphism for $B_{t-1} (\text{mod } p^{t-1})$.

Corollary 3.4: If there exists a cyclic (p,k,λ) design then there exists isomorphic cyclic (p^t,k,λ) designs which are not multiplier equivalent, for all $t > 1$, where $p \equiv 1 \bmod k(k-1)$ and p is a prime.

Corollary 3.5: For all primes $p > c_k$ a constant dependent on k, there exists isomorphic cyclic (p^t,k,λ)-designs, $t > 1$ which are not multiplier equivalent.

There are some composition methods for constructing cyclic designs (nm,k,λ) from cyclic design (n,k,λ) and (m,k,λ) (cf., Jimbo, Kuriki [8]). The simplest is the "direct product" which can be defined when k is a prime and n and m are relatively prime. Using the standard direct product construction we have:

Theorem 3.6: If there exist cyclic (n,q,λ) and cyclic (m,q,λ) designs, where q is a prime and n and m are relatively prime and if there are cyclic isomorphic designs (n,q,λ) which are not multiplier equivalent.

The above construction generalizes the approach of the author for the case $k=3$ [13], which was originally established by N. Brand [3].

4. Cyclic Designs (pq,k,λ)

Since a $(n,2,1)$ design is just the complete graph we assume throughout that $k \geq 3$. P. P. Pálfy ([10],[11]) proved that if n and $\phi(n)$ are relatively prime (where n is a product of distinct primes), then any isomorphic cyclic (n,k,λ) designs must be multiplier equivalent. Thus we restrict our attention to the case $n = pq$ where $q \mid p-1$, for primes p,q.

Alspach and Parsons [1] established that any cyclic $(pq,k,\lambda) p > q$ having a sylow p-subgroup of order p must be multiplier equivalent to any isomorphic cyclic (pq,k,λ) design. Godsil [7] points out that cyclic (pq,k,λ) designs have automorphism groups that are 2-transitive or imprimitive. Combining these observations, we conclude that in order to construct cyclic (pq,k,λ) designs which have sylow p-subgroups of order $p^t, t > 1$. The sylow p-subgroups of such cyclic designs are equivalent to a cyclic code of length q over $GF(p)$. We use these to construct cyclic transversal designs.

Lemma 4.1: There exists cyclic transversal designs (Z_{pq},B) with groups of size p and blocks of size q, where the sylow p-subgroups of the automorphism group has order p^2, whenever $q \mid p-1$ and p is a prime (note: q need not be prime).

Proof: Let P be a cyclic code of length q, dimension 2 over $GF(p)$. Since $q \mid p-1$, the qth roots of unity are in $GF(p)$; then $P = \{s(1,1,\ldots,1) + t(1,r,r^2, \cdots, r^{q-1}) \mid s,t \in Z_p\}$ where $r^q \equiv 1 \bmod p$. P has distance $q-1$, since the roots of the generator polynomial are $r,r^2, \cdots ,r^{q-2}$. Now the groups of the design are $G_i = Z_p \times \{i\}$, $i=0,1,...,q-1$ and for each $(x_0,x_1, \cdots ,x_{q-1}) \in P$, we have a block $\{(x,0),(x,1), \cdots ,(x,q-1)\} \in B$. This obviously gives us a cyclic transversal design with a sylow p-subgroup isomorphic to P. For each $b=(b_0,b_1,...,b_{q-1}) \in P$ define a permutation $\alpha_b:(x,i) \rightarrow (x+b_i,i)(\text{mod } p, \text{mod } q)$. These permutations along with $\beta:(x,i) \rightarrow (x,i+1)$ generate a cyclic group having the desired sylow p-subgroup. Note that for every non-zero $m \in Z_p,(m,1)$ is a multiplier automorphism for the above design.

P. P. Pálfy [10] claims that the automorphism group of the above design contains a subgroup with has non-conjugate cyclic subgroups. However, they are conjugate in the full group. If $(Z_p \times Z_q,B)$ is the design constructed above then:

Lemma 4.2: For primes $p,q,q \mid p-1$, any cyclic design isomorphic to $(Z_p \times Z_q,B)$ as constructed above, must be multiplier equivalent to it.

Proof: If $(Z_p \times Z_q,B')$ is the isomorphic cyclic design with $\alpha:(x,i) \rightarrow (x+1,i+1)$ (mod p,mod q) as the pq-cycle then the automorphism group contains a sylow p-subgroup of order p^2 which is a cyclic code of length q over $GF(p)$ having dimension 2 and distance $q-1$. Since $q \mid p-1$ and $(1,1,1,...,1)$ is in the code then the generator polynomial has roots $\underline{r},\underline{r}^2, \cdots ,\underline{r}^{q-2}$. Since $\underline{r} \equiv r^i \pmod p$ then the two codes are multiplier equivalent. The blocks in B' either correspond to the words in the code or a coset of it. Being cyclic, the latter is impossible. Hence is $\underline{r} \equiv r^i \pmod p$ then $(1,i)$ is a multiplier automorphism for the design.

Corollary 4.3: Isomorphic cyclic $(pq,q,1)$ designs are always multiplier equivalent for primes p,q with $p>q$.

Proof: If $q \nmid p-1$ then pq and $\phi(pq)$ are relatively prime; hence the corollary follows from P. P. Pálfy's result ([10],[11]). If $q \mid p-1$, then we may as well assume by our previous discussion that the sylow p-subgroup has order p^2 and the automorphism group is imprimitive with $Z_p \times \{i\}$ as the imprimitive sets (cf. [14]). This all means that there must exist cyclic $(p,q,1)$ designs on the sets $Z_p \times \{i\}$, $i=0,1, \cdots ,q-1$. The sets $Z_p \times \{i\}$ along with the p^2 blocks cutting across these sets forms a cyclic transversal design with a sylow p-subgroup of order p^2 which is equivalent to a cyclic code. Hence the transversal design is (multiplier) isomorphic to the design constructed in Lemma 3.1. The two isomorphic designs will have isomorphic cyclic subdesigns $(p,q,1)$ which are multiplier isomorphic (mod p) (Bays [2]) and isomorphic cyclic transversal designs which are multiplier isomorphic (mod q). Hence the designs are multiplier isomorphic.

In general, for a cyclic $(pq,k,1)$ design (p,q primes, $p>q$) to have a sylow p-subgroup of order p^t, $t>1$ there must exist cyclic $(p,k,1)$ and $(q,k,1)$ designs. This follows easily from elementary properties of designs along with several remarks by Alspach, Parsons [1].

Lemma 4.4: If a cyclic $(m,k,1)$ design does not exist for either $m=p$ or $m=q$, p,q primes, then any two isomorphic cyclic $(pq,k,1)$ designs will be multiplier equivalent.

Proof: There must be a subdesign on the set of fixed points of any automorphism (subgroup). Assuming $p>q$ and the sylow p-subgroup has order p^t, $t>1$, then each set $Z_p \times \{i\}$ will be the set of fixed points for some subgroup of the sylow p-subgroup [1]. Hence there must exist a cyclic $(p,k,1)$. The orbits under the sylow p-subgroups are the sets $Z_p \times \{i\}$. If one block cuts across k of these sets then exactly p^2 blocks cut across them. The sets of groups which have exactly p^2 blocks cutting across from a cyclic $(q,k,1)$ design. Hence it either of these cyclic subdesigns do not exist then the automorphism group cannot have a sylow p-subgroup of order greater than p. The result now follows from a lemma of Alspach, Parsons [1], (cf. [14]).

We now turn to the main theorem of this section. The author has tried valiantly to prove that isomorphic cyclic $(pq,3,1)$ designs are multiplier equivalent. The basic difficulty was that it simply is not true as we shall see.

Theorem 4.5: For primes $p,q,q \mid p-1$, if there exists a cyclic (p,k,λ) design having multiplier automorphism r, with $r^q \equiv 1 \bmod p$ and there exists a cyclic (q,k,λ) which does not have $-1 (\bmod q)$ as a multiplier automorphism. Then there exist isomorphic cyclic (pq,k,λ) designs which are not multiplier equivalent.

Proof: Let $(Z_p \times Z_q, B)$ be the cyclic transversal design as constructed in Lemma 4.1. We construct a cycle (pq,k,λ) design by placing appropriate copies of the cyclic (p,k,λ) and the cyclic (q,k,λ) designs on the groups and blocks respectively. Define an isomorphism $f:(x,i) \rightarrow (r^{-i}x,i)$. (This is essentially the mapping that P. P. Pálfy uses [10].) The image of this mapping is an isomorphic cyclic (pq,k,λ) design. Note that the cyclic subdesigns on $Z_p \times \{i\}$ and $\{0\} \times Z_q$ remain unchanged. Hence any multiplier isomorphism must be congruent (mod q) to a multiplier automorphism of the cyclic sub (q,k,λ). On the other hand the mapping gives isomorphic cyclic transversal designs which are multiplier equivalent where the multiplier is $(1,-1)$ (mod p,mod q). Hence these designs are not multiplier equivalent, since $-1 (\bmod q)$ is not an automorphism of the cyclic subdesign (q,k,λ).

Corollary 4.6: For all primes $p \equiv 7 \bmod 12$, if $q \mid p-1$, $q>3$ and, $q \equiv 1 \bmod 6$, there exists isomorphic cyclic $(pq,3,1)$ designs which are not multiplier equivalent.

Proof: The well-known block transitive cyclic $(p,3,1)$ design has multiplier automorphism group of order $\frac{(p-1)}{2}$ (cf. Phelps [13]) and -1 (mod q) is never a multiplier automorphism for a cyclic $(q,3,1)$ design, $q>3$.

Note that the proof of the Theorem does not depend on whether q is a prime.

R. M. Wilson's [15] constructions ensure that cyclic (p,k,λ) designs exist where $p \equiv 1$ (mod $qk(k-1)$) (infinitely often). The constructions also ensure that these designs will have the qth roots of unity as multipliers again infinitely often.

5. Conclusions

Because R. M. Wilson's results guarantee the existence of cyclic (p,k,λ) designs of order p, we have:

Theorem 5.1: For any $k>2$, $\lambda>0$ there exists isomorphic cyclic (m,k,λ) designs which are not multiplier equivalent.

Isomorphic circulant graphs on n vertices, are always multiplier equivalent for $n=pq$, ([1], [7], [14]) where p,q are distinct primes. the results of this paper establish that this is not true in general for isomorphic circulant (k-uniform) hypergraphs on pq vertices when $q \mid p-1$.

The most important question remaining is whether isomorphism of cyclic block designs is isomorphism-complete or not (cf. [5]).

References

[1] B. Alspach and T. Parsons, "Isomorphism of circulant graphs and digraphs", *Discrete Math.* 25 (1979) 97 - 108.

[2] S. Bays, "Sur les systèmes cyclique de triples de Steiner differents pour N premier (on puissance du nombre premier) de la forme 6n + 1, II - III", *Comment. Math. Helv.* 3 (1931) 22 - 41.

[3] N. Brand, "On the Bays-Lambossy Theorem", preprint.

[4] N. Brand, "Some combinatorial isomorphism theorems", preprint.

[5] M. J. Colbourn and C. J. Colbourn, "Concerning the complexity of deciding isomorphism of block designs", *Discrete Appl. Math.* 3 (1981) 155 - 162.

[6] M. J. Colbourn and R. Mathon, "On cyclic Steiner 2-designs", *Annals of Discrete Math.* 7 (1980) 215 - 251.

[7] C. Godsil, "On Cayley graph isomorphism", *Ars Combinatoria* 15 (1983) 231 - 246.

[8] M. Jimbo and S. Kuriki, "On a composition of cyclic 2-designs", *Discrete Math.* 43 (1983) 249 - 255.

[9] P. Lambossy, "Sur une manière de differencier les fonctions cycliques d'un form donnée", *Comment. Math. Helv.* 3 (1931) 69 - 102.

[10] P.P. Pálfy, "On regular pronormal subgroups of symmetric groups", *Acta Math. Acad. Sci. Hungar.* 34 (1979) 287 - 292.

[11] P. P. Pálfy, "Isomorphism problems for relational structures with a cyclic automorphism", preprint.

[12] R. Peltesohn, "Eine Lösung der beiden Heffterschen Differenzen probleme", *Compositio Math.* 6 (1939) 251 - 257.

[13] K. T. Phelps, "A construction of Steiner triple systems of order p^n", *Discrete Math.*, to appear.

[14] K. T. Phelps, "Isomorphism of circulant combinatorial structures", preprint.

[15] R. M. Wilson, "Cyclotomy and difference families in elementary abelian groups", *J. Number Theory* 4 (1972) 17 - 47.

Annals of Discrete Mathematics 34 (1987) 393–408
© Elsevier Science Publishers B.V. (North-Holland)

Multiply Perfect Systems of Difference Sets

D.G. Rogers

Fernley House,
The Green
Croxley Green, WD3 3HT
UNITED KINGDOM

TO ALEX ROSA ON HIS FIFTIETH BIRTHDAY

ABSTRACT

A family of difference sets of valency at least 2 is multiply perfect for c if every integer in a run of consecutive integers starting with c occurs a prescribed constant number of times in the difference sets of the family. The theory of perfect systems extends to multiply perfect systems with the advantages that in practical situations we have an economical check on our data and that theoretically we have a possible way of circumventing certain parity constraints. Associations with block designs and graph labelling also either carry over or generalize in a natural way.

1. A general problem in measurement

Quite often in making measurements a_i, $0 \leq i \leq v$, where

$$a_0 \leq a_1 \leq a_2 \leq \cdots \leq a_v,$$

we are most interested in the differences or *relative* measurements $a_i - a_j$, $0 \leq j < i \leq v$: best of all would be if these are all distinct so that there is no duplication of information. Moreover, with several sets of data of this kind, we should like the relative measurements from any one set not only to be distinct one from another, but to differ from those coming from any other set too. We suppose that all relative measurements are integral multiples of some unit. If we measure in this unit, then relative to the smallest measurement a_0 in each set, all the other measurements in the set are integral, and altogether we want none to be duplicated.

At the same time, given the number of sets of measurements available for use and some prescribed overall smallest relative measurement, we may also wish to make the largest relative measurement as small as possible. If we miss any of the integers between the largest and smallest relative measurements, then we may have to try to recover the loss through interpolation. In order to avoid this, our aim is that the relative measurements should in fact form the consecutive run of integers from the smallest measurement.

The problem of arranging measurements so that, if possible, there are no gaps or duplications among the resulting relative measurements arises in, for example, missile guidance (see [12]), coding theory (see [15,16]) and radio astronomy (see [7,8]); for general discussions of this sort of problem, see [9,10]. This is the practical stimulus behind the study of *perfect systems of difference sets* (see [3,5,6]). A survey of these systems is given in [1]. We now define these systems.

A *component* A of *valency* v is a set $A = \{a_i : 0 \leq i \leq v\}$ of integers a_i, $0 \leq i \leq v$, such that

$$a_0 \leq a_1 \leq a_2 \leq \cdots \leq a_v; \tag{1}$$

and the *difference set* $D(A)$ of A, also said to be of *valency* v, is then the set

$$D(A) = \{a_i - a_j : 0 \leq j < i \leq v\}.$$

A family of components A_r, $1 \leq r \leq m$, where m_v are the valency v for $v \geq 2$, is an $(\tilde{m};c)$-system when

$$\bigcup_{r=1}^{m} D(A_r) = \{i : c \leq i < c + l\} \tag{2}$$

where

$$l = \tfrac{1}{2} \sum_{v \geq 2} v(v+1) m_v ; \quad m = \sum_{v \geq 2} m_v ; \tag{3a}$$

$$\tilde{m} = (m_2, m_3, \ldots, m_v, \cdots). \tag{3b}$$

If the components A_r, $1 \leq r \leq m$, form an $(\tilde{m},c)$-system then the family of difference sets $D(A_r)$, $1 \leq r \leq m$, is said to be *perfect* for c or to be a *perfect system of difference sets* with *threshold* c.

$a_v - a_0$

$a_v - a_1$

$a_v - a_2$

$a_2 - a_0$

$a_1 - a_0$ $a_2 - a_1$ $a_v - a_{v-1}$

Figure 1: Difference set $D(A)$ as a difference triangle

It is convenient to display differences as *difference triangles* as in Figure 1. Difference triangles of this sort are characterized by the property that the apex entry of any subtriangle with base on the bottom row is the sum of the entries on the base. Thus Figure 2 displays as difference triangles the difference sets of an $(\tilde{m};6)$-system in which $m_2=6$, $m_3=3$, $m_4=1$ and $m = m_2+m_3+m_4 = 10$ (this example is taken from [2; p.10]).

			51										
		40		43		44		46		50			
	18		32		33	35	38	34	39	36	37		
8		10		22	11	6 29 9		7 27 12		13 23 14			

41	47	42	49	48	45
15 26	16 31	17 25	19 30	20 28	21 24

Figure 2: A $((6,3,1,0,0,\cdots);6)$-system

It is clear that (2) encapsulates the requirement of no gaps or duplications in the series of relative measurements obtained from taking the set of measurements A_r, $1 \leq r \leq m$. Of those relative measurements, the threshold c is the least which it is possible to make, as the name is intended to indicate: operating constraints, as in the application to radio telescopes, may require $c>1$; and indeed a more satisfactory theoretical treatment is possible on considering general c.

It is also readily apparent, however, that (2) is quite stringent: not all runs of consecutive integers admit such partitions into difference sets; and if there are also restrictions on the available valencies of the difference sets, then (2) is still more difficult to meet. For practical purposes it is appropriate to consider ways of relaxing (2) which might yet have uses. One relaxation is to seek families of components A_r, $1 \leq r \leq m$, such that

$$\bigcup_{r=1}^{m} D(A_r) = \{i : c \leq i < c + l'\}$$

where l' is as large as possible. Another is to consider families of components A_r, $1 \leq r \leq m$, such that

$$\bigcup_{r=1}^{m} D(A_r) \subseteq \{i : c \leq i < c + l^*\}$$

where l^* is as small as possible subject to the union of the difference sets containing l distinct differences. We may refer to these two types of families as respectively *maximal* and *minimal* (with respect to c); of course, perfect systems of difference sets come from families which are simultaneously maximal and minimal. Both these relaxations are of some practical interest (see the references given above); but those few results which have been obtained for these sorts of system are largely computational rather than theoretical; and indeed much seems to be unknown.

It should further be observed that (2) or either of the relaxations just described may be only one of the several requirements which we need to meet in a practical application. For example, with radio telescopes, we also wish to keep the union of the components as small as possible; but here too there seems to be little theoretical work (we return to this particular point in Section 3).

A connection between $(\tilde{m};c)$-systems and block designs suggests a further apparently novel relaxation of (2) with attractive practical and theoretical properties and leads us to our subject, *multiply perfect systems.*

2. Block designs and multiply perfect systems

For a component $A = \{a_i : 0 \leq i \leq v\}$ in an $(\tilde{m};c)$-system we form the *blocks*

$$A(j) = \{a_i + j : 0 \leq i \leq v\}, \quad 0 \leq j \leq 2l,$$

as subsets of Z_{2l+1}, the additive group of integers modulo $2l+1$. Condition (1) then ensures that these are the blocks of a cyclic 2-design on Z_{2l+1}, every pair of distinct elements in Z_{2l+1} appearing in a unique block (for $x-y$ or $y-x$ is congruent modulo $2l+1$ to some unique i with $c \leq i < c+l$ when $x \neq y$; in turn, i is in some unique difference set $D(A)$ of the system; and then x and y are both in $A(j)$ for some unique j). As an example, the construction of cyclic Steiner triple systems after this fashion from perfect systems of difference sets all of valency 2 is mentioned in [1; pp. 11-12].

With this in mind, we say that a family of components A_r, $1 \leq r \leq \lambda m$, λm_v of which are of valency v, $v \geq 2$, is an $(\tilde{m};c)_\lambda$-system and that the associated family of difference sets $D(A_r)$, $1 \leq r \leq \lambda m$, is a λ-*perfect system* with *threshold* c when every integer in the set $\{i : c \leq i < c+l\}$ occurs exactly λ times among the difference sets $D(A_r)$, $1 \leq r \leq \lambda m$, where l,m and $\tilde{m}$ are, as before, given by (3). Some examples of $(\tilde{m};c)_2$ systems are shown in Figure 3 displayed in terms of difference triangles.

Clearly an $(\tilde{m};c)_\lambda$-system gives rise to a cyclic 2-design on Z_{2l+1} in which pairs of distinct elements appear in exactly λ blocks. But there are other points of interest to these systems besides this. In terms of our problem of measurement in Section 1, $(\tilde{m};c)_\lambda$-systems uniformly duplicate all of the relative measurements in a given interval and therefore provide an economical and thorough internal check on these measurements. In order to gain some degree of independence between these replications this might be taken further, for instance by requiring that no difference set be repeated, that every difference which appears does so in λ difference sets of the system, that pairs of distinct integers appear in at most one difference set and so on (for example, these criteria are satisfied by the $(\tilde{m};c)_2$-systems in Figure 3 except that those in Figures 3(ii) and (iii) fail on the third criterion, since in both cases, 1 and 4 appear together in two difference sets). Also of note is that the introduction of $(\tilde{m};c)_\lambda$-systems opens up new possibilities in the sense that, for example, in the cases of $(\tilde{m};c)_2$-systems illustrated in Figure 3, it is known that there are no $(\tilde{m};c)$-systems. We comment on this further in Section 3 in the context of graceful graphs, and investigate the matter in some detail in Sections 4, 5, and 6 in the case of systems all of whose difference sets have valency 2.

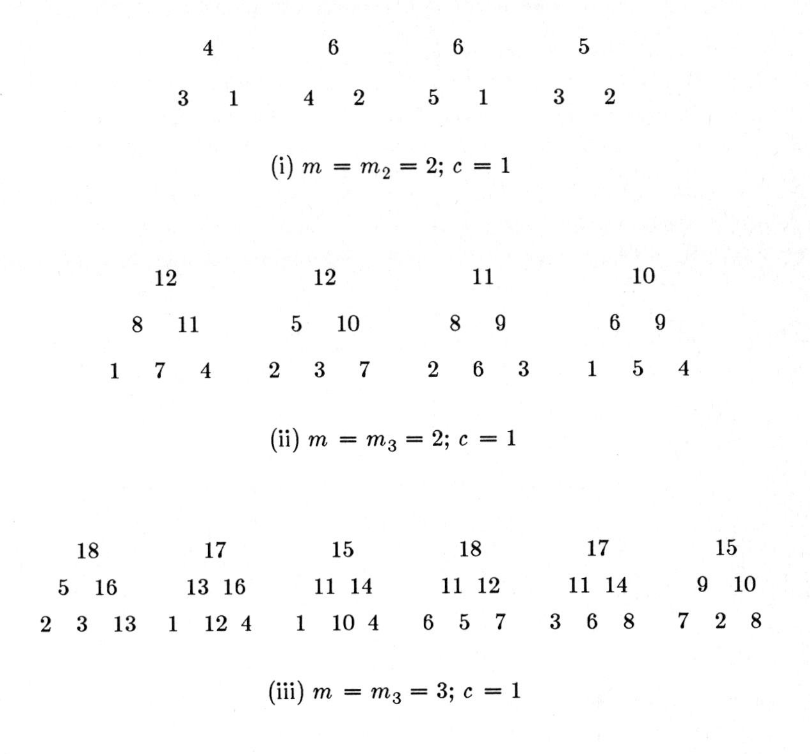

(i) $m = m_2 = 2; c = 1$

(ii) $m = m_3 = 2; c = 1$

(iii) $m = m_3 = 3; c = 1$

Figure 3: Some $(\tilde{m};c)_2$-systems

Naturally much of the theory of perfect systems carries over straightforwardly and largely unchanged to multiply perfect systems: included in this body is the necessity of the *BKT inequality* (see [1 ; p.7] and [6,19]), the *addition theorems* (see [1 ; p.20] and [18]), the *multiplication theorems* (see [1 ; pp.17-18] and [14,17,22]). In connection with the last of these, it is also natural to generalize the notion of a complete (or additive) permutation which figures in the multiplication of perfect systems; this has indeed already been done, in a slightly different setting in [13]. We do not propose to elaborate further on these portions of the general theory, contenting ourselves with a sketch of what we find need for as the occasion for it arises (see especially the last two paragraphs of Section 5).

3. Perfect systems and graceful graphs

The connection between perfect systems and graceful graphs is well known (see [1; pp. 7-9] and [2]). A graph with g edges is said to be c-graceful if we can label the vertices distinctly with some of the numbers $0,1,...,c+g-1$, in such a way that the induced labels of the edges are $c,c+1,...,c+g-1$, where the induced label of an edge is the absolute value of the difference between the labels of its end points. The study of such labellings was pioneered by Rosa in [20] (under the name β-valuation. A motivation in their study was in turn certain special decompositions of complete graphs reminiscent of block designs.

Now suppose that A_r, $1\leq r\leq m$, are the components of an $(\tilde{m};c)$-system and that in addition to (2), we have

$$\bigcup_{r=1}^{m} A_r \subseteq \{0,1,...,c+g-1\}. \tag{4}$$

Then we may regard these components as the vertex labels assigned to various complete graphs which together, by (2), make up a c-graceful graph. Since (2) also implies that, for $r\neq s$, A_r and A_s are either disjoint or intersect in one common element, any graph which arises in this way is decomposable in being the edge disjoint union of complete graphs. It does not appear to be known conversely which decomposable graphs are c-graceful. However, if (4) holds then (2) and (4) remain true on replacing any component $A = \{a_i : 0\leq i\leq v\}$ by $A' = \{a_i - a_0 : 0\leq i\leq v\}$ so that all components now contain 0, in which case the associated c-graceful graph is the windmill graph consisting of m_v copies of the complete graph on $v+1$ vertices all with a unique common vertex (labelled 0). So a necessary condition for a decomposable graph to be c-graceful is that the corresponding windmill graph be c-graceful. (It has in fact become almost customary to stipulate in the definition of an $(\tilde{m};c)$-system that all components contain 0, the better to concentrate on (2); but, as noted towards the end of Section 1, practical applications may require some other configuration.)

A graph may fail to be c-graceful because it has too many or too few edges relative to the number of vertices (the BKT inequality in effect prevents too many edges). But another type of necessary condition arises in the case of Eulerian graphs as first noted by Rosa [20].

Theorem: An Eulerian graph with g edges is c-graceful only if

$$g \equiv 0 \text{ or } 3(2c-1) \pmod 4. \tag{5}$$

Our chief interest here in introducing $(\tilde{m};c)_\lambda$-systems is that they provide a way for dealing with this sort of condition. As we see in Section 4, (5) is not only necessary for c-graceful graphs which are decomposable as the edge disjoint union of triangles, that is for $(\tilde{m};c)$-systems with $m=m_2$, but it is also sufficient among these graphs at least for windmill graphs. On the other hand, when (5) fails then, as we also see, there is an $(\tilde{m};c)_2$-system with $m=m_2$ except possibly for finitely many values of m for each $c\geq 1$.

Of course, we might consider, more generally, an allied notion of multiply graceful graphs. Instead, we close this section more simply with one specific problem which generalizes a question posed by Rosa himself in the Problem Session held at the Australia-

Singapore Joint Conference on Information Processing and Combinatorial Mathematics in Singapore in May 1986 (the case $\lambda=c=1$: are Δ-snakes graceful?).

Problem:

For $\lambda=1$ or 2, is there an $(\tilde{m};c)$-system with $m=m_2$ and components A_r, $1\leq r\leq\lambda m$, such that

$$A_r \cap A_s \neq \emptyset \quad \begin{cases} s=2;\ r=1; \\ s=r\pm 1;\ 1<r<m; \\ s=m-1;\ r=m. \end{cases}$$

4. Regular multiply perfect systems of valency 2

Let c,m and λ be positive integers, and let δ be an integer such that $0\leq\delta<c+3m$. An $(m,3,c)_\lambda^\delta$-system is a family of sets $A_r=\{0,a_1(r),a_2(r)\}$, $1\leq r\leq\lambda m$, with $0<a_1(r)<a_2(r)$, such that every integer in the set

$$\{i:c\leq i\leq c+3m\} - \{c+3m-\delta\}$$

occurs exactly λ times among the difference sets $D(A_r)$ (in all of the constructions to be given here, every integer in this set occurs in exactly λ of the difference sets which may, however, be repeated as sets). If $\lambda=1$ or $\delta=0$, we drop the subscript or superscript respectively. In particular, of course, the difference sets associated with an $(m,3,c)_\lambda$-system form a λ-perfect system with threshold c; this system is *regular* in the sense that all the difference sets have the same valency (namely 2). As shown by Theorem 2, taking $\delta=1$ is one option when (5) fails; and we then show that $\lambda=2$ is another in Theorems 3 and 4.

A necessary condition for $(m,3,c)_\lambda^\delta$-systems can be expressed succinctly in terms of the quantity Δ defined by

$$4\Delta = \lambda(m(m-2c+1) + 2\delta)$$

Theorem 1: If there is an $(m,3,c)_\lambda^\delta$-system then Δ is a non-negative integer.

In the opposite direction we know from [4,11,21] that this condition is sufficient at least when $\lambda=1$ and $\delta=0$ or 1. However, it is convenient and more usual to restate this explicitly in somewhat different terms. Let

$$M_{i,j}^\delta = \{m:m+2\delta \equiv i \text{ or } j \pmod 4\};\ M_{i,j} = M_{i,j}^0.$$

Theorem 2: For $\delta=0$ or 1, there is an $(m,3,c)^\delta$-system if and only if either m is in $M_{0,2c-1}^\delta$ and $m\geq 2c-1$ or $\delta=1$, $c=2$ and $m=1$ or 2.

Note that since m is in $M_{0,2c-1}^1$ exactly when m is not in $M_{0,2c-1}$, the $(m,3,c)^1$-systems provided by Theorem 2 are examples of minimal systems as described in Section 1. It seems likely that Theorem 2 extends to other values of δ (compare [21; p.104]).

Since Δ is always an integer when λ is even, the sufficiency of the condition in Theorem 1 when $\delta=0$ follows in general if it is sufficient for $\lambda=1+\epsilon$ and $m\in M_{0,2c-1}^\epsilon$, for $\epsilon=0$ or 1, by taking appropriate unions of systems for these values of λ. But when $\delta=0$ and $\lambda=2$, the condition in Theorem 1 is just

$$m \geq 2c-1; \tag{6}$$

and it is natural to conjecture that, in this case, (6) is indeed sufficient. We confirm this for all but finitely many exceptional values of m for each $c \geq 1$ as a consequence of Theorem 2 and some special constructions.

Theorem 3: There is an $(m,3,c)_2$-system at least when $m \geq 7(2c-1)$.

For small values of c this is readily improved as follows.

Theorem 4: For $1 \leq c \leq 5$, there is an $(m,3,c)_2$-system for $m \geq 2c-1$ except possibly for

$m = 6,7,11,14$, when $c = 3$;

$m = 9,10,13$, when $c = 4$;

$m = 10,11,14,19,23$, when $c = 5$.

We discuss Theorems 1 and 2 in Section 5 and establish Theorems 3 and 4 in Section 6.

5. Necessary conditions: proofs

Proof of Theorem 1:

Let the sets $A_r = \{0,a_1(r),a_2(r)\}$, $1 \leq r \leq \lambda m$, form an $(m,3,c)^\delta$-system. Then summing the elements of the difference sets $D(A_r)$, $1 \leq r \leq \lambda m$, gives

$$\sum_{r=1}^{\lambda m} (a_1(r)+a_2(r)+a_1(r)) = \lambda \left[\sum_{i=c}^{c+3m} i - 3m - c + \delta \right].$$

Hence

$$\sum_{r=1}^{\lambda m} a_2(r) = \tfrac{1}{2}\lambda \left(\tfrac{1}{2}(3m+2c)(3m+1) - 3m - c + \delta \right). \tag{7}$$

On the other hand, consider the smallest $2\lambda m$ integers which can appear:

$$\sum_{r=1}^{\lambda m} a_2(r) = \sum_{r=1}^{\lambda m} (a_1(r)+a_2(r)-a_1(r))$$

$$\geq \sum_{i=c}^{c+2m-1} i = \tfrac{1}{2}\lambda(2m+2c-1)2m. \tag{8}$$

So the right hand side of (7) and (8) differ by a non-negative integral amount; this difference is the quantity Δ defined earlier.

The alternative form of the necessary condition in Theorem 2 is easily derived from that in Theorem 1 with $\lambda=1$. The sufficiency of this condition is demonstrated by exhibiting partitions $\{a_r,b_r\}$, $1 \leq r \leq m$, of $\{i : 1 \leq i \leq 2m+1\} - \{2m+1-\delta\}$ such that

$$a_r - b_r = c+r-1, \quad 1 \leq r \leq m,$$

for then the sets $A_r = \{0, b_r+c+m-1, a_r+c+m-1\}$, $1 \leq r \leq m$, form an $(m,3,c)^\delta$-system. The necessary condition in Theorem 2 is also necessary for this type of partition too, although not all $(m,3,c)^\delta$-systems arise from partitions in this way. For these partitions, refer to [4,11,21].

It is convenient to mention at this point two technical matters. Again, we let the sets $A_r = \{0,a_1(r),a_2(r)\}$, $1 \le r \le \lambda m$, form an $(m,3,c)_\lambda^\delta$-system. Let x be a fixed positive integer; let p be an arbitrary positive integer; and let

$$f(a) = \begin{cases} a, & a<x; \\ a+p, & x \geq a. \end{cases}$$

Then we say that the $(m,3,c)_\lambda^\delta$-system consisting of the sets A_r, $1 \le r \le \lambda m$, has a *split* at x if every integer in the set

$$\{i : c \le i < x\} \cup \{j : x+p \le j \le c+3m+p\} - \{f(c+3m-\delta)\}$$

occurs exactly λ times among the difference sets $D(A_r^p)$, $1 \le r \le \lambda m$, where

$$A_r^p = \{0, f(a_1(r)), f(a_2(r))\}, 1 \le r \le \lambda m. \tag{9}$$

For example, the $(m,3,c)_\lambda^\delta$-system constructed from partitions as in the previous paragraph have a split at $c+m$; see further the constructions in Section 6.

When $\delta=0$, we are able to use *complete permutations* to generate larger systems from given ones. A permutation π of the set N_c of integers in modulus less than c is *complete* when

$$\{\pi(i)-i : i \in N_c\} = N_c;$$

such permutations exist for all $c \geq 1$. Given a complete permutation π of N_k, if the sets A_r, $1 \le r \le \lambda m$, as above, form an $(m,3,c)_\lambda$-system with a split at z then the sets

$$A_{r,s} = \{0, a_1(r)(2k-1)+s, a_2(r)(2k-1)+\pi(s)\}, \ |s|<c, \ 1 \le r \le \lambda m, \tag{10}$$

form an $(m(2k-1),3,2ck-c-k+1)_\lambda$-system with a split at $2zk-z-k+1$ (this is just an instance of the *multiplication theorem* for perfect systems).

6. Sufficient conditions: constructions

Proof of Theorem 3:

By Theorem 2, for m in $M_{0,2c-1}$ with $m \geq 2c-1$, there is an $(m,3,c)$-system and so by duplication a $(m,3,c)_2$-system.

Now consider the $(2,3,1)_2$-system consisting of the sets, say A_r, $1 \le r \le 4$, whose difference sets are shown in Figure 3(i) (see also the proof of Lemma 1). This system has a split at 3. Following the procedures described in the last two paragraphs of Section 5, form the sets $A_{r,s}^p$ using first (10) and then (9) with $x = 5k-2$; these sets have the property that every integer in the set

$$\{i : k \le i < 5k-2\} \cup \{j : 5k+p-2 \le j < 13k+p-6\}$$

occurs twice in their difference sets (in fact in two distinct difference sets). But, again by Theorem 2, for m' in $M_{0,10k-5}$ with $m' \geq 10k-5$ there is an $(m',3,5k-2)$-system and so by duplication an $(m',3,5k-2)_2$-system. Taking this together with the sets $A_{r,s}^p$ with $p=3m'$ gives an $(m,3,k)_2$-system where $m = m'+4k-2$ (this is just an instance of an addition theorem for perfect systems).

Since $5k-2$ is odd if and only if k is odd, $M_{0,10k-5} = M_{0,2k-1}$; and thus m is in $M^1_{0,2k-1}$ when m' is in $M_{0,2k-1}$. Hence there is an $(m,3,k)_2$-system for m in $M^1_{0,2k-1}$ with $m \geq 10k-5+4k-2 = 14k-7$.

So, at least when $m \geq 14c-7$, there is an $(m,3,c)_2$-system and the theorem is established. □

We now proceed to the proof of Theorem 4 through a series of partial results, the first of which gives rather more information than is stated in Theorem 4 in the case $c=1$.

Lemma 1: For $m \geq 1$, there is an $(m,3,1)_2$ system with no repeated difference sets and with a split at $m+1$.

Proof:

Consider the sets A_r and $A_{r'}$, $1 \leq r \leq m$, given by

$$A_r = \{0, r+m, 2r+m\} \tag{11a}$$

$$A_{r'} = \{0, 3m+1-2r, 3m+1-r\} \tag{11b}$$

Since $\{i : m < i \leq 3m\} = \{2r+m : 1 \leq r \leq m\} \cup \{3m+1-2r : 1 \leq s \leq m\}$, it is clear that these sets form an $(m,3,1)_2$-system with a split at $m+1$ for $m \geq 1$ (see Figure 3(i) for $m=2$, and Figure 4(i) for $m=4$.

6 8 10 12 12 11 10 9

5 1 6 2 7 3 8 4 11 1 9 2 7 3 5 4

(i) $m = m_2 = 4$; $c = 1$

6 8 10 12 12 11 10 9

5 1 5 3 6 4 8 4 11 1 9 2 7 3 7 2

(ii) $m = m_2 = 4$; $c = 1$

Figure 4: Some more $(\tilde{m};c)_2$-systems

However, not all these sets (or their associated difference sets) are distinct (see for example Figure 4(i)). Indeed $A_r = A'_s$ (or equally $D(A_r) = D(A'_s)$ if and only if $r = s = 2k+1$ where $m = 3k+1$. So the $(m,3,1)_2$-system given by (11) has distinct difference sets if and only if $m \not\equiv 1 \pmod 3$.

But, if m=$3k$+1 and r=$2k$+1, we find

$$A_{r-1} = \{0,5k+1,7k+1\};\ A'_{r+1} = \{0,5k,5k+2\};\ A_r = A'_r = \{0,5k+2,7k+3\}.$$

So (see Figure 5) replacing A_{r-1}, A_r, and A'_{r+1} by $\overset{*}{A}_{r-1}$, $\overset{*}{A}_r$, and $\overset{*}{A}_{r+1}$ where

$$\overset{*}{A}_{r-1} = \{0,5k,7k+1\};\ \overset{*}{A}_r = \{0,5k+1,7k+3\};\ \overset{*}{A}_{r+1} = \{0,5k+2,7k+2\},$$

gives a new $(m,3,1)_2$-system with a split at m+1 in which all of the difference sets are distinct. The example in Figure 4(ii) is derived from that in Figure 4(i) in this way.

Alternatively, this results obtains on replacing A'_{r-1}, A'_r, and A_{r+1} by A^+_{r-1}, A^+_r, and A^+_{r+1} where

$$A^+_{r-1} = \{0,5k+2,7k+4\};\ A^+_r = \{0,5k+3,7k+3\};\ A^+_{r+1} = \{0,5k+4,7k+5\}.$$

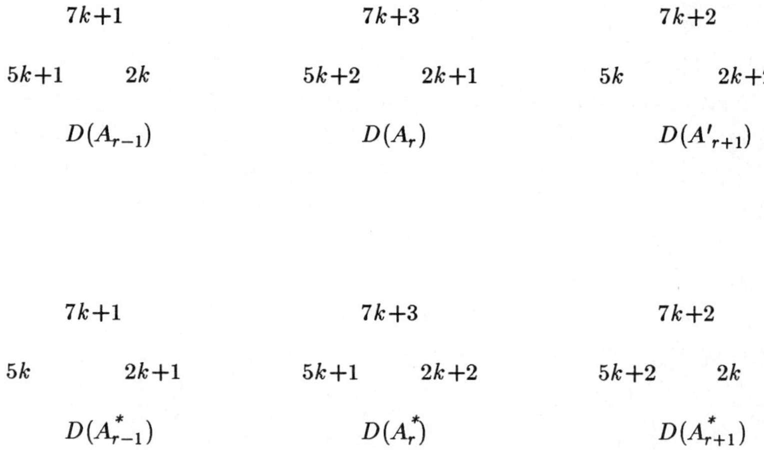

$7k+1$	$7k+3$	$7k+2$
$5k+1$ $2k$	$5k+2$ $2k+1$	$5k$ $2k+2$
$D(A_{r-1})$	$D(A_r)$	$D(A'_{r+1})$

$7k+1$	$7k+3$	$7k+2$
$5k$ $2k+1$	$5k+1$ $2k+2$	$5k+2$ $2k$
$D(\overset{*}{A}_{r-1})$	$D(\overset{*}{A}_r)$	$D(\overset{*}{A}_{r+1})$

Figure 5: Sets of exchangeable difference sets

Another source of $(m,3,1)_2$-systems with splits at m+1 is the construction based on partitions described in Section 5. For instance, we know that, when $m = 4k-1$ or $4k$, $k \geq 1$, there is a partition $\{a_r,b_r\}$ of $\{1,...,2m\}$ such that $a_r - b_r = r+1$, $1 \leq r \leq m$. The sets $\overline{A}_r$ and $\overline{A}'_r$, $1 \leq r \leq m+1$, defined by

$$\overline{A}_r = \{0,b_r+m+1,a_r+m+1\},\ 1 \leq r \leq m;\ \overline{A}_{m+1} = \{0,6m+2,6m+3\};$$

$$\overline{A}'_r = \{0,b_r+m+3,a_r+m+3\},\ 1 \leq r \leq m;\ \overline{A}'_{m+1} = \{0,2,3\},$$

form an $(m,3,1)_2$-system with a split at m+1 in which all of the difference sets are distinct. As a case in point with m=3, the partition

$$\{3,1\},\ \{6,2\},\ \{5,3\}$$

gives rise to the system illustrated in Figure 4(ii) (which in turn in comparison with that in Figure 4(i) then suggested the exchanges described in the previous paragraphs). However, in terms of the problem in measurement, these systems might be suspect because $\overline{A}'_r$ is a 'translate' of $\overline{A}_r$ for $1 \leq r \leq m$.

Our second lemma covers most of the cases $2 \leq c \leq 5$ allowed in Theorem 4 which cannot be obtained simply by duplicating $(m,3,c)$-systems known to exist by Theorem 2.

Lemma 2: For $2 \leq c \leq 5$, there is an $(m,3,c)_2$-system for m in $M^1_{0,2c-1}$ at least when $m \geq 10$, $c=2$; $m \geq 15$, $c=3$; $m \geq 17$, $c=4$; $m \geq 24$, $c=5$.

Proof:

We obtain an $(m,3,c)_2$-system as asserted by taking $(m_i,3,c_i)^{\delta_i}$-systems for $i=1,2$ for the values given in Table 1 (which exist by Theorem 2 under the restrictions mentioned in Table 1) together with further sets as specified in the Appendix for the given value of r.

Table 1

m	c	m_1	c_1	δ_1	m_2	c_2	δ_2	r	Restrictions
$4k+1$	2	$4k-2$	4	1	$4k$	4	0	$12k-6$	$k \geq 3$
$4k+2$	2	$4k$	3	0	$4k+1$	4	1	$12k-3$	$k \geq 2$
m	3	$m-3$	4	0	$m-2$	7	0	$3m-5$	$15 \leq m \equiv 2+\delta \pmod 4$
m	4	$m-4$	7	δ	$m-1$	6	δ	$3m-5-\delta$	$17 \leq m \equiv 1+\delta \pmod 4$
m	5	$m-6$	10	δ	$m-1$	7	δ	$3m-8$	$26 \leq m \equiv 2+\delta \pmod 4$

N.B. $\delta = 0$ or 1.

The multiplicative construction mentioned at the end of Section 5 enables us to give $(m,3,c)_2$-systems when m and c have the following values:

m	6	9	10	14	18
c	2	2	3	4	5

The remaining cases needed to establish Theorem 4 are $(5,3,2)_2$-, $(15,3,5)_2$-, and $(22,3,5)$-systems (note that given the first of these, the second may be obtained on multiplication using a complete permutation of N_2). All three cases may be obtained from our next and final lemma which is of some independent interest as it illustrates a further application of complete permutations.

Lemma 3: If there is an $(m-2k,3,c)$-system with a split at $c+m-2k$ and an $(m-k,3,c)$-system, where

$$2m = 7k + 2c - 1 \tag{12}$$

then there is an $(m,3,c)$-system.

Proof:

Let A_r, $1 \leq r \leq m-2k$, be the sets of an $(m-2k,3,c)$-system with a split at $c+m-2k$ and form the sets A_r^p as in (9) with $x = c+m-2k$ and $p=6k$. Further let $\overline{A}_r$, $1 \leq r \leq m-k$, be the sets of an $(m-k,3,c)$-system.

Now for $d = \frac{1}{2}(3k+1)$ consider the sets A'_s, $|s|<d$, given by

$$A'_s = \{0, c+m+k+d+s-1, c+3m-3k+d+\pi(s)-1\},$$

where π is a complete permutation of N_d. Since π is complete

$$\bigcup_{|s|<d} D(A'_s) = \{c+m+k+d+s-1 : |s|<d\} \cup$$
$$\{c+3m-3k+d+s-1 : |s|<d\} \cup$$
$$\{2m-4k+s : |s|<d\}.$$

But (12) implies that

$$2m-4k = c+m-2k+d-1$$

and so

$$\bigcup_{|s|<d} D(A'_s) = \{c+m-2k+i : 0 \leq i < 6k\} \cup \{c+3m-3k+j : 0 \leq j < 3k\}$$

showing that taking all of the sets A_r, $\overline{A}_s$ and A'_t together gives an $(m,3,c)_2$-system as claimed to exist in the lemma.

Lemma 3 may be applied in the following cases to obtain the systems mentioned just prior to the lemma.

m	5	15	22
c	2	5	5
k	1	3	5

Appendix

We display here the difference sets used to supplement those of the $(m_i,3,c_i)^{\delta_i}$)-systems, $i = 1,2$, in order to prove Lemma 2 (see Table 1).

(i) $m = 4k+1$, $k \geq 3$; $c = 2$; $r = 12k-6$:

$r+2$ $\quad$ $r+7$ $\quad$ $r+6$ $\quad$ $r+7$

r $\;$ 2 $\quad$ $r+5$ $\;$ 2 $\quad$ $r+3$ $\;$ 3 $\quad$ $r+4$ $\;$ 3

(ii) $m = 4k+2$, $k \geq 2$; $c = 2$; $r = 12k-3$:

$r+3$ $\quad$ $r+4$ $\quad$ $r+3$

$r+1$ $\;$ 2 $\quad$ $r+2$ $\;$ 2 $\quad$ r $\;$ 3

(iii) $m \equiv 2,3 \pmod 4$, $m \geq 15$; $c = 3$; $r = 3m-5$:

$$\begin{array}{ccccc} r+6 & r+7 & r+5 & r+7 & r+6 \\ r+3 \quad 3 & r+4 \quad 3 & r+1 \quad 4 & r+2 \quad 5 & r \quad 6 \end{array}$$

(iv) $m \equiv 1+\delta \pmod 4$, $\delta = 0,1$, $m \geq 17$; $c = 4$; $r = 3m-5-\delta$:

$$\begin{array}{ccccc} r+4\delta+5 & r+8 & r+7 & r+8 & r+6 \\ r+4\delta+1 \quad 4 & r+4 \quad 4 & r+2 \quad 5 & r+3 \quad 5 & r \quad 6 \end{array}$$

(v) $m \equiv 2+\delta \pmod 4$, $\delta = 0,1$, $m \geq 26$; $c = 5$; $r = 3m-8$:

$$\begin{array}{ccccccc} r+6 & r+12 & r+10 & r+11 & r+9 & r+8+\delta & r+12-\delta \\ r+1 \quad 5 & r+7 \quad 5 & r+4 \quad 6 & r+5 \quad 6 & r+2 \quad 7 & r \quad 8+\delta & r+3 \quad 9-\delta \end{array}$$

Acknowledgements

I thank J.V. Abrham, J.-C. Bermond, F. Biraud, D.J. Crampin, A. Kotzig, P.J. Laufer, W.H. Mills, F.W. Roush, J.M. Turgeon, and (indirectly) P.R. Wild for supplying much information on perfect systems of difference sets at various times over the past several years. (I was enabled to see the CSIRO research reports of P.R. Wild by R.N. Manchester and T.P. Speed). In particular, especially in connection with this paper, I have much appreciated correspondence with F.W. Roush on the notion of multiply perfect systems.

I have also been grateful over the years to A. Rosa for his characteristically unfailing, unstinting, scholarly, and courteous attention whenever pressed with a query.

I am indebted to various institutions for hospitality during the germination and preparation of this paper, but chiefly to the University of Hawaii and the University of East Anglia.

References

[1] J. Abrham, "Perfect systems of difference sets: a survey", *Ars Combinatoria* 17A (1984) 5-36.

[2] J. Abrham, A. Kotzig, and P.J. Laufer, "Remarks on the minimum number of components in a perfect system of difference sets", *Congressus Numerantium* 52 (1986) 7-19.

[3] J.-C. Bermond, "Graceful graphs, radio antennae, and French windmills", in: *Graph Theory and Combinatorics, Research Notes in Mathematics* 34, Pitman, 1979, pp. 18-37.

[4] J.-C. Bermond, A.E. Brouwer, and A. Germa, "Systèmes de triplets et différences associées", in: *Problèmes combinatoires et théorie des graphes, Proc. Colloque CNRS* (CNRS, Orsay, 1976) pp. 35-48.

[5] J.-C. Bermond and G. Fahri, "Sur un problème combinatoire d'antennes en radioastronomie II", *Ann. Disc. Math.* 12 (1982) 49-53.

[6] J.-C. Bermond, A. Kotzig, and J. Turgeon, "On an combinatorial problem in antennas in radioastronomy", in: *Proc. Eighteenth Hungarian Combinatorial Colloquium, Keszthely, 1976* (North-Holland, Amsterdam, 1976) pp. 135-149.

[7] F. Biraud, E.J. Blum, and J.C. Ribes, "On optimum synthetic linear arrays", *IEEE Trans. Antennas Propagation* AP22 (1974) 108-109.

[8] F. Biraud, E.J. Blum, and J.C. Ribes, "Some new possibilities of optimum synthetic linear arrays for radio astronomy", *Astronomy and Astrophysics* 41 (1975) 409-413.

[9] G.S. Bloom and S.W. Golomb, "Applications of numbered undirected graphs", *Proc. IEEE* 65 (1977) 562-570.

[10] G.S. Bloom and S.W. Golomb, "Numbered complete graphs, unusual rulers, and assorted applications", in: *Theory and applications of graphs* (Proceedings, International Conference, Western Michigan University, Kalamazoo, Michigan, 1976), Lecture Notes in Mathematics 642 (Springer, Berlin, 1978) pp. 53-65.

[11] R.O. Davies, "On Langford's problem II", *Math. Gaz.* 43 (1959) 253-255.

[12] A.E. Eckler, "The construction of missile guidance codes resistant to random interference", *Bell Syst. Tech. J.* 39 (1960) 973-994.

[13] D.F. Hsu and A.D. Keedwell, "Generalized complete mappings, neofields, sequenceable groups, and block designs", *Pacific J. Math.* 111 (1984) 317-322; 117 (1985) 291-312.

[14] A. Kotzig and J. Turgeon, "Perfect systems of difference sets and additive sequences of permutations", *Congressus Numerantium* 24 (1979) 629-636.

[15] J.P. Robinson and A.J. Bernstein, "A class of binary recurrent codes with limited error propagation", *IEEE Trans. Information Theory* IT13 (1967) 106-113 (especially page 107).

[16] J.P. Robinson and A.J. Bernstein, "Error propagation and definite decoding of convolutional codes", *IEEE Trans. Information Theory* IT14 (1968) 121-128 (especially page 126).

[17] D.G. Rogers, "A multiplication theorem for perfect systems of difference sets", *Discrete Math.*, to appear.

[18] D.G. Rogers, "Addition theorems for perfect systems of difference sets", *J. London Math. Soc. (2)* 23 (1981) 385-395.

[19] D.G. Rogers, "Improved bounds for perfect systems of difference sets", to appear.

[20] A. Rosa, "On certain valuations of the vertices of a graph", in: *Theory of Graphs* (Gordon and Black, New York, 1967) pp. 349-355.

[21] J.E. Simpson, "Langford sequences: perfect and hooked", *Discrete Math.* 44 (1983) 97-104.

[22] P.R. Wild, "Combining perfect systems of difference sets", *Bull. London Math. Soc.* 18 (1986) 127-131.

Annals of Discrete Mathematics 34 (1987) 409–418
© Elsevier Science Publishers B.V. (North-Holland)

Some remarks on focal graphs

Gert Sabidussi

Département de Mathématiques et de Statistique
Université de Montréal
C.P. 6128, Succursale "A",
Montréal, Québec, H3C 3J7
CANADA

TO ALEX ROSA ON HIS FIFTIETH BIRTHDAY

ABSTRACT

A graph G is *focal* if for every edge e of G there is exactly one vertex not incident with e which is a fixed point of the stabiliser of e. Some results about the structure of such graphs are given under the additional hypothesis that G contains a cutpoint or a fixed point.

1. Introduction

In 1979, A. Rosa introduced a class of graphs which he called Δ-graphs - we shall use the term focal graphs - and posed the problem of characterising them [2].

(1.1) Definition: A graph G is *focal* if it has no isolated vertex, and for every edge $e = [x,y]$ of G the stabiliser of e fixes exactly one vertex $z \neq x,y$. That is, for any edge $e \in E(G)$ there is exactly one vertex z not incident with e such that every automorphism of G which fixes e also fixes z. The vertex z is called the *focus* of e.

One motivation for introducing these graphs is that they give rise to a class of commutative partial groupoids. On the vertex set of a focal graph a product can be defined for all pairs of adjacent vertices by taking xy to be the focus of the edge $[x,y]$. This notation will be convenient in the sequel. However, beyond using multiplication as a notational device we shall not be concerned with the algebraic properties of these partial groupoids.

Rosa gives two families of examples: odd cycles and the complete bipartite graphs $K_{2,n}$, $n \neq 2$. At first glance the definition of focal graphs appears rather restrictive, and the chances for a characterisation seem good. Nevertheless, focal graphs are very numerous (see Constructions (2.1) and (2.2) below). In fact, the class is so large that it is difficult to pin down any essential structural property of its members. Peter Cameron conjectures (private communication) that focality has the same complexity status as graph isomorphism. Simple standard arguments (attaching pendant vertices) show that deciding whether a graph is focal is isomorphism-easy.

In section 2 we give some examples and constructions (essentially graph products of various kinds) which provide some idea of the extent of the class of focal graphs. The remaining sections deal with focal graphs having cutpoints and/or fixed points (e.g., focal trees). It is not very astonishing that under these additional assumptions some structural results can be obtained.

All graphs will be finite, undirected, and without multiple edges or loops. For a vertex x of G, $V_x := \{y \in V(G) : [x,y] \in E(G)\}$ is the *neighbourhood* of x in G. *Aut G* is the *automorphism group* of G. If $H \subset G$, the *stabiliser* of H in G will be denoted by S_H or $S(H;G)$, i.e., $S_H := \{\sigma \in Aut\ G : \sigma H = H\}$. If H consists of a single edge e we write S_e or $S(e;G)$, and similarly for single vertices. A *fixed point* of G is a vertex which is fixed by all automorphisms of G. An edge $e = [x,y]$ is *involutorial* if there is an automorphism of G which interchanges x and y.

Several other properties of graphs are closely related to focality and will be needed in the sequel.

(1.2) Definition: A graph is *point-focal* if for any $x \in V(G)$ there is a unique $x' \in V(G)$ such that $x \neq x'$ and S_x fixes x', i.e., $S_x \subset S_{x'}$. G is *point-displacing* (or *(0,0)-displacing)* if for any two distinct vertices x,y of G the stabiliser of x displaces y, i.e., there is a $\sigma \in S_x$ such that $\sigma y \neq y$; G is *(1,0)-displacing* if for any $e \in E(G)$ and any vertex x not incident with e the stabiliser of e displaces x. (Isolated vertices are allowed in these definitions.) G is *sharply focal* if it is focal and point-displacing.

(1.3) Remarks.

(1) Even cycles are point-focal and (1,0)-displacing; odd cycles are sharply focal. If G is point-focal, then $S_x \subset S_{x'} \subset S_{(x')'}$ for any $x \in V(G)$, hence $x \rightarrow x'$ is a permutation of order 2. In general it is not an automorphism of G.

(2) *No graph is both focal and point-focal.* Suppose G is such a graph and let $e = [x,y] \in E(G)$. Then $S_e \subset S_{xy} \subset S_{(xy)'}$ and $(xy)' \neq xy$. Hence $(xy)' = x$ or y so that $S_e \subset S_x$, say. Thus e is non-involutorial and therefore $S_e \subset S_x \subset S_{x'}$ as well as $S_e \subset S_y \subset S_{y'}$. This means that $x' = xy = y'$, and hence $x = y$, a contradiction. It follows that G has no edges contrary to the definition of focality.

(3) Let σ be an automorphism of a focal graph G. Then $\sigma(xy) = (\sigma x)(\sigma y)$ for any $[x,y] \in E(G)$.

2. Examples

We begin with two constructions which show that sharply focal graphs as well as bipartite focals form large classes, and then consider the cartesian product of focal graphs. The question whether the product of two focal graphs is focal leads naturally to the class of sharply focal graphs. Throughout this section we consider only *connected* graphs.

(2.1) Construction: Given any connected graph H and a connected focal graph F consider the composition or lexicographic product $L = H[F]$ (i.e., $V(L) = V(H) \times V(F)$, $[(x,u),(y,v)] \in E(L) \Leftrightarrow (x = y$ and $[u,v] \in E(F))$ or $[x,y] \in E(H)$). Let A be the set of all edges of L of the form $e = [(x,u),(y,v)]$, where $x \neq y$. Form G by subdividing every $e \in A$ by a new vertex w_e. Then G is focal with

$(x,u)(x,v) = (x,uv)$, where uv is the product in F, and $(x,u)w_e = (y,v)$, where $[(x,u),(y,v)] = e \in A$.

The second construction also is of lexicographic type.

(2.2) Construction: Let H be any connected bipartite graph with colour classes A,B, and assume that no two distinct vertices in A have the same neighbourhood. Form a new graph G by doubling the vertices in A and at least tripling every $x \in B$. That is, take any family of sets $(J_x)_{x \in B}$ with $|J_x| \geq 3$ for any $x \in B$, and let

$$V(G) = (A \times \{0,1\}) \cup \bigcup_{x \in B} (\{x\} \times J_x),$$

$$[(x,i),(y,j)] \in E(G) \Leftrightarrow x \in A,\ y \in B,\ [x,y] \in E(H).$$

Clearly G is bipartite. From some minor modification of results about the automorphism group of the composition of graphs [3] it follows that G is focal with multiplication given by $(x,i)(y,j) = (x,1-i)$.

The remainder of this section deals with cartesian products. Recall that with respect to cartesian multiplication any connected graph has a unique decomposition into indecomposable factors (see, for example, [1],[4],[5]). As the general situation is rather cumbersome, we shall limit ourselves to the case of indecomposable graphs.

(2.3) Proposition: *Let H_1,H_2 be connected indecomposable graphs. Then $G = H_1 \times H_2$ is focal if and only if H_1 and H_2 are sharply focal, and in that case G is likewise sharply focal.*

The proof hinges on the following general fact about stabilisers of edges in a cartesian product.

(2.4) Lemma: *Let $G = H_1 \times H_2$, where H_1,H_2 are connected indecomposable graphs. Then for any edge $e = [(u,v),(u',v)] \in E(G)$,*

$$S(e;G) = S(e_1;H_1) \times S(v;H_2),$$

where $e_1 = [u,u'] \in E(H_1)$.

Proof: Case 1: H_1 and H_2 are non-isomorphic (actually it suffices that H_1 and H_2 have no common indecomposable factor). Then by [4], Corollary (3.2), any $\alpha \in Aut\ G$ has the form α: $(x,y) \to (\sigma x,\tau y)$, where $\sigma \in Aut\ H_1$, $\tau \in Aut H_2$, and this immediately establishes the claim.

Case 2: $H_1 = H_2$. By [4], Theorem (3.1), any $\alpha \in Aut\ G$ either has the same form as in case 1 or α: $(x,y) \to (\sigma y,\tau x)$, where $\sigma,\tau \in Aut\ H_1$. If $\alpha \in S(e;G)$ is of this type, then

$$\alpha e = [(\sigma v,\tau u),(\sigma v,\tau u')] = [(u,v),(u',v)]$$

so that $u = \sigma v = u'$, a contradiction. Hence we are back in case 1. □

Lemma (2.4) is false without the hypothesis that H_1 be indecomposable.

Proof of (2.3). Necessity: Assume G to be focal. Observe first that H_1 and H_2 are point-displacing. For if H_2, say, is not point-displacing, then there exists a $v \in V(H_2)$ such that $S(v;H_2)$ fixes some $v' \neq v$. Take any $e_1 = [u,u'] \in E(H_1)$ and consider the edge $e = [(u,v),(u',v)]$ of G. If $S_1 := S(e_1;H_1)$ fixes some $u_1 \in V(H_1)$, then by (2.4),

$S(e;G)$ fixes (u_1,v) and (u_1,v'). By focality of G this means that $u_1 = u$ or u'. But then S_1 fixes both u and u', and hence $S(e;G)$ fixes (u,v') and (u',v'), neither of which is incident with e, a contradiction. If S_1 displaces every $x \in V(H_1)$, then $S \times \{1\} \subset S(e;G)$ displaces every $(x,y) \in V(G)$.

To show that H_2 is focal the argument is similar. Take $e_1 = [u,u'] \in E(H_1)$, $v \in V(H_2)$ and consider $e = [(u,v),(u',v)] \in E(G)$. $S_1 := S(e_1;H_1)$ fixes at most one vertex not incident with e_1, otherwise by (2.4), $S(e;G)$ would fix two vertices not incident with e. On the other hand, if S_1 displaces all vertices of H_1 not incident with e_1, then $S \times \{1\} \subset S(e;G)$ displaces all vertices (x,y) of G with $x \neq u,u'$. Since we already know that H_2 is point-displacing, it follows that $\{1\} \times S(v;H_2) \subset S(e;G)$ displaces all vertices of the form (u,y) or (u',y) with $y \neq v$. Therefore $S(e;G)$ displaces all vertices not incident with e, a contradiction.

The proof of the sufficiency is quite similar and is left to the reader. □

The preceding argument depends essentially on the fact that the automorphisms of a cartesian product of indecomposable factors preserves the fibres corresponding to these factors. For products not having this property, e.g., the categorical product, we have not been able to establish similar results apart from isolated examples. Thus, for instance, the categorical product of an odd cycle by itself is focal.

3. Two fixed points

Beginning with this section we shall consider focal graphs having a cutpoint or a fixed point. It will be seen that there is a close relationship between the two types of vertices. Here we determine those focal graphs which have the maximum number of fixed points. This will make it possible to give a description of disconnected focal graphs.

(3.1) Proposition: *Any focal graph G has at most two fixed points. If there are exactly two, say w and z, then either G is the graph of Figure 1, where without loss of generality $|U_z| > |U_w| \geq 2$; or G has the form of Figure 2, where $|U| \geq 3$ and without loss of generality $|U_z| > |U_w| \geq 3$.*

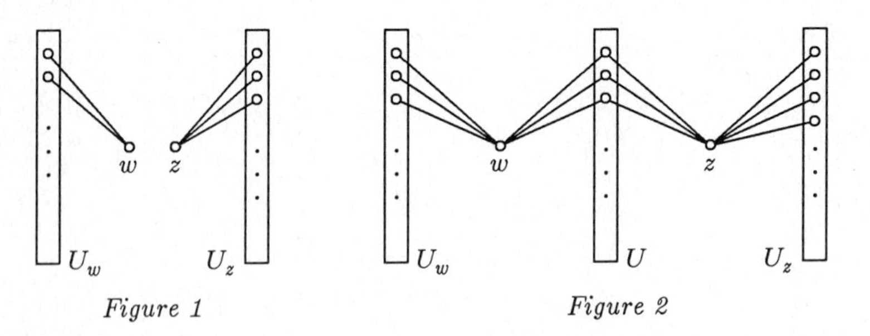

Figure 1 Figure 2

Proof: Let w,z be two distinct fixed points of a focal graph G. Then clearly any edge e must be incident with w or z since S_e fixes both w and z. Now suppose that G has three distinct fixed points u,w,z. Since G has no isolated vertex and any edge incident with one of the three fixed points must also be incident with one of the other two, it follows that $V(G) = \{u,w,z\}$. But no graph of order 3 has three fixed points.

Again, let w,z be two fixed points. Then $[w,z] \notin E(G)$, otherwise $S_{[w,z]} = Aut\ G$ would fix $wz \neq w,z$, so that G would have three fixed points. It follows that every edge of G is incident with exactly one of w,z. In other words, if we put $U = V_w \cap V_z$, $U_w = V_w \backslash U$, $U_z = V_z \backslash U$, then $U_w \cup U \cup U_z = V(G) \backslash \{w,z\}$ is an independent set. Thus G has the form given in Figure 1 or Figure 2 according as U is empty or not. □

Note that in the case where G is connected the two fixed points are also cutpoints of G.

Now consider a disconnected focal graph G. As can be seen from Figure 1 the components of G need not be focal. Suppose G has a non-focal component H_0. By definition of focality, H_0 has at least one edge, say $[x_0,y_0]$. Let H_1 ($\neq H_0$) be the component of G containing $z := x_0y_0$. Since any automorphism of H_1 can be extended to an automorphism of G it follows that z is a fixed point of H_1 (and hence of G), and that z is the focus of every edge of H_0. Moreover, the stabiliser of any edge e of H_0 cannot fix any vertex of H_0 except possibly the two endpoints of e, i.e., H_0 is (1,0)-displacing. We sum this up in:

(3.2) Proposition: *Any disconnected focal graph G is of one of the following three types:*

(1) *All components of G are focal.*

(2) *There are both focal and non-focal components. In this case, G has a unique fixed point z, and z lies in one of the focal components; every non-focal component is (1,0)-displacing.*

(3) *No component is focal. In this case G has two fixed points and is the graph of Figure 1.*

We conclude this section with a characterisation of focal trees.

(3.3) Proposition: *Any focal tree has one of the forms given in Figure 3.*

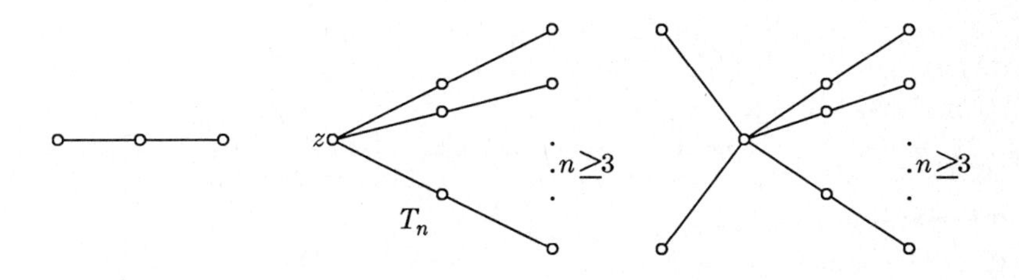

Figure 3

Proof: Let T be a focal tree. If T has a central edge $e_0 = [z_1,z_2]$, then e_0 is involutorial, otherwise z_1,z_2 are two adjacent fixed points, contrary to (3.1). However, if e_0 is involutorial, then S_{e_0} displaces all vertices of T, again a contradiction. It follows that

T has a central vertex z and that all edges of T are non-involutorial. Hence given any edge e, S_e fixes all vertices between e and z, which means that every vertex is at distance at most 2 from z. If $a \in V_z$, then any permutation of $V_a \setminus \{z\}$ belongs to $S_{[a,z]}$. Hence a is of degree at most 2. If z is incident with a pendant edge e_1, then it is incident with exactly two such edges since the stabiliser of e_1 permutes all other pendant edges incident with z. Hence T has one of the forms given in Figure 3 depending on whether there are no vertices at distance 2 from z, no pendant vertices adjacent to z, or both types of vertices exist. In the last two cases the number of vertices at distance 2 from z is ≥ 3, otherwise any S_e fixes more than one vertex not on e. □

4. The block structure of focal graphs

In this section we consider the block-cutpoint tree T_G of a connected focal graph G. Recall that the block-cutpoint tree of any connected graph is of even diameter and has a central vertex. An *extremal block* of G is one which is an endpoint of a longest path in T_G. An *extremal cutpoint* is the unique cutpoint of G belonging to an extremal block. A block is *trivial* if it consists of a single edge.

(4.1) Lemma: *If G has a non-trivial extremal block, then diam T_G = 2 or 4.*

Proof: Let $B_0 a_1 B_1 \cdots a_n B_n$ be a longest path in T_G, where each B_i is a block, and each a_i is a cutpoint of G, and suppose that $n \geq 2$. By hypothesis we may also assume that B_0 is not a single edge. Hence there is an edge $e \in E(B_0)$ which is not incident with a_1. Then S_e fixes B_0 and hence also a_1. This implies that $S_{a_1} \subset S_{B_1}$. For if there is a $\sigma \in S_{a_1}$ with $\sigma B_1 \neq B_1$, then

$$B_n a_n \cdots a_2 B_1 a_1 (\sigma B_1)(\sigma a_2) \cdots (\sigma a_n)(\sigma B_n)$$

is a path in T_G of length $2(2n-1)$, whence $n \leq 1$, a contradiction. On the other hand, S_{a_1} displaces all vertices of G not in B_0. In particular, there is a $\tau \in S_{a_1}$ which displaces a_2. Hence

$$B_n a_n \cdots B_2 a_2 B_1 (\tau a_2)(\tau B_2) \cdots (\tau a_n)(\tau B_n)$$

is a path of length $2(2n-2)$ and therefore $n \leq 2$ as claimed. □

If *diam* $T_G = 2$, i.e., if T_G is a star, then the center of T_G is a cutpoint of G, say z. Clearly z is a fixed point of G. If *diam* $T_G = 4$, then G has a central block B and all other blocks meet B. Moreover:

(4.2) Lemma: *If G has a non-trivial extremal block, and if diam $T_G = 4$, then the central block B is likewise non-trivial; and if G has a fixed point z, then z is a cutpoint belonging to a non-trivial extremal block.*

Proof: If B is an edge, say $[a,a']$, then the stabiliser of any edge not incident with a or a' fixes both a and a', a contradiction.

Now assume that G has a fixed point z. Let $B_1, \cdots, B_n$ be the extremal blocks of G, a_i the unique cutpoint in B_i, $i = 1, \cdots, n$. Suppose that z is not a cutpoint. If z belongs to some B_i, take any edge $e \in E(G) \setminus E(B_i)$ not incident with a_i (such an edge exists because *diam* $T_G = 4$). Since z is a fixed point, S_e must fix both z and a_i which is impossible. If z does not belong to any B_i, choose e in some non-trivial block B_j such that e is not incident with a_j. Then S_e fixes the two distinct vertices a_j and

z, the same contradiction as before.

It follows that $z = a_i$ for some i. At least one of the extremal blocks containing z must be non-trivial, for if not, take any e in a non-trivial block B_k, e not incident with a_k. Then S_e fixes both a_k and z, and by assumption the two vertices are distinct. □

An example illustrating the situation described in Lemma (4.2) is the graph of Figure 4.

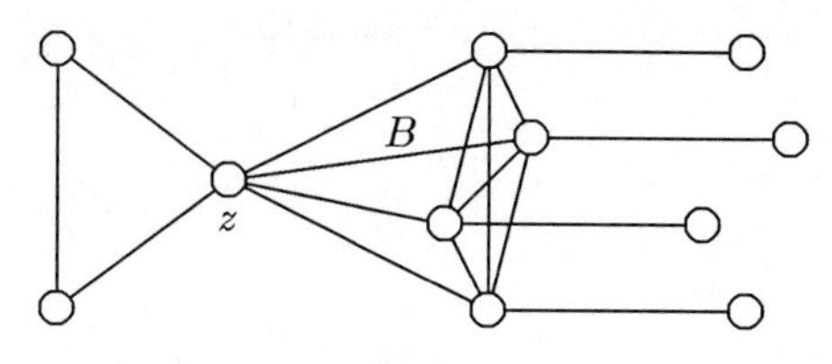

Figure 4

(4.3) Proposition: *diam* $T_G \leq 8$ *for any connected focal graph* G.

Proof: Let $B_0a_1B_1 \cdots a_nB_n$ be a longest path in T_G. By Lemma (4.1) we may assume that every extremal block of G is trivial. Let A be the set of pendant edges of G incident with a_1 (i.e., the extremal blocks containing a_1).

Case 1: $|A| = 2$, say $A = \{e,e'\}$, where $e = [a_1,u]$, $e' = [a_1,u']$. Then $ua_1 = u'$ and S_e displaces all vertices other than a_1,u,u'; in particular, S_e displaces a_i, $i \geq 2$. Hence by the argument used in the proof of (4.1), T_G contains a path of length $2(2n-2)$, i.e., $n \leq 2$.

Case 2: $|A| \neq 2$. Take any $e = [a_1,u] \in A$. Then ua_1 is not incident with any edge in A. If $ua_1 \in V(B_1)\backslash\{a_2\}$, then S_e fixes B_1 but displaces a_2. Hence T_G contains a path of length $2(2n-2)$, so that $n \leq 2$ as before. If $ua_1 \in V(B_i)$ with $i \geq 2$, then S_e fixes a_2, and hence $ua_1 = a_2$. On the other hand, if $n \geq 3$, then S_e displaces a_3. Hence T_G contains a path of length $2(2n-4)$ or $2(2n-3)$, depending on whether S_e does or does not stabilise B_2. In either case $n \leq 4$. □

Note that the case $|A| = 1$ can occur only if B_1 is trivial. For if $A = \{[a_1,u]\}$, then $S_{a_1} = S_u$ and hence the stabiliser of any edge of B_1 incident with a_1 would fix both u and a_2, and this is possible only if B_1 consists of the single edge $[a_1,a_2]$. We therefore obtain: *if diam* $T_G = 6$ *or 8, then the extremal blocks of* G *are single edges, and any extremal cutpoint* a *is either incident with at least three pendant edges or with exactly one. In the latter case the unique non-extremal block containing* a *is likewise trivial. If diam* $T_G = 8$*, then the central block of* G *is non-trivial.*

The cases that *diam* $T_G = 6$ or 8 can actually occur. An example for diameter 6 is the focal tree T_n of Figure 3. For diameter 8, take any connected, sharply focal graph H and at each $x \in V(H)$ attach a copy of T_n using z as the vertex of attachment. This yields a focal graph G with *diam* $T_G \geq 8$. It follows that *diam* $T_G = 8$ and hence that H consists of a single block. We state this separately as:

(4.4) Corollary: *Any connected, sharply focal graph is 2-connected.*

5. Structure near a fixed point

In this section we consider connected focal graphs *having exactly one fixed point* z. In such a graph all edges not incident with z have the same focus, viz. z. One cannot hope, therefore, to obtain much information about G except in the vicinity of z. We shall derive some simple structural details about the part of G lying within distance 2 of z, but except in the bipartite case they are far from adding up to a characterisation of even this small portion of G. The bipartite case, on the other hand, is close to trivial. We shall need some terminology and notation.

For $i \in \mathcal{N}$ put

$$A_i := \{x \in V(G) : dist(x,z) = i\}, \ B_i := \{x \in A_i : V_x \cap A_{i+1} \neq \emptyset\}.$$

Given $x \in A_i$, $i \geq 1$, define the set of *upper* [*lower*] *neighbours* of x to be $U_x := V_x \cap A_{i+1}$ [$L_x := V_x \cap A_{i-1}$]. If $U_x \neq \emptyset$, x is *terminal.*

G has the following simple properties:

(1) $A_2 = \{xz : x \in B_1\}$. *More precisely, if x is a neighbour of z, then U_x is empty or consists of the single vertex xz.* Let $y \in U_x$ and consider $e = [x,y]$. Then $S_e \subset S_x = S_{[x,z]} \subset S_{xz}$, i.e., S_e fixes z and $xz \neq x,z$. The only way this can happen is if $xz = y$.

(2) *Let $x,y \in V_z$. If x is terminal and y is not, then x and y are nonadjacent.* Otherwise $[x,y]$ is a non-involutorial edge and $S_{[x,y]} = S_x \cap S_y$ would fix both z and yz.

It is an immediate consequence of (2) that *z is a cutpoint of G if and only if it has terminal neighbours* (i.e., if $A_1 \backslash B_1 \neq \emptyset$).

Let G_0, G_1 be the subgraphs of G induced by $\{z\} \cup B_1 \cup \bigcup \{A_i : i \geq 2\}$ and $\{z\} \cup (A_1 \backslash B_1)$, respectively. Clearly $G_0 \cup G_1 = G$ and $G_0 \cap G_1 = z$. Both G_0 and G_1 may consist of several blocks.

We shall consider G_1 first, assuming that $A_1 \backslash B_1 \neq \emptyset$. Note that if $e = [x,y] \in E(G_1)$, then $xy \in V(G_1)$. Otherwise $xy \in V(G_0) \backslash \{z\}$ and since the action of S_e on G_0 is the same as that of the full automorphism group of G, it would follow that xy is a fixed point of G, a contradiction.

Put $D := G_1 - z$. Since $V(D)$ is the neighbourhood of z in G_1, the structure of G_1 is given by that of D. The vertices of D are invariantly defined, hence $\sigma|_D \in Aut\ D$ for every $\sigma \in Aut\ G$. Conversely, every automorphism of D can be extended to an automorphism of G by taking the identity on $V(G) \backslash V(D)$. This says in particular that $S(x;D) = \{\sigma|_D : \sigma \in S_{[x,z]}\}$ for any $x \in V(D)$. Since $xz \in V(G_1) \backslash \{z\} = V(D)$ and $xz \neq x$, it follows that $S(x;D)$ fixes exactly one vertex of D different from x, i.e., D is point-focal. Similarly, one shows that D is (1,0)-displacing, using that $xy = z$ for every $[x,y] \in E(D)$. We sum this up as:

(5.1) Lemma: *$G_1 - z$ is (1,0)-displacing and point-focal, the focal map being $x \rightarrow xz$. Conversely, the Zykov-sum of any point-focal, (1,0)-displacing graph and a one-point graph can play the role of G_1.*

We now turn to the consideration of G_0, continuing the enumeration of simple facts begun above.

(3) $|L_x| = 1$ *or* ≥ 3 *for any* $x \in A_i$, $i \geq 2$; $|U_x| \geq 3$ *for any* $x \in B_i$, $i \geq 2$. Let $x \in B_i$ and suppose x has a unique upper neighbour u. Then for any $w \in L_x$ the stabiliser of $[w,x]$ fixes both u and z. Hence $|U_x| \geq 2$. If $U_x = \{u,v\}$, then $S_{[x,u]}$ would fix v and z; similarly for L_x.

(4) *If* $x \in A_i$, $i \geq 2$, *and* $|L_x| = 1$, *then* x *is terminal.* For if there is a $y \in U_x$, then $S_{[x,y]}$ fixes z as well as the unique lower neighbour of x.

(5) Partition B_1 into equivalence classes by setting $x \sim y \Leftrightarrow xz = yz$, where $x,y \in B_1$. By (1) the equivalence class containing x is $L_{xz} = V_{xz} \cap V_z = V_{xz} \cap B_1$. Hence: *any two distinct vertices in* A_2 *have disjoint lower neighbourhoods.*

(6) *If* $x \in B_1$, *then* S_x *does not stabilise any* L_u, $u \in A_2$, $u \neq xz$. If it did, then $S_{[x,z]} = S_x$ would fix both u and xz.

Properties (3)-(6) refer to edges of G whose endpoints have different distances from z. For a bipartite graph this is the entire set of edges. In this case the information obtained above, in particular (5),(3),(6), describes G to within distance 2 from z.

(5.2) Proposition: *Given a connected, bipartite, focal graph* G *having a unique fixed point* z, *let* G_2 *be the subgraph of* G *induced by the vertices of* G *with* $dist(x,z) \leq 2$. *Then* G_2 *is the sum of a family of rooted graphs* $(H_i,z_i)_{i \in I}$ *(sum modulo identification of the roots) such that (i) at most one* H_i *is a focal tree and* z_i *its central vertex; (ii) all other* H_i*'s are complete bipartite graphs of the form* K_{2,n_i}, $n_i \neq 2$, *and* z_i *belongs to the two-element colour class; and (iii) every block of the form* K_{2,n_i} *occurs at least in triplicate, i.e., if* $H_i \simeq K_{2,n_i}$, *then there are at least two other indices* $j,k \in I$ *for which* $H_j \simeq H_k \simeq K_{2,n_i}$. *Conversely, any such graph is focal and if it is not a tree, can be extended to a bipartite focal graph of any diameter.*

The extension to arbitrary diameter ≥ 3 can be carried out as shown in Figure 5.

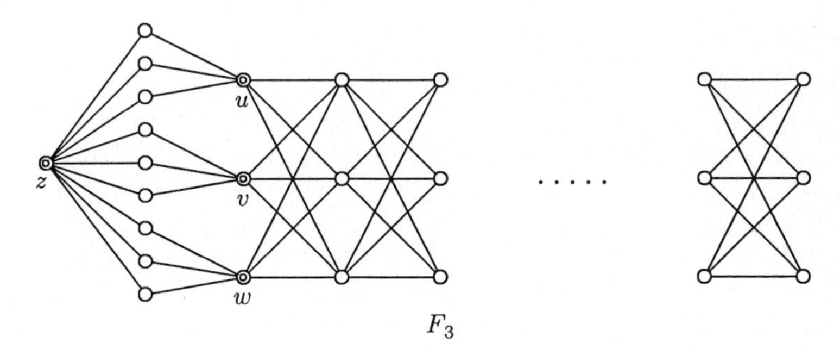

F_3

Figure 5

Suppose that for a given $n \neq 2$, G_2 has exactly r blocks $H_{i_1}, \cdots, H_{i_r}$ isomorphic to $K_{2,n}$. Take the composition $\tilde{P} = P[\overline{K}_r]$ of the discrete graph $\overline{K}_r$ with a path P of any length and identify the top vertices of $H_{i_1}, \cdots, H_{i_r}$ with the "endpoints" of $\tilde{P}$ (u,v,w in Figure 5). This yields a bipartite focal graph whose diameter depends solely on the length of P. F_3 is the simplest such graph.

The graph of Figure 5 shows incidentally that the number of foci in a focal graph is not tied to the diameter of the graph (it may be as low as 4). Also, at least for small diameters, there is no dependence on the connectivity. For any $k \geq 3$ one can construct k-connected focal graphs of diameter 3 or 4 with exactly 5 or 7 foci, respectively. On the other hand, it follows from property (1) earlier in this section that in the presence of a unique fixed point z, any k-connected focal graph of diameter ≥ 5 must have at least $k+1$ foci (viz. z and A_2).

Acknowledgements

I wish to thank Alex Rosa for having pointed out to me the existence of focal graphs and to have listened patiently to numerous false conjectures of mine about them. Support of the Natural Sciences and Engineering Research Council of Canada, Grant A7315, is gratefully acknowledged.

References

[1] J. Feigenbaum, J. Herschberger, and A. Schaffer, "A polynomial time algorithm for finding the prime factors of cartesian product graphs", *Discrete Appl. Math.* 12 (1985) 123-138.

[2] A. Rosa, in: *Combinatorics 79 (Part II)* (M. Deza and I.G. Rosenberg, eds.) *Annals of Discrete Math.* 9 (1980) 307.

[3] G. Sabidussi, "The composition of graphs", *Duke Math. J.* 26 (1959) 693-696.

[4] G. Sabidussi, "Graph multiplication", *Math. Z.* 72 (1960) 446-457.

[5] P. Winkler, "Factoring a graph in polynomial time", *European J. Combin.*, to appear.

Annals of Discrete Mathematics 34 (1987) 419–436
© Elsevier Science Publishers B.V. (North-Holland)

Some Perfect One-Factorizations of K_{14}

E. Seah and D.R. Stinson

Department of Computer Science
University of Manitoba
Winnipeg, Manitoba, R3T 2N2
CANADA

TO ALEX ROSA ON HIS FIFTIETH BIRTHDAY

ABSTRACT

We use orderly algorithms to generate 16 new perfect one-factorizations of K_{14}. This provides a lower bound of 20 nonisomorphic perfect one-factorizations of K_{14} (the previous bound was 4).

1. Introduction

Let Gr be an r-regular graph on n vertices. A *one-factorization* of Gr is a partition of the edge-set of Gr into r *one-factors,* each of which contains $\frac{n}{2}$ edges that partition the vertex set of Gr. A *complete graph* on n vertices, K_n, is $(n-1)$-regular.

A *perfect* one-factorization (or P1F) is a one-factorization in which every pair of distinct one-factors forms a Hamiltonian cycle of the graph. P1Fs are known to exist on K_{p+1} and K_{2p}, where p is an odd prime. These are called GK_{p+1} and GA_{2p}, respectively; see [7] and [2]. P1Fs are also known to exist on K_{16}, K_{28}, K_{244}, and K_{344} (see [1]). Recently, the authors discovered P1Fs for K_{36} (in [10]) and K_{50} (unpublished). No examples of a P1F are known for any other values of n. It has been conjectured that every K_{2n} has a P1F, but this appears to be a very difficult problem. A summary of known results on P1Fs is given in [8].

Denote by $N(n)$ the number of nonisomorphic P1Fs of K_n. The following is well known:

Theorem 1.1 [9]: $N(4) = N(6) = N(8) = N(10) = 1$, and $N(12) = 5$. □

In this paper, we investigate the number of nonisomorphic P1Fs of K_{14}. Four nonisomorphic P1Fs of K_{14} were shown to exist in [1]. In this paper, we construct 16 new P1Fs and hence improve the lower bound to 20.

Theorem 1.2: $N(14) \geq 20$.

We also compute the automorphism groups of these P1Fs. We find examples where the automorphism group has order 2, 3, 4, 6, 12, 84, and 156. It is interesting to note that of all the 20 P1Fs known so far, none is automorphism-free. The existence of an automorphism-free P1F for K_{12} would lead one to suspect that there might be some of these for K_{14}.

The basic methods used to establish the results of this paper are orderly algorithms. We consider only nonisomorphic P1Fs, by eliminating isomorphic structures as they are being constructed. These algorithms are described in the remainder of the paper.

2. Orderly algorithms for generating P1Fs of a complete graph

In this section, we outline two orderly algorithms that can be used to generate all the (nonisomorphic) P1Fs of a complete graph K_{2n}. These are modifications of the orderly algorithm described in [11]. We first need to define orderings on edges, one-factors, etc., of K_{2n}. All orderings are defined lexicographically, as follows.

- Suppose that the vertices are numbered 1, ..., $2n$. An edge e will be written as an ordered pair (p,p') with $1 \leq p < p' \leq 2n$.
- For any two edges $e_1 = (p_1,p'_1)$ and $e_2 = (p_2,p'_2)$, we say $e_1 < e_2$ if either of the following is true: (i) $p_1 < p_2$, (ii) $p_1 = p_2$ and $p'_1 < p'_2$.
- A one-factor f can be written as a set of ordered edges, i.e. $f = (e_1, e_2, \cdots e_n)$, where $e_1 < e_2 < \cdots < e_n$.
- For two one-factors $f_i = (e_{i1}, e_{i2}, \cdots e_{in})$ and $f_j = (e_{j1}, e_{j2}, \cdots e_{jn})$, we say $f_i < f_j$ if there exists a k $(1 \leq k \leq n)$ such that $e_{il} = e_{jl}$ for all $l < k$, and $e_{ik} < e_{jk}$.
- A one-factorization F of K_{2n} can be written as an ordered set of $2n-1$ one-factors, i.e. $F = (f_1, f_2, \cdots f_{2n-1})$, where $f_i < f_j$ whenever $i < j$. Note that we are only interested in P1Fs. We use F, G, H to denote P1Fs, and f_i, g_i, h_i the corresponding one-factors.
- For two P1Fs F and G, we say that $F < G$ if there exists some i, $1 \leq i \leq 2n-1$, such that $f_i < g_i$ and $f_j = g_j$ for all $j < i$.
- For $1 \leq i \leq 2n-1$, $F_i = (f_1, f_2, \cdots f_i)$ will denote a partial P1F consisting of an ordered set of i one-factors. We say that i is the *rank* of the partial P1F. Note that $F_{2n-1} = F$, a (complete) P1F. We can also extend our ordering to partial P1Fs of rank i, in an analogous manner.

Given a partial P1F F_i (of rank i), we can rename the $2n$ points using a permutation α and obtain another partial P1F, denoted F_i^α. We say that F_i is *canonical* if $F_i^\alpha \geq F_i$ for all permutations α. It is easy to see that if two partial P1Fs of rank i, F_i and G_i, are distinct and are both canonical, then F_i and G_i are nonisomorphic. Also, if $F_i = (f_1, f_2, \cdots f_i)$ is canonical, and $1 \leq j \leq i$, then $F_j = (f_1, f_2, \cdots f_j)$ is also canonical.

We need the following definitions to help explain the canonicity testing and the algorithms:

- Let S_i denote the set of all one-factors containing the edge $(1\ i+1)$.
- If an F_i of rank i contains one one-factor from each of $S_1,...,S_i$, then we say that F_i is *proper*. Note that if $F_i = (f_1,f_2, \cdots f_i)$ is proper, then $F_j = (f_1,f_2, \cdots f_j)$ is proper, for $1 \le j \le i$. Also, note that any P1F is proper.
- Let $\mathcal{F}_i$ denote the set of canonical proper P1Fs of rank i.

It is interesting to note that a canonical partial P1F need not be proper. For example, there are exactly 32 (nonisomorphic) canonical partial P1Fs of rank 3 (see [6]), but only 24 of these are proper.

In testing whether F_i is canonical, we are interested only in those α's that map a one-factor of $F_i \cup \{f\}$ $(f \in S_{i+1})$ into $((1\ 2)(3\ 4)(5\ 6) \cdots (2n{-}1\ 2n))$, the very first (smallest) one-factor of S_1. The number of such mappings is $(i+1)2^n n!$. For $n = 7$, the maximum number would be $(12+1)2^7 7! = 8{,}386{,}560$, which is much smaller than the 14! permutations of the vertex set.

A further improvement can be achieved by testing only those α's which map two one-factors of $F_i \cup \{f\}$ into a fixed set of two one-factors, which is the approach we use. Since we require any two disjoint one-factors of K_{14} to form a Hamiltonian cycle, we can map them into $f_a = ((1\ 2)(3\ 4)(5\ 6) \cdots (13\ 14))$ and $f_b = ((1\ 3)(2\ 5)(4\ 7)(6\ 9)(8\ 11)(10\ 13)(12\ 14))$, where f_b is the smallest one-factor in S_2 that forms a Hamiltonian cycle with f_a. There are 28 different ways to map one Hamiltonian cycle of length 14 into another, and hence the maximum number of mappings needed is $\binom{13}{2} 28 = 2184$.

There are two ways that one can go about generating the P1Fs: (A) breadth-first algorithm, and (B) depth-first algorithm.

(A) Breadth-first algorithm

A *breadth-first* algorithm generates each set $\mathcal{F}_i$ of canonical proper partial P1Fs of rank i in turn, starting with $i{=}1$ and ending with $i{=}2n{-}1$. Once the whole process is through, $\mathcal{F}_{2n-1}$ is the set of all of the nonisomorphic P1Fs of K_{2n} (in canonical form). The following pseudocode describes how to generate $\mathcal{F}_{i+1}$ from $\mathcal{F}_i$ (step $i+1$):

$\mathcal{F}_{i+1} = \emptyset$;
For each $F_i \in \mathcal{F}_i$ do
    For each one-factor $f \in S_{i+1}$ that is disjoint from and forms Hamiltonian cycles with all one-factors of F_i do
        For each permutation α do
            (1) compute f^α and F_i^α;
            (2) if $F_i^\alpha \cup \{f^\alpha\} < F_i \cup \{f\}$, then
                $F_i \cup \{f\}$ is not canonical, so discard it and go on to the next f;
    {Here $F_i^\alpha \cup \{f^\alpha\} \ge F_i \cup \{f\}$ for all α. Hence $F_i \cup \{f\}$ is canonical and proper, so save it for the next step.}
    $\mathcal{F}_{i+1} = \mathcal{F}_{i+1} \cup \{F_i \cup \{f\}\}$.

(B) Depth-first algorithm

A *depth-first* algorithm uses backtracking. Instead of generating all canonical proper partial P1Fs of each rank in turn, a depth-first algorithm tries all possible ways of extending each given F_i to a P1F, before trying the next F_i. The following recursive pseudocode describes how to generate, from a given F_i, all F_{2n-1} extending F_i, where $2 \leq i \leq 2n-1$. Note that $\mathcal{F}_1$ consists of one one-factor, $((1\ 2)(3\ 4) \cdots (2n-1\ 2n))$, the very first (smallest) one-factor of S_1.

```
Procedure Depth-first (F_i, i):
If i = 2n-1 then
    F_i is a P1F
else
    For each f ∈ S_{i+1} that is disjoint from and forms a Hamiltonian cycle with
    each of the one-factors of F_i do
        If F_i ∪ {f} is canonical then
            Depth-first (F_i ∪ {f}, i+1);
End Depth-first.
```

It is not difficult to see that both the depth-first and the breadth-first algorithms will enumerate all proper partial P1Fs of each rank. Since all (complete) P1Fs are proper, we can determine the number of nonisomorphic P1Fs by either method.

The depth-first method has the following two advantages over a breadth-first method.

(1) We can incorporate *pruning,* by showing that some F_i cannot be extended to a P1F. Given an F_i, we determine the sets of the one-factors in each of the sets $S_{i+1}, S_{i+2}, \cdots S_{2n-1}$, which are disjoint from and form Hamiltonian cycles with each of the one-factors of F_i. If any of these sets is empty, then the given F_i cannot be completed to a P1F of K_{14} so we do not have to investigate any extensions of F_i.

(2) No storage is required for the partial P1Fs at each step as in the case of the breadth-first method.

3. Results on K_{12} and K_{14}

All of the computer work in this paper was implemented in Pascal/VS and run on the University of Manitoba Amdahl 580 computer. We started out by implementing the breadth-first algorithm, since this method tells us the number of nonisomorphic proper partial P1Fs at each intermediate level before proceeding to the next level. It took approximately 132 minutes of CPU time to construct the 5 P1Fs of K_{12}. Using the depth-first method, incorporating pruning, the number of intermediate proper partial P1Fs is significantly reduced, and the enumeration took only 23 minutes of CPU time. Table 1 gives the number of canonical proper partial P1Fs of each rank, both with and without pruning.

Table 1
Nonisomorphic Canonical Proper Partial P1Fs of K_{12}

i	# of canonical proper partial P1Fs of rank i without pruning	with pruning
3	24	24
4	395	395
5	2679	2679
6	10987	10906
7	13791	3542
8	3491	14
9	209	7
10	6	6
11	5	5

When we used the breadth-first method to attempt to determine $N(14)$, it did not take long for us to conclude that the complete enumeration is out of reach at this time. The number of partial P1F structures generated, and the CPU time taken, increase dramatically from one step to the next, as indicated by Table 2.

Table 2
Nonisomorphic Canonical Proper Partial P1Fs of K_{14}
Using the Breadth-first Method

i	# of canonical proper partial P1Fs of rank i
3	174
4	23704
5*	34272

* using only the first 464 sets $F_4 \in \mathcal{F}_4$

Consequently, we decided to improve the lower bound on $N(14)$ by constructing as many P1Fs of K_{14} as possible. We used the breadth-first method to construct all partial P1Fs in $\mathcal{F}_4$. Then, given a partial P1F in $\mathcal{F}_4$, the depth-first method was used to generate all extensions to complete P1Fs.

4. New P1Fs of K_{14} and their Automorphism Groups

We found 16 new P1Fs of K_{14}. The 20 nonisomorphic P1Fs, and their automorphism groups, are listed in the Appendix.

The four previously known P1Fs of K_{14} are sets 1, 13, 14, and 15. Sets 1 and 13 are GA_{14} and GK_{14} respectively. Sets 13, 14, and 15 are constructed from even starters in Z_{12}; and set 13 can also be generated by a starter in Z_{13}.

The 20 P1Fs have automorphism groups as follows:

Z_2	is the automorphism group for sets 7 and 8;
Z_3	is the automorphism group for sets 16-20;
$Z_2 \times Z_2$	is the automorphism group for set 2;
Z_6	is the automorphism group for sets 4-6, 11, and 12;
$Z_2 \times Z_6$	is the automorphism group for set 3;
Q_6	(a dicyclic group) is the automorphism group for sets 9 and 10;
Z_{12}	is the automorphism group for sets 14 and 15;
$[Z_{13}]Z_{12}$	(semidirect product) is the automorphism group for set 13;
$[Z_{14}]Z_6$	(semidirect product) is the automorphism group for set 1.

In [4] and [5], Ihrig studies automorphism groups of P1Fs. The next two theorems give several properties such a group must have, if it contains an automorphism of order 2 having fixed points.

Theorem 4.1: ([4], Theorem 5.5) If a P1F for K_{2n} contains a noncentral automorphism of order 2 having fixed points, then the P1F is either GA_{2p} or GK_{p+1}. □

Theorem 4.2: ([4], Theorem 5.9) If a P1F of K_{2n} contains a central automorphism of order 2 having fixed points, then the following statements hold:

(a) the order of the automorphism group divides $2n-2$;

(b) there are at most 3 automorphisms of order 2, and only one of these has fixed points. □

We note that the P1Fs of sets 1-15 all have an automorphism of order 2 containing fixed points. Our examples illustrate every group order allowed by Theorem 4.2(a) (namely, orders 2, 4, 6, and 12). Also, note that sets 2 and 3 each contain three automorphisms of order 2, while sets 4-12, 14, and 15 each contain only one such automorphism.

If the automorphism group of a P1F does not contain an automorphism of order 2 containing fixed points, then the following results hold.

Theorem 4.3: ([5], Theorem 3.10) If a P1F of K_{2n} contains no automorphism of order 2 having fixed points, then the order of the automorphism group is $m_0 m_1 m_2$, where $m_0 \mid 2n$, $m_1 \mid (2n-1)$, and $m_2 \mid (n-1)$. Further, m_2 is odd; and at least one of m_0, m_1, and m_2 is equal to 1. □

In the case of K_{14}, we obtain $m_0 \mid 14$, $m_1 \mid 13$, and $m_2 \mid 3$. If $m_1 = 13$, then the P1F must be generated by a starter in Z_{13}. These were enumerated in [2], and set 13 (GK_{14}) is the only example. Hence, we can ignore this case and assume that $m_1 = 1$. Then, the order of the automorphism group must divide 42.

We have enumerated all P1Fs of K_{14} having an automorphism of order 7, and set 1 is the only example. Consequently, the order of the automorphism group must divide 6, and 1 and 3 are the only new possibilities. As mentioned already, sets 16-20 have automorphism groups isomorphic to Z_3, and we have no examples with trivial automorphism groups.

Hence, we have examples for every possible group order, except 1. In fact, we have exhaustively enumerated all P1Fs with groups of order at least 3, and the list given in this paper is complete (the algorithms used will be described in a forthcoming paper).

Two of the new P1Fs (sets 9 and 10) have the property that their automorphism groups are the dicyclic group Q_6 of order 12. The dicyclic group Q_{2n} of order $4n$ is the group defined by

$$Q_{2n} = \{a^i b^j : 0 \leq i \leq 2n-1,\, 0 \leq j \leq 1,\, a^{2n} = e,\, b^2 = a^n,\, bab^{-1} = a^{-1}\}$$

A *sequencing* of a finite group G of order $2n$ is an ordering $e, a_2, a_3, \ldots, a_{2n}$ of all elements of G so that the partial products $e, ea_2, ea_2a_3, \ldots, ea_2a_3 \cdots a_{2n}$ are distinct and hence also all of G. The sequencing is *symmetric* if in addition the following are true: (1) G has a unique element z of order 2, (2) $a_{n+1} = z$, and (3) $a_{n+i+1} = (a_{n+1-i})^{-1}$. In [3], Anderson observed that these two P1Fs give rise to symmetric sequencings in the group Q_6. (A sequencing of this group was previously unknown). Subsequently, he showed that for any odd $n \geq 3$, the dicyclic group Q_{2n} can be symmetrically sequenced.

Given a symmetric sequencing of a group G, one can construct a one-factorization (not necessarily perfect) of $K_{|G|+2}$ (see [3]). It thus seems hopeful that symmetric sequencings of Q_{2n} can be used to construct P1Fs of K_{4n+2}. However, it remains to be seen whether symmetric sequencings will give us a new class of P1Fs.

Added in Proof:

The P1F of K_{50} is presented in "A perfect one-factorization of K_{50}", by E.C. Ihrig, E. Seah and D.R. Stinson, *J. Comb. Math. Comb. Comput.* 1 (1987) 217-219. As well, an exhaustive enumeration of P1Fs of K_{14} having non-trivial automorphism groups has recently been done by the authors. This work is contained in "On the enumeration of one-factorizations of the complete graph containing prescribed automorphism groups", *Math. Comp.*, to appear. There are a total of 21 such P1Fs, including the 20 presented in this paper.

Acknowledgements

We would like to thank Bruce Anderson and Edwin Ihrig for their useful comments. As well, we are indebted to the referee for pointing out reference [6].

References

[1] B.A. Anderson, "Some perfect one-factorizations", *Proc. Seventh Southeastern Conf. Combinatorics, Graph Theory, Computing,* 1976, pp. 79-91.

[2] B.A. Anderson, "Symmetry groups of some perfect one-factorizations of complete graphs", *Discrete Math.* 18 (1977) 227-234.

[3] B.A. Anderson, "Sequencings of dicyclic groups", *Ars Combinatoria* 23 (1987) 131-142.

[4] E.C. Ihrig, "Symmetry groups related to the construction of perfect one factorizations of K_{2n}", *J. Comb. Theory* B40 (1986) 121-151.

[5] E.C. Ihrig, "The structure of symmetry groups of perfect one factorizations of K_{2n}", to appear.

[6] N.P. Korovina, "On the construction of closed triples of pair groups of order 12" (in Russian), *Kombinatornyi Anal.* 2 (1972) 42-45.

[7] E. Lucas, *Recréations Mathématiques, Vol. 2,* Gauthier-Villars, Paris, 1883 (Sixième Recréation, Les Jeux des Demiselles, 161-197).

[8] E. Mendelsohn and A. Rosa, "One-factorizations of the complete graph -- a survey", *J. Graph Theory* 9 (1985) 43-65.

[9] L.P. Petrenyuk and A.Y. Petrenyuk, "Intersection of perfect one-factorizations of complete graphs", *Cybernetics* 16 (1980) 6-9.

[10] E. Seah and D.R. Stinson, "A perfect one-factorization of K_{36}", *Discrete Math.*, to appear.

[11] E. Seah and D.R. Stinson, "An enumeration of non-isomorphic one-factorizations and Howell designs for the graph K_{10} minus a one-factor", *Ars Combinatoria* 21 (1986) 145-161.

Appendix

Set 1:

$|A| = 84$ (GA_{14})

$A = < g_1, g_2 >$

$g_1 = (1\ 12\ 11\ 8\ 3\ 7)(2\ 14\ 13\ 10\ 5\ 9)$

$g_2 = (1\ 13\ 4\ 9\ 8\ 5\ 12\ 2\ 11\ 6\ 7\ 10\ 3\ 14)$

g_1 induces $(f_2\ f_4\ f_8\ f_{12}\ f_{11}\ f_7\ f_3)(f_5\ f_6)(f_9\ f_{10})(f_{13}\ f_{14})$

g_2 induces $(f_2\ f_3\ f_4\ f_{11}\ f_7\ f_{12})(f_5\ f_{10}\ f_{14}\ f_6\ f_9\ f_{13})$

1 2	3 4	5 6	7 8	9 10	11 12	13 14
1 3	2 5	4 7	6 9	8 11	10 13	12 14
1 4	2 6	3 8	5 10	7 12	9 14	11 13
1 5	2 4	3 9	6 8	7 13	10 12	11 14
1 6	2 3	4 10	5 7	8 14	9 11	12 13
1 7	2 9	3 5	4 11	6 13	8 12	10 14
1 8	2 10	3 12	4 6	5 14	7 11	9 13
1 9	2 8	3 13	4 5	6 12	7 14	10 11
1 10	2 7	3 6	4 14	5 11	8 13	9 12
1 11	2 13	3 7	4 12	5 9	6 14	8 10
1 12	2 14	3 11	4 8	5 13	6 10	7 9
1 13	2 12	3 14	4 9	5 8	6 11	7 10
1 14	2 11	3 10	4 13	5 12	6 7	8 9

Set 2:

$|A| = 4.$

$A \quad Z_2 \times Z_2$

$g_1 = (1\ 2)(3\ 5)(4\ 6)(7\ 9)(8\ 10)(11\ 13)(12\ 14)$

$g_2 = (1\ 2)(3\ 6)(4\ 5)(7\ 10)(8\ 9)(11\ 14)(12\ 13)$

g_1 induces $(f_5\ f_6)(f_9\ f_{10})(f_{13}\ f_{14})$

g_2 induces $(f_3\ f_4)(f_7\ f_8)(f_{11}\ f_{12})$

1 2	3 4	5 6	7 8	9 10	11 12	13 14
1 3	2 5	4 7	6 9	8 11	10 13	12 14
1 4	2 6	3 8	5 10	7 12	9 14	11 13
1 5	2 4	3 9	6 8	7 13	10 12	11 14
1 6	2 3	4 10	5 7	8 14	9 11	12 13
1 7	2 9	3 13	4 6	5 11	8 12	10 14
1 8	2 10	3 5	4 14	6 12	7 11	9 13
1 9	2 8	3 12	4 5	6 13	7 14	10 11
1 10	2 7	3 6	4 11	5 14	8 13	9 12
1 11	2 13	3 14	4 8	5 12	6 10	7 9
1 12	2 14	3 7	4 13	5 9	6 11	8 10
1 13	2 12	3 11	4 9	5 8	6 14	7 10
1 14	2 11	3 10	4 12	5 13	6 7	8 9

Set 3:

$|A| = 12.$

$A \simeq Z_2 \times Z_6$

$A = < g_1, g_2 >$

$g_1 = (3\ 11\ 10\ 4\ 12\ 9)(5\ 13\ 8\ 6\ 14\ 7)$

$g_2 = (1\ 2)(3\ 6)(4\ 5)(7\ 10)(8\ 9)(11\ 14)(12\ 13)$

g_1 induces $(f_3\ f_{11}\ f_{10}\ f_4\ f_{12}\ f_9)(f_5\ f_{13}\ f_8\ f_6\ f_{14}\ f_7)$

g_2 induces $(f_3\ f_4)(f_{11}\ f_{12})(f_9\ f_{10})$

1 2	3 4	5 6	7 8	9 10	11 12	13 14
1 3	2 5	4 7	6 9	8 11	10 13	12 14
1 4	2 6	3 8	5 10	7 12	9 14	11 13
1 5	2 4	3 9	6 8	7 13	10 12	11 14
1 6	2 3	4 10	5 7	8 14	9 11	12 13
1 7	2 10	3 6	4 11	5 14	8 13	9 12
1 8	2 9	3 12	4 5	6 13	7 14	10 11
1 9	2 7	3 13	4 6	5 11	8 12	10 14
1 10	2 8	3 5	4 14	6 12	7 11	9 13
1 11	2 13	3 14	4 8	5 12	6 10	7 9
1 12	2 14	3 7	4 13	5 9	6 11	8 10
1 13	2 12	3 11	4 9	5 8	6 14	7 10
1 14	2 11	3 10	4 12	5 13	6 7	8 9

Set 4:

$|A| = 6.$

$A \simeq Z_6$

$A = < g_1 >$

$g_1 = (3\ 10\ 7\ 4\ 9\ 8)(5\ 14\ 12\ 6\ 13\ 11)$

g_1 induces $(f_3\ f_{10}\ f_7\ f_4\ f_9\ f_8)(f_5\ f_{14}\ f_{12}\ f_6\ f_{13}\ f_{11})$

1 2	3 4	5 6	7 8	9 10	11 12	13 14
1 3	2 5	4 7	6 9	8 11	10 13	12 14
1 4	2 6	3 8	5 10	7 12	9 14	11 13
1 5	2 4	3 9	6 8	7 13	10 12	11 14
1 6	2 3	4 10	5 7	8 14	9 11	12 13
1 7	2 12	3 11	4 5	6 13	8 9	10 14
1 8	2 11	3 6	4 12	5 14	7 10	9 13
1 9	2 13	3 10	4 6	5 11	7 14	8 12
1 10	2 14	3 5	4 9	6 12	7 11	8 13
1 11	2 7	3 14	4 8	5 13	6 10	9 12
1 12	2 8	3 7	4 13	5 9	6 14	10 11
1 13	2 10	3 12	4 14	5 8	6 11	7 9
1 14	2 9	3 13	4 11	5 12	6 7	8 10

Set 5:

$|A| = 6.$

$A \simeq Z_6$

$A = < g_1 >$

$g_1 = (3\ 10\ 7\ 4\ 9\ 8)(5\ 13\ 11\ 6\ 14\ 12)$

g_1 induces $(f_3\ f_{10}\ f_7\ f_4\ f_9\ f_8)(f_5\ f_{13}\ f_{11}\ f_6\ f_{14}\ f_{12})$

1 2	3 4	5 6	7 8	9 10	11 12	13 14
1 3	2 5	4 7	6 9	8 11	10 13	12 14
1 4	2 6	3 8	5 10	7 12	9 14	11 13
1 5	2 4	3 9	6 8	7 14	10 11	12 13
1 6	2 3	4 10	5 7	8 13	9 12	11 14
1 7	2 11	3 12	4 6	5 13	8 9	10 14
1 8	2 12	3 5	4 11	6 14	7 10	9 13
1 9	2 14	3 10	4 5	6 11	7 13	8 12
1 10	2 13	3 6	4 9	5 12	7 11	8 14
1 11	2 8	3 7	4 14	5 9	6 13	10 12
1 12	2 7	3 13	4 8	5 14	6 10	9 11
1 13	2 9	3 14	4 12	5 11	6 7	8 10
1 14	2 10	3 11	4 13	5 8	6 12	7 9

Set 6:

$|A| = 6.$

$A \simeq Z_6$

$A = < g_1 >$

$g_1 = (3\ 14\ 12\ 4\ 13\ 11)(5\ 9\ 8\ 6\ 10\ 7)$

g_1 induces $(f_3\ f_{14}\ f_{12}\ f_4\ f_{13}\ f_{11})(f_5\ f_9\ f_8\ f_6\ f_{10}\ f_7)$

1 2	3 4	5 6	7 8	9 10	11 12	13 14
1 3	2 5	4 7	6 9	8 11	10 13	12 14
1 4	2 6	3 8	5 10	7 12	9 14	11 13
1 5	2 4	3 9	6 8	7 14	10 11	12 13
1 6	2 3	4 10	5 7	8 13	9 12	11 14
1 7	2 12	3 10	4 14	5 11	6 13	8 9
1 8	2 11	3 13	4 9	5 14	6 12	7 10
1 9	2 13	3 7	4 11	5 12	6 10	8 14
1 10	2 14	3 12	4 8	5 9	6 11	7 13
1 11	2 7	3 14	4 6	5 8	9 13	10 12
1 12	2 8	3 5	4 13	6 7	9 11	10 14
1 13	2 10	3 11	4 5	6 14	7 9	8 12
1 14	2 9	3 6	4 12	5 13	7 11	8 10

Set 7:

$|A| = 2.$

$A \simeq Z_2$

$A = < g_1 >$

$g_1 = (3\ 4)(5\ 6)(7\ 8)(9\ 10)(11\ 12)(13\ 14)$

g_1 induces $(f_3\ f_4)(f_5\ f_6)(f_7\ f_8)(f_9\ f_{10})(f_{11}\ f_{12})(f_{13}\ f_{14})$

1 2	3 4	5 6	7 8	9 10	11 12	13 14
1 3	2 5	4 7	6 9	8 11	10 13	12 14
1 4	2 6	3 8	5 10	7 12	9 14	11 13
1 5	2 4	3 9	6 14	7 13	8 12	10 11
1 6	2 3	4 10	5 13	7 11	8 14	9 12
1 7	2 12	3 6	4 13	5 11	8 9	10 14
1 8	2 11	3 14	4 5	6 12	7 10	9 13
1 9	2 8	3 7	4 11	5 14	6 10	12 13
1 10	2 7	3 12	4 8	5 9	6 13	11 14
1 11	2 13	3 10	4 14	5 12	6 8	7 9
1 12	2 14	3 13	4 9	5 7	6 11	8 10
1 13	2 9	3 11	4 6	5 8	7 14	10 12
1 14	2 10	3 5	4 12	6 7	8 13	9 11

Set 8:

$|A| = 2.$
$A \simeq Z_2$
$A = < g_1 >$
$g_1 = (3\ 4)(5\ 6)(7\ 8)(9\ 10)(11\ 12)(13\ 14)$
g_1 induces $(f_3\ f_4)(f_5\ f_6)(f_7\ f_8)(f_9\ f_{10})(f_{11}\ f_{12})(f_{13}\ f_{14})$

1 2	3 4	5 6	7 8	9 10	11 12	13 14
1 3	2 5	4 7	6 9	8 11	10 13	12 14
1 4	2 6	3 8	5 10	7 12	9 14	11 13
1 5	2 4	3 11	6 13	7 9	8 12	10 14
1 6	2 3	4 12	5 14	7 11	8 10	9 13
1 7	2 9	3 6	4 13	5 8	10 12	11 14
1 8	2 10	3 14	4 5	6 7	9 11	12 13
1 9	2 11	3 10	4 8	5 12	6 14	7 13
1 10	2 12	3 7	4 9	5 13	6 11	8 14
1 11	2 14	3 12	4 6	5 9	7 10	8 13
1 12	2 13	3 5	4 11	6 10	7 14	8 9
1 13	2 8	3 9	4 14	5 7	6 12	10 11
1 14	2 7	3 13	4 10	5 11	6 8	9 12

Set 9:

$|A| = 12.$
$A \simeq Q_6$ (dicyclic group)
$A = < g_1, g_2 >$
$g_1 = (3\ 5\ 4\ 6)(7\ 13\ 8\ 14)(9\ 11\ 10\ 12)$
$g_2 = (3\ 13\ 10\ 4\ 14\ 9)(5\ 11\ 7\ 6\ 12\ 8)$
g_1 induces $(f_3\ f_5\ f_4\ f_6)(f_7\ f_{13}\ f_8\ f_{14})(f_9\ f_{11}\ f_{10}\ f_{12})$
g_2 induces $(f_3\ f_{13}\ f_{10}\ f_4\ f_{14}\ f_9)(f_5\ f_{11}\ f_7\ f_6\ f_{12}\ f_8)$

1 2	3 4	5 6	7 8	9 10	11 12	13 14
1 3	2 5	4 7	6 9	8 11	10 13	12 14
1 4	2 6	3 8	5 10	7 12	9 14	11 13
1 5	2 4	3 11	6 13	7 9	8 12	10 14
1 6	2 3	4 12	5 14	7 11	8 10	9 13
1 7	2 9	3 14	4 8	5 11	6 10	12 13
1 8	2 10	3 7	4 13	5 9	6 12	11 14
1 9	2 8	3 13	4 6	5 12	7 14	10 11
1 10	2 7	3 5	4 14	6 11	8 13	9 12
1 11	2 14	3 6	4 9	5 8	7 13	10 12
1 12	2 13	3 10	4 5	6 7	8 14	9 11
1 13	2 11	3 12	4 10	5 7	6 14	8 9
1 14	2 12	3 9	4 11	5 13	6 8	7 10

Set 10:

$|A| = 12.$

$A \simeq Q_6$ (dicyclic group)

$A = < g_1, g_2 >$

$g_1 = (3\ 5\ 4\ 6)(7\ 12\ 8\ 11)(9\ 13\ 10\ 14)$

$g_2 = (3\ 7\ 14\ 4\ 8\ 13)(5\ 10\ 11\ 6\ 9\ 12)$

g_1 induces $(f_3\ f_5\ f_4\ f_6)(f_7\ f_{12}\ f_8\ f_{11})(f_9\ f_{13}\ f_{10}\ f_{14})$

g_2 induces $(f_3\ f_7\ f_{14}\ f_4\ f_8\ f_{13})(f_5\ f_{10}\ f_{11}\ f_6\ f_9\ f_{12})$

1 2	3 4	5 6	7 8	9 10	11 12	13 14
1 3	2 5	4 7	6 9	8 11	10 13	12 14
1 4	2 6	3 8	5 10	7 12	9 14	11 13
1 5	2 4	3 13	6 12	7 11	8 9	10 14
1 6	2 3	4 14	5 11	7 10	8 12	9 13
1 7	2 10	3 11	4 5	6 13	8 14	9 12
1 8	2 9	3 6	4 12	5 14	7 13	10 11
1 9	2 7	3 12	4 8	5 13	6 10	11 14
1 10	2 8	3 7	4 11	5 9	6 14	12 13
1 11	2 13	3 5	4 9	6 8	7 14	10 12
1 12	2 14	3 10	4 6	5 7	8 13	9 11
1 13	2 12	3 14	4 10	5 8	6 11	7 9
1 14	2 11	3 9	4 13	5 12	6 7	8 10

Set 11:

$|A| = 6.$

$A \simeq Z_6$

$A = < g_1 >$

$g_1 = (3\ 7\ 11\ 4\ 12\ 8)(5\ 10\ 14\ 6\ 9\ 13)$

g_1 induces $(f_3\ f_{11}\ f_7\ f_4\ f_{12}\ f_8)(f_5\ f_{10}\ f_{14}\ f_6\ f_9\ f_{13})$

1 2	3 4	5 6	7 8	9 10	11 12	13 14
1 3	2 5	4 7	6 9	8 11	10 13	12 14
1 4	2 6	3 8	5 10	7 12	9 14	11 13
1 5	2 8	3 10	4 13	6 12	7 14	9 11
1 6	2 7	3 14	4 9	5 11	8 13	10 12
1 7	2 14	3 9	4 11	5 13	6 10	8 12
1 8	2 13	3 12	4 10	5 9	6 14	7 11
1 9	2 4	3 5	6 11	7 10	8 14	12 13
1 10	2 3	4 6	5 12	7 13	8 9	11 14
1 11	2 10	3 7	4 12	5 14	6 8	9 13
1 12	2 9	3 11	4 8	5 7	6 13	10 14
1 13	2 12	3 6	4 14	5 8	7 9	10 11
1 14	2 11	3 13	4 5	6 7	8 10	9 12

Set 12:

$|A| = 6.$

$A \simeq Z_6$

$A = < g_1 >$

$g_1 = (3\ 8\ 13\ 4\ 7\ 14)(5\ 9\ 12\ 6\ 10\ 11)$

g_1 induces $(f_3\ f_8\ f_{13}\ f_4\ f_7\ f_{14})(f_5\ f_9\ f_{12}\ f_6\ f_{10}\ f_{11})$

1 2	3 4	5 6	7 8	9 10	11 12	13 14
1 3	2 5	4 7	6 9	8 11	10 13	12 14
1 4	2 6	3 8	5 10	7 12	9 14	11 13
1 5	2 8	3 13	4 6	7 11	9 12	10 14
1 6	2 7	3 5	4 14	8 12	9 13	10 11
1 7	2 10	3 12	4 5	6 14	8 13	9 11
1 8	2 9	3 6	4 11	5 13	7 14	10 12
1 9	2 13	3 11	4 8	5 14	6 12	7 10
1 10	2 14	3 7	4 12	5 11	6 13	8 9
1 11	2 3	4 10	5 9	6 7	8 14	12 13
1 12	2 4	3 9	5 8	6 10	7 13	11 14
1 13	2 12	3 14	4 9	5 7	6 11	8 10
1 14	2 11	3 10	4 13	5 12	6 8	7 9

Set 13:

$|A| = 156.$

1 2	3 4	5 6	7 8	9 10	11 12	13 14
1 3	2 5	4 7	6 9	8 11	10 13	12 14
1 4	2 6	3 8	5 10	7 12	9 14	11 13
1 5	2 9	3 7	4 11	6 13	8 14	10 12
1 6	2 10	3 12	4 8	5 14	7 13	9 11
1 7	2 13	3 11	4 14	5 9	6 12	8 10
1 8	2 14	3 13	4 12	5 11	6 10	7 9
1 9	2 12	3 14	4 10	5 13	6 8	7 11
1 10	2 11	3 9	4 13	5 7	6 14	8 12
1 11	2 8	3 10	4 6	5 12	7 14	9 13
1 12	2 7	3 5	4 9	6 11	8 13	10 14
1 13	2 4	3 6	5 8	7 10	9 12	11 14
1 14	2 3	4 5	6 7	8 9	10 11	12 13

Set 14:

$|A| = 12.$

$A \simeq Z_{12}$

$A = < g_1 >$

$g_1 = (3\ 5\ 9\ 8\ 14\ 12\ 4\ 6\ 10\ 7\ 13\ 11)$

g_1 induces $(f_3\ f_5\ f_9\ f_8\ f_{14}\ f_{12}\ f_4\ f_6\ f_{10}\ f_7\ f_{13}\ f_{11})$

1 2	3 4	5 6	7 8	9 10	11 12	13 14
1 3	2 5	4 7	6 9	8 11	10 13	12 14
1 4	2 6	3 8	5 10	7 12	9 14	11 13
1 5	2 9	3 14	4 12	6 13	7 11	8 10
1 6	2 10	3 11	4 13	5 14	7 9	8 12
1 7	2 13	3 10	4 8	5 9	6 12	11 14
1 8	2 14	3 7	4 9	5 11	6 10	12 13
1 9	2 8	3 13	4 6	5 12	7 14	10 11
1 10	2 7	3 5	4 14	6 11	8 13	9 12
1 11	2 3	4 5	6 7	8 14	9 13	10 12
1 12	2 4	3 6	5 8	7 13	9 11	10 14
1 13	2 11	3 12	4 10	5 7	6 14	8 9
1 14	2 12	3 9	4 11	5 13	6 8	7 10

Set 15:

$|A| = 12.$

$A \simeq Z_{12}$

$A = < g_1 >$

$g_1 = (3\ 5\ 14\ 9\ 7\ 11\ 4\ 6\ 13\ 10\ 8\ 12)$

g_1 induces $(f_3\ f_5\ f_{14}\ f_9\ f_7\ f_{11}\ f_4\ f_6\ f_{13}\ f_{10}\ f_8\ f_{12})$

1 2	3 4	5 6	7 8	9 10	11 12	13 14
1 3	2 5	4 7	6 9	8 11	10 13	12 14
1 4	2 6	3 8	5 10	7 12	9 14	11 13
1 5	2 14	3 9	4 12	6 11	7 13	8 10
1 6	2 13	3 11	4 10	5 12	7 9	8 14
1 7	2 11	3 5	4 9	6 12	8 13	10 14
1 8	2 12	3 10	4 6	5 11	7 14	9 13
1 9	2 7	3 12	4 8	5 13	6 10	11 14
1 10	2 8	3 7	4 11	5 9	6 14	12 13
1 11	2 4	3 13	5 14	6 7	8 9	10 12
1 12	2 3	4 14	5 8	6 13	7 10	9 11
1 13	2 10	3 14	4 5	6 8	7 11	9 12
1 14	2 9	3 6	4 13	5 7	8 12	10 11

Set 16:

$|A| = 3.$

$A \simeq Z_3$

$A = < g_1 >$

$g_1 = (1\ 5\ 14)(3\ 11\ 13)(4\ 8\ 7)(6\ 12\ 10)$

g_1 induces $(f_2\ f_3\ f_{10})(f_4\ f_9\ f_{12})(f_5\ f_{13}\ f_{14})(f_6\ f_7\ f_{11})$

1 2	3 4	5 6	7 8	9 10	11 12	13 14
1 3	2 5	4 7	6 9	8 11	10 13	12 14
1 4	2 6	3 8	5 10	7 12	9 14	11 13
1 5	2 4	3 10	6 12	7 9	8 13	11 14
1 6	2 3	4 12	5 13	7 10	8 14	9 11
1 7	2 11	3 14	4 6	5 12	8 10	9 13
1 8	2 9	3 12	4 14	5 7	6 13	10 11
1 9	2 12	3 13	4 10	5 8	6 14	7 11
1 10	2 14	3 6	4 8	5 11	7 13	9 12
1 11	2 13	3 9	4 5	6 7	8 12	10 14
1 12	2 10	3 11	4 13	5 9	6 8	7 14
1 13	2 8	3 7	4 9	5 14	6 11	10 12
1 14	2 7	3 5	4 11	6 10	8 9	12 13

Set 17: $|A| = 3.$

$A \simeq Z_3$

$A = < g_1 >$

$g_1 = (1\ 8\ 3)(2\ 9\ 14)(4\ 11\ 5)(7\ 13\ 10)$

g_1 induces $(f_2\ f_{11}\ f_{10})(f_3\ f_8\ f_4)(f_5\ f_{14}\ f_9)(f_6\ f_{12}\ f_{13})$

1 2	3 4	5 6	7 8	9 10	11 12	13 14
1 3	2 5	4 7	6 9	8 11	10 13	12 14
1 4	2 6	3 8	5 10	7 13	9 12	11 14
1 5	2 4	3 9	6 13	7 12	8 14	10 11
1 6	2 14	3 12	4 5	7 11	8 10	9 13
1 7	2 11	3 10	4 14	5 9	6 12	8 13
1 8	2 12	3 5	4 9	6 14	7 10	11 13
1 9	2 8	3 11	4 13	5 14	6 7	10 12
1 10	2 13	3 14	4 12	5 8	6 11	7 9
1 11	2 10	3 13	4 6	5 12	7 14	8 9
1 12	2 9	3 7	4 11	5 13	6 8	10 14
1 13	2 7	3 6	4 10	5 11	8 12	9 14
1 14	2 3	4 8	5 7	6 10	9 11	12 13

Set 18:

$|A| = 3.$
$A \simeq Z_3$
$A = < g_1 >$
$g_1 = (1\ 5\ 9)(2\ 7\ 14)(4\ 6\ 8)(10\ 13\ 11)$
g_1 induces $(f_2\ f_{13}\ f_4)(f_3\ f_8\ f_{14})(f_5\ f_{10}\ f_9)(f_7\ f_{11}\ f_{12})$

1 2	3 4	5 6	7 8	9 10	11 12	13 14
1 3	2 5	4 7	6 9	8 11	10 13	12 14
1 4	2 6	3 8	5 11	7 10	9 14	12 13
1 5	2 4	3 13	6 12	7 14	8 10	9 11
1 6	2 13	3 12	4 9	5 8	7 11	10 14
1 7	2 10	3 14	4 11	5 12	6 8	9 13
1 8	2 12	3 5	4 10	6 14	7 9	11 13
1 9	2 7	3 10	4 12	5 13	6 11	8 14
1 10	2 14	3 11	4 13	5 9	6 7	8 12
1 11	2 3	4 8	5 14	6 10	7 13	9 12
1 12	2 9	3 7	4 6	5 10	8 13	11 14
1 13	2 11	3 6	4 14	5 7	8 9	10 12
1 14	2 8	3 9	4 5	6 13	7 12	10 11

Set 19:

$|A| = 3.$
$A \simeq Z_3$
$A = < g_1 >$
$g_1 = (1\ 4\ 7)(2\ 11\ 5)(3\ 12\ 10)(6\ 8\ 14)$
g_1 induces $(f_2\ f_{14}\ f_{10})(f_3\ f_7\ f_4)(f_5\ f_9\ f_{12})(f_6\ f_{11}\ f_{13})$

1 2	3 4	5 6	7 8	9 10	11 12	13 14
1 3	2 5	4 7	6 9	8 11	10 13	12 14
1 4	2 6	3 8	5 11	7 10	9 14	12 13
1 5	2 13	3 14	4 10	6 11	7 9	8 12
1 6	2 7	3 10	4 13	5 12	8 14	9 11
1 7	2 11	3 13	4 12	5 14	6 10	8 9
1 8	2 12	3 5	4 14	6 7	9 13	10 11
1 9	2 4	3 7	5 8	6 12	10 14	11 13
1 10	2 3	4 6	5 7	8 13	9 12	11 14
1 11	2 14	3 12	4 8	5 9	6 14	7 13
1 12	2 14	3 6	4 9	5 13	7 11	8 10
1 13	2 9	3 11	4 5	6 8	7 14	10 12
1 14	2 8	3 9	4 11	5 10	6 13	7 12

Set 20:

$|A| = 3.$

$A \simeq Z_3$

$A = < g_1 >$

$g_1 = (1\ 8\ 11)(2\ 9\ 10)(3\ 5\ 4)(6\ 12\ 14)$

g_1 induces $(f_2\ f_{14}\ f_7)(f_3\ f_{11}\ f_8)(f_5\ f_{13}\ f_6)(f_9\ f_{12}\ f_{10})$

1 2	3 4	5 6	7 8	9 10	11 12	13 14
1 3	2 5	4 7	6 9	8 11	10 13	12 14
1 4	2 6	3 8	5 11	7 13	9 12	10 14
1 5	2 7	3 6	4 10	8 12	9 14	11 13
1 6	2 12	3 11	4 14	5 9	7 10	8 13
1 7	2 9	3 14	4 5	6 8	10 11	12 13
1 8	2 14	3 10	4 11	5 7	6 12	9 13
1 9	2 8	3 12	4 13	5 10	6 11	7 14
1 10	2 11	3 9	4 6	5 13	7 12	8 14
1 11	2 13	3 7	4 9	5 8	6 14	10 12
1 12	2 4	3 13	5 14	6 7	8 10	9 11
1 13	2 3	4 8	5 12	6 10	7 9	11 14
1 14	2 10	3 5	4 12	6 13	7 11	8 9

Annals of Discrete Mathematics 34 (1987) 437—440
© Elsevier Science Publishers B.V. (North-Holland)

A Construction for Orthogonal Designs with Three Variables

Jennifer Seberry

Department of Computer Science
University College
The University of New South Wales
Australian Defence Forces Academy
Canberra, A.C.T. 2600
AUSTRALIA

TO ALEX ROSA ON HIS FIFTIETH BIRTHDAY

ABSTRACT

We show how orthogonal designs $OD(48p^2t;16p^2t, 16p^2t,16p^2t)$ can be constructed from an Hadamard matrix of order $4p$ and an $OD(4t;t,t,t,t)$. This allows us to assert that $OD(48p^2t; 16p^2t,16p^2t,16p^2t)$ exist for all $t,p \le 102$ except possibly for $t \in \{67,71,73,77,79,83,86,89,91,97\}$. These designs are new.

1. Introduction

Let $H = (h_{ij})$ be a matrix of order h with $h_{ij} \in \{1,-1\}$. H is called an *Hadamard matrix* of order n, if $HH^T = hI_h$ where I_h denotes the identity matrix order of h.

An *orthogonal design* A, of order n, type $(p_1,p_2, \ldots ,p_u)$, denoted $OD(n;p_1,p_2, \ldots ,p_u)$, on the commuting variables $(\pm x_1,\pm x_2, \ldots , \pm x_u,0)$ is a square matrix of order n with entries $\pm x_k$ where each x_k occur p_k times in each row and column such that the rows are pairwise orthogonal.

In other words

$$AA^T = (p_1 x_1^2 + \cdots + p_u x_u^2)I_n .$$

It is known that the maximum number of variables is an orthogonal design is $\rho(n)$, the Radon number, where for $n = 2^a b$, b odd, set $a = 4c + d, 0 \le d < 4$, then $\rho(n) = 8c + 2^d$.

$OD(4t;t,t,t,t)$, otherwise called Baumert-Hall arrays, and $OD(2^p;a,b,2^p-a-b)$ have been extensively used to construct Hadamard matrices and weighing matrices. For details see Geramita and Seberry (1979).

Geramita, Geramita and Wallis (=Seberry) observed (see Geramita and Seberry [1979, §4.3]) that if A,B,C,D are four circulant or type 1 matrices of order n satisfying

$$AA^T + BB^T + CC^T + DD^T = (\sum_{i=1}^{u} p_i x_i^2) I_n$$

then A,B,C,D can be used in the Goethals-Seidel array or (J. Seberry Wallis-Whiteman array)

$$\begin{bmatrix} A & BR & CR & -DR \\ -BR & A & D^TR & -C^TR \\ CR & -D^TR & A & B^TR \\ DR & C^TR & -B^TR & A \end{bmatrix} \quad (1).$$

to form an orthogonal design $OD(4n;p_1,p_2,...,p_u)$.

2. Background

Kharaghani (1985) defined $C_k = [h_{ij}.h_{kj}]$ and applying that to Hadamard matrices of order $4p$, obtained matrices satisfying

$$C_i C_j = 0,\ i \neq j \quad (2).$$
$$\sum_{i=1}^{4p} C_i^2 = (4p)^2 I_{4p}.$$

He then used this to show there are Bush-type (blocks J_{4p} down the diagonal) and Szekeres-type ($h_{ij} = -1 \Rightarrow h_{ji} = 1$ and *not* necessarily vice versa) Hadamard matrices. By using a symmetric Latin square he could also have shown that regular symmetric Hadamard matrices with constant diagonal of order $(4p)^2$ could be constructed by his method.

Hammer, Sarvate and Seberry applied Kharaghani's method to $OD(n;s_1, \ldots, s_u)$ and in particular $OD(4t;t,t,t,t)$ and $OD4s;s,s,s,s)$ obtaining existence of $OD(48s^2t$ $12s^2t,\ 12s^2t,\ 12s^2t,\ 12s^2t)$ and $OD(80s^2t;\ 20s^2t,\ 20s^2t,\ 20s^2t,\ 20s^2t)$ from $OD(4s;s,s,s,s)$ and $0D(4t;t,t,t,t)$. Seberry (to appear) extended this further to obtain $OD(16ks^2t;4ks^2t,4ks^2t,4ks^2t,4ks^2t)$ for {1,3,5,...}.

We modify their techniques to obtain new orthogonal designs.

3. Construction

Let $C_1,C_2, \ldots, C_{4p}$ be the Kharaghani matrices of order $4p$ obtained from an orthogonal design $OD(4p;p,p,p,p)$.

Let a,b,c be commuting variables and write

$$[aC_1 : aC_2 :\ldots: aC_p : bC_{p+1} :\ldots: bC_{2p} : cC_{2p+1} :\ldots: cC_{3p}]$$
$$[aC_{3p+1} :\ldots: aC_{4p} : cC_{p+1} :\ldots: cC_{2p} : bC_{2p+1} :\ldots: bC_{3p}]$$

for the first rows of two block circulant matrices W_1 and W_2. It can be checked that $W_1 W_2^T = W_2 W_1^T$. Write

$$[aC_{2p+1}:....:aC_{3p} : bC_1:...:bC_p : cC_{3p+1}:...:cC_{4p}]$$

$$[aC_{p+1}:....:aC_{2p} : bC_1:...bC_p : bC_{3p+1}:....bC_{4p}]$$

for the first rows of two block back-circulant matrices W_3 and W_4. It can be checked that $W_3 W_4^T = W_4 W_3^T$.

Now by virtue of being circulant and back-circulant

$$W_i W_j^T = W_j W_i^T \qquad i \in \{1,2\}, \quad j \in \{3,4\}.$$

Example. Let $p = 3$ so there are 12 matrices of order 12, $C_1, \ldots, C_{12}$. Then

$$W_1 = \begin{bmatrix} aC_1 & aC_2 & aC_3 & bC_4 & bC_5 & bC_6 & cC_7 & cC_8 & cC_9 \\ aC_3 & aC_1 & aC_2 & bC_6 & bC_4 & bC_5 & cC_9 & cC_7 & cC_8 \\ aC_2 & aC_3 & aC_1 & bC_5 & bC_6 & bC_4 & cC_8 & cC_9 & cC_7 \\ cC_7 & cC_8 & cC_9 & aC_1 & aC_2 & aC_3 & bC_4 & bC_5 & bC_6 \\ cC_9 & cC_7 & cC_8 & aC_3 & aC_1 & aC_2 & bC_6 & bC_4 & bC_5 \\ cC_8 & cC_9 & cC_7 & aC_2 & aC_3 & aC_1 & bC_5 & bC_6 & bC_4 \\ bC_4 & bC_5 & bC_6 & cC_7 & cC_8 & cC_9 & aC_1 & aC_2 & aC_3 \\ bC_6 & bC_4 & bC_5 & cC_9 & cC_7 & cC_8 & aC_3 & aC_1 & aC_2 \\ bC_5 & bC_6 & bC_4 & cC_8 & cC_9 & cC_7 & aC_2 & aC_3 & aC_1 \end{bmatrix}$$

$$W_2 = \begin{bmatrix} aC_{10} & aC_{11} & aC_{12} & cC_4 & cC_5 & cC_6 & bC_7 & bC_8 & bC_9 \\ aC_{12} & aC_{10} & aC_{11} & cC_6 & cC_4 & cC_5 & bC_9 & bC_7 & bC_8 \\ aC_{11} & aC_{12} & aC_{10} & cC_5 & cC_6 & cC_4 & bC_8 & bC_9 & bC_7 \\ bC_7 & bC_8 & bC_9 & aC_{10} & aC_{11} & aC_{12} & cC_4 & cC_5 & cC_6 \\ bC_9 & bC_7 & bC_8 & aC_{12} & aC_{10} & aC_{11} & cC_6 & cC_4 & cC_5 \\ bC_8 & bC_9 & bC_7 & aC_{11} & aC_{12} & aC_{10} & cC_5 & cC_6 & cC_4 \\ cC_4 & cC_5 & cC_6 & bC_7 & bC_8 & bC_9 & aC_{10} & aC_{11} & aC_{12} \\ cC_6 & cC_4 & cC_5 & bC_9 & bC_7 & bC_8 & aC_{12} & aC_{10} & aC_{11} \\ cC_5 & cC_6 & cC_4 & bC_8 & bC_9 & bc_7 & aC_{11} & aC_{12} & aC_{10} \end{bmatrix}$$

$$W_1 W_2^T = I_q \times bc \sum_{i=4}^{q} C_i c_i^T$$

but $C_i C_i^T$ is symmetric and so $W_1 W_2^T = W_2 W_1^T$.

W_3 and W_4 are formed similarly but are back-circulant in blocks (i.e., type 2) in the language of Seberry-Wallis and Whiteman).

Thus W_1, W_2, W_3, W_4 are Williamson-type matrices of order $12p^2$ they can be used to replace the variables of an $OD(4;1,1,1,1)$ to get an $OD(48p^2;\ 16p^2,\ 16p^2,\ 16p^2,\ 16p^2)$. In general they can be used to replace the variables of an $OD(4t;t,t,t,t)$ so we have

Theorem: If there is an Hadamard matrix of order $4p$ and an $OD(4t;t,t,t,t)$ then there is an $OD(48p^2t;16p^2t,16p^2t,16p^2t,16p^2t)$.

Since Hadamard matrices of order $4p$ exist for all $p \leq 102$ and $OD(4t;t,t,t,t)$ exist for all $t \leq 102$ except possibly for $t \in S$, $S = \{67,71,73,77,79,83,86,89,91,97\}$ (see Seberry (1986a)) we have

Corollary: *$OD(48p^2t;16p^2t,16p^2t,16p^2t,16p^2t)$ exist for all $t,p \leq 102$ except possibly for $t \in S$.*

Acknowledgement

The author's research is supported in part by grants from ACRB and ATERB.

References

[1] J. Cooper and J. Seberry Wallis, "A construction of Hadamard arrays", *Bull. Austral Math. Soc.,* 7 (1972) 269-278.

[2] A.V. Geramita and J. Seberry, *Orthogonal Designs: Quadratic Forms and Hadamard Matrices* Marcel Dekker, New York-Basel (1986).

[3] J. Hammer, D.G. Sarvate, J. Seberry, "A note on orthogonal designs", *Ars Combinatoria,* submitted.

[4] H. Kharagani, "New class of weighing matrices", *Ars Combinatorica,* Vol. 19, pp. 69-72.

[5] J. Seberry, "More towards proving the Hadamard conjecture", (preprint).

[6] J. Seberry-Wallis, "Hadamard matrices, Part IV of W.D. Wallis, A. P. Street, and J. Seberry-Wallis", *Combinatorics: Room Square, sum free sets and Hadamard matrices,* Lecture Notes in Mathematics, Vol. 292, Springer-Verlag, Berlin-Heidelberg New York.

Annals of Discrete Mathematics 34 (1987) 441–448
© Elsevier Science Publishers B.V. (North-Holland)

Isomorphism Classes of Small Covering Designs with Block Size Five

R.G. Stanton

Department of Computer Science
University of Manitoba
Winnipeg, Manitoba, R3T 2N2
CANADA

TO ALEX ROSA ON HIS FIFTIETH BIRTHDAY

ABSTRACT

The number of nonisomorphic covering designs with block size five is determined for $v < 12$. The corresponding excess graphs include the Petersen graph.

1. Introduction

For balanced incomplete block designs with parameters (v,b,r,k,λ), various authors have determined the number of nonisomorphic designs for small values of v; see, for example, [2,3,5,11,13], and the references there cited. In a covering design on v elements with block size k (see [12]), every pair of elements occurs at least once; however, in order to cover all pairs, it may be necessary to have a few repetitions. We always use the smallest number of blocks possible in such a cover, and this number is denoted by $\mathrm{N}(2,k,v)$.

In a BIBD, one can establish by elementary counting that $bk = rv$; the same counting argument trivially establishes for a covering design the well known bound

$$\mathrm{N}(2,k,v) \geq \mathrm{N}(1,k-1,v-1)v = \lceil (v-1)/(k-1) \rceil v,$$

which should probably, in consequence, be called the Fisher bound. Usually, the value of $\mathrm{N}(2,k,v)$ is very close to the bound.

If we set $N = \mathrm{N}(2,k,v)$ for brevity, then there are $Nk(k-1)/2$ pairs in the blocks of the covering design. Of these, $v(v-1)/2$ are distinct, and hence there must be a total of $Nk(k-1)/2 - v(v-1)/2$ excess pairs. The excess pairs can be represented by a graph on v points called the *excess graph*. This graph will be empty if the covering design is a BIBD.

Not a great deal is known about nonisomorphic solutions for covering designs. The case for N(2,4,9) is discussed in [10]; see also [1] for a correction; the case $N_\lambda(3,4,7)$ is discussed in [9].

In this paper, we consider the case for $N(2,5,v)$ for $v<12$. The values of $N(2,5,v)$ for $v<24$ have been found by various authors, and a summary is given in Mills [6], who includes a complete list of references to earlier results. Many general results for $N(2,5,v)$ have been recently established by Mullin and others; see [4,7,8].

We henceforth employ the abbreviation $N(v) = N(2,5,v)$ throughout this paper.

2. The Trivial Cases for $N(v)$, $4 < v < 10$.

For $v = 5$, $N(v) = 1$; the cover 12345 is, of course, unique, and the excess graph is empty.

For $v = 6$, $N(v) = 3$; up to isomorphism, the cover is 12345, 12346, 12356. The excess graph is shown in Figure 1.

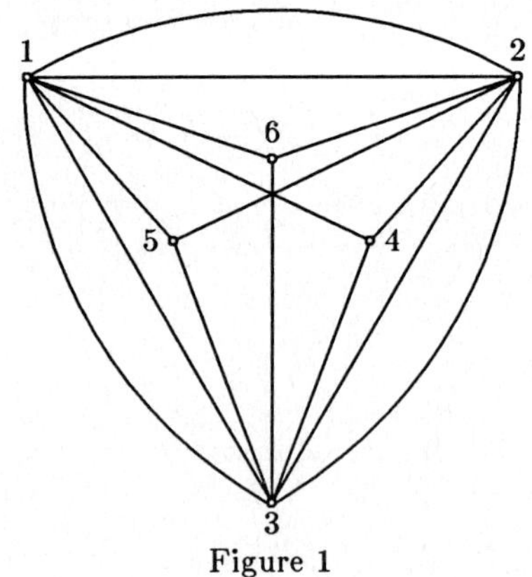

Figure 1

For $v = 7$, $N(v) = 3$; every symbol has frequency at least 2, and hence there is a single symbol that has frequency 3. We immediately get the unique cover 12345, 12367, 14567. The excess graph is shown in Figure 2.

If $v = 8$, $N(v) = 4$; since $N(2,4,7) = 5$, it is not possible to have a symbol of frequency 4. Hence there are four symbols of frequency 3 and four symbols of frequency 2; we let the symbols of frequency 3 be 1,2,3,4. Now the blocks 123XX, 124XX, 134XX, 234XX, cannot be completed; hence there must be a block 12345. Some pair from 1,2,3,4, must occur twice more; so we may take the blocks as 12XXX, 12XXX, XXXXX. Since $f(5) = 2$, the last block must be X5678. Then we obtain, up to isomorphism, the unique cover 12345, 12346, 12378, 45678, and have established

Lemma 1: The cover for $v = 8$ is unique, and consists of the four blocks 12345, 12346, 12378, 45678. □

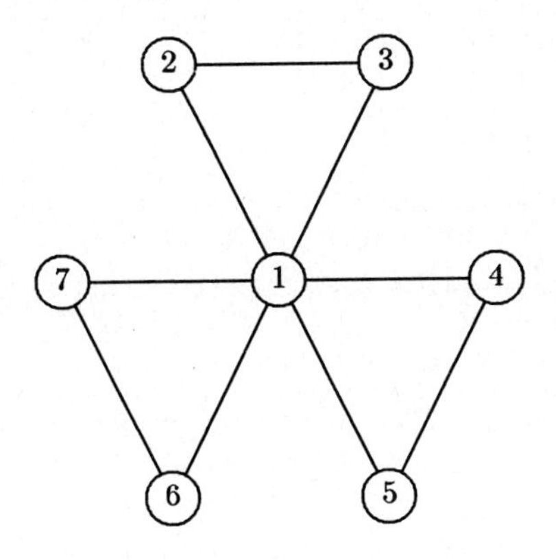

Figure 2

The excess graph is shown in Figure 3.

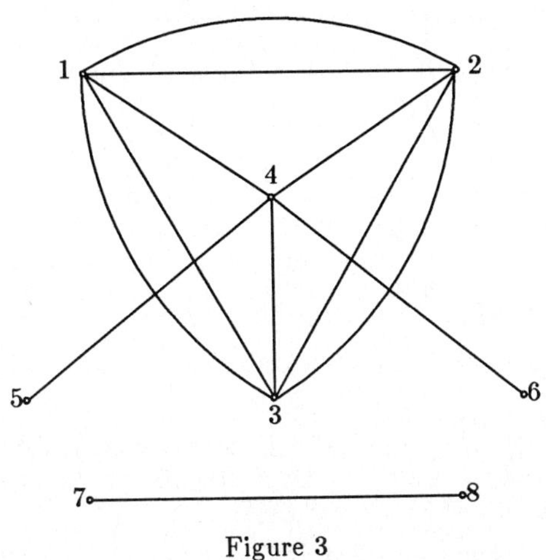

Figure 3

We now suppose that $v = 9$; then $N(v) = 5$, and every symbol has frequency at least 2. Since $N(2,4,8) > 5$, no symbol can have frequency 5. Also, there must be at least two symbols of frequency 2; so we may let these be 1 and 2. Then we obtain the blocks 12345, 16789, 26789. The use of a block 36789 leads to a contradiction; hence the frequency is 3 for all of the symbols 3,4,5,6,7,8,9. Thus we obtain

Lemma 2: If $v = 9$, there is a unique cover given by 12345, 16789, 26789, 34567, 34589. □

The excess graph is shown in Figure 4.

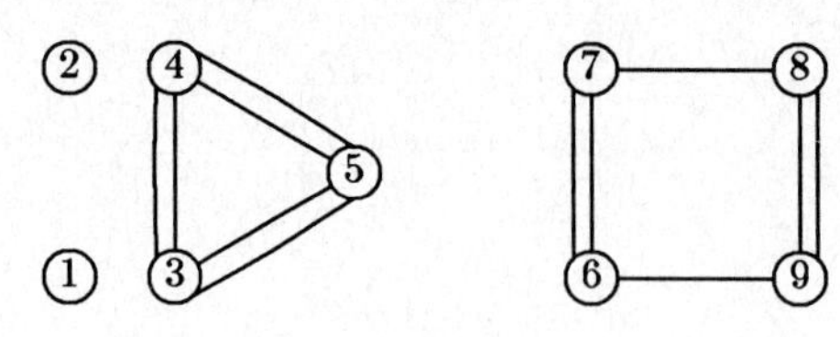

Figure 4

3. The Case v = 10.

Here N(v) = 6, and each symbol has frequency at least 3. If we assume that there are two blocks 12345 and 12346 that have four elements in common, then we are forced to have blocks i789T (i = 1,2,3,4), and this is not possible. Hence, no two blocks can have four elements in common.

Similarly, if two blocks are disjoint, say 12345 and 6789T, then we need 1678X and 19TXX; this requires 6,7,8, to occur with 2,3,4,5, and we cannot achieve this. Hence it is not possible to have two disjoint blocks.

Now designate the block 12345 as the first block B_1, and let x_i represent the number of blocks having intersection number i with B_1. From

$$x_1 + x_2 + x_3 = 5,\ x_1 + 2x_2 + 3x_3 = 10,$$

we deduce that $x_2 + 2x_3 = 5$. Hence $x_2 > 0$, and we may take the first two blocks to be 12345, 12678.

First, suppose that no pair ij ever occurs three times; then the blocks can be written in the form 12345, 12678, 19TXX, 29TXX, 9XXXX, TXXXX. If B_3 is 19T34, then the last blocks would have to be 93678 and T4678; this is not possible. Hence there is no loss of generality in taking B_3 as 19T36 and B_4 as 29T47. This forces the final blocks to be 95837 and T5846.

The excess graph is shown in Figure 5; it is simply the ubiquitous Petersen graph.

Now suppose that some pair, say 12, can occur three times. We may then start with the blocks 12345, 12678, 129T3. The remaining blocks must be 3678X, XXXXX, XXXXX. Since (45) and (9T) are equivalent pairs, we may take B_4 to be 34678; then the remaining blocks must be 9T546 and 9T578.

The excess graph is shown in Figure 6.

We sum up the results of this section in

Theorem 1: If v = 10, then N(v) = 6, and there are two covering designs, namely,

12345, 12678, 1369T, 2479T, 35789, 4568T, and

12345, 12678, 1239T, 34678, 4569T, 5789T. □

We note that the first design is embeddable in the standard BIBD with v = 11, k = 5, and λ = 2 (add the blocks 1489E, 157TE, 238TE, 2569E, 3467E).

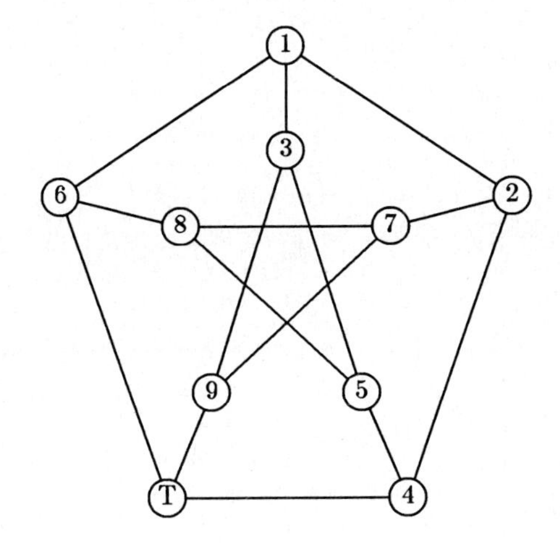

Figure 5

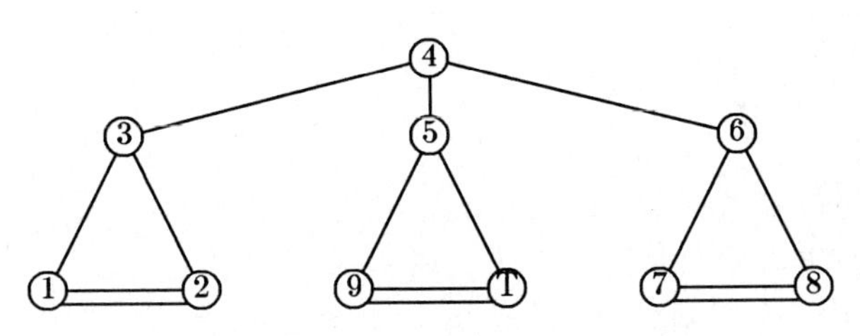

Figure 6

4. The Case v = 11.

If $v = 11$, $\mathrm{N}(v) = 7$, and $f(i) > 2$ for all i. Suppose, if possible, that $f(\mathrm{E}) = 5$; then the blocks are $(\mathrm{EXXXX})^5$, 12345, XXXXX. Consider now the sets $A = \{1,2,3,4,5\}$ and $B = \{6,7,8,9,\mathrm{T}\}$; then the number of pairs $a_i b_j$ ($a_i \in A$; $b_j \in B$) is 25. Each of the first five blocks contributes 0, 3, or 4 such pairs; the last block contributes 0, 4, or 6. We at once see that the only possibilities are $25 = 6 + 4(4) + 3$, or $26 = 6 + 5(4)$. However, we also need ten pairs $b_i b_j$, and (in both cases) there are only four pairs from the four blocks that have a (2,2) split from A and B. It follows that the last block must have the form $aabbb$, and the fifth block must have the form $\mathrm{E}abbb$. We thus have a total of 16 elements from A and 14 from B in the cover; this is a contradiction, since $f(i) > 2$ for i selected from B. We thus see that no element can have frequency 5; hence there must be nine elements of frequency 3, and two elements of frequency 4.

We now consider the possibility that $x_4>0$; let the first two blocks be 12345, 12346. At most two of the elements 1,2,3,4, can have frequency 4; so we may let $f(1) = 3$. Then 1 must appear with 7,8,9,T,E in a single block, and this is certainly impossible. Thus $x_4 = 0$ in all cases.

If $x_0>0$, we can write the blocks as 12345, 6789T, $(\mathrm{EXXXX})^3$, $(\mathrm{XXXXX})^2$. (It is possible that a symbol E appears in one of the last two blocks.) If we set $A = \{1,2,3,4,5\}$ and $B = \{6,7,8,9,\mathrm{T}\}$, then there must be 25 pairs $a_i b_j$. But the maximum number of such pairs is $3(4) + 2(6) < 25$. Hence we conclude that $x_0 = 0$.

Now let T and E be the symbols of frequency 4, and consider the block TE123. The intersection numbers of this block satisfy the equations

$$x_1 + x_2 + x_3 = 6,\ x_1 + 2x_2 + 3x_3 = 12.$$

Either $x_3>0$ or $x_2 = 6$. In the latter case, we can write the six other blocks in the form $aabbb$, where $A = \{\mathrm{T,E},1,2,3\}$ and $B = \{4,5,6,7,8\}$. Thus we have to have six triples that cover all pairs of the form bb; these can be written uniquely as 456, 478, 495, 578, 769, 869. Now we need to add the symbol 3 to two of the blocks so that all of the pairs 34, 35, 36, 37, 38, 39, occur. This is impossible, and so we conclude that $x_3>0$.

Consequently, we may assume two initial blocks TE123 and TE145. Since there can be only two symbols of frequency 4, we may take $f(1) = 3$; hence 16789 is a block. Since $x_4 = 0$, we see that $f(\mathrm{T}) = f(\mathrm{E}) = 4$.

Now let $A = \{\mathrm{T,E},2,3,4,5\}$ and $B = \{6,7,8,9\}$. There must be 24 pairs of the form ab, and this requires us to have two elements from B in each of the last four blocks. The distribution 67XXX, 68XXX, 79XXX, 89XXX, cannot be completed; so the last four blocks must be taken as 67XXX, 67XXX, 89XXX, 89XXX.

If T and E occur together in these last four blocks, the skeleton is readily completed to be 67TE2, 67345, 89TE3, 89245. The corresponding excess graph is shown in Figure 7.

On the other hand, if T and E do not occur together, then the last four blocks may be taken as 67T24, 67E35, 89T34, 89E25, and the corresponding excess graph is shown in Figure 8. (Note that the completion 67T24, 67E35, 89T25, 89E34, is an isomorphic cover under application of the permutation (TE)(25)(34).)

Both of these covers are obtainable by the "star" method of Mullin (cf. [8]). Take the geometry 124, 235, 346, 457, 561, 672, 713, and leave the points 4,5,7, fixed. The blocks of Mullin's "star" cover then are

$$457,\ 5616^*1^*,\ 6726^*2^*,\ 7131^*3^*,\ 1241^*2^*,\ 2352^*3^*,\ 3463^*6^*.$$

If we inflate 457 to be the block 22^*457, we obtain the cover corresponding to Figure 7 (the isomorphism is easily obtained from the excess graph). On the other hand, if we inflate 457 to be the block 23457, then we have the cover corresponding to Figure 8 (again, the isomorphism is easily obtained from the excess graph).

The two cases come from the fact that the excess graph of a "star" cover is merely made up of four double links. In Figure 7, we use two points from the same double link to inflate the triple, whereas in Figure 8 we use points from different double links. So the Mullin "star" cover is unique, but there are two nonisomorphic standard covers. We state this result as

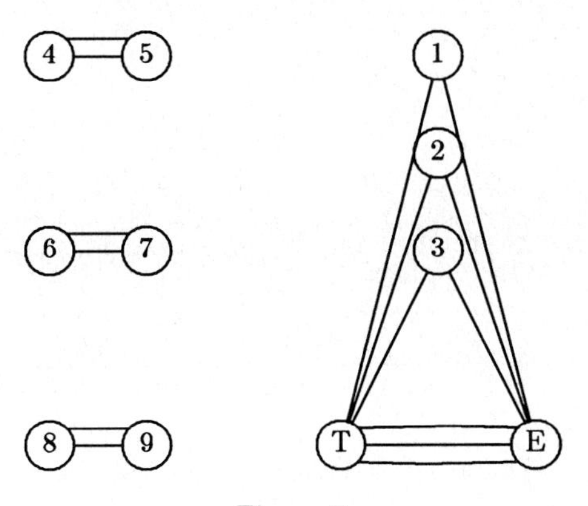

Figure 7

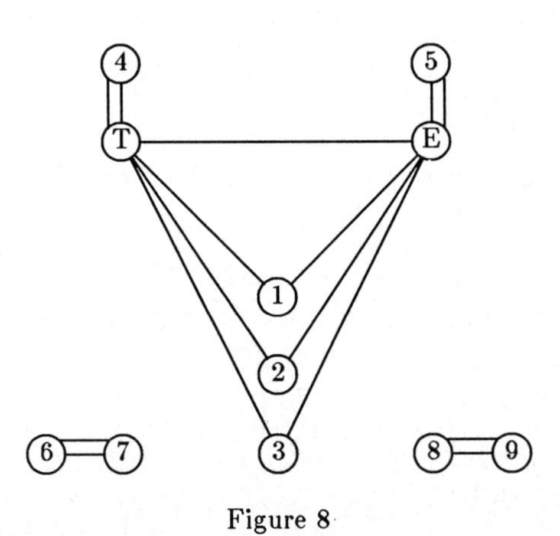

Figure 8

Theorem 2: If $v = 11$, we have two covers given by

TE123, TE145, TE267, TE389, 16789, 24589, 34567, and

TE123, TE145, 16789, T2467, T3489, E2589, E3567. □

5. Conclusion.

All nonisomorphic covers by quintuples have been determined for $v<12$. The only regular excess graph with a connectivity greater than unity is the Petersen graph.

References

[1] J.A. Bate and G.H.J. van Rees, "Some results on N(4,6,10), N(4,6,11), and related coverings", *Congressus Numerantium* 48 (1985) 25-45.

[2] R.A. Fisher, "An examination of the different possible solutions of a problem in incomplete blocks", *Annals of Eugenics* 10 (1940) 52-75.

[3] P.B. Gibbons, R.A. Mathon, and D.G. Corneil, "Computing techniques for the construction and analysis of block designs", *Utititas Mathematica* 11 (1977) 161-192.

[4] E. Lamken, W.H. Mills, R.C. Mullin, and S.A. Vanstone, "On covering pairs by quintuples", *J. Comb. Theory* A, to appear, 1987.

[5] R.A. Mathon, K.T. Phelps, and A. Rosa, "Small Steiner triple systems and their properties", *Ars Combinatoria* 15 (1983) 3-110.

[6] W.H. Mills, "A covering of pairs by quintuples", *Ars Combinatoria* 18 (1984) 21-31.

[7] W.H. Mills and R.C. Mullin, "On covering pairs by quintuples: the case $v \equiv 3$ modulo 4", submitted.

[8] R.C. Mullin, "On covering pairs by quintuples: the cases $v \equiv 3$ or 11 modulo 20", *Utilitas Mathematica* 31 (1987).

[9] R.C. Mullin, E. Nemeth, R.G. Stanton, and W.D. Wallis, "Isomorphism properties of some small covers", *Utilitas Mathematica* 29 (1986) 269-283.

[10] R.G. Stanton, J.L. Allston, D.D. Cowan, and W.D. Wallis, "The number of nonisomorphic solutions to a problem in covering designs", *Utilitas Mathematica* 21A (1982) 119-136.

[11] R.G. Stanton and R.J. Collens, "A computer system for research on the family classification of BIBDs", *Proc. International Conference on Combinatorial Theory,* Academia dei Lincei, Rome, 1973, pp. 133-169.

[12] R.G. Stanton, J.G. Kalbfleisch, and R.C. Mullin, "Covering and packing designs", *Proc. Second Chapel Hill Conference on Combinatorial Mathematics and its Applications,* University of North Carolina, Chapel Hill, 1970, pp. 428-450.

[13] R.G. Stanton, R.C. Mullin, and J.A. Bate, "Isomorphism classes of a set of prime BIBD parameters", *Ars Combinatoria* 2 (1976) 251-264.

Annals of Discrete Mathematics 34 (1987) 449–460
© Elsevier Science Publishers B.V. (North-Holland)

Graphs Which are not Leaves of Maximal Partial Triple Systems

D.R. Stinson

Department of Computer Science
University of Manitoba
Winnipeg, Manitoba, R3T 2N2
CANADA

W.D. Wallis

Department of Mathematics
Southern Illinois University
Carbondale, Illinois, 62901
U.S.A.

TO ALEX ROSA ON HIS FIFTIETH BIRTHDAY

ABSTRACT

The *leave* of a set of edge-disjoint triangles is the complement of the union of the triangles. The leave of a maximal partial Steiner triple system is therefore a triangle-free graph, and it satisfies certain restrictions on its number of vertices and edges and its degrees. A *pseudoleave* is defined to be a graph which is triangle-free and satisfies the parameter restrictions but is not the leave of any partial triple system. Upper and lower bounds for the number of edges in a pseudoleave on n vertices are obtained, and it is proven that most values between these bounds can be attained.

1. Introduction

A *Steiner triple system* of order n (denoted STS(n)) is a pair $(X,\mathcal{A})$ where X is a set of n points, and $\mathcal{A}$ is a collection of 3-subsets (called *blocks),* such that every pair of points occurs in a unique block. It is well known that an STS(n) exists if and only if $n \equiv 1$ or 3 modulo 6.

A *partial* STS(n) (denoted PTS(n)) is similar, except that we require that every pair of points occurs in at most one block (such designs are also called *packings).* If $D = (X,\mathcal{A})$ is a PTS(n), then we define a graph $L(D)$, which we call the *leave* of D, as follows. $L(D)$ has vertex set X, and xy is an edge of $L(D)$ if and only if there is no block $A \in \mathcal{A}$ such that $\{x,y\} \subseteq A$. Alternatively, if we interpret each block of D as a triangle, then $L(D)$ is the complement (in the complete graph based on X) of the union of those triangles.

Suppose we are given a graph G, and we want to determine whether G is the leave of a PTS(n), where n is the number of vertices of G. This is equivalent to asking whether $K_n - G$ can be edge-partitioned into triangles corresponding to a block of the PTS. If G is a leave, it must satisfy some numerical conditions. Since every block (=triangle) uses up three pairs of points (=edges of $K_n - G$), we must have

$$\epsilon \equiv \binom{n}{2} \text{ modulo } 3, \tag{1}$$

where ϵ denotes the number of edges in G. Also, whenever a point x occurs in a block, it occurs in two pairs. Hence, if d_x denotes the degree of x in G, we require

$$d_x \equiv (n-1) \text{ modulo } 2, \text{ for every point } x. \tag{2}$$

Unfortunately, these two necessary conditions are not sufficient for G to be a leave. Indeed, it is unlikely that one can determine in polynomial time whether a graph is a leave of a PTS, since this question has been shown to be NP-complete by Holyer [3].

An interesting case is to consider *maximal* PTS(n) (denoted MPTS(n)). A PTS(n) $(X,\mathcal{A})$ is said to be maximal if there is no PTS(n) $(X,\mathcal{A}')$ with $\mathcal{A} \subset \mathcal{A}'$. Equivalently, we require the property that

$$G \text{ is triangle-free}, \tag{3}$$

where $G = L(X,\mathcal{A})$.

We suspect that it is also an NP-complete problem to recognize leaves of MPTS, but this has not been proved.

Leaves of MPTS are studied in [1] and [2], where the following existence result is proved.

Theorem 1.1: If a graph G on n vertices satisfies conditions (1), (2), and (3), and every vertex of G has degree 0 or 2, then G is a leave of an MPTS if and only if G is not the disjoint union $C_4 \cup C_5$ (with no isolated vertices). □

In this paper, we study graphs which satisfy conditions (1), (2), and (3), but which are not leaves of MPTS. We define a *pseudoleave* (of an MPTS) to be a graph which satisfies conditions (1), (2), and (3), as well as

$$K_n - G \text{ cannot be edge-partitioned into triangles.} \tag{4}$$

We are interested in how many edges a pseudoleave on n vertices can have. Define $S(n) = \{\epsilon$: there is a pseudoleave on n vertices having ϵ edges$\}$. We refer to $S(n)$ as the *spectrum* of n. Also, we define $Max(n) = \max\{\epsilon \in S(n)\}$ and $Min(n) = \min\{\epsilon \in S(n)\}$.

In Section 2, we derive some lower bounds on $Min(n)$ and some upper bounds on $Max(n)$. In Sections 3 and 4, we give some explicit constructions for pseudoleaves. These enable us to show that a wide range of values of ϵ are in the spectrum for n. In particular, we determine $Max(n)$ exactly for $n \equiv 7$ or 11 (mod 12), $n \geq 11$.

We summarize our results in Tables 1 and 2.

Table 1
The maximum and minimum number of edges in pseudoleaves

n	$Min(n)$	$Max(n)$
$\equiv 0,2$ mod 6	$\geq n/2+3$	$\leq n^2/4-3$
$\equiv 4$ mod 6	$\geq n/2+4$	$\leq n^2/4-1$
$\equiv 1$ mod 12, $n \geq 13$	≥ 9	$\leq (n^2-2n+1)/4$
$\equiv 3$ mod 12, $n \geq 15$	≥ 9	$\leq (n^2-2n-3)/4$
$\equiv 5$ mod 12, $n \geq 17$	≥ 10	$\leq (n^2-2n+1)/4$
$\equiv 7$ mod 12, $n \geq 7$	≥ 9	$\leq (n^2-2n+1)/4$
$\equiv 9$ mod 12, $n \geq 21$	≥ 9	$\leq (n^2-2n-3)/4$
$\equiv 11$ mod 12	≥ 10	$\leq (n^2-2n+1)/4$

Table 2
Constructions for pseudoleaves

n (mod 12)	Number of edges $\epsilon \in S(n)$	Construction (see §4)
1	$\equiv 3$ mod 6, $(3n+3)/2 \leq \epsilon \leq (n^2-2n-11)/4$	(i)
	$\equiv 0$ mod 6, $2n-2 \leq \epsilon \leq (n^2-6n+5)/4$	(iii)
5	$\equiv 1$ mod 6, $(3n+11)/2 \leq \epsilon \leq (n^2-2n-11)/4$	(i)
	$\equiv 4$ mod 6, $2n-6 \leq \epsilon \leq (n^2-6n+21)/4$	(iii)
9	$\equiv 3$ mod 6, $(3n+3)/2 \leq \epsilon \leq (n^2-2n-27)/4$	(i)
	$\equiv 0$ mod 6, $2n-6 \leq \epsilon \leq (n^2-6n+21)/4$	(iii)
3	$\equiv 3$ mod 6, $(3n-3)/2 \leq \epsilon \leq (n^2-2n-15)/4$	(ii)
	$\equiv 0$ mod 6, $2n-6 \leq \epsilon \leq (n^2-6n+9)/4$	(iii)
7	$\equiv 3$ mod 6, $(3n-3)/2 \leq \epsilon \leq (n^2-2n+1)/4$	(ii)
	$\equiv 0$ mod 6, $2n-2 \leq \epsilon \leq (n^2-6n-7)/4$	(iii)
11	$\equiv 1$ mod 6, $(3n+5)/2 \leq \epsilon \leq (n^2-2n+1)/4$	(ii)
	$\equiv 4$ mod 6, $2n-6 \leq \epsilon \leq (n^2-6n+9)/4$	(iii)
0	$\equiv 0$ mod 6, $n+6 \leq \epsilon \leq (n^2-2n)/4$	(iv)
	$\equiv 3$ mod 6, $2n-3 \leq \epsilon \leq (n^2-4n-12)/4$	(v)
4	$\equiv 0$ mod 6, $n+2 \leq \epsilon \leq (n^2-2n-8)/4$	(iv)
	$\equiv 3$ mod 6, $2n+1 \leq \epsilon \leq (n^2-4n-28)/4$	(v)
8	$\equiv 4$ mod 6, $n+2 \leq \epsilon \leq (n^2-2n-8)/4$	(iv)
	$\equiv 1$ mod 6, $2n-3 \leq \epsilon \leq (n^2-4n-12)/4$	(v)
2	$\equiv 4$ mod 6, $n+2 \leq \epsilon \leq (n^2-2n-8)/4$	(iv)
	$\equiv 1$ mod 6, $2n-3 \leq \epsilon \leq (n^2-4n+8)/4$	(v)
6	$\equiv 0$ mod 6, $n+6 \leq \epsilon \leq (n^2-2n-24)/4$	(iv)
	$\equiv 3$ mod 6, $2n-3 \leq \epsilon \leq (n^2-4n-24)/4$	(v)
10	$\equiv 0$ mod 6, $n+2 \leq \epsilon \leq (n^2-2n-8)/4$	(iv)
	$\equiv 3$ mod 6, $2n+1 \leq \epsilon \leq (n^2-4n-24)/4$	(v)

2. Bounds on the number of edges in a pseudoleave

In this section, we make some observations on the number of edges in a pseudoleave on n vertices.

Theorem 2.1: If n is even, then any pseudoleave on n vertices has at least $n/2+c$ edges, where

$$c = \begin{cases} 3 & \text{if } n \equiv 0 \text{ or } 2 \text{ modulo } 6 \\ 4 & \text{if } n \equiv 4 \text{ modulo } 6. \end{cases}$$

Proof:

Since n is even, every vertex in a pseudoleave has positive degree, and hence $\epsilon \geq n/2$. If $n \equiv 4$ modulo 6, then $\epsilon \geq n/2+1$, since $\epsilon \equiv \binom{n}{2}$ modulo 3.

If $\epsilon = n/2$ ($n \equiv 0$ or 2 modulo 6), then the pseudoleave must consist of $n/2$ disjoint edges. But such a graph is a leave (see [6]). The next possible value for ϵ is $n/2+3$.

If $\epsilon = n/2+1$ ($n \equiv 4$ modulo 6), then the pseudoleave would consist of $(n-4)/2$ disjoint edges and a 3-star. However, this graph is also a leave (see [6]), whence we obtain $\epsilon \geq n/2+4$. □

For odd n, we can prove only constant lower bounds.

Theorem 2.2: If $n \geq 11$ is odd, then any pseudoleave on n vertices has at least c edges, where

$$c = \begin{cases} 9 & \text{if } n \equiv 1 \text{ or } 3 \text{ modulo } 6 \\ 10 & \text{if } n \equiv 5 \text{ modulo } 6. \end{cases}$$

Proof:

From Theorem 1.1, we can assume that there is a vertex of degree ≥ 4. Since a pseudoleave is triangle-free, it is easy to check that there must be at least eight edges. Then since $\epsilon \equiv \binom{n}{2}$, the result follows. □

We now turn our attention to upper bounds. First, we prove

Theorem 2.3: If n is even, then any pseudoleave on n vertices contains at most $n^2/4-c$ edges, where

$$c = \begin{cases} 3 & \text{if } n \equiv 0 \text{ or } 2 \text{ modulo } 6 \\ 1 & \text{if } n \equiv 4 \text{ modulo } 6. \end{cases}$$

Proof:

We use the well known result [7] that a triangle-free graph on n vertices contains at most $n^2/4$ edges, and equality holds if and only if the graph is $K_{n/2,n/2}$. Since $\epsilon \equiv \binom{n}{2}$ modulo 3 in a pseudoleave, the stated bound follows if $n \equiv 4$ modulo 6. If $n \equiv 0$ or 2 modulo 6, then $n^2/4 \equiv \binom{n}{2}$ modulo 3. However, $K_{n/2,n/2}$ can never be a pseudoleave: if $n/2$ is odd, then $K_{n/2,n/2}$ is a leave (since an STS($n/2$) exists); if $n/2$ is even then the vertices have even degree, which is not allowed. Hence we must reduce the number of edges by at least three. □

Finally, we consider the maximum number of edges in a pseudoleave on an odd number of vertices. We use bounds on the number of edges in a triangle-free graph on an odd number of vertices, in which every vertex has even degree (i.e., the graph is Eulerian). These bounds are due to Novák [4].

Lemma 2.4: If xy is any edge in a triangle-free Eulerian graph on n vertices, where n is odd, then $d_x + d_y \leq n-1$ where d_x (resp. d_y) denotes the degree of x (resp. y).

Proof:

Since xy is an edge and the graph contains no triangles, x and y cannot have a common neighbour. Hence $d_x + d_y \leq n$. However, n is odd, and d_x and d_y are both even, so $d_x + d_y \leq n-1$. □

Lemma 2.5: If a triangle-free Eulerian graph on n vertices, where n is odd, does not contain an edge xy where $d_x + d_y = n-1$, then the number ϵ of edges satisfies

$$\epsilon \leq \frac{n^2 - 3n}{4}.$$

Proof:

From Lemma 2.4, we may assume that $d_x + d_y < n-1$ for every edge xy. Since d_x and d_y are both even, we must have $d_x + d_y \leq n-3$. We calculate

$$\sum_x d_x^2 = \sum_{xy \text{ an edge}} (d_x + d_y) \leq (n-3)\epsilon.$$

Also, we have:

$$\sum_x d_x = 2\epsilon,$$

and

$$\sum_x 1 = n.$$

The average value of the degrees d_x is $\bar{d} = 2\epsilon/n$. If we now calculate the variance of the d_x, we obtain:

$$\begin{aligned} 0 &\leq \sum_x (d_x - \bar{d})^2 \\ &= \sum_x d_x^2 - 2\bar{d}\sum_x d_x + n\bar{d}^2 \\ &\leq (n-3)\epsilon - 2(\frac{2\epsilon}{n})(2\epsilon) + n(\frac{2\epsilon}{n})^2. \end{aligned}$$

This simplifies to $\epsilon \leq \dfrac{n(n-3)}{4}$, as desired. □

We now consider the situation where, for some edge xy, $d_x+d_y = n-1$. Let A denote the set of neighbours of x (excluding y), and let B denote the set of neighbours of y (excluding x). We have noted that $A \cap B = \emptyset$; hence there is precisely one more vertex in the graph, say z. Let C denote the neighbours of z in A, and let D denote the neighbours of z in B. Finally, let $a = |A|$, $b = |B|$, $c = |C|$, and $d = |D|$.

We observe that a and b must both be odd, and $a+b = n-3$. Also, there can be no edge uv with $u \in C$, $v \in D$, since this would create a triangle. Hence, we obtain a bound on ϵ:

$$\begin{aligned}\epsilon &\leq n-2+ab+c+d-cd \\ &= n-1+ab-(c-1)(d-1).\end{aligned}$$

This quantity is maximized when $c = 1$ or $d = 1$; and a and b are as equal as possible (subject to the conditions that $a+b = n-3$, and a and b are both odd). Thus, if $n \equiv 1 \bmod 4$, we take $a = b = \dfrac{n-3}{2}$, and if $n \equiv 3 \bmod 4$, we take $a = \dfrac{n-5}{2}$, $b = \dfrac{n-1}{2}$.

Hence, we have:

Lemma 2.6 (Novák [4]): In a triangle-free Eulerian graph on n vertices, where n is odd, the number of edges $\epsilon \leq (n^2-2n+c)/4$, where

$$c = \begin{cases} 5 & \text{if } n \equiv 1 \bmod 4 \\ 1 & \text{if } n \equiv 3 \bmod 4. \end{cases}$$

□

We are now in a position to prove

Theorem 2.7: If a pseudoleave on n vertices, where n is odd, contains an edge xy, where $d_x+d_y = n-1$, then $\epsilon \leq (n^2-2n+c)/4$, where

$$c = \begin{cases} -3 & \text{if } n \equiv 3,9 \bmod 12 \\ 1 & \text{if } n \equiv 1,5,7,11 \bmod 12. \end{cases}$$

Proof:

Apply Lemma 2.6, noting that $\epsilon \equiv \binom{n}{2} \bmod 3$. □

3. Even subgraphs

In this section we discuss the existence of some subgraphs of the complete regular bipartite graph $K_{n,n}$, and of related graphs that we call even subgraphs.

It will be useful to recall the relationship between bipartite graphs and Latin rectangles and squares. A one-factor in $K_{n,n}$ can be interpreted as a permutation π of $\{1,...,n\}$ by the following rule: the ith entry of π is j if and only if the one-factor joins the ith vertex in the left-hand set to the jth vertex in the right-hand set. The rows of a Latin rectangle clearly correspond to pairwise edge-disjoint one-factors, and a Latin square is equivalent to a one-factorization of $K_{n,n}$. No confusion will arise if we treat a Latin rectangle as equivalent to the union of the one-factors which correspond to its rows.

It is well known [5] that any Latin rectangle can be extended. This means that, given any set of h pairwise edge-disjoint one-factors of $K_{n,n}$, we can find a further set of k edge-disjoint one-factors disjoint from them, provided $h+k \leq n$.

We shall say that a graph is *even* if it has an even number of edges and every vertex has even degree. For our constructions in the next section, we need certain bipartite graphs to have even subgraphs. When n is even, the bipartite graphs concerned are complete. When n is odd, we consider the bipartite graphs L_n, derived from $K_{n,n}$ by deleting a one-factor, and the graphs M_n, derived from $K_{n,n+2}$ by deleting $n-1$ independent edges and a 3-edge star independent of all of those edges.

Lemma 3.1:

(i) If n and α are even then $K_{n,n}$ has an even subgraph with α edges whenever $4 \leq \alpha \leq n^2-4$;

(ii) If n and α are even then $K_{n,n+2}$ has an even subgraph with α edges whenever $4 \leq \alpha \leq n(n+2)-4$;

(iii) If n is odd and α is even then L_n has an even subgraph with α edges whenever $4 \leq \alpha \leq n(n-1)-4$;

(iv) If n is odd and α is even then M_n has an even subgraph with α edges whenever $4 \leq \alpha \leq (n+2)(n-1)-4$.

It is easy to see that no even subgraphs with two edges exist; and, since all four classes of supergraphs are even graphs, no even subgraph of them could have a complement with two edges either. The Lemma states that there are proper even subgraphs for all other numbers of edges.

Proof of Lemma 3.1:

(i) Let us write $\alpha = 2tn+2r$, where $0 \leq 2t < n$, and $0 \leq 2r < 2n$. We distinguish cases where $2r = 0$, $2r = 2$, $4 \leq 2r \leq 2n-4$ and $2r = 2n-2$.

If $2r = 0$, then we require a Latin rectangle with $2t$ rows, which is easy.

If $4 \leq 2r \leq 2n-4$, first select two one-factors whose union is the disjoint union of a $2r$-cycle and a $(2n-2r)$-cycle. Call the $(2n-2r)$-cycle F. Then extend the corresponding 2-row Latin rectangle to a $(2t+2)$-row Latin rectangle. Finally, delete F from the Latin rectangle.

If $2r = 2$ and $n \geq 6$, start by taking four factors: two of them have union $C_4 \cup C_{2n-4}$ and the other two have union $C_6 \cup C_{2n-6}$. Complete a Latin rectangle as before and delete the C_{2n-6} and the C_4. A similar construction works for the case $2r = 2n-2$; the C_{2n-4} and the C_6 are deleted. (When $2r = 2$ we clearly require that $t>0$.)

The restriction "$n \geq 6$" in the last case is needed so that C_{2n-6} will be a cycle. However, the only cases needed for part (i) which are excluded by this requirement are even graphs on 6 and 10 edges in $K_{4,4}$, and these are easily constructed: one is a 6-cycle, and the other is its complement.

(ii) We interpret $K_{n,n+2}$ as the union of a $K_{n,n}$ with a $K_{n,2}$, where the two graphs have one n-set of vertices in common and the other n-set in the $K_{n,n}$ is disjoint from the 2-set in the $K_{n,2}$. The $K_{n,n}$ contains even subgraphs with α edges for all α satisfying $4 \leq \alpha \leq n^2-4$, so there are even subgraphs of $K_{n,n+2}$ with these numbers of edges. Also, if we take unions of these subgraphs with the $K_{n,2}$, we obtain examples for $2n+4 \leq \alpha \leq n(n+2)-4$. This covers all cases unless $n^2-4 < 2n+4$; but this is true only when $n < 4$, and no cases of the Lemma occur then.

(iii),(iv) Suppose that n is odd. It is easy to see that all of the constructions of part (i) are compatible with first taking a one-factor from $K_{n,n}$ (except that the upper bound on α will of course become $n(n-1)-4$). So part (iii) may be proven analogously to part (i). In the same way part (iv) is an easy extension of part (ii). □

4. Construction of pseudoleaves

We now present constructions for pseudoleaves with a wide spectrum of numbers of edges, as promised in the introduction.

Theorem 4.1: There exists a pseudoleave with n vertices and ϵ edges for every ϵ in Table 2.

Proof:

It is easy to check that the numbers presented in Table 2 satisfy the congruential conditions on degrees and number of edges. We give the solutions as five separate cases, although it will be seen that the methods are closely related, so only the first case is presented in full detail.

(i) Suppose that n is odd and $h = \frac{1}{2}(n-3)$ is also odd. We construct an n-vertex graph G_1, with vertices labelled $x, y, z, y_1, \cdots y_h, z_1, \cdots z_h$ with adjacencies defined as follows (see also Figure 1):

$$x \sim y, \quad x \sim z;$$

$$y \sim y_1, y_2, \cdots y_h;$$

$$z \sim z_1, z_2, \cdots z_h;$$

$$y_1 \sim z_1,\ y_2 \sim z_2\ ,\ \cdots\ ,\ y_h \sim z_h.$$

It will be observed that the complement of this graph is not partitionable into triangles -- in particular there would be no triangle containing yz -- and that it contains no triangle itself. These properties are preserved if edges of the form $y_i z_j$ are added. If we add α edges joining $\{y_i\}$ and $\{z_j\}$ the graph resulting will be a pseudoleave provided its total number of edges -- $(\frac{3}{2}n - \frac{5}{2} + \alpha)$ -- is divisible by 3 and every vertex has even degree. The graph induced by $(y_1,y_2, \cdots y_h, z_1,z_2, \cdots z_h)$ is the complement of L_n, and the congruential necessary conditions are seen to be that the α edges chosen form an even subgraph of L_n and that

(a) $\alpha \equiv 1 \pmod 3$ if $n \equiv 1 \pmod 6$,

(b) $\alpha \equiv 1 \pmod 3$ if $n \equiv 3 \pmod 6$,

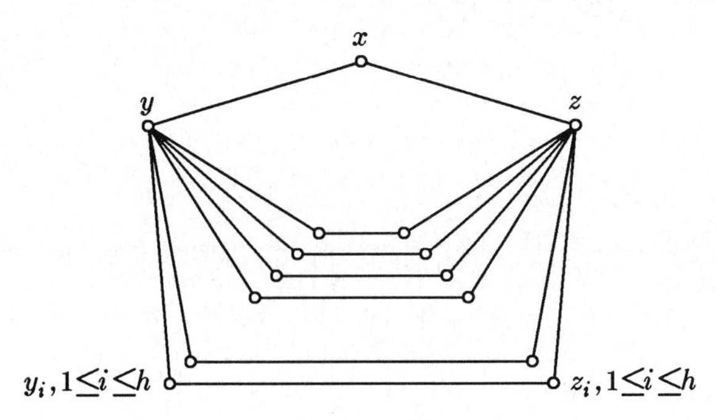

Figure 1. G_1

(c) $\alpha \equiv 2 \pmod 3$ if $n \equiv 5 \pmod 6$,

(since $n-3 \equiv 2 \pmod 4$, these conditions amount to $n \equiv 1$, 9, and 5 (mod 12), respectively). Since α must be even, we can rewrite the first congruence as $\alpha \equiv 4 \pmod 6$; we can reinterpret (a) as

> If $n \equiv 1 \pmod{12}$ then there is a pseudoleave with $\frac{3}{2}n + \frac{3}{2} + 6k$ edges whenever $K_{\frac{n-3}{2},\frac{n-3}{2}}$ has an even subgraph with $4+6k$ edges.

Since this even subgraph exists for $0 \le k \le (n^2-8n-17)/24$, we have the first line of Table 2. The third and fifth lines follow from (c) and (b), respectively. These three lines are marked (i) in the Table.

(ii) If n is odd but $h = \frac{1}{2}(n-3)$ is even the above construction will not work because the vertices y and z would have odd degree. So we modify G_1 to construct a graph G_2 with vertices x, y, z, $y_1, \cdots y_{h-1}$, $z_1, \cdots z_{h+1}$. The adjacencies are as before, but y_{h-1} is adjacent to z_{h-1}, z_h, and z_{h+1}. See Figure 2.

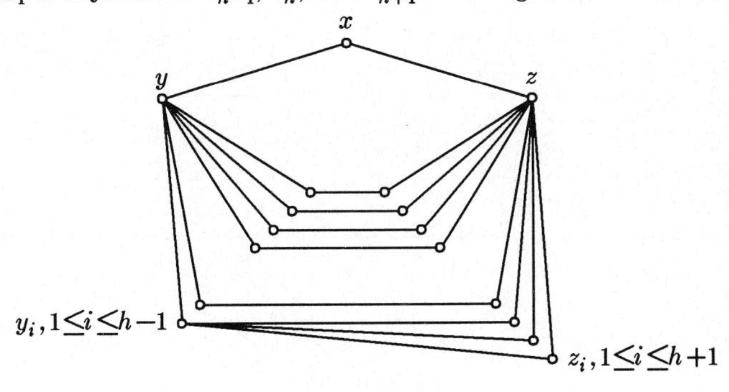

Figure 2. G_2

In this case the graph induced by $\{y_1, \cdots y_{h-1}, z_1, \cdots z_{h+1}\}$ is the complement of M_n. Lemma 3.1 tells us that suitable even subgraphs of M_n exist to provide the families indicated in lines 7, 9, and 11 of the table (marked (ii)) and again yz lies in no triangle.

(iii) All of the graphs from constructions (i) and (ii) have an odd number of edges. To obtain an even number of edges we use a graph G_3, depicted in Figure 3, with vertices

$$x,y,z,y^1,y^2,z^1,z^2,y_1, \cdots y_l,z_1, \cdots z_m,$$

where $l = m = ½(n-7)$ if $½(n-7)$ is even, and $l = ½(n-5), m = ½(n-9)$ if $½(n-7)$ is odd.

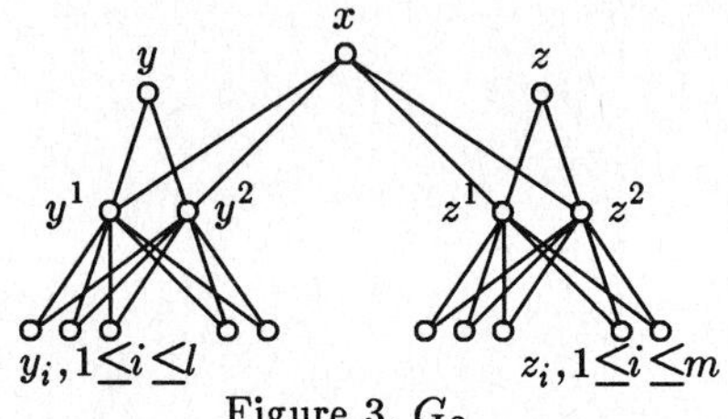

Figure 3. G_3

The adjacencies are

$$x \sim y^1,\ x \sim y^2,\ x \sim z^1,\ x \sim z^2,$$

$$y \sim y^1,\ y \sim y^2,\ z \sim z^1,\ z \sim z^2,$$

$$y^i \sim y_j,\ z^i \sim z_j \text{ for all } i,j.$$

This graph is triangle-free. The four edges y^1z^1, y^1z^2, y^2z^1, y^2z^2 cannot all lie in triangles in the complement of G_3: if y^iz^j lies in a triangle then the third vertex must be y^{3-i} or z^{3-j}, and hence a set of triangles involving all four of these edges would necessarily be a decomposition of the K_4 with vertices $\{y^1,y^2,z^1,z^2\}$ into two triangles, which is impossible. By choosing suitable even subgraphs of the complete bipartite graph on $\{y_i\}$ and $\{z_i\}$ we obtain the remaining pseudoleaves when n is odd (marked "(iii)" in the table).

(iv) Suppose that n is even; write $h = ½(n-2)$. The graph G_4, depicted in Figure 4, has vertices y, z, $y_1, \cdots y_l$, $z_1, \cdots z_m$, where $l = m = h$ if h is odd, and $l = h-1$, $m = h+1$ if h is even.

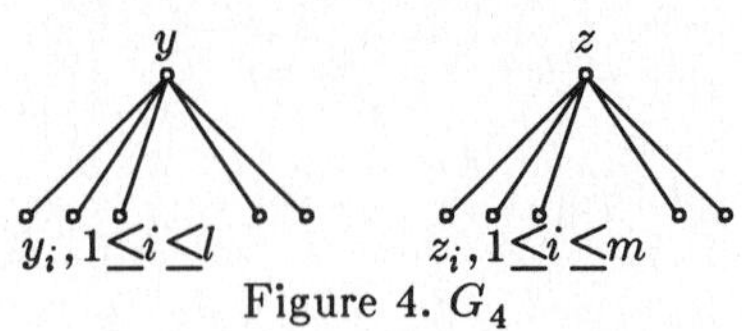

Figure 4. G_4

The adjacencies are

$$y \sim y_i,\ z \sim z_i \quad \text{for all } i,j.$$

The examples are constructed by adding to G_4 an even subgraph of $K_{l,m}$. From Lemma 3.1 we see that all numbers of edges marked (iv) in Table 2 are obtained; and yz does not belong to any triangle in the complement.

(v) When n is even, we construct pseudoleaves with an odd number of edges by starting with a graph G_5, depicted in Figure 5, having vertices y, z, y^1, y^2, z^1, z^2, $y_1, \cdots y_l, z_1, \cdots z_m$.

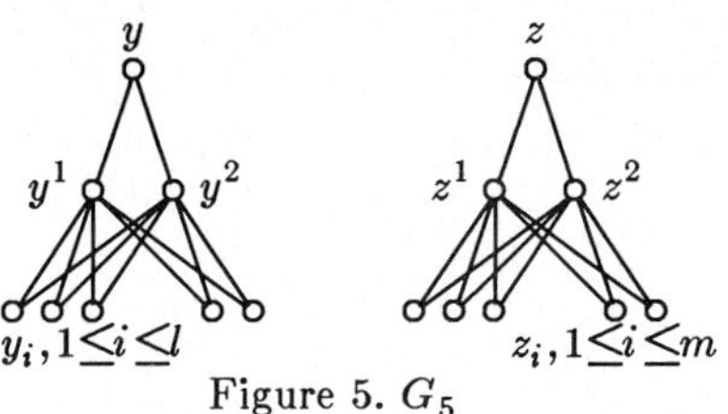

Figure 5. G_5

The edges are

$$y \sim z,\ y \sim y^1,\ y \sim y^2,\ z \sim z^1,\ z \sim z^2,$$

$$y^i \sim y_j \quad \text{for } i=1,2,\ 1 \leq j \leq l,$$

$$z^i \sim z_j \quad \text{for } i=1,2,\ 1 \leq j \leq m.$$

If $h = \frac{1}{2}(n-6)$ is even, then $l = m = h$; if h is odd then $h = l-1$ and $m = h+1$. If an even subgraph of $K_{l,m}$ is added, graphs with ϵ edges are obtained in all of the cases marked (v) in Table 2. Again, not all of y^1z^1, y^1z^2, y^2z^1 and y^2z^2 can lie in triangles in the complements of these graphs. □

Added In Proof

Using results contained in "Realizing small leaves of partial triple systems" (by C.J. Colbourn, *Ars Combinatoria* 23A (1987) 91-94), it is possible to show that any pseudoleave on n vertices must have at least $c\sqrt{n}$ edges for some positive constant c.

Acknowledgements

We would like to thank Alex Rosa and Charlie Colbourn for their helpful comments and suggestions. In particular, Rosa brought Novák's work to our attention, and Colbourn provided the $c\sqrt{n}$ lower bound mentioned above.

References

[1] C.J. Colbourn and A. Rosa, "Quadratic leaves of maximal partial triple systems", *Graphs and Combinatorics* 2 (1986) 317-337.

[2] A.J.W. Hilton and C.A. Rodger, "Triangulating nearly complete graphs of odd order", to appear.

[3] I. Holyer, "The NP-completeness of edge-colouring", *SIAM J. Computing* 10 (1981) 718-720.

[4] J. Novák, "Eulerovske grafy bez trojuhelniku s maximalnim poctem hran", Sbornik vedeckych praci VSST, Liberec, 1971.

[5] H.J. Ryser, *Combinatorial Mathematics* (Carus Math. Monograph 14), MAA *per* Wiley, New York, 1963.

[6] J. Schönheim, "On maximal systems of k-tuples", *Studia Sci. Math. Hung.* 1 (1966) 363-368.

[7] P. Turán, "An extremal problem in graph theory", *Mat. Fiz. Lapok* 48 (1941) 436-452.

Annals of Discrete Mathematics 34 (1987) 461–464
© Elsevier Science Publishers B.V. (North-Holland)

Symmetric 2-(31,10,3) Designs with Automorphisms of Order Seven

Vladimir D. Tonchev

Institute of Mathematics
P.O. Box 373
Sofia 1090
BULGARIA

TO ALEX ROSA ON HIS FIFTIETH BIRTHDAY

ABSTRACT

The greatest prime which can divide the order of the automorphism group of a symmetric 2-(31,10,3) design is 7. It is shown that there exist precisely four nonisomorphic 2-(31,10,3) designs having an automorphism of order 7. One of these designs does not possess any ovals and yields an extremal doubly-even self-dual code of length 64.

The four symmetric 2-(31,10,3) designs

Let us acknowledge at the outset the inspiration provided by R. Mathon and A. Rosa's survey on block designs with small parameters [8]. We assume that the reader is familiar with the basic definitions and facts from design theory [4,5] and coding theory [6,7].

As seen from [8], the smallest parameters of a non-Hadamard symmetric design for which a full enumeration up to isomorphism is not yet completed, are 2-(31,10,3). In fact, it seems that besides the existence, very little is known about such designs. Since a cyclic (31,10,3) difference set does not exist [3], it can be easily deduced using a theorem of Aschbacher [1], that the greatest prime which can be an order of the automorphism group of a 2-(31,10,3) design is 7. The 2-(31,10,3) design listed in Hall's book possesses an automorphism of order 7. It is our goal in this note to demonstrate that there are precisely 4 nonisomorphic 2-(31,10,3) designs admitting automorphisms of order 7.

Our approach is similar to that from [9]. Suppose that a symmetric 2-(31,10,3) design D with an automorphism β of order 7 is given. It is easy to see that β must fix exactly three points and blocks. Consequently, we can assume that D has an incidence matrix of the following form:

I	3_A	3_B	3_C	000 ... 000
3_D	2_a	2_b	2_c	100 ... 100
3_E	2_d	2_e	2_f	010 ... 010
3_F	2_g	2_h	2_i	001 ... 001
0...0 0...0 0...0	1...1 0...0 0...0	0...0 1...1 0...0	0...0 0...0 1...1	111 111 111

where I is the identity matrix of order 7, and 3_X (resp., 2_x) is a circulant of order 7 containing 3 ones (resp., 2 ones) in a row. It is clear that:

(i) the submatrix of (1) $3_A3_B3_C$ (as well as $3_D3_E3_F$) is an incidence matrix of a cyclic 2-(7,3,3) design;

(ii) any horizontal or vertical submatrix of (1) consisting of four consecutive circulants (e.g., $3_D2_a2_b2_c$ or $3_A2_a2_d2_g$) is an incidence matrix of a cyclic pairwise balanced design on 7 points such that any two points occur together in two blocks.

Up to permutation of columns, we have the following possibilities for the first row of a circulant of type 2_x or 3_X:

	2_1: 1100000	3_1: 1110000
	2_2: 1010000	3_2: 1101000
(2)	2_3: 1001000	3_3: 1100100
		3_4: 1100010
		3_5: 1010100

As a first step, we have to fill up the matrix (1) with circulants from (2) so that conditions (i) and (ii) are satisfied. In order to obtain all nonisomorphic solutions, we examine further all 31×31 matrices obtained from a matrix of the form (1) by cyclic permutations of the rows of the circulants $2_a,...,2_i$.

It can be shown that up to permutation equivalence there are exactly 21 matrices of the form (1) satisfying conditions (i) and (ii). Only four of these matrices yield 2-(31,10,3) designs: two matrices lead to unique solutions, while each of the remaining two yields a pair of isomorphic solutions. Base blocks of these solutions are listed in Table I, where we assume that β acts on the points and blocks as follows:

$$\beta = (1,2,...,7)(8,9,...,14)(15,16,...,21)(22,23,...,28)(29)(30)(31)$$

Table I. Symmetric 2-(31,10,3) designs with automorphisms of order 7

Design	Base Blocks	Group Order	Orbit Lengths
1	$B_1 = (1,8,12,14,15,19,21,22,26,28)$	63	3,7,21
	$B_8 = (2,3,5,8,10,18,21,26,27,29)$		
	$B_{15} = (2,3,5,11,14,19,20,22,24,30)$		
	$B_{22} = (2,3,5,12,13,15,17,25,28,31)$		
2	$B_1 = (1,8,13,14,15,18,21,22,25,27)$	42	3,7,21
	$B_8 = (1,6,7,8,11,16,18,24,28,29)$		
	$B_{15} = (1,4,7,9,11,15,20,26,27,30)$		
	$B_{22} = (1,4,6,10,14,19,20,22,28,31)$		
3	$B_1 = (1,8,13,14,15,18,21,22,25,27)$	21	1,1,1,7,21
	$B_8 = (1,5,7,8,10,19,20,25,28,29)$		
	$B_{15} = (1,3,7,9,12,16,18,25,26,30)$		
	$B_{22} = (1,3,7,11,14,17,19,22,23,31)$		
4	$B_1 = (1,8,12,14,15,17,21,22,24,28)$	21	3,7,7,7,7
	$B_8 = (1,6,7,8,10,18,21,23,26,29)$		
	$B_{15} = (1,4,7,12,13,17,19,23,25,30)$		
	$B_{22} = (1,4,6,11,14,15,16,25,26,31)$		
	Common Base Blocks		
	$B_{29} = (8,9,10,11,12,13,14,29,30,31)$		
	$B_{30} = (15,16,17,18,19,20,21,29,30,31)$		
	$B_{31} = (22,23,24,25,26,27,28,29,30,31)$		

The four designs from table I are distinguished by the orders and orbit lengths of their automorphism groups. Design 1 is isomorphic to the design given in [5].

An interesting feature of the 2-(31,10,3) designs is the possibility to construct a binary self-dual doubly-even (i.e., with all weights divisible by 4) code of length 64 from any such design (a good reference for self-dual codes is [7, Ch. 19]). More concretely, the following statement is true:

Proposition 1: The rows of the matrix

$$\left|\begin{array}{ccc} & 1 \quad . \quad . \quad . & 10 \\ & & 1 \\ & & . \\ I_{32} & A & . \\ & & . \\ & & 1 \end{array}\right|$$

where A is an incidence matrix of a symmetric 2-(31,10,3) design, generate a binary self-dual doubly-even code of length 64 and minimum Hamming distance $d \geq 8$.

The minimum distance d of a doubly-even self-dual code of length 64 is at most 12; a code with $d=12$ is called *extremal.*

An oval in a symmetric 2-(31,10,3) design is a set of 4 points such that each block contains at most 2 of them [2]. In addition to proposition 1, the following holds:

Proposition 2 [10]: A code constructed as in proposition 1 is extremal if and only if the corresponding design does not possess any oval.

We checked by computer that only design 2 from table I does not possess any ovals, and thus it yields an extremal doubly-even code.

In view of proposition 2, it is of interest to find other 2-(31,10,3) designs without ovals. Applying the same technique, one may try to construct designs with automorphisms of order 5. However, an exhaustive computer search showed that no such designs exist. Therefore, the full automorphism group of a 2-(31,10,3) design without automorphisms of order 7 must be a (2,3)-group.

References

[1] M. Aschbacher, "On collineation groups of symmetric block designs", *J. Comb. Thy.* A11 (1971) 272-281.

[2] E.F. Assmus, Jr., and J.H. Van Lint, "Ovals in projective designs", *J. Comb. Thy.* A27 (1977) 307-324.

[3] L.D. Baumert, *Cyclic Difference Sets,* Lecture Notes in Mathematics 182, Springer-Verlag, Berlin-Heidelberg-New York, 1971.

[4] Th. Beth, D. Jungnickel, and H. Lenz, *Design Theory,* Bibliographisches Institut, Mannheim-Wien-Zürich, 1985.

[5] M. Hall, Jr., *Combinatorial Theory,* Blaisdell, Waltham MA, 1967.

[6] J.H. Van Lint, *Introduction to Coding Theory,* Springer-Verlag, Berlin-Heidelberg-New York, 1982.

[7] F.J. MacWilliams and N.J.A. Sloane, *The Theory of Error-correcting Codes,* North-Holland, Amsterdam, 1978.

[8] R. Mathon and A. Rosa, "Table of parameters of BIBDs with $r \leq 41$ including existence, enumeration, and resolvability results", *Ann. Discrete Math.* 26 (1985) 275-308.

[9] V.D. Tonchev, "Hadamard matrices of order 28 with automorphisms of order 7", *J. Comb. Thy.* A40 (1985) 62-81.

[10] V.D. Tonchev, "Symmetric designs without ovals and extremal self-dual codes", *Proc. of the Passo della Mendola Conference, "Combinatorics 86",* to appear.

Annals of Discrete Mathematics 34 (1987) 465–470
© Elsevier Science Publishers B.V. (North-Holland)

Embeddings of Steiner Systems $S(2,4,v)$

R. Wei and L. Zhu

Department of Mathematics, Suzhou University,
Suzhou,
People's Republic of China

TO ALEX ROSA ON HIS FIFTIETH BIRTHDAY

1. Introduction

A pairwise balanced design *(PBD)* of index unity, denoted by $B[K,v]$, is a pair (X,A) where X is a v-set and A is a collection of some subsets (called blocks) of X such that any two distinct elements of X are contained in exactly one block in A and every block in A contains k elements where $k \in K$. A Steiner system $S(2,k,v)$ (also called a $(v,k,1)$–*BIBD*) is a $B[k,v]$ where we write $B[k,v]$ in place of $B[\{k\},v]$.

If (X,A) and (Y,B) are two *PBDs* such that $X \subseteq Y$ and $A \subseteq B$, we say that (X,A) is embedded in (Y,B) and that (Y,B) contains (X,A) as a subdesign. It is necessary for an $S(2,k,v)$ to contain an $S(2,k,u)$ as a subdesign that $v \geq (k-1)u+1$. It is well known that an $S(2,3,v)$ exists if and only if $v \equiv 1$ or $3 (mod\ 6)$, any positive integer satisfying this congruence is called admissible. Doyen and Wilson [6] have proved that any $S(2,3,u)$ can be embedded in some $S(2,3,v)$ for every admissible $v \geq 2u+1$. In this paper, we shall investigate the case $k=4$. Hanani [see e.g., 8] has proved that any $(S,4,v)$ exists if and only if $v \equiv 1$ or $4 (mod\ 12)$, such a positive integer will also be called admissible here. In 1981, Brouwer and Lenz [3] proved that if $u \geq 10{,}000$, any $S(2,4,u)$ can be embedded in some $S(2,4,v)$ for every admissible $v \geq 4u-12$. In this paper, we shall show that the same result still holds for small values of u, and that for infinitely many values of u the necessary condition $v \geq 3u+1$ is also sufficient. That is, we shall prove the following

Theorem 1.1 If $u \geq 85$, then any $S(2,4,u)$ can be embedded in some $S(2,4,v)$ for every admissible $v \geq 4u-12$.

Theorem 1.2. Let $u \geq 85$. If $u \equiv 4 (mod\ 12)$ or $u \equiv 1$ or $13 (mod\ 48)$, then an $S(2,4,u)$ can be embedded in some $S(2,4,v)$ if and only if v is admissible and $v \geq 3u+1$.

2. Some Preliminaries

For general background see Denes and Keedwell [5], Hall [7], Hanani [8] and Wilson [13].

A parallel class of blocks of a $PBD(X,A)$ is a subset A' of A such that A' is a partition of X.A transversal design $T(m,n)$ is a $PBD(X,A)$ with mn points and a distinguished parallel class of blocks A' (called groups) such that every group contains n

points and every block of $A \backslash A'$ contains m points. Evidently, every block intersects each group in exactly one point. It is well known that the existence of a $T(m,n)$ is equivalent to the existence of $m-2$ pairwise orthogonal Latin squares *(POLS)* of order n. Therefore, a $T(4,n)$ exists for every $n \neq 2,6$. Moreover, from [1, Example 8.27], [2], [10] and [12] we have

Lemma 2.1. A $T(5,n)$ exists for every $n \neq 2,3,6,10$ and a $T(6,n)$ exists for every $n \neq 2,3,4,6,10,14,18,20,22,26,28,30,34,38,42,44,52$.

A subdesign S of a transversal design T is a *TD* whose points and blocks are respectively points and blocks of T, and whose groups are subsets of the groups of T. If S is a $T(k,u)$ and T is a $T(k,v)$ with $u<v$, then we denote by $T(k,v)-T(k,u)$ or $IA_{k-2}(v,u)$ a design obtained by deleting all blocks of S from T. In fact, the missing subdesign S is not necessarily to exist. The following two lemmas are the direct corollaries of proposition 3.4 and corollary 1.3 in [4].

Lemma 2.2. Suppose there exist $T(6,k), T(5,m)$ and $T(5,m+1)$. Then there exists an $IA_3(mk+t,t)$ where $0 \leq t \leq k$.

Lemma 2.3. Suppose there exist a $T(6,k)$ and an $IA_3(m+l_i,l_i)$ where $l_i \geq 0, 1 \leq i \leq k$ and $l_1+l_2+\cdots+l_k=n$. Then there exists an $IA_3(km+n,n)$.

From [11] we have

Lemma 2.4. There exists an $IA_3(16+d,d)$ for $d=0,1,3,4,5$.

Now we may prove the following

Lemma 2.5. Let $k \equiv 0,1$ or $3 (mod\ 4)$. If $k>4$ and $k \geq t \geq 0$, then there exists an $IA_3(4k+t,t)$.

Proof:

Suppose $k \equiv 0,1$ or $3(mod\ 4)$ and a $T(6,k)$ exists. Taking $m=4$ in lemma 2.2 we obtain an $IA_3(4k+t,t)$. Otherwise, we have from lemma 2.1 that $k \in \{20,28,44,52\}$. Let $k'={}^k/4$ for such values of k, so that a $T(6,k')$ exists. Now take $m=16$ in lemma 2.3 and select certain values of $l_i \in \{0,1,3,4,5\}$ such that an $IA_3(m+l_i,l_i)$ exists from lemma 2.4 and $l_1+l_2+\cdots+l_{k'}=t$. We then obtain an $IA_3(16k'+t,t)$, i.e., a required $IA_3(4k+t,t)$.

Let $K=\{4,5,8,9,12\}$. It has been proved in [8] that a $B[K,u]$ exists if and only if $u \equiv 0$ or $1 (mod\ 4)$ and $u \geq 4$. The following lemma 2.6 shows that embeddings of $B[K,u]$ will yield some embeddings of $S(2,4,v)$. □

Lemma 2.6. Let $u,v \equiv 0$ or $1(mod\ 4)$. If there is a $B[K,v]$ containing a $B[K,u]$ as a subdesign, then there is an $S(2,4,3v+1)$ containing an $S(2,4,3u+1)$ as a subdesign.

The proof of this lemma is immediate and omitted here. The next lemma 2.7 is useful in proving theorem 1.1, which is essentially the lemma 3.1 in [3].

Lemma 2.7. Suppose there exists an $IA_3(m+n,n)$. Suppose also there exist a $B[K \cup \{n^*\}, m+n]$ and a $B[K \cup \{(n')^*\}, m'+n']$ where $0 \leq m' \leq m$ and $0 \leq n' \leq n$. Then there is a $B[K \cup \{u^*\}, v]$ where $u=4n+n'$ and $v=4m+m'+u$.

Here, by $B[K\cup\{t^*\},v]$ we mean a *PBD* with v points and block size in K except a unique block of size t. To use this lemma we need the following

Lemma 2.8. Let $k\equiv 0,1$ or $3(mod\ 4)$. If $k>3$ and $k\geq t\geq 0$, then there is a $B[K\cup\{t^*\},4k+t]$.

Proof:

$t=0$ is obvious and we suppose $k\geq t\geq 1$. If $K\equiv 0$ or $1(mod\ 4)$, delete $k-t$ points from one group of a $T(5,k)$ and construct a $B[K,k]$ on each of other groups.

If $k\equiv 3(mod\ 4)$, delete $k-(t-1)$ points from one group of a $T(5,k)$ and add a new point ∞ to each group. Then construct a $B[K,k+1]$ on the groups (with ∞) remaining one group of size $t-1$, which forms together with ∞ a block of size t. The conclusion then follows. □

From [3] we also have

Lemma 2.9. Let $v\equiv 0$ or $1(mod\ 4)$. A $B[K,v]$ containing a block of size 4 or 5 exists whenever $v\geq 13$ or 17, respectively.

3. Lower bound of u

In this section we shall give the proof of theorem 1.1. In view of lemma 2.6, we shall investigate some embeddings of $B[K,u]$ first.

Proposition 3.1 Let $u,v\equiv 0$ or $1(mod\ 4)$. If $4u-3\leq v\leq 5u$, then any $B[K,u]$ can be embedded in some $B[K,v]$ as a subdesign.

Proof:

If $v=4u-3$, add a point ∞ to every group of a $T(4,u-1)$ and then construct the $B[K,u]$ on each one.

If $v=4u+d$ where $0\leq d\leq u$, then $d\equiv 0$ or $1(mod\ 4)$. Delete $u-d$ points from one group of a $T(5,u)$ and construct a $B[K,d]$ on the size d group. On the other groups of size u construct the given $B[K,u]$. This completes the proof. □

Proposition 3.2 Let $u,v\equiv 0$, or $1(mod\ 4)$ and $u\geq 28$. If $v\geq 5u$, then any $B[K,u]$ can be embedded in some $B[K,v]$ as a subdesign.

Proof:

Suppose $u=4t$ where $t\geq 7$. We make use of lemma 2.7 by taking $n'=0$ and $n=t$. We can find some k such that $k\geq t$, $k\equiv 0,1$ or $3(mod\ 4)$ and $0\leq m'=v-4m-u\leq m$ where $m=4k$. In fact, we may take k as large as possible such that $m'\leq 32$. Since $v\geq 5u$ and the difference between two consecutive k is at most 2, this is always possible except $t\equiv 2(mod\ 4)$ and $5u\leq v\leq 5u+16$. The required $IA_3(4k+t,t), B[K\cup\{t^*\},4k+t]$ and $B[K,m']$ in lemma 2.7 come from lemma 2.5, lemma 2.8 and the fact that $m'=v(mod\ 4)$, respectively. For the exceptional case $t\equiv 2(mod\ 4)$ and $5u\leq v\leq 5u+16$, take $n'=4$ and $n=k=t-1$ in lemma 2.7. We further take $m=4k$ and $m'=v-5u+16$ such that $16\leq m'\leq 32$. The required $B[K\cup\{4^*\},m'+4]$ comes from lemma 2.9 Thus the case $u=4t$ is complete.

Now we suppose $u=4t+1$ and $t\geq 7$. We still make use of lemma 2.7 but taking $n'=1$ $n=t$ this time. Then we can similarly obtain the required conclusion except when $t\equiv 2(mod\ 4)$ and $5u\leq v\leq 5u+12$. In this case, take $n'=5$ and $n=k=t-1$ in lemma 2.7. We further take $m=4k$ and $m'=v-5u+20$ such that $20\leq m'\leq 32$. The required $B[K\cup\{5^*\},m'+5]$ comes from lemma 2.9. This completes the case $u=4t+1$ and then the proof. □

Proof of theorem 1.1.

For any $u\equiv 1$ or $4(mod\ 12)$ and $u\geq 85$ we write $u=3u^*+1$, where $u^*\equiv 0$ or $1(mod\ 4)$ and $u^*\geq 28$. For any admissible $v\geq 4u-12$ we write $v=3v^*+1$ where $v^*\equiv 0$ or $1(mod\ 4)$ and $v^*\geq 4u^*-3$. By propositions 3.1 and 3.2 we know that any $B[K,u^*]$ can be embedded in some $B[K,v^*]$ as a subdesign. From lemma 2.6 we then know that any $S(2,4,u)$ can be embedded in some $S(2,4,v)$ as a subdesign for any admissible u and v, where $u\geq 85$ and $v\geq 4u-12$. The proof is complete. □

4. Necessary and sufficient condition

In this section we shall give the proof of theorem 1.2. We need some results on resolvable designs. A $PBD(X,A)$ is resolvable if A can be partitioned into some parallel classes of blocks.

From [9] we have

Lemma 4.1. A resolvable $(v,4,1)$–*BIBD* exists if and only if $v\equiv 4(mod\ 12)$.

Theorem 4.2. Let $u\equiv 4(mod\ 12)$. Then any $S(2,4,u)$ can be embedded in some $S(2,4,v)$ as a subdesign for every admissible v satisfying $3u+1\leq v\leq 4u$.

Proof:

Let (X,A) be a resolvable $(u,4,1)$–*BIBD* where $u=12t+4$ and A can be partitioned into $4t+1$ parallel classes, say $A_e, 1\leq e\leq 4t+1$.

Let $X'=X\times I_3, I_3=\{1,2,3\}$. For every block $\{a,b,c,d\}$ in some A_e, we may use either one of the following two ways to form new blocks which contain a copy of the block, i.e., $\{a_1,b_1,c_1,d_1\}$. We briefly write $(x,j)=x_j$ where $x\in X$ and $j\in I_3$. Let

(A) the blocks on $\{a,b,c,d\}\times I_3$ be

$$a_1b_1c_1d_1 \quad a_1b_2c_3d_4 \quad a_1b_3c_2d_3$$

$$a_2b_2c_2d_1 \quad a_3b_1c_2d_2 \quad a_2b_1c_3d_3$$

$$a_3b_3c_3d_1 \quad a_2b_3c_1d_2 \quad a_3b_2c_1d_3$$

or

(B) The blocks on $\{a,b,c,d\}\times I_3\cup\{g_{e1},g_{e2},g_{e3}\}$ be

$$a_1b_1c_1d_1 \quad b_1c_2d_3g_{e1} \quad c_1d_2b_3g_{e2} \quad d_1b_2c_3g_{e3}$$

$$a_2b_2c_2d_2 \quad a_1d_2c_3g_{e1} \quad d_1c_2a_3g_{e2} \quad c_1a_2d_3g_{e3}$$

$a_3b_3c_3d_3 \quad d_1a_2b_3g_{e1} \quad a_1b_2d_3g_{e2} \quad b_1d_2a_3g_{e3}$

$c_1b_2a_3g_{e1} \quad b_1a_2c_3g_{e2} \quad a_1c_2b_3g_{e3}$.

For every other block in A_e we form the new blocks in the same way as we did for $\{a, b, c, d\}$. Denote all these new blocks by B_e, which has a copy of A_e i.e., $\{a_1b_1c_1d_1 | \{a,b,c,d\} \in A_e\}$.

If $v=3u+d$ where $1 \le d \le u$ and $d \equiv v \equiv 1$ or $4 (mod\ 12)$, let $d^*=(d-1)/3$. For $1 \le e \le d^*$, we form B_e from A_e in the way of (B). For other e, $d^*+1 \le e \le 4t+1$, let B_e be the blocks obtained by the way of (A). Denote by G the set of all points g_{ei}, surely $|G|=d-1$.

Let B_∞ be the set of all blocks $\{a_1,a_2,a_3,\infty\}$, $a \in X$. Let B' be the blocks of an $S(2,4,d)$ on the set $G \cup \{\infty\}$.

Let $Y=X' \cup G \cup \{\infty\}$ and $B=(\bigcup_{e=1}^{4t+1} B_e) \cup B_\infty \cup B'$, then (Y,B) is an $S(2,4,v)$ which contains a copy of (X,A), i.e., an $S(2,4,u)(X^*,A^*)$ where $X^*=X \times \{1\}$ and $A^*=\{A \times \{1\} | A \in A\}$. The remaining verification is straightforward . Since the subdesign $S(2,4,u)$ can be replaced by any other $S(2,4,u)$ on the same set, the proof is complete. □

Theorem 4.3. Let $u \equiv 1$ or $13 (mod\ 48)$. Then any $S(2,4,u)$ can be embedded in some $S(2,4,v)$ as a subdesign for every admissible v satisfying $3u+1 \le v \le 4u-3$.

Proof:

First let $u=48t+1$ and $v=3u+1+d$. Construct a $T(4,12t)$ on a set X such that its blocks consist of $12t$ parallel classes denoted by A_e where $1 \le e \le 12t$. Let $X'=X \times I_3$ and as in the proof of theorem 4.2, we may use (A) or (B) to construct new blocks for every A_e. It is clear that the maximum value of $|G|$ is $36t$.

Case 1. $0 \le d \le 36t$.

Add a point α to every group of the $T(4,12t)$ on X and construct an $S(2,4,12t+1)$ on each one, thus we obtain an $S(2,4,u)$. For the blocks other than A_e in the $S(2,4,u)$, we use (A) on $(X \cup \{\alpha\}) \times I_3$ to get new blocks. Denote all these blocks by B^*. Finally, construct an $S(2,4,d+4)$ having blocks B' on $G \cup \{\infty\} \cup (\{\alpha\}) \times I_3)$. Let B_∞ be the blocks of $(a \times I_3) \cup \{\infty\}$ for all $a \in X$. Let $Y=((X \cup \{\alpha\}) \times I_3) \cup \{\infty\} \cup G, B=B^* \cup (\bigcup_{e=1}^{12t} B_e) \cup B_\infty \cup B'$. *(Y,B)* is the required $S(2,4,v)$ containing a copy of the $S(2,4,u)$.

Case 2. $36t \le d \le 4u-3$.

Let $X=\bigcup_{1 \le i \le 4} S_i$ where S_i are the groups of the $T(4,12t)$ on X. Let S be a set of size $12t$. For $1 \le i \le 4$, take $S_i \times \{j\}, j=1,2,3$ and S as groups of a $T(4,12t)$ and denote its blocks by $B^{(i)}$. Let $B^*=\bigcup_{1 \le i \le 4} B^{(i)}$. On $(S_i \times \{j\}) \cup \{\infty\}$ construct an $S(2,4,12t+1)$ where $1 \le i \le 4$ and $1 \le j \le 3$. Denote all these blocks by B_∞. Finally, construct an $S(2,4,d+4)$ on $G \cup S \cup \{\infty\}$ and denote its blocks by B'. We then obtain an $S(2,4,v)$ (Y,B) containing an $S(2,4,u)$ on $(X \times \{1\}) \cup \{\infty\}$, where

$Y=(X\times I_3)\cup S\cup G\cup\{\infty\}$ and $B=B^*\cup(\bigcup_{e=1}^{12t} B_e)\cup B_\infty\cup B'$.

When $u=48t+13$, the proof is similar. In this case, we consider $T(4,12t+3)$ instead of $T(4,12t)$. This completes the proof. □

Proof of theorem 1.2. The conclusion follows immediately from theorem 1.1, theorem 4.2 and theorem 4.3. □

References

[1] T. Beth, D. Jungnickel and H. Lenz, *Design Theory,* Bibliographisches Institut, Zürich, 1985.

[2] A. Brouwer, "The number of mutually orthogonal Latin squares-a table up to order 10,000", *Math. Centrum report ZW 123,* Amsterdam, June 1979.

[3] A. Brouwer and H. Lenz, "Subspaces of linear spaces of line size 4", *Europ. J. Combinatorics* 2(1981) 323-330.

[4] A. Brouwer and G. Van Rees, "More mutually orthogonal Latin squares", *Discrete Math.* 39(1982) 263-281.

[5] J. Denes and A.D. Keedwell, *Latin Squares and Their Applications,* London 1974.

[6] J. Doyen and R.M. Wilson, "Embeddings of Steiner triple systems", *Discrete Math.* 5(1973) 229-239.

[7] M. Hall Jr., *Combinatorial Theory,* (Blaisdell, Waltham, MA, 1967).

[8] H. Hanani, "Balanced incomplete block designs and related designs", *Discrete Math.* 11(1975) 255-369.

[9] H. Hanani, D.K. Ray-Chaudhuri and R.M. Wilson, "On resolvable designs", *Discrete Math.* 3 (1972) 343-357.

[10] R. Roth and M. Peters, "Four pairwise orthogonal Latin squares of order 24", *J. Combinatorial Theory (A)* 44 (1987) 152-155.

[11] E. Seiden and C.J. Wu, "Construction of three mutually orthogonal Latin squares by the method of sum composition", in: *Essays in probability and statistics, S. Ikeda e.a. (eds.),* 1976, Shinko Tsusho Co. Ltd. (dist.), Tokyo.

[12] D.T. Todorov, "Three mutually orthogonal Latin squares of order 14", *Ars Combinatoria* 20 (1985) 45-48.

[13] R.M. Wilson, "Construction and uses of pairwise balanced designs", *Math. Centre Tracts* 55 (1974) 18-41.

Science Libr